INTRODUCTION TO CHEMICAL ENGINEERING FLUID MECHANICS

Designed for undergraduate courses in fluid mechanics for chemical engineers, this textbook illustrates the fundamental concepts and analytical strategies in a rigorous and systematic, yet mathematically accessible, manner. Dimensional analysis and order-of-magnitude estimation are presented as tools to help students identify which forces are important in different settings. The friction factors for pipes and other conduits, the terminal velocities of particles, drops, and bubbles, and flow in porous media, packed beds, and fluidized beds are explained from an experimental viewpoint. The physical and mathematical distinctions among major flow regimes, including unidirectional flow, lubrication flow, creeping flow, pseudosteady flow, irrotational flow, laminar boundary layers, turbulent shear flow, and compressible flow are described. Including a full solutions manual for instructors available at www.cambridge.org/deen, this is the ideal text for a one-semester course.

William M. Deen is the Carbon P. Dubbs Professor in the Department of Chemical Engineering at MIT. He is an author of some 200 research publications in bioengineering, colloid science, membrane science, quantitative physiology, and toxicology, most involving aspects of diffusion or fluid flow. During his 40 years of teaching at MIT, he has focused on undergraduate and graduate fluid mechanics, heat transfer, and mass transfer. He is the author of *Analysis of Transport Phenomena* (2012), which is used internationally in graduate-level transport courses. Among his awards are the 2012 Bose Award for Excellence in Teaching from the MIT School of Engineering and the 2012 Warren K. Lewis Award for Contributions to Chemical Engineering Education from the AIChE.

CAMBRIDGE SERIES IN CHEMICAL ENGINEERING

Books in Series

Baldea and Daoutidis, *Dynamics and Nonlinear Control of Integrated Process Systems*

Chau, *Process Control: A First Course with MATLAB*

Cussler, *Diffusion: Mass Transfer in Fluid Systems, Third Edition*

Cussler and Moggridge, *Chemical Product Design, Second Edition*

De Pablo and Schieber, *Molecular Engineering Thermodynamics*

Deen, *Introduction to Chemical Engineering Fluid Mechanics*

Denn, *Chemical Engineering: An Introduction*

Denn, *Polymer Melt Processing: Foundations in Fluid Mechanics and Heat Transfer*

Duncan and Reimer, *Chemical Engineering Design and Analysis: An Introduction*

Fan and Zhu, *Principles of Gas–Solid Flows*

Fox, *Computational Models for Turbulent Reacting Flows*

Franses, *Thermodynamics with Chemical Engineering Applications*

Leal, *Advanced Transport Phenomena: Fluid Mechanics and Convective Transport Processes*

Lim and Shin, *Fed-Batch Cultures: Principles and Applications of Semi-Batch Bioreactors*

Marchisio and Fox, *Computational Models for Polydisperse Particulate and Multiphase Systems*

Mewis and Wagner, *Colloidal Suspension Rheology*

Morbidelli, Gavriilidis, and Varma, *Catalyst Design: Optimal Distribution of Catalyst in Pellets, Reactors, and Membranes*

Nicoud, *Chromatographic Processes*

Noble and Terry, *Principles of Chemical Separations with Environmental Applications*

Orbey and Sandler, *Modeling Vapor–Liquid Equilibria: Cubic Equations of State and their Mixing Rules*

Petyluk, *Distillation Theory and its Applications to Optimal Design of Separation Units*

Rao and Nott, *An Introduction to Granular Flow*

Russell, Robinson, and Wagner, *Mass and Heat Transfer: Analysis of Mass Contactors and Heat Exchangers*

Schobert, *Chemistry of Fossil Fuels and Biofuels*

Shell, *Thermodynamics and Statistical Mechanics*

Sirkar, *Separation of Molecules, Macromolecules and Particles: Principles, Phenomena and Processes*

Slattery, *Advanced Transport Phenomena*

Varma, Morbidelli, and Wu, *Parametric Sensitivity in Chemical Systems*

Introduction to Chemical Engineering Fluid Mechanics

William M. Deen
Massachusetts Institute of Technology

CAMBRIDGE
UNIVERSITY PRESS

CAMBRIDGE
UNIVERSITY PRESS

Shaftesbury Road, Cambridge CB2 8EA, United Kingdom

One Liberty Plaza, 20th Floor, New York, NY 10006, USA

477 Williamstown Road, Port Melbourne, VIC 3207, Australia

314–321, 3rd Floor, Plot 3, Splendor Forum, Jasola District Centre, New Delhi – 110025, India

103 Penang Road, #05–06/07, Visioncrest Commercial, Singapore 238467

Cambridge University Press is part of Cambridge University Press & Assessment, a department of the University of Cambridge.

We share the University's mission to contribute to society through the pursuit of education, learning and research at the highest international levels of excellence.

www.cambridge.org
Information on this title: www.cambridge.org/9781107123779

First published 2016

A catalogue record for this publication is available from the British Library

ISBN 978-1-107-12377-9 Hardback

Additional resources for this publication at www.cambridge.org/9781107123779

To Meredith and Michael

Contents

Preface

WHAT IS CHEMICAL ENGINEERING FLUID MECHANICS?

Quantitative experimentation with fluids began in antiquity, and the foundations for the mathematical analysis of fluid flow were well established by the mid 1800s. Although a mature subject, fluid mechanics remains a very active area of research in engineering, applied mathematics, and physics. As befits a field that is both fascinating and useful, it has been the subject of innumerable introductory textbooks. However, only a few have focused on the aspects of fluid mechanics that are most vital in chemical engineering.

Certain results that stem from conservation of mass and momentum in fluids cut across all fields. However, the kinds of flow that are of greatest interest differ considerably among the various branches of engineering. One thing that distinguishes fluid mechanics in chemical engineering from that in, say, aeronautical or civil engineering, is the central importance of viscosity. Viscous stresses are at the heart of predicting flow rates in pipes, which has always been the main application of fluid mechanics in process design. Moreover, chemical engineering encompasses many technologies that involve bubbles, drops, particles, porous media, or liquid films, where small length scales amplify the effects of viscosity. Surface tension, usually not a concern in other engineering disciplines, also can be important at such length scales. In addition, in chemical engineering applications even gases usually can be idealized as incompressible. Another feature of chemical engineering fluid mechanics is an emphasis on microscopic analysis to calculate velocity fields. Determining velocities and pressures, and finding the resulting forces or torques, is often not an end in itself. Detailed velocity fields are needed to predict concentration and temperature distributions, which in turn are essential for the analysis and design of reactors and separation devices. Of lesser concern than in some other disciplines are the fluid dynamics of rotating machinery, flow in open channels, and flow at near-sonic velocities (where gas compressibility is important). Thus, chemical engineering fluid mechanics is characterized by a heightened interest in the microscopic analysis of incompressible viscous flows. Biomedical and mechanical engineers share some of the same concerns.

PURPOSE AND ORGANIZATION

This book is designed mainly as a text for chemical engineering undergraduates. The intention is to present fundamental concepts in a rigorous but mathematically accessible manner. A recurring theme is how to identify what is important physically in a novel situation and how to use such insights in modeling. That is illustrated by examples both within and outside the traditional domain of chemical engineering. The end-of-chapter problems tend to be challenging. They are intended not just to provide practice in certain kinds of calculations, but to build confidence in analyzing physical systems and to help develop engineering judgment.

The essential prerequisites are introductory mechanics, multivariable calculus, and ordinary differential equations. The information on vectors and tensors that is needed to understand certain derivations is summarized in an appendix, thereby making that part of the mathematics self-contained. Familiarity with a few numerical methods (e.g., solving first-order differential equations) is helpful but not necessary. No prior experience with partial differential equations is assumed; solution methods are explained as the need arises. A basic background in thermodynamics is presumed only in the last chapter.

The book has four parts. Part I, "Use of Experimental Data" (Chapters 1–3), discusses fluid properties, representative magnitudes of velocities and forces, and certain kinds of design. The information in these chapters is largely empirical. After surveying gas and liquid properties, Chapter 1 introduces dimensional analysis and the several uses of dimensionless groups, with an emphasis on groups that indicate the relative importance of different kinds of forces. Chapter 2 focuses on pressure–flow relationships in long pipes or other conduits. Chapter 3 discusses drag forces, terminal velocities of particles, porous media, packed beds, and fluidized beds. While presenting various experimental results and explaining certain engineering calculations, Chapters 2 and 3 introduce phenomena and relationships that are revisited later from more fundamental viewpoints.

Part II, "Fundamentals of Fluid Dynamics" (Chapters 4–6), lays the groundwork for predictive modeling. Chapter 4, on static fluids, explains the interactions among pressure, gravity, and surface tension and begins to make force calculations more precise. Chapter 5 introduces the continuity equation (the differential equation that embodies conservation of mass), the concept of rate of strain, and other aspects of kinematics, the description of fluid motion. Chapter 6 provides a general description of viscous stresses and combines that with conservation of linear momentum. The main result is the Navier–Stokes equation, the differential equation that ordinarily governs momentum changes within fluids. As aspects of vectors, tensors, and analytical geometry become relevant, the reader is referred to specific sections of the Appendix.

Part III, "Microscopic Analysis" (Chapters 7–10), illustrates how to use the governing equations of Chapter 6 to predict velocity and pressure fields, and how then to calculate fluid forces and torques. Chapter 7 is devoted to unidirectional flow, the simplest set of applications. Chapter 8 discusses how to anticipate and justify simplifications of the Navier–Stokes equation when viscous stresses are much more important than the inertia of the fluid. Introduced there are the lubrication, creeping flow, and pseudosteady approximations. Chapter 9 extends the discussion of approximations to laminar flows where inertia is prominent, as in boundary layers. Concepts unique to turbulent flow (Kolmogorov scales, time-smoothing, Reynolds stress, and mixing lengths) are presented in Chapter 10. Numerous connections are made between results derived in Part III and the experimental observations in Part I.

Part IV, "Macroscopic Analysis" (Chapters 11–12), focuses on flow problems that are too complex for the approaches in Part III, but where a less detailed kind of analysis is useful. Integral forms of the conservation equations are derived in Chapter 11 and simplified to algebraic equations that are practical for applications, such as the engineering Bernoulli equation. Key assumptions are justified by referring to the microscopic results of Part III. The simplified macroscopic balances are applied to a variety of systems in Chapter 11. Chapter 12 revisits pipe flow, including now resistances due to entrance regions and pipe fittings. It concludes with an introduction to compressible flow.

Although proofs of all key theoretical results are provided, some derivations are put in "additional notes" at the ends of sections. Any subsection so labeled can be skipped without loss of continuity. In contrast, all the examples illustrate core material and merit study. Many of the end-of-chapter problems present additional theoretical or experimental results or describe new kinds of applications. It is recommended that all the problem statements be read as part of the chapter, even if solutions are not to be worked out.

In manuscript form, the book has been used successfully as the text in a one-semester course for chemical engineering undergraduates. In a fast-paced course with four contact hours per week over 14 weeks, approximately 80% of the material was covered. Overall, the content provides a reasonable foundation for practicing chemical engineers and good preparation for graduate-level study of fluid mechanics. If supplemented by a comparable introduction to heat and mass transfer, it would be good preparation for graduate study of transport phenomena.

ACKNOWLEDGMENTS

Some of the examples and problems originated with MIT colleagues or faculty elsewhere. I have identified such unpublished sources with footnotes that state "this problem was suggested by," a phrase intended to give credit without blame. I have revised what others had written in homework or exam problems, often extensively, or elaborated on what they and I discussed, and therefore take all responsibility for errors or confusion.

I have learned a great deal over the years from MIT faculty with whom I have taught fluid mechanics or otherwise discussed the subject. Among those present or former colleagues are Robert C. Armstrong, Martin Z. Bazant, Robert A. Brown, Fikile R. Brushett, Arup K. Chakraborty, Patrick S. Doyle, Kenneth A. Smith, James W. Swan, and Preetinder S. Virk. As a graduate student at Stanford long ago, I was inspired by a course in viscous flow theory taught by Andreas Acrivos. My education has been advanced no less by interactions with generations of MIT students, who have always impressed me with their curiosity and determination to learn. Responding to their questions has continually sharpened my own understanding of the subject.

Although undertaken after retirement, the writing of this book was supported in part by the Carbon P. Dubbs Chair in the Department of Chemical Engineering at MIT. I am appreciative of that support and the other encouragement I have received from the Department.

Last but not least are those at Cambridge University Press who helped improve the book and make it a reality, including acquisition editor Michelle Carey, copy-editor Steven Holt, and production editor Charles Howell.

W. M. D.

Symbols

Following is a list of the more commonly used symbols. Omitted are coordinates (defined in Section A.5) and many quantities that appear in just one chapter. In general, scalars are italic Roman or Greek (e.g., a or δ), vectors are bold Roman (e.g., $\mathbf{v}$), and tensors are bold Greek (e.g., $\boldsymbol{\tau}$). Magnitudes of vectors and tensors are denoted usually by the corresponding italic letter (e.g., v or τ), and vector and tensor components are represented using subscripted italics (e.g., v_x or τ_{yx}). Natural and base-10 logarithms are written as "ln" and "log," respectively.

ROMAN LETTERS

A	Surface area or cross-sectional area.
Ar	Archimedes number.
Bo	Bond number.
C_D	Drag coefficient [Eq. (3.2-1)].
C_f	Friction factor or drag coefficient for a flat plate [Eq. (3.2-13)].
Ca	Capillary number.
D	Diameter (also d).
D_H	Hydraulic diameter [Eq. (2.4-1)].
$\mathbf{e_i}$	Unit vector associated with coordinate i.
E_c	Rate of mechanical energy loss due to compression [Eq. (11.4-2)].
E_v	Rate of mechanical energy loss due to viscous dissipation [Eq. (11.4-3)].
f	Friction factor for a tube or other conduit [Eq. (2.2-4)].
F_D	Drag force.
$\mathbf{F}$	Force vector.
$\mathbf{F_0}$	Force due to static pressure variations.
$\mathbf{F_B}$	Net buoyancy force, $\mathbf{F_0} - \mathbf{F_G}$.
$\mathbf{F_G}$	Gravitational force.
$\mathbf{F_P}$	Pressure force.
$\mathbf{F}_{\mathcal{P}}$	Flow-dependent part of pressure force.
$\mathbf{F}_{\boldsymbol{\tau}}$	Viscous force.
Fr	Froude number.
$\mathbf{g}$	Gravitational acceleration vector.
$\mathbf{G}$	Torque vector.
h	Height (usually).
k	Wall roughness parameter (usually) or Darcy permeability (Chapter 3).
K_i	Loss coefficient for event or device i [Eq. (12.3-1)].
L	Length as a dimension.

L	Object length or characteristic length.
L_E	Entrance length for tubes or other conduits.
m	Mass.
M	Mass as a dimension.
Ma	Mach number.
$\mathbf{n}$	Unit vector normal to a surface, directed outward from a control volume or from phase 1 to phase 2.
P	Absolute or thermodynamic pressure.
$\mathcal{P}$	Dynamic pressure (also called modified pressure or equivalent pressure).
q	Volume flow rate per unit width in two-dimensional flows.
Q	Volume flow rate.
$\mathbf{r}$	Position vector.
R	Radius (usually) or universal gas constant (Sections 1.2, 12.4, and 12.5).
Re	Reynolds number.
$\mathbf{s}$	Stress vector.
S	Surface area.
t	Time.
$\mathbf{t}$	Unit vector tangent to a surface.
T	Absolute temperature.
T	Time as a dimension.
$\mathbf{u}$	Interfacial velocity (Chapter 6), turbulent velocity fluctuation (Chapter 10), or control-surface velocity (Chapter 11).
u	In boundary-layer flows, the outer velocity evaluated at the surface (Chapters 9 and 10).
u_τ	Friction velocity [Eq. (10.2-4)].
U	Characteristic velocity, usually a mean fluid velocity or particle velocity.
v_s	Superficial velocity [Eq. (3.4-1)].
$\mathbf{v}$	Fluid velocity.
V	Volume (usually) or velocity.
w	Mass flow rate (Chapters 11 and 12).
$\mathbf{w}$	Vorticity vector.
W_m	Rate of work done on a system by moving surfaces (shaft work).

GREEK LETTERS

α	Contact angle at a three-phase contact line.
β	At a contraction or expansion, the smaller diameter divided by the larger one.
δ	Boundary-layer thickness (usually).
δ_{ij}	Kronecker delta [Eq. (A.3-6)].
$\boldsymbol{\delta}$	Identity tensor.
Δ	Difference along the direction of flow (downstream value minus upstream value) or differential change.
ε	Volume fraction of fluid (void fraction) in porous media or packed beds.
ε_{ijk}	Permutation symbol [Eq. (A.3-15)].
ϕ	Volume fraction of solids in porous media (Chapter 3) or velocity potential (Chapters 5 and 9).
$\boldsymbol{\Gamma}$	Rate-of-strain tensor [Eq. (6.5-1)].
γ	Surface tension (usually) or heat-capacity ratio [Eq. (12.4-1)].
μ	Viscosity.

ν	Kinematic viscosity, μ/ρ.
ρ	Density.
$\boldsymbol{\sigma}$	Total stress tensor.
$\boldsymbol{\tau}$	Viscous stress tensor.
τ_w	Shear stress at a wall or other solid surface (also τ_0).
$\boldsymbol{\Omega}$	Vorticity tensor [Eq. (6.5-3)].
ψ	Stream function.

SPECIAL SYMBOLS

D/Dt	Material derivative, $\partial/\partial t + \mathbf{v} \cdot \nabla$.
∇	Gradient operator.
∇^2	Laplacian operator, $\nabla \cdot \nabla$.
$\sim$	Order-of-magnitude (OM) equality.

Part I

Use of experimental data

1

Properties, dimensions, and scales

1.1 INTRODUCTION

Fluids, including both gases and liquids, are materials that deform continuously when subjected to shearing forces. If a flowing fluid contains a dye or other tracer, the labeled region tends to change shape from instant to instant. The viscosity of a fluid is a reflection of its resistance to such deformations. This chapter begins with definitions of viscosity and three other material properties that are important in fluid mechanics, namely density, kinematic viscosity, and surface tension. Representative values of each are presented, and some of the differences between Newtonian and non-Newtonian fluids are described.

Physical quantities have *dimensions*, such as mass (M), length (L), and time (T). The concepts of mass, length, and time are more fundamental than the *units* in a particular system of measurement, such as the kilogram (kg), meter (m), and second (s) that underlie the SI system. To have general validity, a physical relationship must be independent of the observer, including the observer's choice of units. That can be achieved by making each variable or parameter in an equation *dimensionless*, which means that the physical quantities are grouped in such a way that their dimensions cancel. Dimensionless parameters are ratios, and often the numerator and denominator are each the *scale* of a variable. A scale is the maximum value of something, such as a velocity or force. Thus, the numerical value of such a group reveals how two things compare, and thereby offers insight into what is important in the process or phenomenon under consideration and what might be negligible. Several dimensionless groups that arise in fluid mechanics are discussed.

Dimensional analysis, the last major topic of the chapter, provides a systematic way to identify the dimensionless groups that are involved in something of interest. In addition to yielding relationships that are more universal, grouping quantities in this manner minimizes the number of independent variables. This greatly facilitates the design of experiments and interpretation of data, as discussed here and explored further in Chapters 2 and 3.

1.2 FLUID PROPERTIES

Viscosity

A definition of viscosity is provided by the idealized experiment in Fig. 1.1. Imagine that a fluid fills a space of thickness H between parallel plates, each of which has an area A. Suppose that the upper plate is pulled to the right (in the x direction) at a constant velocity U, while the lower one is held in place. Because fluids tend to adhere to surfaces in a way that prevents relative motion or "slip," the fluid contacting the top plate also moves at velocity U, while that next to the bottom remains stationary. Once enough time

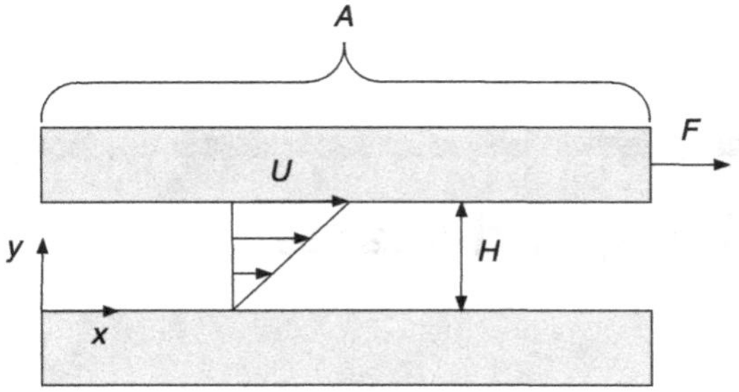

Figure 1.1 Shear flow between parallel plates. A force F applied to the top plate moves it to the right at velocity U, while the bottom plate is kept stationary. The resulting fluid velocity varies linearly from zero at the bottom to U at the top.

has elapsed, the fluid velocity varies linearly with height (the y coordinate), as shown by the arrows. As elements of the fluid at different heights move to the right at different speeds, the distance between any two of them increases. Such elongation of imaginary lines connecting pairs of points indicates that the fluid is being deformed.

A constant force F, to the right, must be applied to the top plate to keep it moving, and the absence of slip transfers that force to the adjacent fluid. The fluid itself may be thought of as having innumerable layers, each of which resists moving at a velocity different from the adjacent ones. This internal friction, which is proportional to the viscosity, transmits the horizontal force through successive fluid layers and to the bottom plate. (Not shown in Fig. 1.1 is the restraining force of magnitude F, to the left, which would be needed to keep the bottom plate stationary.)

A *stress* is a force per unit area. In principle, a stress may be computed at any point on a surface by finding the differential force ΔF acting on a differential area ΔA, and calculating the ratio $\Delta F/\Delta A$ in the limit $\Delta A \to 0$. Intensive variables such as this (i.e., ones independent of system size) are central to the analysis of fluid mechanics. *Shear* stresses act parallel to (tangent to) surfaces; *normal* stresses act perpendicular to surfaces. The shear stress that a solid exerts on a fluid, called the *wall shear stress*, is denoted as τ_w.[1] In the idealized experiment in Fig. 1.1, τ_w happens to be the same everywhere on the top plate. Accordingly, its value at any point equals the shear stress averaged over the entire surface. That is, $\tau_w = F/A$.

Experiments like that in Fig. 1.1 reveal that τ_w for any fluid depends on U/H, but not on U or H separately. The velocity-to-thickness ratio is the *shear rate*, $s = U/H$.[2] It is found that τ_w always increases with s. The shear stress and shear rate are related as

$$\tau_w = \frac{\mu U}{H} = \mu s \tag{1.2-1}$$

where μ is the *viscosity*. In a *Newtonian* fluid, μ is independent of s and the shear stress is exactly proportional to the shear rate. Numerous fluids of practical interest are Newtonian, including all gases, liquids with molecular weights (relative molecular masses) less than about 10^3 (such as water and common organic solvents), and suspensions of non-aggregating particles. For a given Newtonian fluid, μ depends mainly on temperature. In *non-Newtonian* fluids, the value of μ (i.e., the ratio τ_w/s) depends in some way on s.

[1] This symbol is used in Part I of this book, but a more systematic way to identify stress components is needed later. In the notation introduced in Part II, τ_w at the top plate in Fig. 1.1 would be written as τ_{yx}. The first subscript indicates that the surface (the plane $y = H$) is perpendicular to the y axis, and the second refers to a resultant force in the x direction.

[2] A symbol for shear rate that appears very widely in the literature is $\dot{\gamma}$. Because of the potential for confusing that with surface tension (to be denoted as γ), s is used here instead.

Non-Newtonian fluids also have other distinctive features, as discussed later in this section. Among such fluids are polymer melts, concentrated solutions of flexible polymers, and concentrated suspensions of particles that deform, orient, or aggregate.

A flow that closely approximates that in Fig. 1.1 is realized in a *Couette viscometer*, in which a liquid sample fills the annular space between long, concentric cylinders. If the gap between cylinders is small compared with their radii, the surfaces will appear to the liquid to be planar, just as the Earth seems flat from human eye level. One cylinder is rotated while the other is held in place. The linear velocity of the rotating cylinder is its radius times the angular velocity, and s is the linear velocity divided by the gap width. The torque measured on either cylinder is proportional to τ_w. This and other viscometers avoid practical problems that would occur with the system in Fig. 1.1, such as keeping the gap width H and fluid–solid contact area A constant.

Density and kinematic viscosity

The *density* ρ of a fluid is its mass per unit volume. In that density times velocity is momentum per unit volume, ρ is needed to describe inertia. Moreover, ρg is the gravitational force per unit volume, g being the gravitational acceleration. In principle, the density at any point in a fluid could be found by repeatedly determining the mass Δm in a volume ΔV, and finding the limit of $\Delta m/\Delta V$ as $\Delta V \to 0$. In liquids, where ρ is nearly independent of pressure (P), it is spatially uniform unless there are significant variations in temperature (T). In gases, ρ is related to P and T by an equation of state, and pressure gradients will cause it to vary with position, even if the gas is isothermal. The equation of state for an *ideal gas*, which is adequate at low to moderate pressures, is

$$\rho = \frac{MP}{RT} \tag{1.2-2}$$

where M is the average molecular weight, R is the universal gas constant, and T is the absolute temperature. Assuming ρ to be constant, even in gases, usually leads to negligible error. Exceptions include gases spanning large heights, flowing in very long pipes, or moving at velocities that approach the speed of sound. Those situations are unusual enough in chemical engineering that constancy of ρ will be our default assumption. The corresponding idealization is called *incompressible flow*. Although such flows are our main concern, an introduction to *compressible flow*, where variations in ρ must be taken into account, is provided in Chapter 12.

The ratio of viscosity to density occurs frequently enough to have its own name, the *kinematic viscosity*. It is denoted by the Greek "nu,"

$$\nu = \frac{\mu}{\rho}. \tag{1.2-3}$$

Units and values

The SI system, in which the unit of force is the newton ($\mathrm{N = kg\,m\,s^{-2}}$), is used in this book. The units of energy and power are the joule ($\mathrm{J = N\,m = kg\,m^2\,s^{-2}}$) and watt ($\mathrm{W = J\,s^{-1} = kg\,m^2\,s^{-3}}$), respectively. The unit of pressure (or stress) is the pascal ($\mathrm{Pa = N\,m^{-2} = kg\,m^{-1}\,s^{-2}}$). One atmosphere equals 1.013×10^5 Pa = 101.3 kPa. The standard gravitational acceleration is $g = 9.807\ \mathrm{m\ s^{-2}}$ and the gas constant is $R = 8314\ \mathrm{kg\ m^2\ s^{-2}\ kg\text{-}mol^{-1}\ K^{-1}}$.

When making engineering estimates, one should have in mind typical values for the various fluid properties. Table 1.1 lists properties of several gases at near-ambient

Table 1.1 Viscosity, density, and kinematic viscosity of gases (at 27 °C and 100 kPa)[a]

Gas	M(daltons)	$\mu(10^{-5}$ Pa · s)	ρ(kg m^{-3})	$\nu(10^{-5}$ m^2 s^{-1})
H_2	2	0.90	0.0819	11.0
CH_4	16	1.12	0.652	1.72
N_2	28	1.79	1.14	1.57
Air	29	1.86	1.18	1.58
C_2H_6	30	0.95	1.22	0.78
O_2	32	2.08	1.30	1.60
CO_2	44	1.50	1.79	0.84

[a] The viscosities are from Lide (1990), the densities are from Eq. (1.2-2), and the kinematic viscosities were calculated from those viscosities and densities. "Air" is dry air.

Table 1.2 Viscosity, density, and kinematic viscosity of Newtonian liquids[a]

Liquid	T(°C)	$\mu(10^{-3}$ Pa · s)	$\rho(10^3$ kg m^{-3})	$\nu(10^{-6}$ m^2 s^{-1})
Water	20	1.002	0.9982	1.004
Benzene	20	0.652	0.8765	0.744
Ethanol	20	1.200	0.7893	1.520
Mercury	20	1.554	13.55	0.115
Phenol	18	12.7	1.0576	12.0
Olive oil	20	84.0	0.918	91.5
Machine oils	16	114–661	0.87	130–760
Glycerol	20	1490	1.2613	1181
Honey	20	19,000	1.42	13,400

[a] All viscosities and densities are from Lide (1990), except as noted below, and ν was calculated from μ and ρ. The density for the "machine oils" is a value typical of motor oils. The properties of a representative honey are from www.airborne.co.nz.

conditions. The entries are arranged in order of increasing molecular weight. Despite 20-fold variations in M, the viscosities fall within a narrow range, from about 1×10^{-5} to 2×10^{-5} Pa · s. With little difference in μ and with ρ proportional to M, the variations in ν are due largely to the differences in M. Given that most gas densities are roughly 1 kg m^{-3}, a typical kinematic viscosity at ambient temperature and pressure is about 1×10^{-5} m^2 s^{-1}.

The properties of a number of Newtonian liquids are shown in Table 1.2. The viscosity of water at room temperature is 1×10^{-3} Pa · s, its density is 1×10^3 kg m^{-3}, and its kinematic viscosity is therefore 1×10^{-6} m^2 s^{-1}. Although the viscosity of water is about 100 times that for gases, its kinematic viscosity is about 10 times smaller because its density is 1000 times larger. The viscosity, density, and kinematic viscosity of benzene and ethanol, which are representative of common organic solvents, are each the same order of magnitude as those for water. The variations in μ and ν among the other liquids are noteworthy; they range over factors of about 10^4 and 10^5, respectively. Differences in viscosity should not be confused with differences in density, as occurs sometimes in casual speech or writing. Despite its high density, mercury has an ordinary viscosity, whereas the liquids in the last five lines of Table 1.2 have increasingly large viscosities, but ordinary densities.

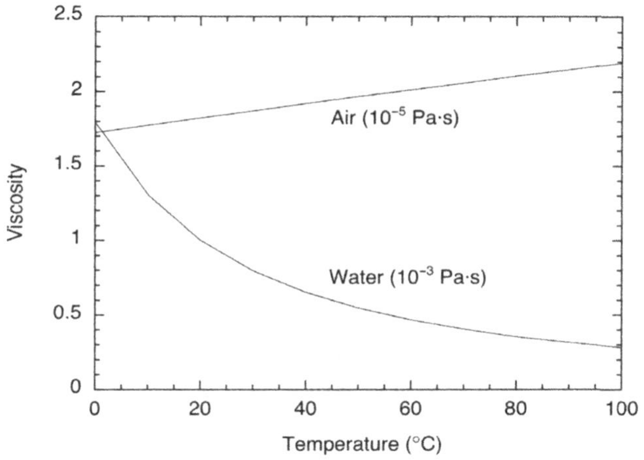

Figure 1.2 Viscosities of air and water as a function of temperature. The values for air are from a correlation in Kadoya *et al.* (1985) for dry air and those for liquid water are from Lide (1990).

Gas viscosities increase with temperature (approximately as $T^{1/2}$), whereas those for liquids decrease, as exemplified by the results for air and water in Fig. 1.2. Liquid viscosities are more temperature-sensitive. As T increases from 0 to 100 °C, μ for liquid water decreases by 84%, whereas μ for air increases by only 27%.

Both for gases and for liquids, the effects of pressure on viscosity are generally negligible (Blevins, 2003). That is true for gases if $P/P_c < 0.2$ and $T/T_c > 1$, where P_c and T_c are the critical temperature and pressure, respectively. For air, $P_c = 36.4$ atm and $T_c = 78.6$ K. For water at 50 atm, the kinematic viscosity is only about 1% less than that at 1 atm.

Estimation methods to use when specific data are unavailable may be found in Reid *et al.* (1987). Gas viscosities are explained well by kinetic theory (Hirschfelder *et al.*, 1954; Kennard, 1938). Although there are no comparably simple and predictive theories for liquids, molecular-dynamics simulations provide a way to relate liquid viscosities and other transport properties to intermolecular forces (Allen and Tildesley, 1987).

Non-Newtonian liquids

As already mentioned, a dependence of viscosity on shear rate is the hallmark of non-Newtonian liquids. The experimental and theoretical investigation of stresses in such fluids constitutes the field of *rheology*. Aside from the dependence of μ on s, the behavior of polymeric liquids, in particular, can differ strikingly from that of low-molecular-weight liquids. For example, when a Newtonian liquid is stirred, the air–liquid interface is depressed in the vicinity of the stir rod; with various polymer solutions the interface is elevated, and the solution may even climb the rod as stirring proceeds. Also, a steady stream of Newtonian liquid leaving a small tube reaches a diameter that does not differ greatly from that of the tube, whereas jets of polymeric liquids can swell to several times the tube diameter. Further, flexible polymer molecules resist elongation and, because of their size, can require several seconds to adjust shape in response to applied forces. This can give the fluid an elastic character and cause viscous stresses to depend not just on instantaneous shear rates, but also the recent history of the sample. Fluids in which μ is time-dependent are *viscoelastic*. These and other special characteristics of polymeric

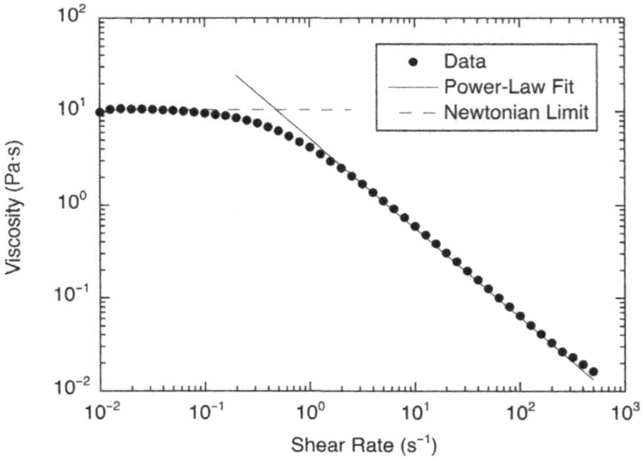

Figure 1.3 Viscosity as a function of shear rate for a solution containing 66 mM cetylpyridinium chloride and 40 mM sodium salicilate (Haward and McKinley, 2012).

liquids are surveyed in Chapter 2 of Bird *et al.* (1987). Colloidal dispersions and other particle suspensions in low-molecular-weight solvents can also be non-Newtonian, especially when very concentrated. For example, whole blood, which is almost half red cells by volume, has a viscosity that depends on shear rate when s is small.

Many features of polymeric liquids are beyond the scope of this book. The dependence of μ on s, and its consequences for predicting the flow rates of non-Newtonian liquids in pipes or other conduits, tends to be of greatest practical interest and will be our focus. A relationship that describes data for many liquids over practical ranges of s is

$$\mu = ms^{n-1} \tag{1.2-4}$$

where m and n are positive constants. The values of m and n are specific to a given material and temperature. In polymer solutions, m and n both depend on concentration. They are obtained by substituting Eq. (1.2-4) into Eq. (1.2-1) and fitting the resulting equation to data for τ_w as a function of s. A material that obeys Eq. (1.2-4) is called a *power-law fluid*, and such liquids can follow this relationship for values of s spanning several orders of magnitude. For polymer solutions, n often ranges from 0.2 to 0.8 (Bird *et al.*, 1987, pp. 172–175; Tanner, 2000, pp. 18–19). A fluid with $n < 1$, such that μ decreases with increasing s, is referred to as *shear-thinning*. In the rarer situations where $n > 1$, such that μ increases with increasing s, it is *shear-thickening*. For $n = 1$, the viscosity is independent of s and the Newtonian case is recovered, with $\mu = m$.

Equation (1.2-4) implies that $\mu \to \infty$ as $s \to 0$ for a shear-thinning fluid. In reality, μ tends to be constant at very low shear rates, as exemplified by the data in Fig. 1.3. What was studied in this case was a surfactant solution that forms highly elongated micelles. However, the overall trends are like those found for numerous polymer solutions and polymer melts. At low shear rates the solution was Newtonian with a viscosity of about 10 Pa · s, as indicated by the dashed line. At higher shear rates there was a power-law region where

$$\mu = \mu_0 \left(\frac{s}{s_0} \right)^{n-1} \tag{1.2-5}$$

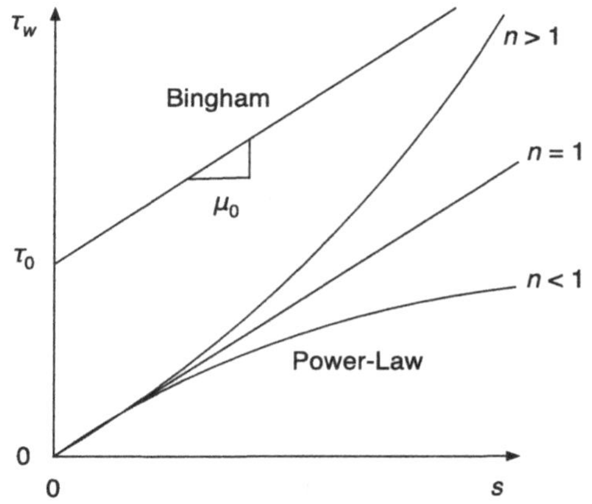

Figure 1.4 Shear stress as a function of shear rate for power-law and Bingham fluids.

a relationship that is obtained by setting $m = \mu_0 s_0^{1-n}$ in Eq. (1.2-4). The curve fit shown by the solid line corresponds to $\mu_0 = 10.5$ Pa · s, $n = 0.04$, and $s_0 = 0.48$ s^{-1}. Although beyond the range of these experiments, at extremely high shear rates there would be another Newtonian plateau at a lower viscosity. To describe the upper and lower Newtonian limits, the power-law region, and the transitions in a single expression, an equation with more degrees of freedom is needed. Such an expression is the *Carreau–Yasuda equation*, which involves five fitted constants (Bird *et al.*, 1987, p. 171). Their inaccuracy for extreme values of s notwithstanding, Eqs. (1.2-4) and (1.2-5) are adequate for most pipe-flow calculations.

Another class of materials flows only when the shear stress exceeds a certain threshold, called the *yield stress*. One way to describe their viscosity is

$$\mu = \begin{cases} \infty & \text{for } \tau_w < \tau_0 \\ \mu_0 + \dfrac{\tau_0}{s} & \text{for } \tau_w \geq \tau_0 \end{cases} \tag{1.2-6}$$

where τ_0 is the yield stress and μ_0 is a second material-specific constant [not to be confused with μ_0 in Eq. (1.2-5)]. A material that obeys Eq. (1.2-6) is called a *Bingham fluid* or *Bingham plastic*. The infinite viscosity for $\tau_w < \tau_0$ means simply that the material behaves as a rigid solid when subjected to insufficient shear. However, for $\tau_w \geq \tau_0$ the viscosity becomes finite and decreases with increasing s. If s greatly exceeds τ_0/μ_0, the fluid becomes Newtonian with $\mu = \mu_0$. House paint and various foods behave much like this.

The qualitative dependence of shear stress on shear rate for Newtonian, power-law, and Bingham fluids is shown in Fig. 1.4. For a Newtonian fluid ($n = 1$), the straight line through the origin indicates that μ (which in this case equals the slope) is constant and that there is no yield stress. For a power-law fluid with $n \neq 1$, the curvature of the function $\tau_w(s)$ signifies a variable μ, but the passing of the curve through the origin indicates the absence of a yield stress. For the Bingham model, the nonzero intercept is τ_0 and the constant slope is μ_0.

Equations (1.2-4)–(1.2-6) are examples of rheological *constitutive equations*, which relate viscous stresses to rates of shear or deformation. The constitutive equations for Newtonian, power-law, and Bingham fluids are presented in more general form in Section 6.5. Many other non-Newtonian constitutive equations have been proposed, some

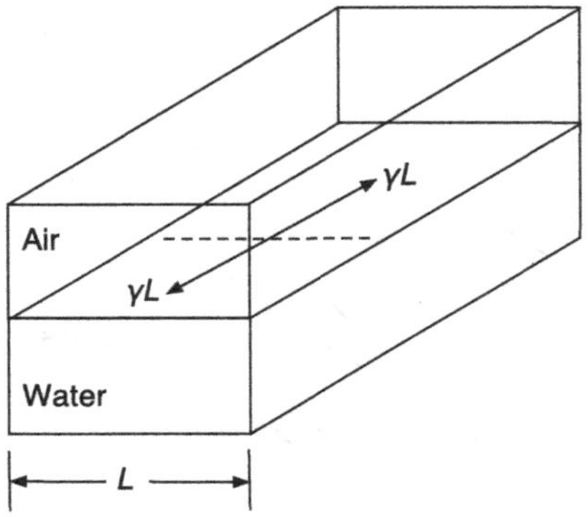

Figure 1.5 Surface tension at a flat air–water interface. A force of magnitude γL pulls away from each side of the imaginary cut shown by the dashed line.

purely empirical and others derived from molecular-level mechanics (Bird *et al.*, 1987; Tanner, 2000).

Surface tension

There is a surface tension (γ) associated with the interface between a gas and a liquid, or between immiscible liquids. Surface tension may be viewed either as an energy per unit area or as a force per unit length (J/m^2 or N/m in SI units). In thermodynamics, it is the energy required to increase the interfacial area; in mechanics, the usual perspective in this book, it is a force that acts on an imaginary line or contour within an interface. The direction of the force is tangent to the interface and away from the line or contour. The surface tension force at a planar interface is illustrated in Fig. 1.5, which shows a container of width L that is partly filled with water. At an imaginary cut through the air–water interface (dashed line), there are opposing forces of magnitude γL acting horizontally on each side of the line of length L. The forces are *tensile*, meaning that each pulls on the imaginary line.

If an interface is flat and γ is constant, the surface tension forces will balance and not be observable. However, surface tension can become very evident when an interface is nonplanar, when γ varies with position (due to gradients in temperature or surfactant concentration), or when an interface ends at a three-phase *contact line*. A contact line is created, for example, at the edge of a liquid drop resting on a solid surface, where solid, liquid, and gas all meet. The surroundings exert a force that pulls away from the contour that corresponds to the contact line. That force may cause the drop to either spread or contract, depending on the angle the gas–liquid interface makes with the surface, which is called the *contact angle*. Contact angles and other aspects of surface tension are discussed in much more detail in Chapter 4. Surface tension affects the dynamics of drops, bubbles, and foams, causes wetting of surfaces, and leads to fluid uptake by porous media. These effects are most prominent at small length scales. Many such phenomena are described in Adamson and Gast (1997) and De Gennes *et al.* (2003).

Table 1.3 shows values of γ for several liquids in contact with air. The surface tension for water (about 7×10^{-2} N/m or 70 mN/m) is about three times that for benzene, ethanol, or other common organic solvents (20–30 mN/m). Mercury is unusual in that γ is about 20 times that for the organics (almost 500 mN/m). Surface tensions decrease with increasing temperature, as exemplified by the entries for water and mercury.

The surface tension for water in contact with various immiscible liquids is less than that for water and air. For example, at room temperature $\gamma = 45.7$ mN/m for decane/water and 16.4 mN/m for olive oil/water (Than *et al.*, 1988).

Table 1.3 Surface tension for liquid–air interfaces[a]

Liquid	T(K)	γ(mN/m)
Water	293	72.88
	303	71.40
	313	69.92
Benzene	293	28.88
Ethanol	293	22.39
Mercury	293	486.5
	303	484.5

[a] From Jasper (1972).

Continuum approximation

Little more is said in this book about molecules, because in continuum mechanics they are ignored. Indeed, continuum descriptions of fluid mechanics were already well advanced by the mid 1700s, long before the existence of molecules was firmly established. Continuum modeling is natural in that, in our everyday experiences with air and water, we are unaware of individual molecules. Of great importance, postulating that velocities, stresses, and material properties are smoothly varying functions of position allows us to use the tools of calculus. In general, this approach hinges on there being enough molecules present in a system to average their effects. That is easiest to visualize with density. When we speak of the local value of ρ as the ratio $\Delta m/\Delta V$ in the limit $\Delta V \to 0$, what is really meant is that there exists a ΔV that is small compared with the system volume, but still large enough to contain a predictable number of molecules. If ΔV were so small that it enclosed only a few molecules, ρ would fluctuate wildly, according to the thermal motion of molecules into or out of the imaginary box. Thus, ρ becomes unpredictable if the system is too small. The same is true for the other properties we have discussed. If the number of molecules present is inadequate, flow models must be stochastic (based on probabilities) rather than deterministic.

To have continuum behavior with property values that are independent of system size, the smallest linear dimension of the system (ℓ) must greatly exceed whichever is the larger of two molecular length scales. One is the intermolecular spacing, which may be expressed as an average center-to-center distance (r). The other is the mean distance traveled in a random "jump." Single molecular displacements in liquids are typically a small fraction of r, whereas the average distance between collisions in a gas (λ, the mean free path) greatly exceeds r. This makes r limiting in liquids and λ limiting in gases. The need to have $\ell \gg r$ is evident already from the discussion of density, and $\ell \gg \lambda$ is required to make collisions among the gas molecules much more frequent than their collisions with boundaries. Collisions among the gas molecules are what transmit momentum in a gas, and their frequency determines μ. If a container is so small that collisions occur mainly with the walls, the usual values of μ will be inapplicable.

A typical value of r for water or organic solvents is 0.4 nm. Allowing for a factor of 10, this suggests $\ell > 4$ nm as a criterion for the continuum modeling of such liquids. From kinetic theory (Kennard, 1938), the mean free path in a gas is

$$\lambda = \frac{k_B T}{\sqrt{2}\, \pi d^2 P} \tag{1.2-7}$$

where k_B is Boltzmann's constant (1.38×10^{-23} J K^{-1}) and d is the molecular diameter. For $d = 0.3$ nm and ambient conditions ($P = 101$ kPa and $T = 293$ K), $\lambda = 0.1$ μm. Again allowing a factor of 10, this suggests $\ell > 1$ μm for gases. Because λ varies as P^{-1}, in rarefied gases the minimum system dimension increases markedly. It is assumed throughout this book that the criteria for continuum analysis are met. Additional discussion of the continuum approximation, including experimental support for the suggested criteria, is presented in Deen (2012, pp. 19–21).

1.3 SCALES AND DIMENSIONLESS GROUPS

Scales

As mentioned already, the *scale* of a variable is its maximum value. Order-of-magnitude precision suffices, and we are concerned only with absolute values, not algebraic signs. When discussing scales a distinction is made between independent variables (position or time) and dependent ones (such as velocity, pressure, or shear stress). For an independent variable, what is sought usually is not its maximum per se, but rather the increment in it that gives a significant change in a dependent variable. For example, a human heart beats approximately once per second. Whether the cardiovascular function of an individual is monitored for minutes or hours, the time scale for oscillations in arterial pressure is one second. Thus, length or time scales are usually *changes* in position or time that correspond to something of interest.

As shown in Parts II and III of this book, the differential equations that govern fluid dynamics are complex enough that approximations are almost always needed to obtain useful results. Including all possible forces in an analytical or computational model is ordinarily impractical. However, such inclusiveness also tends to be unnecessary, in that certain effects are usually dominant in a given system, while others are negligible. Comparing force scales is a way to identify what is most important in a particular problem. Although the utility of various approximations will not be fully evident until later, we can begin now to think about what is "large" or "small" and what kinds of simplifications might be possible.

The magnitudes of forces in fluids are determined by the material properties discussed in Section 1.2 and by the length and velocity scales in particular systems. A sampling of length scales (L) and velocity scales (U) is given in Table 1.4. The systems are as follows: a colloidal particle settling in water; a bacterium swimming in water; blood plasma flowing in a capillary; water flowing in a microfluidic device; a bubble rising in water; blood flowing in a large artery; a liquid flowing through a pipe in a chemical plant; crude oil flowing in a large pipeline; and air movement relative to automobiles or trains that are traveling at highway or passenger-train speeds. In each case L is a diameter or other cross-sectional dimension and U is either a mean velocity in a channel or the velocity difference between an object and the surrounding fluid. Except possibly for the transportation vehicles, each system has been subjected to analysis or design by chemical engineers. Even though fields such as meteorology, ship or aircraft design, and geophysics have been excluded from Table 1.4, the length and velocity ranges are remarkably broad, spanning 7 and 10 orders of magnitude, respectively. The range for the product LU covers some 17 orders of magnitude, reflecting the fact that L and U tend to vary in parallel. The range for Re (to be defined shortly) is similarly broad.

Table 1.4 Length and velocity scales in fluid mechanics[a]

System	L(m)	U(m s^{-1})	LU(m^2 s^{-1})	Re
Colloidal particle	1×10^{-7}	5×10^{-9}	5×10^{-16}	5×10^{-10}
Bacterium	1×10^{-6}	1×10^{-5}	1×10^{-11}	1×10^{-5}
Blood capillary	1×10^{-5}	1×10^{-3}	1×10^{-8}	1×10^{-2}
Microfluidic channel	1×10^{-4}	1×10^{-2}	1×10^{-6}	1
Air bubble	1×10^{-3}	0.3	3×10^{-4}	3×10^{2}
Major artery	3×10^{-2}	0.1	3×10^{-3}	8×10^{2}
Process pipe	0.1	1	0.1	1×10^{5}
Oil pipeline	1	3	3	3×10^{5}
Automobile or train	3	30	90	6×10^{6}

[a] Except as noted below, the values of ν used to calculate Re $= LU/\nu$ were those for either water or air at room temperature. The kinematic viscosities for arterial blood in the body and crude oil in a heated pipeline were assumed to be 4 and 10 times that for room-temperature water, respectively. For blood plasma at body temperature (37 °C), ν is nearly the same as for water at 20 °C. For the air bubble, ν for water was used.

Table 1.5 Dimensions of quantities involving mass, length, and time

Quantity	Dimension	Quantity	Dimension
Mass	M	Energy	ML^{2}T^{-2}
Length	L	Power	ML^{2}T^{-3}
Time	T	Viscosity	ML^{-1}T^{-1}
Velocity	LT^{-1}	Density	ML^{-3}
Acceleration	LT^{-2}	Kinematic viscosity	L^{2}T^{-1}
Force	MLT^{-2}	Surface tension	MT^{-2}
Stress or pressure	ML^{-1}T^{-2}	Angle	None

Dimensions

In mechanics, a sufficient set of primary dimensions is mass (M), length (L), and time (T). Various dimensions derived from those are shown in Table 1.5. In equations that involve dimensions, the quantities themselves are placed inside square brackets. For example, the dimension of pressure is written as

$$[P] = \mathrm{ML}^{-1}\mathrm{T}^{-2}. \tag{1.3-1}$$

The singular "dimension" is used, not the plural "dimensions," because what is being referred to is a specific combination of the primary dimensions. Dimensions are denoted by Roman letters (e.g., L), whereas variables or parameters are italicized (e.g., L).

Stress scales

Four independent groups that have the dimension of stress can be constructed from μ, ρ, γ, g, L, and U, as will be shown next. Each is a stress scale associated with a particular kind of force. The corresponding force scales are obtained by multiplying each by the area scale, L^2. Accordingly, ratios of forces are the same as ratios of stresses.[3]

The *inertial stress* scale is the force required to effect a significant change in the momentum of a fluid. It may be inferred by calculating the constant force that would either bring an element of fluid to a stop, or double its momentum, within a certain time. Suppose that the fluid element is a cube with edge length ℓ. Its initial velocity is U and the time we choose is how long it takes to move one length at that speed, or ℓ/U. The mass of the fluid element is $\rho\ell^3$ and its initial momentum is $\rho\ell^3 U$. The force that will *change* the momentum by that amount is expressed in terms of a stress acting on one face of the cube. Equating the change in momentum with the product of stress, area, and time gives $\rho\ell^3 U = (\text{stress})(\ell^2)(\ell/U)$, or stress $= \rho U^2$.

The inertial stress may be inferred also from a dimensional argument. In that ρU is the momentum per unit volume, ρ and U must be involved, and possibly also L. The remaining quantities in the full list (μ, γ, and g) are associated with other kinds of stresses and not relevant to inertia. What we seek then is a product of ρ, U, and L, each raised to some power, that has the dimension of stress. The dimensional requirement is written as

$$\mathrm{ML^{-1}T^{-2}} = [\rho]^a[U]^b[L]^c = [\mathrm{ML^{-3}}]^a[\mathrm{LT^{-1}}]^b[\mathrm{L}]^c \tag{1.3-2}$$

where the exponents a, b, and c are to be determined. For this relationship to be satisfied, the exponents of M, L, and T each must match. Thus,

$$\text{M:}\quad 1 = a \tag{1.3-3a}$$

$$\text{L:}\quad -1 = -3a + b + c \tag{1.3-3b}$$

$$\text{T:}\quad -2 = -b. \tag{1.3-3c}$$

The solution to Eq. (1.3-3) is $a = 1$, $b = 2$, and $c = 0$, indicating again that the inertial stress scale is ρU^2. Notice that L is excluded on dimensional grounds.

The other three stress scales are easily identified. As shown by Eq. (1.2-1), a shear stress is a viscosity times a velocity, divided by a length. Accordingly, $\mu U/L$ is the *viscous stress* scale. In that ρg is the force per unit volume due to gravity, it must be multiplied by L to have the dimension of stress, giving ρgL as the *gravitational stress* scale. Because γ is a force per unit length, it must be divided by L. Thus, the stress associated with *surface tension* is γ/L. That each of these scales has the correct dimension ($\mathrm{ML^{-1}T^{-2}}$) may be confirmed by writing expressions similar to Eq. (1.3-2).

Dimensionless groups

Dimensionless groups provide a way to compare scales. Considering all possible ratios of the four stress (or force) scales, six independent dimensionless groups or "numbers" can be constructed. Five of these arise often enough to have been named, most in honor of a pioneering researcher. The dimensionless stress ratios are given in Table 1.6, along with their two-letter symbols and names.

The most important of these dimensionless groups is the *Reynolds number* (Re), which is the ratio of inertial to viscous stresses.[4] As shown in Chapters 2 and 3, it is

[3] Although it is a type of stress that will receive much attention later, pressure can be ignored at present because it is mainly a consequence of other forces. As stated in Kármán (1954, p. 75), "In an incompressible fluid, the pressure is a kind of passive reaction, whose magnitude is just sufficient to balance the other forces acting on a fluid element."

[4] Osborne Reynolds (1842–1912) was an engineering professor at the University of Manchester, England. His experiments with water flow in pipes, reported in 1883, established the importance of the dimensionless

Table 1.6 Dimensionless groups corresponding to stress or force ratios

Stress or force ratio	Definition	Name
$\dfrac{\text{Inertial}}{\text{Viscous}}$	$\text{Re} = \dfrac{\rho U L}{\mu} = \dfrac{UL}{\nu}$	Reynolds
$\dfrac{\text{Inertial}}{\text{Gravitational}}$	$\text{Fr} = \dfrac{U^2}{gL}$	Froude
$\dfrac{\text{Inertial}}{\text{Surface Tension}}$	$\text{We} = \dfrac{\rho U^2 L}{\gamma}$	Weber
$\dfrac{\text{Viscous}}{\text{Gravitational}}$	$\dfrac{\text{Fr}}{\text{Re}} = \dfrac{\mu U}{\rho g L^2} = \dfrac{\nu U}{g L^2} = \dfrac{\text{Ca}}{\text{Bo}}$	–
$\dfrac{\text{Viscous}}{\text{Surface Tension}}$	$\text{Ca} = \dfrac{\mu U}{\gamma}$	Capillary
$\dfrac{\text{Gravitational}}{\text{Surface Tension}}$	$\text{Bo} = \dfrac{\rho g L^2}{\gamma}$	Bond

central in correlating experimental data on forces and flow rates. It is also the most important determinant of the amount of mathematical complexity encountered in theoretical analyses, as seen in Part III. When Re is much smaller than unity, calculations are greatly simplified because the differential equations that describe conservation of momentum become linear. On the other hand, when Re exceeds a certain critical value, there is typically a transition from laminar flow, which is orderly, to turbulent flow, which is chaotic. The critical value of Re depends on the system, but is often at least in the thousands. Values of Re for various systems are given in Table 1.4. The examples include flow at low Re (the first three entries), laminar flow at moderate to large Re (the next three), and turbulent flow (the last three).

Dimensionless groups need not be ratios of just stresses or forces. Three not listed in Table 1.6 arose implicitly in Section 1.2. The ratio of the mean free path in a gas to the system dimension is the *Knudsen number*, Kn. Thus, what is needed for continuum modeling of gas flows in small spaces is $\text{Kn} \ll 1$. The ratio of fluid velocity to the speed of sound is the *Mach number*, Ma. Neglecting the compressibility of gases requires that $\text{Ma} \ll 1$. The relaxation time of a polymer molecule divided by the experimental time scale is the *Deborah number*, De. The effects of viscoelasticity are negligible when $\text{De} \ll 1$.

Dimensionless groups often can be viewed as ratios of time scales. A *process* time scale (t_p) is either the overall duration of a process or a time scale imposed on a system. The time required to empty a tank and the period of a heart are examples of an overall duration and an imposed time scale, respectively. The process time scale may be compared with one derived from the characteristic length and velocity, which is the *convective* time scale $t_c = L/U$. The ratio of process to convective time scales is the *Strouhal number*, $\text{Sr} = t_p U/L$. Any ratio of rates is also a ratio of time scales, the smaller the time scale, the faster the rate. Of course, a force is a rate of momentum transfer. The Reynolds number can be interpreted as $\text{Re} = t_v/t_c$, where $t_v = L^2/\nu$ is the *viscous* time scale. Comparing time scales is especially informative when a process involves a series of events. If

group that would be named for him. He also pioneered the theory of hydrodynamic lubrication (Chapter 8) and established a procedure for time-averaging that is still used in the analysis of turbulent flow (Chapter 10).

one of the obligatory steps has a much larger time scale than the others, it is what will limit how fast the process can be completed.

Given the multitude of topics encountered in the physical sciences and engineering, the potential number of dimensionless groups is nearly limitless. Some 200 have appeared in the chemical engineering literature (Catchpole and Fulford, 1966). This daunting total notwithstanding, most dimensionless groups of importance in fluid mechanics are either listed in Table 1.6 or were mentioned in the two preceding paragraphs.

Example 1.3-1 Deep-water waves. Which forces most affect waves in large bodies of water? Consider a series of waves that is characterized by a wavelength (λ), a wave height (h), and a wave speed (c).

One pertinent comparison is between gravity and surface tension, in that both tend to level the surface of a liquid. Gravity does this by causing downward flow from peaks to troughs, and surface tension acts much as stretching smooths a wrinkled piece of cloth. The *Bond number* (Table 1.6) provides the desired ratio. Setting $L = h = 0.1$ m gives

$$\text{Bo} = \frac{\rho g h^2}{\gamma} = \frac{(1 \times 10^3)(9.81)(0.1)^2}{7 \times 10^{-2}} = 1.4 \times 10^3.$$

Because waves in oceans or large lakes are unlikely to be shorter than the 0.1 m assumed here, and can be much taller, it is safe to conclude that surface tension is negligible relative to gravity. In that $\lambda > h$ typically, Bo would be even larger with $L = \lambda$. The length at which Bo = 1, namely $[\gamma/(\rho g)]^{1/2}$, is called the *capillary length*. Thus, Bo may be viewed not just as a stress or force ratio, but as the system size divided by the capillary length (each squared). Surface tension competes well with gravity only when L is comparable to or smaller than the capillary length, which is about 3 mm for water at room temperature. Thus, the conclusion that Bo $\gg 1$ in deep-water waves is conservative.

The value of the Reynolds number is always germane. Using $L = h = 0.1$ m again and assuming that $U = c = 1$ m/s,

$$\text{Re} = \frac{hc}{\nu} = \frac{(0.1)(1)}{1 \times 10^{-6}} = 1 \times 10^5.$$

This suggests that viscous stresses are negligible relative to inertial ones. The conclusion again is conservative, in that both the length and the velocity are more likely to exceed the assumed values than to be much smaller than them. A caveat is that, when solid surfaces are present, viscous stresses are not necessarily negligible when Re is large. However, solid surfaces are not a concern with deep-water waves.

A third comparison, between inertia and gravity, is provided by the *Froude number* (Table 1.6). Using the same length and velocity as before,

$$\text{Fr} = \frac{c^2}{gh} = \frac{(1)^2}{(9.81)(0.1)} = 1.0.$$

That Fr = 1, almost exactly, is coincidental, given that the values of c and h are only order-of-magnitude estimates. What is significant is that Fr does not differ greatly from unity, which supports the view that inertia and gravity are of comparable importance. The overall conclusion is that only gravity and inertia significantly affect the motion of deep-water waves. Wave motion is discussed further in Example 1.4-1.

Example 1.3-2 Inkjet printing. The breakup of a liquid jet into drops is crucial for the performance of an inkjet printer. Which forces most affect the flow, once the jet has left the print nozzle? Reasonable length and velocity scales are the nozzle diameter and jet speed, respectively. Based on representative data in Wijshoff (2010), we assume that $L = 32\ \mu\text{m}$, $U = 7$ m/s, $\mu = 1 \times 10^{-2}$ Pa · s, $\gamma = 3 \times 10^{-2}$ N/m, $\rho = 1 \times 10^{3}$ kg/m^3, and $\nu = 1 \times 10^{-5}$ m^2/s. Notice the unusual combination of a small length and a large velocity. The ink viscosity is 10 times that of water and its surface tension is about 40% that of water.

We begin again by comparing gravity with surface tension. The Bond number is

$$\mathrm{Bo} = \frac{\rho g L^2}{\gamma} = \frac{(1 \times 10^3)(9.81)(3.2 \times 10^{-5})^2}{3 \times 10^{-2}} = 3.4 \times 10^{-4}$$

which indicates that gravity is negligible in this case. For the ink, the capillary length is about 2 mm, which greatly exceeds the nozzle diameter. A comparison of viscous vs. surface-tension forces is provided by the *capillary number* (Table 1.6),

$$\mathrm{Ca} = \frac{\mu U}{\gamma} = \frac{(1 \times 10^{-2})(7)}{3 \times 10^{-2}} = 2.3.$$

This finding that Ca does not differ greatly from unity supports the idea that drop formation involves a competition between surface tension (which favors jet breakup) and viscous stresses (which resist it). The Reynolds number,

$$\mathrm{Re} = \frac{LU}{\nu} = \frac{(3.2 \times 10^{-5})(7)}{1 \times 10^{-5}} = 22$$

is large enough to indicate that inertia is important in the jet, but not so large as to preclude a role for viscous stresses. When working only with order-of-magnitude estimates such as these, caution is needed in excluding forces from further consideration. The overall conclusion is that inertial, viscous, and surface tension forces are all important in this system, whereas gravity is negligible. This is supported by the experimental and computational results reviewed in Wijshoff (2010). The Ohnesorge number (Problem 1.7) directly compares the relative importance of inertia and viscosity in resisting the drop breakup that is caused by surface tension.

1.4 DIMENSIONAL ANALYSIS

Some of the insights that can be gained by combining physical quantities into dimensionless groups were illustrated in Section 1.3. We now consider more generally what can be learned from the dimensions of the quantities involved in a process or phenomenon. *Dimensional analysis* has two main objectives: to express physical relationships in the most universal manner, and to minimize the number of variables or parameters that must be considered in experiments or calculations. For anyone new to this topic but interested in more depth than what follows, good sources are Bridgman (1931) and Chapter 1 of Barenblatt (1996).

Pi theorem

The key result, called the *Pi theorem*, may be stated as follows: *If a process or phenomenon is governed by n quantities that contain m independent dimensions, then the number of independent dimensionless groups that govern the process or phenomenon is $n - m$.* Put another way, if one of the dimensionless groups is an unknown that we wish to determine, then it is a function of at most $n - m - 1$ others. If a quantity in the list of what is involved is already dimensionless, then it is a "group" in itself. A proof of the Pi theorem may be found, for example, in Bridgman (1931, pp. 37–40).

A key word in the Pi theorem is "independent." If any dimension in a set can be constructed as a product of the others (perhaps each raised to some power), then it is not independent. The most obvious form of dependence is if two or more quantities have the same dimension, in which case only one is independent. Moreover, simply inverting something or raising it to any power does not produce an independent dimension or an independent dimensionless group.

In discussions of how to use the Pi theorem, m is sometimes equated with the number of primary dimensions (e.g., from among M, L, and T). That is usually, but not always, the same as the number of independent dimensions. The need for a distinction is illustrated by a simple example. Suppose that some phenomenon involves just two velocities, u and v. In that both have the dimension LT^{-1}, there is only one independent dimension among the two governing quantities. Clearly, there is also one independent dimensionless group. That group is u/v or, more generally, $(u/v)^a$, where a can differ from unity. This is consistent with the Pi theorem as stated, for which $n - m = 2 - 1 = 1$. Using the number of primary dimensions (L and T) would have given $m = 2$ along with $n = 2$ and incorrectly predicted an absence of dimensionless groups.

Example 1.4-1 Speed of water waves. It was concluded in Example 1.3-1 that waves in large bodies of water are affected only by inertia and gravity. Putting viscosity and surface tension aside, the list of quantities that might be important in wave dynamics includes ρ, g, the wave speed (c), the wavelength (λ), the mean depth of the water (H), and the wave height (h). The objective is to use dimensional analysis to relate c to the other five governing quantities. In other words, c will be viewed as an unknown that depends on five parameters.

Of the six quantities, ρ, c, and λ are dimensionally independent of one another. Among these three, only ρ contains mass, only c contains time, and no product of powers of ρ and c could then cancel mass and time to give only the length in λ. In that H and h also have the dimension of length, they are not independent of λ. The dimension of g (LT^{-2}) is the same as that of c^2/λ, so that it too is not independent of the others. Thus, the six quantities involve three independent dimensions, which happens to equal the number of primary dimensions.

With $n = 6$ and $m = 3$, the Pi theorem indicates that there are just three independent dimensionless groups, which will be denoted as N_1, N_2, and N_3. It is convenient to isolate the unknown in the first of these. That can be done by defining N_1 as the ratio of c to a velocity scale derived from ρ, g, and λ. (We could use H or h here instead of λ, but including more than one length would add no dimensional information.) The velocity scale is determined by[5]

$$LT^{-1} = [\rho]^a[g]^b[\lambda]^c = [ML^{-3}]^a[LT^{-2}]^b L^c. \tag{1.4-1}$$

Because M is absent from the left-hand side and is contained only in ρ, it is evident that $a = 0$. The remaining exponents in Eq. (1.4-1) are determined from

$$\text{L:} \quad 1 = b + c \tag{1.4-2a}$$

$$\text{T:} \quad -1 = -2b. \tag{1.4-2b}$$

[5] The exponent c should not be confused with the wave speed.

Accordingly, $b = c = 1/2$ and the first dimensionless group is

$$N_1 = \frac{c}{(g\lambda)^{1/2}} \tag{1.4-3}$$

which is the square root of a Froude number based on c and λ (Table 1.6).

In that ρ is the only one of the six quantities that contains M, it is excluded not just from N_1, but also from the other dimensionless groups. Thus, ρ must not be involved at all, and actually $n = 5$ and $m = 2$. The total number of dimensionless groups is still three. A second and third independent group are created by using one quantity in each that was not present in N_1, namely H and h. The simplest choices are ratios of lengths,

$$N_2 = \frac{H}{\lambda} \tag{1.4-4}$$

$$N_3 = \frac{h}{\lambda}. \tag{1.4-5}$$

These choices are not unique. For example, either could be replaced by h/H.

With only three dimensionless variables, the wave speed is governed by a function that might be written as

$$F(N_1, N_2, N_3) = 0. \tag{1.4-6}$$

Although dimensional analysis tells us nothing about the form of this function, in principle we could solve for N_1 and rewrite the relationship as

$$N_1 = f(N_2, N_3) \tag{1.4-7}$$

where $f(x, y)$ is another function. Thus, the conclusion is that

$$\frac{c}{(g\lambda)^{1/2}} = f\left(\frac{H}{\lambda}, \frac{h}{\lambda}\right). \tag{1.4-8}$$

Although dimensional analysis did not reveal the form of $f(x, y)$, it was very useful nonetheless. That is, a problem that involved five dimensional variables (plus the constant g) was reduced to one with just three dimensionless variables. Suppose that we wanted to investigate wave speeds experimentally, and had a way to measure c for various λ, H, and h. Using Eq. (1.4-8) as guidance and doing experiments involving combinations of 10 values each of H/λ and h/λ, a total 100 experiments might provide a reasonable representation of $f(x, y)$. However, without dimensional analysis we might think it necessary to combine 10 levels each of λ, H, h, and ρ, entailing 10,000 experiments!

The more variables that can be eliminated using experiment, theory, or physical insight, the more robust are the conclusions from dimensional analysis. Two suppositions that prove to be true for waves are: (i) for $H/\lambda \to \infty$, the bottom is too distant to affect the motion near the surface; and (ii) for $h/\lambda \to 0$, the wave height is too small to influence c. In these limits, the precise values of H/λ and h/λ no longer matter. Put another way, $f(x, y)$ evidently approaches a constant as $x \to \infty$ and $y \to 0$.[6] Denoting that constant as K, Eq. (1.4-8) becomes

$$c = K(g\lambda)^{1/2} \quad (H/\lambda \to \infty \text{ and } h/\lambda \to 0). \tag{1.4-9}$$

[6] Conversely, if a function such as $f(x, y)$ is known to be constant under certain conditions, it usually indicates that at least one of its arguments is very small or very large.

The detailed theoretical analysis in Example 9.2-3 confirms that c for small-amplitude waves in deep water varies as $(g\lambda)^{1/2}$. Further, it indicates that $K = (2\pi)^{-1/2} = 0.399$.

Example 1.4-2 Shear stress in pipe flow. Steady (time-independent) flow of a Newtonian fluid in a cylindrical pipe is the most common application of fluid mechanics in chemical engineering. The objective is to use dimensional analysis to see what determines the wall shear stress τ_w in a long, liquid-filled pipe with a diameter D and a mean fluid velocity U.

Experiment and theory show that, in a sufficiently long pipe with a smooth wall, τ_w is determined only by U, D, ρ, and μ. It is found that τ_w in such a pipe is uniform over the surface, so that pipe length and position along the pipe do not require consideration. In the absence of a fluid–fluid interface, surface tension can play no role. Moreover, in a filled pipe gravity has no independent significance, although g can affect U (Chapter 2).

The dimensions of D, U, and ρ are independent; because M appears only in ρ and T is present only in U, no combination of ρ and U can cancel both M and T, as would be needed to obtain the dimension L of D. However, μ is not independent, as both ρUD and $\tau_w D/U$ also have the dimension $\mathrm{ML^{-1}T^{-1}}$. Because the inertial stress (ρU^2) and viscous stress ($\mu U/D$) are both involved, τ_w also does not have an independent dimension.

With five governing quantities and three independent dimensions, the Pi theorem indicates that there are $5 - 3 = 2$ independent dimensionless groups. Because τ_w is the unknown, it will be isolated in N_1. The simplest possibilities for N_1 are to divide τ_w by either ρU^2 or $\mu U/D$. Following a widely used convention for pipe flow, we choose one-half of the inertial stress scale and let $N_1 = 2\tau_w/(\rho U^2)$. Aside from τ_w, the quantities in the list are what is needed to form a Reynolds number. Thus, we set $N_2 = \mathrm{Re} = UD\rho/\mu = UD/\nu$. The conclusion is then

$$\frac{2\tau_w}{\rho U^2} = f(\mathrm{Re}). \tag{1.4-10}$$

The function f for pipe flow is called the *friction factor*, and how it is evaluated from data and used in design calculations is the focus of Chapter 2. In that a set of five dimensional quantities has been reduced to just two dimensionless ones, the benefits of dimensional analysis are even greater than in the preceding example.

Dimensional analysis does not reveal which is the better choice for the denominator of N_1, the inertial stress or the viscous one. Highlighting inertia, as done in Eq. (1.4-10), reflects a bias toward the large values of Re that are often encountered (Table 1.4). That is, when Re is large, inertia is prominent. However, if we were interested only in pipe flow at small Re, it would be reasonable to suppose that inertia is negligible and to adopt the viscous stress. The alternate definition of N_1 would give

$$\frac{\tau_w D}{\mu U} = K \quad (\mathrm{Re} \ll 1) \tag{1.4-11}$$

where K is a constant. The underlying assumption is that the precise value of Re no longer matters, if it is small enough. With one variable (Re) eliminated, the remaining one (the new N_1) must equal a constant, as shown. Dimensionally, neglecting inertia is equivalent to removing ρ from the list of governing quantities, which reduces the number of groups to one. It is known from both experiment and theory that $f = 16/\mathrm{Re}$ for $\mathrm{Re} < 2100$ (Chapter 2), which indicates that $K = 8$. A more detailed analysis of pipe flow (Chapter 7) is needed to understand why Eq. (1.4-11) does not actually require that $\mathrm{Re} \ll 1$, and why $\mathrm{Re} < 2100$ is sufficient.

Example 1.4-3 Energy of an atomic blast. In what became a very famous paper, the British fluid dynamicist G. I. Taylor estimated the energy of an atomic explosion (E) from a time series of photographs of an early US bomb test (Taylor, 1950). Although his calculations were based on a detailed analysis of the dynamics of a spherical shock wave created by an intense explosion, his principal result can be obtained by dimensional analysis alone (Barenblatt, 1996, pp. 47–50). Images that had been released in 1947 showed the radius of a hemispherical fireball (R) as a function of time (t), with $t = 0$ corresponding to the nearly instantaneous detonation. In that an explosion pushes against the surrounding atmosphere, it may be reasoned that the rate of expansion of the hemisphere (the boundary of which corresponded to a sharp decrease in pressure) was affected by the ambient air density (ρ). Following this reasoning, the governing quantities consist of two variables (R and t) plus two constants (E and ρ).

It is readily confirmed that R, t, and ρ are dimensionally independent of one another. With four quantities and three independent dimensions, the Pi theorem indicates that there will be just one dimensionless group. That group can be constructed by dividing E by an energy scale derived from R, t, and ρ. The energy scale is identified using

$$\mathrm{ML^2T^{-2}} = [R]^a[t]^b[\rho]^c \tag{1.4-12}$$

from which it follows that

$$\mathrm{M}: \quad 1 = c \tag{1.4-13a}$$

$$\mathrm{L}: \quad 2 = a - 3c \tag{1.4-13b}$$

$$\mathrm{T}: \quad -2 = b. \tag{1.4-13c}$$

Thus, $a = 5$, $b = -2$, and $c = 1$. In that there are no other groups to affect it, the dimensionless energy evidently is a constant. Denoting that constant as K, the conclusion is that

$$E = \frac{KR^5\rho}{t^2} \tag{1.4-14}$$

in agreement with what was found by Taylor (1950). Given that E and ρ are constants, the analysis predicts that $R \propto t^{2/5}$.

To evaluate E from the photographs, Taylor took the square root of both sides of Eq. (1.4-14) and rearranged the result to give a straight line in a log–log plot. The equation of the line was[7]

$$\frac{5}{2}\log R = \frac{1}{2}\log\left(\frac{E}{K\rho}\right) + \log t. \tag{1.4-15}$$

When measurements from the photographs were plotted as $(5/2)\log R$ versus $\log t$, a straight line with the predicted slope of unity was obtained, and the intercept provided the value of $E/(K\rho)$. The value of K is not revealed by dimensional analysis, but detailed analysis of the gas dynamics suggested that $K = 0.86$ (Taylor, 1950). This gave $E = 7.1 \times 10^{20}$ erg $= 7.1 \times 10^{13}$ J.

The foregoing examples included three situations in which a dimensionless unknown was inferred to equal a constant, denoted as K. Using experimental or theoretical information to supplement what could be learned from dimensional analysis, the values of K in Eqs. (1.4-9), (1.4-11), and (1.4-14) were found to be 0.399, 8, and 0.86, respectively. Obtaining an expression in which the unknown constant is within a factor of 10 of unity

[7] In this book the natural logarithm of x is denoted as $\ln x$ and the base-10 logarithm as $\log x$.

is a regular outcome of this kind of analysis. Thus, even without supplementary information, one can usually find the correct order of magnitude for the unknown physical quantity by assuming simply that $K = 1$. This ability to get the correct order of magnitude is evidence that whatever was large or small in the system being analyzed was properly accounted for in the scales used to define the dimensionless groups.

Dynamic similarity

In addition to minimizing the number of independent variables, dimensional analysis provides a way to extrapolate experimental results to systems that are larger or smaller, have flow that is faster or slower, or have different fluid properties. In general, it indicates that a dimensionless group N_1 is a function of two other kinds of groups. One kind, consisting of N_2, N_3, etc., involves fluid properties and/or velocities. The other involves only length ratios, also called *aspect ratios*. These are denoted now as ϕ_i, where $i = 1$, 2, etc. For example, the shape of an ellipse is fully characterized by one aspect ratio, the minor axis divided by the major axis. The more complicated the shape of an object, the greater the number of ϕ_i values needed to describe it. In general, the group of interest is related to the others as

$$N_1 = f(N_2, N_3, \ldots, \phi_1, \phi_2, \ldots). \tag{1.4-16}$$

In a complex system that has not yet been studied, the form of the function f is typically unknown.

Suppose that experiments are done using a scale model. By definition, all ϕ_i in the model are the same as in the system it represents. Any two such objects are said to be *geometrically similar*. If, in addition, the fluid properties and characteristic velocity in the experiments are chosen so that N_2, N_3, etc., are each the same as in the real system, the two systems are *dynamically similar*. Because the value of each argument of f in Eq. (1.4-16) is then the same, so is the value of N_1. Thus, N_1 for the full-size system can be predicted from that measured with the model, even if the equations that govern the flow are unknown or too difficult to solve. Experiments with land-vehicle or aircraft designs in wind tunnels, or ship designs in towing tanks, seek to exploit this concept. However, because dynamic similarity is much more difficult to achieve than geometric similarity, compromises are often needed in designing and interpreting such experiments. Extrapolating data for pipe flow (as in Chapter 2) is more straightforward, because it is relatively easy to make the flow in one pipe dynamically similar to that in another.

1.5 CONCLUSION

The viscosity of a fluid (μ), which is a measure of its resistance to flow, is the shear stress at a test surface divided by the shear rate. If μ is independent of the shear rate, the fluid is Newtonian; otherwise, it is non-Newtonian. All gases are Newtonian, as are water and many organic solvents. For water and such solvents at room temperatures, μ ($\sim 10^{-3}$ Pa · s) is roughly 100 times that for air and other common gases ($\sim 10^{-5}$ Pa · s). There are other Newtonian liquids, such as lubricating oils, with much larger viscosities. Non-Newtonian liquids include polymer melts, various polymer solutions, and certain concentrated suspensions. For common liquids the density (ρ) is roughly 1000 times that for gases at ambient temperature and pressure ($\sim 10^3$ kg/m^3 vs. 1 kg/m^3). Liquids with large viscosities often have ordinary densities, and a dense liquid need not be highly viscous.

A stress is a force per unit area. The stresses in a fluid are influenced by its material properties (μ, ρ, and the surface tension γ), the gravitational acceleration (g), and the length scale (L) and velocity scale (U) of the system. Identifying L and U is often straightforward, so that the various kinds of stresses can be estimated without detailed knowledge of a flow. There are four characteristic stress scales: ρU^2 for inertia, $\mu U/L$ for viscosity, γ/L for surface tension, and $\rho g L$ for gravity. Each is the order of magnitude of stresses with a particular origin. Examining ratios of these scales is helpful in deciding which kinds of forces are most influential in a particular system. The Reynolds number (Re), which is the inertial stress scale divided by the viscous stress scale, is the most important such ratio.

In general, using dimensionless variables yields relationships that are more universal, because they are independent of units, and ones that are more efficient for experimental design and data analysis, because they contain as few variables as possible. Dimensional analysis indicates how many variables are needed and provides guidance on how to define them. If two systems are governed by the same set of dimensionless groups and the corresponding groups are all equal, they are said to be dynamically similar. Dynamic similarity allows predictions to be made for situations that might be impractical to explore experimentally.

REFERENCES

Adamson, A. W. and A. P. Gast. *Physical Chemistry of Surfaces*, 6th ed. Wiley, New York, 1997.

Allen, M. P. and D. J. Tildesley. *Computer Simulation of Liquids*. Clarendon Press, Oxford, 1987.

Barenblatt, G. I. *Scaling, Self-Similarity, and Intermediate Asymptotics*. Cambridge University Press, Cambridge, 1996.

Bates, R. L., P. L. Fondy, and R. R. Corpstein. An examination of some geometric parameters of impeller power. *Ind. Eng. Chem. Proc. Des. Develop.* 2: 310–314, 1963.

Bird, R. B., R. C. Armstrong, and O. Hassager. *Dynamics of Polymeric Liquids*, Vol. 1, 2nd ed. Wiley, New York, 1987.

Blevins, R. D. *Applied Fluid Dynamics Handbook*. Krieger, Malabar, FL, 2003.

Bridgman, P. W. *Dimensional Analysis*. Yale University Press, New Haven, CT, 1931.

Catchpole, J. P. and G. Fulford. Dimensionless groups. *Ind. Eng. Chem.* 58(3): 46–60, 1966.

Deen, W. M. *Analysis of Transport Phenomena*, 2nd ed. Oxford University Press, New York, 2012.

De Gennes, P.-G., F. Brochard-Wyart, and D. Quéré. *Capillarity and Wetting Phenomena*. Springer, New York, 2003.

Eggers, J. and E. Villermaux. Physics of liquid jets. *Rep. Prog. Phys.* 71: 036601, 2008.

Haward, S. J. and G. H. McKinley. Stagnation point flow of wormlike micellar solutions in a microfluidic cross-slot device: effects of surfactant concentration and ionic environment. *Phys. Rev. E* 85: 031502, 2012.

Hirschfelder, J. O., C. F. Curtiss, and R. B. Bird. *Molecular Theory of Gases and Liquids*. Wiley, New York, 1954.

Jasper, J. J. The surface tension of pure liquid compounds. *J. Phys. Chem. Ref. Data* 1: 841–1010, 1972.

Kadoya, K., N. Matsunaga, and A. Nagashima. Viscosity and thermal conductivity of dry air in the gaseous phase. *J. Phys. Chem. Ref. Data* 14: 947–970, 1985.

Kármán, T. von. *Aerodynamics*. Cornell University Press, Ithaca, New York, 1954.

Kennard, E. H. *Kinetic Theory of Gases*. McGraw-Hill, New York, 1938.

Kleiber, M. Body size and metabolic rate. *Physiol. Rev.* 27: 511–541, 1947.

Lide, D. R., Ed. *Handbook of Chemistry and Physics*, 71st ed. Chemical Rubber Co., Cleveland, OH, 1990.

Reid, R. C., J. M. Prausnitz, and B. E. Poling. *The Properties of Gases and Liquids*, 4th ed. McGraw-Hill, New York, 1987.
Tanner, R. I. *Engineering Rheology*, 2nd ed. Oxford University Press, Oxford, 2000.
Taylor, G. I. The formation of a blast wave by a very intense explosion. II. The atomic explosion of 1945. *Proc. Roy. Soc. Lond.* A201: 175–186, 1950.
Than, P., L. Preziosi, D. D. Joseph, and M. Arney. Measurement of interfacial tension between immiscible liquids with the spinning rod tensiometer. *J. Colloid Int. Sci.* 124: 552–559, 1988.
White, D. A. and J. A. Tallmadge. Theory of drag out of liquids on flat plates. *Chem. Eng. Sci.* 20: 33–37, 1965.
Wijshoff, H. The dynamics of the piezo inkjet printhead operation. *Phys. Rep.* 491: 77–177, 2010.

PROBLEMS

1.1. Falling body* It is desired to predict the time t_h that is needed for an object of mass m to travel downward a distance h, if gravity is the only force.

(a) If the object is stationary before release, what could you conclude about t_h from dimensional analysis alone?
(b) How is the dimensional analysis changed if the object is propelled downward at an initial velocity U?
(c) Derive from basic mechanics the expression for t_h for $U \geq 0$. When put in dimensionless form, how does it compare with what you found in parts (a) and (b)?

* This problem was suggested by H. A. Stone.

1.2. Pendulum A simple pendulum consists of a bob of mass m attached to a rigid rod, with R being the distance from the pivot to the bob center. The rod mass and bob diameter are each small (relative to m and R, respectively), the pivoting is frictionless, and the effects of the surrounding air are negligible. The bob is initially at rest, with a small angle θ_0 between the rod and a vertical line. It is desired to predict the period t_p of the oscillations, once the bob is released.

(a) What can you conclude about t_p from dimensional analysis alone?
(b) Derive from basic mechanics the expression for t_p. When put in dimensionless form, how does it compare with what you found in part (a)?

1.3. Salad dressing* A simple salad dressing can be prepared by adding one part vinegar to three parts vegetable oil and beating the mixture vigorously using a whisk. The result is an emulsion in which vinegar droplets are dispersed in the oil. The palatability of the dressing depends on the volume V of an average vinegar droplet. If it is supposed that V depends on the whisking velocity (hand speed) U, the diameter D of the whisk wires, the oil viscosity μ, and the oil–water surface tension γ, identify a set of dimensionless groups that might be used to quantify this process.

* This problem was suggested by P. S. Doyle.

1.4. Heat transfer coefficient Suppose that a sphere of diameter D moves at a velocity U relative to a liquid, and that the sphere temperature T_S exceeds the liquid temperature

T_L far from the sphere. The rate of heat transfer from sphere to liquid (energy per unit time) may be expressed as

$$Q = hA(T_S - T_L) \quad \text{(P1.4-1)}$$

where A is the surface area of the sphere and h is the heat transfer coefficient. The value of h depends on the thermal conductivity of the liquid (k, SI unit W m^{-1} K^{-1}) and its heat capacity per unit mass at constant pressure ($\hat{C}_P$, SI unit J kg^{-1} K^{-1}), in addition to D, U, ρ, and μ. Temperature adds to (M, L, T) a fourth primary dimension, denoted as Θ.

(a) It is customary to embed h in a *Nusselt number*,

$$\text{Nu} = \frac{hD}{k}. \quad \text{(P1.4-2)}$$

Confirm that this is dimensionless.

(b) If $N_1 = \text{Nu}$, U appears only in N_2, and μ appears only in N_3, what are suitable choices for the dimensionless groups?

1.5. Oscillating drops Liquid drops are generally nonspherical when formed (e.g., upon release from a pipette), and if they are in a certain size range their shape oscillates in time. It is desired to predict the period t_o of the oscillations.

(a) Assuming that t_o depends only on the surface tension γ, the liquid density ρ, and the linear dimension R (the radius of a sphere of equal volume), what can you conclude from dimensional analysis?

(b) Implicit in part (a) is that inertia and surface tension are important, but the oscillations are unaffected by viscosity or gravity. Estimate the range of R for which that will be true for a water drop.

1.6. Dip coating Various industrial processes require that a flat sheet be coated with a uniform liquid film. *Dip coating* is a procedure in which an immersed substrate is withdrawn through a liquid–gas interface. In the process in Fig. P1.6, in which a substrate is pulled upward at a constant velocity V, the coating achieves a uniform thickness H at a certain height above the liquid. It is desired to predict H from V and the liquid properties (μ, ρ, and γ).

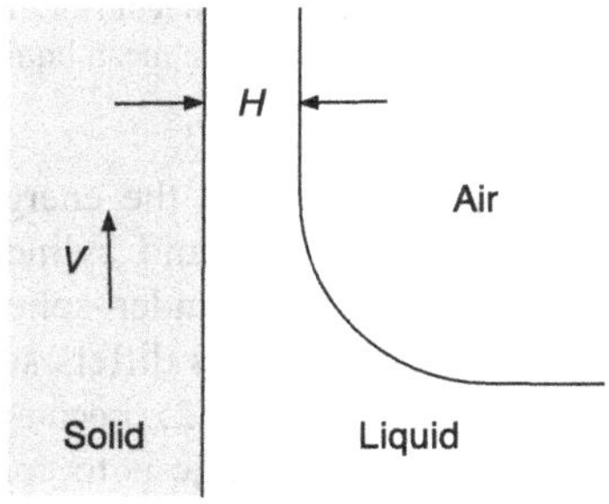

Figure P1.6 One side of a solid sheet being withdrawn upward from a liquid.

(a) Representative values are $H = 0.02$ to 0.2 cm, $V = 0.4$ to 3 cm/s, $\mu = 0.3$ to 2 Pa · s, $\rho = 900$ kg/m^3, and $\gamma = 30$ mN/m. By examining the ranges of dimensionless groups, show that viscosity, gravity, and surface tension are all important, but inertia is not.

(b) A theoretical result that is applicable here is (White and Tallmadge, 1965)

$$H\left(\frac{\rho g}{\gamma}\right)^{1/2} = 0.944\left(\frac{\mu V}{\gamma}\right)^{2/3}\left(1 - \frac{H^2 \rho g}{\mu V}\right)^{2/3}. \qquad \text{(P1.6-1)}$$

Show that this may be rewritten as a relationship between Bo and Ca.

(c) If one group is chosen now as $N = (\text{Bo}/\text{Ca})^{1/2} = [H^2 g/(\nu V)]^{1/2}$ and Ca is the second group, show that

$$\frac{N}{(1 - N^2)^{2/3}} = 0.944\text{Ca}^{1/6}. \qquad \text{(P1.6-2)}$$

Use this to derive explicit expressions for N for Ca $\to 0$ and Ca $\to \infty$. How does H vary with V in these limits?

1.7 Breakup of liquid jets Circular jets of liquid are notoriously unstable. The breakup of one into drops may be observed by placing a finger under a thin stream of water from a faucet and noticing the rapid series of impacts at a certain distance from the nozzle. The objective is to explore why liquid cylinders break up and to estimate how fast that occurs in a water jet. The jet radius is R and its velocity in a lab reference frame is U, as shown in Fig. P1.7(a). The enlargement in Fig. P1.7(b) shows the breakup process in a reference frame moving with the jet. Transforming a segment of length L into a sphere of radius a involves relative velocities on the order of v, which is an unknown. To form the drop, a typical element of liquid must travel a distance in the jet reference frame that is a few times R.

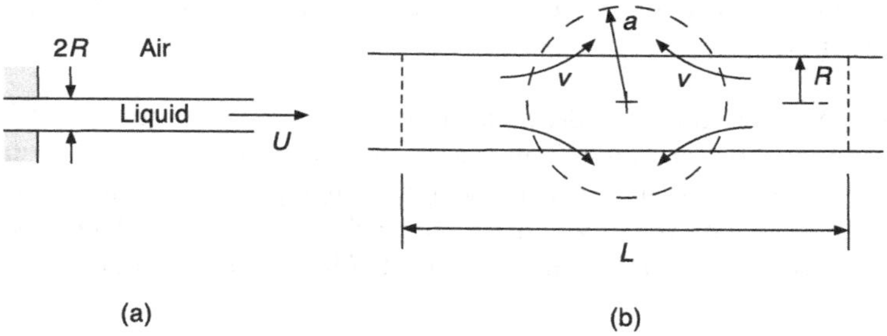

Figure P1.7 Liquid jet in air: (a) intact jet in a lab reference frame, and (b) breakup process in a reference frame moving at the mean liquid velocity.

(a) In that γ is the energy per unit area of interface and systems seek lower energy states, a liquid cylinder will be unstable if it can reduce its surface area. Show that the cylinder–sphere transformation in Fig. P1.7(b) will reduce the area if $L/R > 9/2$. This differs somewhat from the actual criterion for instability ($L/R > 2\pi$, Problem 4.12), because small deformations must reduce the area immediately if the shape change is to continue; the cylinder does not "know" at the outset that it can become a sphere.

(b) For low-viscosity liquids such as water, $L/R = 9.0$ results in the fastest breakup (Eggers and Villermaux, 2008) and is what is usually seen. Accordingly, what drop size would you expect from the breakup of a water jet with $R = 1$ mm?

(c) Jet breakup due to surface tension is resisted by inertia and viscosity. If solely inertia were involved, the time for breakup would be t_{is}, the inertial–surface-tension time scale. The corresponding velocity scale is $v = R/t_{is}$. By equating the inertial force scale to that for surface tension, derive an expression for t_{is}. Evaluate t_{is} for a water jet with $R = 1$ mm.

(d) The viscous–surface-tension time scale (t_{vs}) is calculated similarly. Show that the ratio of time scales is

$$\frac{t_{vs}}{t_{is}} = \frac{\mu}{(\rho R \gamma)^{1/2}} = \text{Oh} \qquad \text{(P1.7-1)}$$

and that Oh, the *Ohnesorge number*, is small for the water jet of part (c). This implies that inertia slows the breakup much more than does viscosity. Accordingly, the actual time for the water jet to break into drops is roughly t_{is}, and the jet will travel a distance of approximately Ut_{is} before drops form.

1.8. Valve scale-up A valve manufacturer has received an inquiry from a company seeking large units for an oil pipeline. Certain types currently in production might be suitable, but they are much smaller than what is needed. Before investing in expensive prototypes, the manufacturer wants to know what the flow resistances would be for enlarged versions of current designs. The measure of resistance is the pressure drop $|\Delta P|$ across an open valve (upstream minus downstream value). The manufacturer can readily measure $|\Delta P|$ as a function of the mean fluid velocity U in any existing unit, using either air or water at ambient conditions. The pipe diameter is D. You agree to provide guidance on extrapolating such data to valves for the oil pipeline.

(a) A good starting point is to assume that $|\Delta P|$ depends on D, U, ρ, μ, and, depending on the valve type, k internal dimensions, $d_1, d_2, \ldots, d_k$. What can you conclude about $|\Delta P|$ from dimensional analysis?

(b) Suppose that for the oil pipeline $D = 1.0$ m, $U = 2.0$ m s^{-1}, $\rho = 900$ kg m^{-3}, and $\mu = 0.025$ Pa · s. If the tests are done with 1/10-size versions of what is proposed, which is the better test fluid (air or water) and what would you recommend for U?

(c) A staff member points out that, for Re > 10^5 (based on D), the company's experience is that $|\Delta P|$ is independent of μ. How would this alter your recommendations, if at all?

(d) If it were desired to predict $|\Delta P|$ for *water* in a pipe with $D = 1.0$ m and $U = 2.0$ m s^{-1}, how would your analysis and experimental recommendations change?

1.9. Ship scale-up Experiments with scale models are sometimes used to guide ship design. One quantity of interest is the *drag force* (F_D) that opposes the ship's forward motion. It can be evaluated for a model by towing at a steady speed in a test tank and measuring the tension on the tow line. It is reasonable to suppose that F_D depends on the ship speed U, ship length L, k other ship dimensions ($d_1, d_2, \ldots, d_k$), μ and ρ of water, and g. Gravity has a role because surface ships are affected by waves, including the ones they generate. You may assume that the model size greatly exceeds the capillary length, making surface tension negligible.

(a) What can you conclude about F_D from dimensional analysis?

(b) The largest practical model of a modern ship might be 1/100 scale, with a length of about 1 m. Ideally, the model in a towing tank would be dynamically similar to the full-size ship. To what extent can that be achieved?

1.10. Power input in a stirred tank The effectiveness of mixing within a batch reactor or other stirred tank is determined mainly by the stirring power. The power Φ is the rate at which work is done on the liquid by the moving impeller. It is affected by the angular velocity ω and diameter d of the impeller, ρ, μ, and sometimes g. A baffled tank with a turbine impeller is shown in Fig. P1.10. Lengths less critical than d, but still influential, include the width w_i of the impeller blades, the width w_b of the baffles that project in from the wall, the height h of the impeller above the bottom, the liquid depth H, and the tank diameter D. Geometric parameters that are ignored here for simplicity include the number and pitch of the impeller blades and the number of baffles.

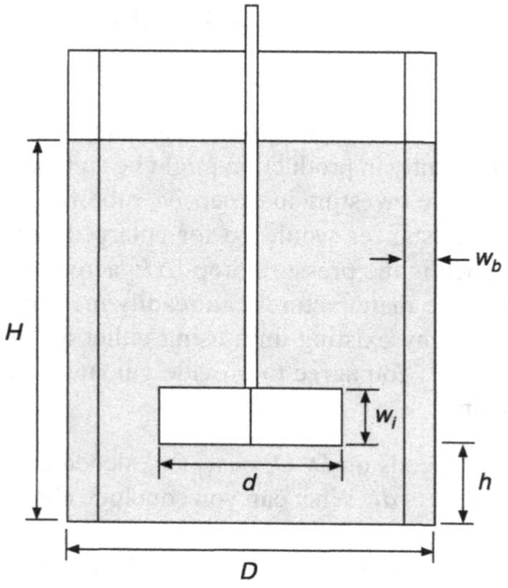

Figure P1.10 Geometric parameters for a stirred cylindrical tank with baffles.

(a) Considering the complete list of quantities (Φ, ω, d, ρ, μ, g, w_i, w_b, h, H, D), what can you conclude about Φ from dimensional analysis? It is helpful to arrange the groups so that Φ, μ, and g each appear in only one group. Show that

$$N_1 = \frac{\Phi}{\omega^3 d^5 \rho} = \text{Po} \qquad \text{(P1.10-1)}$$

where Po is the *power number*.

(b) Baffles improve mixing by preventing a purely swirling motion of the liquid. They also tend to flatten the gas–liquid interface and thereby minimize the influence of gravity on the flow pattern. If g is unimportant and only geometrically similar tanks are to be considered, how can the dimensional analysis be simplified?

(c) Bates *et al.* (1963) found that Po $\rightarrow A/\text{Re}_i$ for small Re_i and Po $\rightarrow B$ for large Re_i, where Re_i is the impeller Reynolds number and A and B are each constant for a given impeller type. Could you have anticipated the forms of these relationships?

1.11. Underwater swimming This problem concerns the power Φ that an organism of length L must expend to swim underwater at a constant velocity U. In general, Φ will be affected by ρ and μ of water, in addition to L and U.

(a) If N_1 has Φ in the numerator and ρ in the denominator, what can you conclude from dimensional analysis?

(b) It has been found that, from microbes to whales, the overall rate of metabolic energy production varies as body weight to the 3/4 power. Observed originally for mammals ranging from mice (20 g) to cattle (600 kg) (Kleiber, 1947), this is called *Kleiber's law*. When combined with certain assumptions, Kleiber's law suggests that $\Phi = CL^{9/4}$, where C is constant within an aquatic family. What are the necessary assumptions?

(c) Suppose that a modern fish with $L = 0.5$ m and $U = 1$ m/s had a prehistoric relative with $L = 5$ m. Show that Re for the modern fish is already quite large. For many objects at such large Re, the *drag force* F_D (the fluid-dynamic force that opposes the motion) is approximately $F_D = K_1 L^2 \rho U^2$, where K_1 is a dimensionless coefficient that depends only on the shape of the object (Chapter 3). Assume that this holds at any Re greater than or equal to that of the modern fish. Given that $\Phi \propto F_D U$, what might have been the swimming speed of the now-extinct giant?

(d) Suppose that a bacterium with $L = 1$ μm propels itself at $U = 20$ μm/s by movement of flagella. For Re $\ll 1$, the drag force is $F_D = K_2 \mu U L$, where K_2 is another dimensionless constant that depends only on the shape of the object. If there were a geometrically similar bacterium with $L = 2$ μm, how fast would you expect it to swim?

2

Pipe flow

2.1 INTRODUCTION

The prediction of flow rates in tubes or other fluid-filled conduits is the most common application of fluid mechanics in chemical engineering. This topic is important also historically, as experiments with liquid flow through tubes in the early 1800s were among the first to quantify viscosity. This chapter focuses on the steady (time-independent) flow of Newtonian fluids through conduits of uniform cross-section. It begins with definitions of various quantities involved in pipe flow, principally the *friction factor*, which is a dimensionless wall shear stress. The observed dependence of the friction factor on the Reynolds number is detailed both for laminar and for turbulent flow. Discussed then is the relationship between the wall shear stress and the pressure drop along a tube. Although most of the data to be discussed were obtained using cylindrical tubes, some of the results can be extended to noncircular cross-sections by using the concept of a *hydraulic diameter*. After that is described, the effects of wall roughness on the friction factor are discussed. Such effects can be quite important in turbulent flow.

2.2 SHEAR STRESS

Fundamental quantities

A typical objective in piping design is to achieve a certain *volume flow rate* Q, which is the volume passing through the pipe per unit time (e.g., m^3/s). Closely related is the *mass flow rate*, ρQ (kg/s). Another measure of flow rate is the *mean velocity* U, which is Q divided by the cross-sectional area. In a cylindrical tube of diameter D that area is $A = \pi D^2/4$. Accordingly,

$$U = \frac{Q}{A} = \frac{4Q}{\pi D^2}. \tag{2.2-1}$$

A key aspect of the flow of a constant-density fluid in any completely filled conduit with rigid, impermeable walls, not necessarily cylindrical, is that Q does not vary with position.[1] If the cross-sectional area is uniform, U is also constant.

Flow along a solid surface, such as a pipe wall, creates a shear stress there, τ_w. As described in Section 1.2, τ_w equals the viscosity times the rate of shear at the surface. The shear rate, in turn, is the rate at which the velocity varies with distance from the surface. Thus, τ_w for a given fluid is determined by the variation in velocity over the

[1] If Q were not the same everywhere, then the mass flow rate entering a given segment of the pipe would differ from that leaving it, implying continual creation or destruction of matter. For constant ρ, conservation of mass implies conservation of volume.

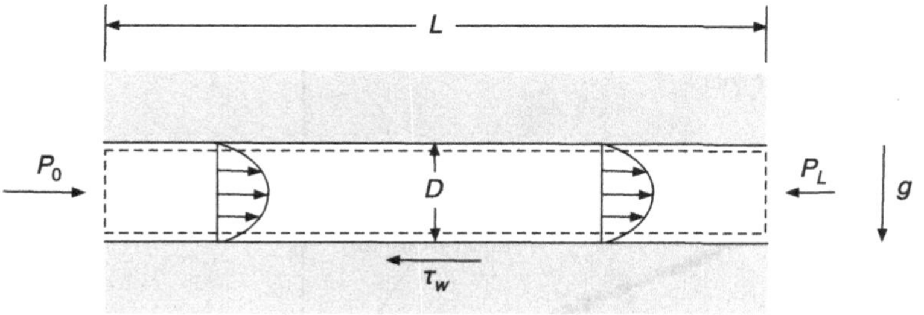

Figure 2.1 Velocity profile and stresses in a long, horizontal tube. The local velocity is greatest at the center and zero at the walls. Acting on the control volume shown by the dashed rectangle are horizontal forces arising from the wall shear stress (τ_w) and the pressures at the ends (P_0 and P_L). The direction of gravity is indicated by the arrow "g."

cross-section, or *velocity profile*. In a tube of sufficient length L, the constancy of U and D allows the velocity profile to become independent of position, as depicted qualitatively in Fig. 2.1. When the final form of the profile is reached, the flow is said to be *fully developed*. This occurs at a predictable distance from the tube inlet, called the *entrance length*, which is often short relative to L. In this chapter we will neglect the transitional region near the inlet and assume that τ_w does not vary with position along the tube. Entrance lengths and the additional flow resistances associated with entrance regions are discussed in Section 12.2.

Figure 2.1 also shows the stresses acting on the fluid inside a horizontal tube. The dashed rectangle, which encloses all of the fluid, defines the system to be examined. Such a system, chosen for analytical purposes, is called a *control volume*. The only *horizontal* forces exerted by the surroundings on this control volume are due to the pressures at the ends and the shear stress at the wall. Each force is a stress times an area. For steady flow, the net force on the control volume must be zero. Defining forces in the flow direction as positive and opposing ones as negative, the horizontal force balance is

$$(P_0 - P_L)\frac{\pi D^2}{4} - \tau_w \pi D L = 0. \tag{2.2-2}$$

Accordingly, the shear stress and pressure difference are related as

$$\tau_w = \frac{D}{4}\left(\frac{P_0 - P_L}{L}\right) = -\frac{D}{4}\frac{\Delta P}{L} = \frac{D}{4}\frac{|\Delta P|}{L} \quad \text{(horizontal tube)}. \tag{2.2-3}$$

It is seen that a pressure decrease in the direction of flow is needed to offset the viscous stress at the wall. A convention we will use is that the change in a quantity in the flow direction, denoted using Δ, is its downstream minus its upstream value. Because $\Delta P = P_L - P_0 < 0$, the pressure *drop* is $|\Delta P|$.

Friction factor

A choice to be made when correlating experimental data on tube flow is whether to view τ_w or $|\Delta P|$ as the primary variable. As shown by Eq. (2.2-3), if either is known the other is readily calculated. The wall shear stress has the merit not only of being constant along a tube, but also of being unaffected by whether the tube is horizontal or vertical. That is, for a given U and D, the velocity profile that determines τ_w turns out to be independent of the direction of gravity. In contrast, as shown in Section 2.3, for a given τ_w the value of $|\Delta P|$ is very sensitive to the orientation of the tube. Thus, in this section we will relate

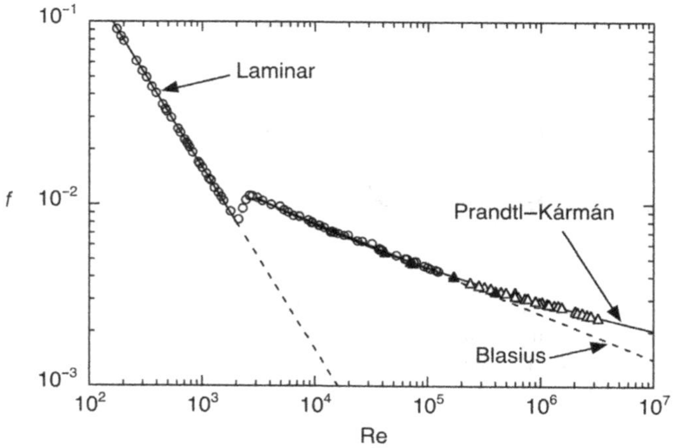

Figure 2.2 Friction factor for smooth cylindrical tubes. The curves labeled Laminar, Prandtl–Kármán, and Blasius are based on Eqs. (2.2-6), (2.2-7), and (2.2-9), respectively. The dashed lines are extrapolations. The symbols show data for water: circles, $D = 1$ cm from Koury (1995); filled and open triangles, $D = 5$ cm and 10 cm, respectively, from Nikuradse (1932).

τ_w to U, D, ρ, and μ. We omit Q because it is not independent of U and D, and exclude L because, in fully developed flow, it affects τ_w only via its influence on U.

As discussed in Example 1.4-2, our list of five governing quantities, three of which are dimensionally independent, implies that there are just two independent dimensionless groups. Thus, one group is a function only of the other. In defining a dimensionless wall shear stress, it is customary to divide τ_w by a modified inertial stress scale, $\rho U^2/2$. A second group, independent of τ_w, is the Reynolds number based on D and U. Accordingly,

$$f(\text{Re}) = \frac{2\tau_w}{\rho U^2} \tag{2.2-4}$$

$$\text{Re} = \frac{UD\rho}{\mu} = \frac{UD}{\nu}. \tag{2.2-5}$$

The dimensionless shear stress (f) is called the *friction factor*.[2]

The symbols in Fig. 2.2 show representative data for f obtained over a range of Re spanning four orders of magnitude. In this log–log plot there is initially a linear decline in f. At a critical value of the Reynolds number ($\text{Re}_c \cong 2100$) a sharp rise in f occurs, followed by a decline that is more gradual than the earlier one. What happens at the critical Reynolds number is a transition from laminar to turbulent flow. In laminar flow all elements of the fluid move in straight lines that are parallel to the tube wall, whereas in turbulence the pattern is chaotic, with eddies of varying size continually forming and decaying. Such eddies greatly increase τ_w, as detailed in Chapter 10. Although $\text{Re}_c = 2100$ is typical of ordinary laboratory or industrial conditions, Re_c can be increased significantly if special precautions are taken to avoid vibrations or other flow disturbances. Thus, where the laminar–turbulent transition will occur is not entirely predictable, and in piping design it is prudent to avoid values of Re between 2000 and 4000.

[2] In this book we use the *Fanning* friction factor. Also found in the literature is the *Darcy* friction factor, which equals $4f$.

Also shown in Fig. 2.2 are curves that fit the data within certain ranges of Re. In the laminar regime, the friction factor is given exactly by

$$f = \frac{16}{\text{Re}} \quad (\text{Re} \leq 2100). \tag{2.2-6}$$

This is a dimensionless form of *Poiseuille's law* (also called the *Hagen–Poiseuille equation*) [see also Eq. (2.3-7)].[3] The finding that $f \propto \text{Re}^{-1}$ indicates that inertial effects are entirely absent for $\text{Re} < \text{Re}_c$. That is, if both sides of Eq. (2.2-6) are multiplied by Re, the equation becomes $\text{Re}f = 2\,\tau_w D/(\mu U) = 16$ or $\tau_\omega = 8\mu U/D$. The inertial stress scale is eliminated and only the viscous one ($\mu U/D$) remains.

For turbulent flow in smooth tubes, the most widely accepted expression is the *Prandtl–Kármán equation*,[4]

$$\frac{1}{\sqrt{f}} = 4.0 \log(\text{Re}\sqrt{f}) - 0.4 \quad (\text{Re} \geq 3 \times 10^3). \tag{2.2-7}$$

This is accurate to within about 4% for Re up to 3×10^6. In chemical engineering applications, Re in pipes rarely exceeds that (Table 1.4). For smooth tubes at higher Re (for which there are limited data up to 3.5×10^7), it has been found that Eq. (2.2-7) begins to underestimate f, and other expressions have been proposed (Zagarola and Smits, 1998). The theoretical reasons for choosing $\sqrt{f}$ and $\text{Re}\sqrt{f}$ as variables are explained in Chapter 10.

The Prandtl–Kármán equation can be awkward to use because f appears on both sides. An explicit expression that gives results within 1% of Eq. (2.2-7) for $5 \times 10^3 \leq \text{Re} \leq 5 \times 10^7$ is

$$f = \left[3.6 \log\left(\frac{\text{Re}}{6.9}\right)\right]^{-2} \quad (\text{Re} \geq 5 \times 10^3). \tag{2.2-8}$$

This fairly simple expression, which was due originally to Colebrook (1939) and modified slightly by Haaland (1983), has largely been overlooked. If included in Fig. 2.2, the corresponding curve would be visually indistinguishable from that for the Prandtl–Kármán equation. Even more convenient, but valid over a more limited range of Re, is the *Blasius equation*,

$$f = 0.0791\,\text{Re}^{-1/4} \quad (3 \times 10^3 \leq \text{Re} \leq 1 \times 10^5). \tag{2.2-9}$$

The impact of turbulence may be judged by comparing the values of f calculated from Eqs. (2.2-6) and (2.2-9). At $\text{Re} = 1 \times 10^4$, f in turbulent flow is approximately five times

[3] Jean Poiseuille (1797–1869) was a French physician whose interest in blood flow led him to measure pressure–flow–diameter relationships for water and other liquids in glass capillary tubes in the 1830s and 1840s. He is also the inventor of the U-tube mercury manometer, which he used to measure arterial blood pressures in animals (Sutera and Skalak, 1993). Gotthilf Hagen (1797–1884), a German engineer, independently and concurrently quantified flow in tubes.

[4] Equation (2.2-7) dates from the early 1930s. Ludwig Prandtl (1875–1953) was a German fluid dynamicist who made numerous seminal contributions. He was responsible for introducing the idea of control volumes into engineering analysis (Vincenti, 1982). He also originated the boundary-layer concept (Chapter 9) and first exploited the idea of a turbulent mixing length (Chapter 10). His research group at Göttingen established many of the principles that came to underlie modern aircraft design. Theodore von Kármán (1881–1963) also made major contributions to aerodynamics. Hungarian-born, he did his doctoral research under Prandtl and then established his own research program at Aachen. They remained collaborators and friendly competitors (Kármán, 1954). Von Kármán emigrated to the United States in 1930 and co-founded what became the Jet Propulsion Laboratory at Caltech.

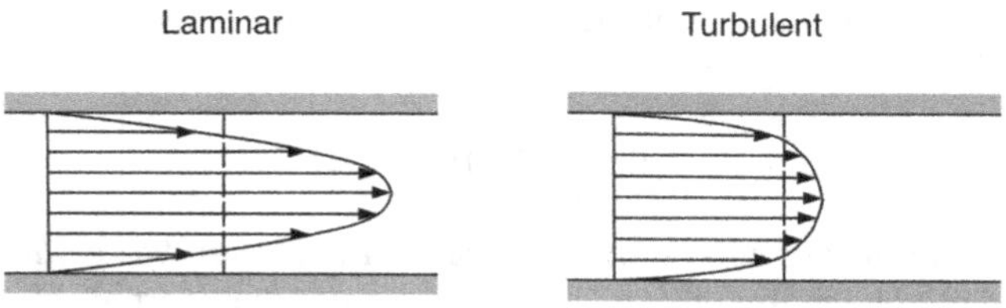

Figure 2.3 Velocity profiles for laminar or turbulent flow in a tube. The mean value in each case is indicated by the dashed line. In turbulent flow the local velocity fluctuates from moment to moment; what is shown is the time-smoothed velocity.

what it would be if laminar flow could be maintained. At $\text{Re} = 1 \times 10^5$, there is a 28-fold increase in f. The increased shear stress is caused by a change in the velocity profile, as depicted in Fig. 2.3. In laminar flow the variation in velocity with radial position is parabolic (Chapter 7), whereas in turbulent flow the profile is more blunted (Chapter 10). Turbulence causes fluid elements at different radial positions to intermingle, such that most move at very similar velocities. Given the relatively uniform velocity in the central part of the tube, there must be a larger gradient near the wall, if the mean velocity is the same as for laminar flow. The higher shear rate at the wall is what increases τ_w in turbulent flow.

Turbulence can be costly in terms of energy usage. In a horizontal pipe, the required pumping power equals $Q|\Delta P|$ (Section 11.4) and $|\Delta P|$ is proportional to τ_w [Eq. (2.2-3)]. The effects of turbulence on τ_w in liquids can be greatly lessened by small concentrations of dissolved polymers, a phenomenon called *drag reduction* (McComb, 1990, pp. 130–141; Virk, 1975). For example, 10 ppm (parts per million, by weight) polyethylene oxide with an average molecular weight of 5×10^6 can reduce f by as much as 70%, although f remains above that for laminar flow. Such additives may be justified when pumping costs are large and high purity is unnecessary, as in oil pipelines. Drag reduction is discussed further in Problem 2.7. For certain other purposes, turbulence is advantageous. For example, the increased mixing due to the turbulent eddies facilitates heat transfer between the fluid and the pipe wall. That is helpful whenever a process stream must be heated or cooled.

2.3 PRESSURE DROP AND DYNAMIC PRESSURE

Friction factor and pressure drop

Although τ_w is arguably the more fundamental quantity, $|\Delta P|$ is much easier to measure. Also, because pumps are essentially devices that boost the pressure at selected places in piping networks, values of $|\Delta P|$ are needed in design. For these reasons, it is desirable to rewrite the friction factor in terms of $|\Delta P|$. Substituting Eq. (2.2-3) into Eq. (2.2-4) gives

$$f = \frac{D}{2\rho U^2}\frac{|\Delta P|}{L} \quad \text{(horizontal tubes)} \tag{2.3-1}$$

as a more practical form of the friction factor, but one limited to horizontal flow. What is needed (and derived next) is a relationship that is valid for pipes inclined at any angle.

Two inclined tubes are shown in Fig. 2.4. As before, the control volume consists of all fluid within a tube. For downward flow at an angle α, as in Fig. 2.4(a), the gravitational

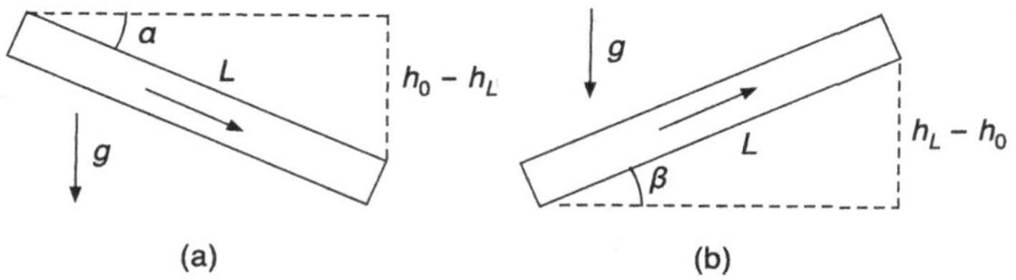

Figure 2.4 Inclined tubes, each of length L: (a) downward flow at an angle α and (b) upward flow at an angle β. The heights of the tube inlet and outlet are h_0 and h_L, respectively.

force in the direction of flow is the fluid mass times the corresponding component of the gravitational vector, or

$$\left(\frac{\pi D^2 L\rho}{4}\right) g \sin\alpha = \left(\frac{\pi D^2 L}{4}\right)\rho g\left(\frac{h_0 - h_L}{L}\right) = -\rho g\,\Delta h\frac{\pi D^2}{4}. \tag{2.3-2}$$

It is seen that the tube length cancels and that the key quantity is the height difference between the ends, Δh. Including gravity within the force balance in Eq. (2.2-2) gives

$$-\Delta P\frac{\pi D^2}{4} - \rho g\,\Delta h\frac{\pi D^2}{4} - \tau_w \pi D L = 0 \tag{2.3-3}$$

or, solving for the shear stress,

$$\tau_w = -\frac{D}{4}\frac{\Delta(P+\rho g h)}{L} = -\frac{D}{4}\frac{\Delta\mathcal{P}}{L}. \tag{2.3-4}$$

For upward flow at an angle β, as in Fig. 2.4(b), the component of gravity in the flow direction is $-g\sin\beta = -g\,\Delta h/L$ and the final result is the same.

The last expression in Eq. (2.3-4) makes use of the definition

$$\mathcal{P} \equiv P + \rho g h. \tag{2.3-5}$$

We will refer to $\mathcal{P}$ as the *dynamic pressure*; it is also called the *equivalent pressure* or *modified pressure*.[5] This variable neatly combines the effects of pressure and gravity. It is useful not just in pipe-flow calculations, but in many other contexts, and is encountered very frequently in this book. As exemplified by Eq. (2.3-4), it is only height *differences* that matter. Accordingly, h can be measured from any convenient horizontal plane. Whereas the absolute value of P is physically meaningful, that of $\mathcal{P}$ (which depends on how h is chosen) is not. Only variations in $\mathcal{P}$ are significant.

What is apparent now is that, in general, the shear stress is balanced by a difference not in P, but in $\mathcal{P}$. Using Eq. (2.3-4) to evaluate τ_w in Eq. (2.2-4) gives

$$f = \frac{D}{2\rho U^2}\frac{|\Delta\mathcal{P}|}{L}. \tag{2.3-6}$$

This is the starting point for most design calculations involving pipe flow. For a horizontal pipe, $\Delta h = 0$, $\Delta P = \Delta\mathcal{P}$, and Eq. (2.3-1) is recovered. In general, the power that must be supplied by a pump is $Q|\Delta\mathcal{P}|$ (Section 11.4).

[5] As discussed in Section 6.6, $\mathcal{P}$ is constant throughout a static fluid. In that variations in $\mathcal{P}$ are either the cause or consequence of flow, "dynamic pressure" is very descriptive. However, the terminology differs among authors. In particular, our "modified inertial stress scale," $\rho U^2/2$, is often called the dynamic pressure.

Using $f = 16/\mathrm{Re}$ in Eq. (2.3-6) and replacing U by Q, the result for laminar flow is

$$Q = \frac{\pi D^4}{128\mu} \frac{|\Delta\mathscr{P}|}{L}. \tag{2.3-7}$$

This is a dimensional form of Poiseuille's law. Although presented here as an empirical finding, it is derived from first principles in Chapter 7.

Circuit analogy

Equation (2.3-7) shows that, for laminar flow, Q is proportional to $|\Delta\mathscr{P}|$. That relationship is analogous to Ohm's law for circuits, with Q serving as the current and $|\Delta\mathscr{P}|$ replacing the voltage drop. The corresponding resistance is $r = |\Delta\mathscr{P}|/Q = 128\mu L/(\pi D^4)$. The electrical analogy can be exploited when designing complex flow networks for low-to-moderate Re, as in microfluidic devices (Oh *et al.*, 2012). For laminar flow through n tubes in series with resistances $r_1, r_2, \ldots, r_n$, the overall resistance is $r_T = r_1 + r_2 + \cdots + r_n$. For n tubes in parallel, $1/r_T = (1/r_1) + (1/r_2) + \cdots + (1/r_n)$.

In turbulent flow the relationship between Q and $|\Delta\mathscr{P}|$ is nonlinear. For example, in the Blasius range, where $f \propto \mathrm{Re}^{-1/4}$, Q varies as $|\Delta\mathscr{P}|^{4/7}$. If f were entirely independent of Re, as occurs for turbulent flow in very rough tubes (Section 2.5), then Q would vary as $|\Delta\mathscr{P}|^{1/2}$. The electrical analogy is less useful in turbulent flow because the resistance ($|\Delta\mathscr{P}|/Q$) depends on the current (Q).

Pipe-flow problems have many variations, according to what is given and what is to be calculated. Whenever possible, Re should be found first, because it determines which of Eqs. (2.2-6)–(2.2-9) govern f. The following examples involve cylindrical tubes with walls that are assumed to be functionally smooth.

Example 2.3-1 Pressure drop for water in process pipes. It is desired to find the values of $|\Delta P|$ for water flow at 20 °C through either of two pipes. One is entirely horizontal and the other has horizontal and vertical segments. In each, the flow rate is $Q = 8.0 \times 10^{-3}$ m^3 s^{-1}, the diameter is $D = 0.1$ m, and the total length is $L = 30$ m. The second pipe has a bend, such that flow in the first 20 m is horizontal and that in the last 10 m is upward. As given in Table 1.2, $\rho = 1.00 \times 10^3$ kg m^{-3} and $\nu = 1.00 \times 10^{-6}$ m^2 s^{-1}.

In this problem Re can be calculated at the outset. Using Eq. (2.2-1), the mean velocity that corresponds to the specified Q and D is

$$U = \frac{4Q}{\pi D^2} = \frac{4(8 \times 10^{-3})}{\pi (0.1)^2} = 1.02 \text{ m s}^{-1}.$$

Accordingly,

$$\mathrm{Re} = \frac{DU}{\nu} = \frac{(0.1)(1.02)}{1 \times 10^{-6}} = 1.02 \times 10^5$$

which indicates that the flow is turbulent and that the Blasius equation is applicable. From Eq. (2.2-9), the friction factor is

$$f = 0.0791(1.02 \times 10^5)^{-1/4} = 4.43 \times 10^{-3}.$$

Rearranging Eq. (2.3-6) gives

$$|\Delta\mathscr{P}| = \frac{2\rho U^2 L}{D} f = \frac{2(1 \times 10^3)(1.02)^2(30)}{0.1}(4.43 \times 10^{-3}) = 2.76 \times 10^3 \text{ Pa}.$$

Because the flow rate, dimensions, and fluid properties are the same for the two pipes, the values of $|\Delta\mathcal{P}|$ are the same. For the first case, where the pipe ends are at the same level, $|\Delta P| = |\Delta\mathcal{P}|$. For the second, where the outlet is 10 m above the inlet,

$$|\Delta P| = |\Delta\mathcal{P}| + \rho g\,\Delta h = (2.76 \times 10^3) + (1 \times 10^3)(9.81)(10) = 1.01 \times 10^5 \text{ Pa}.$$

It is seen that the vertical segment in the second pipe creates a nearly 40-fold increase in $|\Delta P|$.

Pipe fittings, such as the 90° elbow that would join the horizontal and vertical segments of the second pipe, themselves create additional pressure drops. The entrance region also causes the total pressure drop to exceed that for fully developed flow. Such corrections, which are negligible for sufficiently long pipes, will be discussed in Sections 12.2 and 12.3.

Example 2.3-2 Pressure drop in an oil pipeline. The objective is to calculate the pressure drop at a given flow rate in a large oil pipeline. The mean velocity is $3.0\,\text{m s}^{-1}$, the pipe diameter is 1.0 m, the length is 100 km, $\rho = 850$ kg m^{-3}, and $\nu = 1.06 \times 10^{-5}$ m^2/s^{-1}. There is no net change in elevation. Although the volume flow rate is not needed in what follows, we note that the given values of U and D correspond to $Q = 2.4\,\text{m}^3\,\text{s}^{-1}$.

Again, the Reynolds number can be calculated immediately:

$$\text{Re} = \frac{DU}{\nu} = \frac{(1)(3)}{1.06 \times 10^{-5}} = 2.83 \times 10^5.$$

This is large enough to require either Eq. (2.2-7) or Eq. (2.2-8). The value of $|\Delta P|$ from the Prandtl–Kármán equation will be compared with that from the Colebrook equation.

Because f, which is unknown, appears on both sides, the Prandtl–Kármán equation requires iteration. An initial guess for f can be obtained either from Fig. 2.2 or by using the Blasius equation. From Eq. (2.2-9),

$$f = 0.0791(2.83 \times 10^5)^{-1/4} = 3.43 \times 10^{-3}.$$

Using this on the right-hand side of Eq. (2.2-7), an improved value for f is

$$\begin{aligned} f &= [4.0\log(\text{Re}\sqrt{f}) - 0.4]^{-2} \\ &= \{4\log[(2.83 \times 10^5)(3.43 \times 10^{-3})^{1/2}] - 0.4\}^{-2} = 3.68 \times 10^{-3}. \end{aligned}$$

Using this new value on the right-hand side of Eq. (2.2-7) gives $f = 3.66 \times 10^{-3}$. In that f has apparently converged to within 1%, and that the Prandtl–Kármán equation is not more accurate than that, there is little to be gained by continuing. (Indeed, an additional iteration gives the same value of f to three digits.) With no height difference between the ends of the pipeline, the pressure drop is

$$|\Delta P| = \frac{2\rho U^2 L}{D} f = \frac{2(8.5 \times 10^2)(3)^2(1.0 \times 10^5)}{1}(3.66 \times 10^{-3}) = 5.59 \times 10^6 \text{ Pa}.$$

Alternatively, f can be calculated in one step using Eq. (2.2-8):

$$f = \left[3.6\log\left(\frac{2.83 \times 10^5}{6.9}\right)\right]^{-2} = 3.63 \times 10^{-3}.$$

This value from the Colebrook equation is about 1% smaller than that from the Prandtl–Kármán equation. This difference is within the scatter of typical data for turbulent flow.

Example 2.3-3 Flow rate in an oil pipeline. For the pipeline in Example 2.3-2, suppose that a breakdown at the pumping facility halves $|\Delta P|$, to 2.80×10^6 Pa. How much will the flow rate be affected?

Although the new value of U is unknown, Re will certainly remain large enough for the flow to be turbulent. A quick estimate can be obtained by recalling that the Blasius equation implies that $U \propto (|\Delta \mathcal{P}|/L)^{4/7}$ for a given diameter and fluid properties. This suggests that cutting $|\Delta P|$ in half will reduce U and Re each by only about one third [i.e., $(0.5)^{4/7} = 0.67$]. Because the original Re was well within the turbulent range (Re $\cong 3 \times 10^5$ in Example 2.3-2), we can be confident that the Prandtl–Kármán equation will remain applicable.

Although Re is unknown, $\text{Re}\sqrt{f}$ is independent of U and can be calculated as

$$\text{Re}\sqrt{f} = \frac{UD}{\nu}\left(\frac{D|\Delta P|}{2\rho U^2 L}\right)^{1/2} = \left(\frac{D^3|\Delta P|}{2\nu^2 \rho L}\right)^{1/2}$$
$$= \left[\frac{(1)^3(2.80 \times 10^6)}{2(1.06 \times 10^{-5})^2(8.5 \times 10^2)(1.0 \times 10^5)}\right]^{1/2} = 1.21 \times 10^4.$$

Substituting this into Eq. (2.2-7) gives

$$f = [4 \log(1.21 \times 10^4) - 0.4]^{-2} = 3.94 \times 10^{-3}$$

without the need for iteration. The Reynolds number and mean velocity are then

$$\text{Re} = \frac{\text{Re}\sqrt{f}}{f^{1/2}} = \frac{1.21 \times 10^4}{(3.94 \times 10^{-3})^{1/2}} = 1.93 \times 10^5$$
$$U = \frac{\nu \text{Re}}{D} = \frac{(1.06 \times 10^{-5})(1.93 \times 10^5)}{1} = 2.04 \text{ m s}^{-1}.$$

The values of Re and U are each 68% of those in Example 2.3-2, in excellent agreement with the quick estimate. Because D is unchanged, Q is also 68% of its former value.

In contrast to Example 2.3-2, the Colebrook equation is less convenient here than is the Prandtl–Kármán equation. Because Eq. (2.2-8) does not contain a combination of f and Re that is independent of U, it would require iteration.

Example 2.3-4 Capillary viscometer. A simple way to determine the viscosity of a Newtonian liquid is to measure its volume flow rate through a tube of precisely known dimensions, at a given $|\Delta \mathcal{P}|$. Devices for this, called *capillary viscometers*, typically contain a U-shaped glass tube, in one leg of which is a small-diameter test section. The volumetric flow rate can be found by timing the movement of a meniscus between a pair of marks above the test section, and $|\Delta \mathcal{P}|$ can be calculated from the difference in liquid heights between the legs. Such viscometers operate in the laminar regime, allowing μ to be found from Poiseuille's law. Suppose that it is desired to find the viscosity of a liquid that has a density of 900 kg m^{-3}, that 0.40 ml passes through the viscometer in 20 s when $|\Delta \mathcal{P}| = 880$ Pa, and that the test-section diameter and length are 0.5 mm and 5 cm, respectively.

The volume flow rate is $Q = (4.0 \times 10^{-7}\text{ m}^3)/(20\text{ s}) = 2.0 \times 10^{-8}\text{ m}^3\text{ s}^{-1}$. From Eq. (2.3-7),

$$\mu = \frac{\pi D^4 |\Delta \mathcal{P}|}{128 Q L} = \frac{\pi(5 \times 10^{-4})^4(8.8 \times 10^2)}{128(2 \times 10^{-8})(5 \times 10^{-2})} = 1.35 \times 10^{-3}\text{ Pa}\cdot\text{s}.$$

Table 2.1 Hydraulic diameters for various conduits

Cross-sectional shape	Dimensions	Hydraulic diameter
Circular	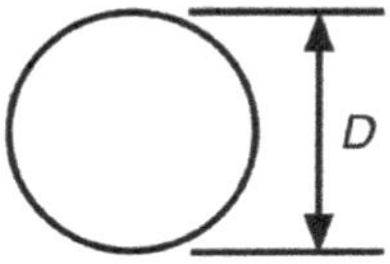	$D_H = D$
Rectangular	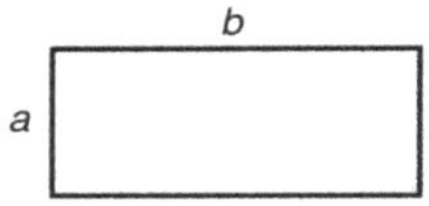	$D_H = \dfrac{2ab}{a+b}$
Parallel plate (slit)	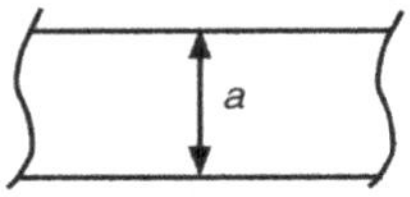	$D_H = 2a$
Annular	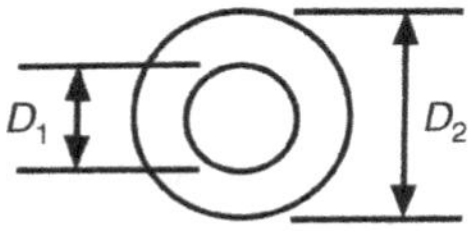	$D_H = D_2 - D_1$
Triangular	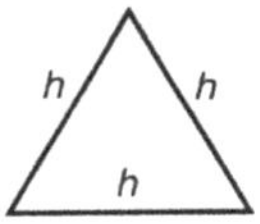	$D_H = \dfrac{h}{\sqrt{3}}$

The viscosity of this liquid is seen to be similar to that of water. The Reynolds number is

$$\mathrm{Re} = \frac{UD\rho}{\mu} = \frac{4Q\rho}{\pi D\mu} = \frac{4(2\times10^{-8})(9\times10^{2})}{\pi(5\times10^{-4})(1.35\times10^{-3})} = 34$$

which confirms that the flow is laminar. As exemplified here, iteration is unnecessary in laminar pipe-flow calculations, no matter which quantity is the unknown.

2.4 NONCIRCULAR CROSS-SECTIONS

Turbulent flow

Many conduits are not cylindrical. For example, heating and ventilating ducts often have rectangular cross-sections, and the flow in certain heat exchangers is in the annular space between two tubes. A remarkable finding is that, for turbulent flow, the results for circular cross-sections can be applied to noncircular ones, provided that an appropriate effective diameter is used. What works is the *hydraulic diameter*,

$$D_H = \frac{4A}{C} \tag{2.4-1}$$

where A is the cross-sectional area and C, the *wetted perimeter*, is the length of wall within a cross section that is in contact with the fluid. For a cylindrical tube, $A = \pi D^2/4$, $C = \pi D$, and the 4 in the numerator of Eq. (2.4-1) ensures that $D_H = D$.

Hydraulic diameters for several cross-sections are shown in Table 2.1. The flow direction is into or out of the page. The *parallel-plate* (or *slit*) geometry is an idealization in which the width of a rectangular channel (b) is so much larger than its height (a) that the side walls can be ignored. In the *annular* channel the fluid is between concentric, cylindrical tubes, each of which contributes to the wetted perimeter.

For noncircular cross-sections, D is replaced by D_H both in the friction factor and in the Reynolds number. Thus,

$$f(\text{Re}_H) = \frac{D_H}{2\rho U^2}\frac{|\Delta\mathcal{P}|}{L}, \quad \text{Re}_H = \frac{UD_H}{\nu} \tag{2.4-2}$$

where $U = Q/A$. The key assumption is that f is given still by Eqs. (2.2-7)–(2.2-9). This proves to be quite accurate for turbulent flow, as shown in Schlichting (1968, p. 576) for square, rectangular, triangular, and annular channels.

One reason this approach works well is that, for any conduit of uniform cross-section, D_H arises naturally from an overall force balance. For a control volume that includes all the fluid within the conduit, the pressure at each end acts on the area A, the wall shear stress acts on the area CL, and gravity acts on the volume AL, where L is the channel length. Accordingly, a force balance like that in Section 2.3 leads to

$$\tau_w = \frac{A\,|\Delta\mathcal{P}|}{CL}. \tag{2.4-3}$$

Dividing by $\rho U^2/2$ and setting $A/C = D_H/4$ gives the expression for f in Eq. (2.4-2). In other words, D_H is the cross-sectional dimension to use when relating the shear stress to the pressure drop. That is true whether the flow is laminar or turbulent. What is special about turbulent flow is that the velocity profile very close to the wall has a nearly universal form (Chapter 10). Because the profile near the wall is insensitive to the channel shape, so is the relationship between f and Re_H.

Laminar flow

Although accurate for turbulent flow, the assumption that $f(\text{Re}_H)$ is the same for all channel shapes fails in the laminar regime. That is, if Eq. (2.2-6) is generalized as $f\,\text{Re}_H = c$, then c for non-cylindrical conduits usually has a value other than 16. The resulting difficulty is lessened by the fact that c often can be predicted from first principles. That is true for all the shapes in Table 2.1, and several such results are derived in Chapter 7. Consequently, exhaustive sets of experiments are usually unnecessary for laminar flow.

For a parallel-plate channel $c = 24$, and for an equilateral triangular channel $c = 40/3$. For a rectangular conduit with $\alpha = a/b \le 1$,

$$f\,\text{Re}_H = c = \frac{24}{(1+\alpha)^2}\left[1 - 6\alpha\sum_{n=0}^{\infty}\frac{\tanh(\lambda_n/\alpha)}{\lambda_n^5}\right]^{-1}, \quad \lambda_n = \left(n+\frac{1}{2}\right)\pi \tag{2.4-4}$$

where n has integer values. This expression is less cumbersome than it might appear, because only three terms in the sum ($n = 0, 1, 2$) are needed for four-digit accuracy. For an annular channel with $\kappa = D_1/D_2$,

$$f\,\text{Re}_H = c = \frac{16(1-\kappa)^2 \ln\kappa}{[1-\kappa^2+(1+\kappa^2)\ln\kappa]}. \tag{2.4-5}$$

As shown in Fig. 2.5, c for rectangular channels falls from 24 for $\alpha \to 0$ to 14.23 for $\alpha = 1$. For annular channels the extremes are $c \to 16$ for $\kappa \to 0$ and $c \to 24$ for $\kappa \to 1$. Notice that the parallel-plate result ($c = 24$) corresponds to either of two limits: a rectangular channel with distant side walls, or an annular one with a narrow gap. Thus,

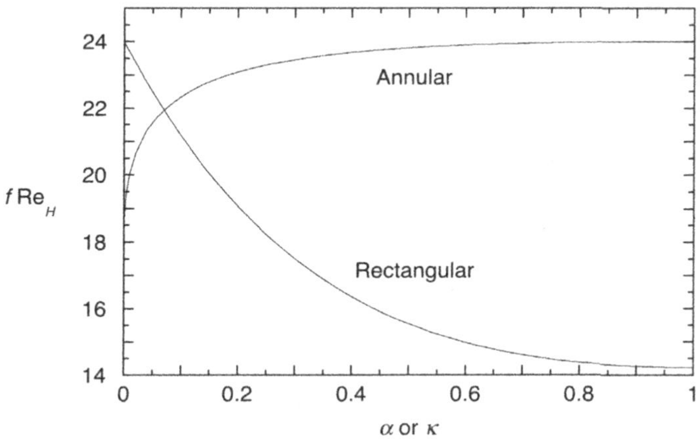

Figure 2.5 Friction factors for laminar flow in rectangular or annular channels. Referring to the diagrams in Table 2.1, $\alpha = a/b$ for a rectangle and $\kappa = D_1/D_2$ for an annulus.

there are two ways in which the parallel-plate idealization can be realized in practice. Among the examples just discussed, c ranges from 17% smaller to 50% larger than the cylindrical value of 16.

Example 2.4-1 Pressure drop for air in a triangular duct. It is desired to determine $|\Delta\mathcal{P}|$ for air flow through an equilateral triangular duct, given that $h = 0.5$ m (side length), $L = 20$ m, $Q = 0.25$ m^3 s^{-1}, $\rho = 1.2$ kg m^{-3}, and $\nu = 1.6 \times 10^{-5}$ m^2 s^{-1}.

The cross-sectional area is $A = (\sqrt{3}/4)h^2$, the mean velocity is $U = (4/\sqrt{3})Q/h^2 = 2.31$ m/s, and the hydraulic diameter is $D_H = h/\sqrt{3} = 0.289$ m. Accordingly,

$$\mathrm{Re}_H = \frac{UD_H}{\nu} = \frac{(2.31)(0.289)}{1.6 \times 10^{-5}} = 4.17 \times 10^4.$$

This indicates that the flow is turbulent and that the Blasius equation is applicable. Combining Eqs. (2.2-9) and (2.4-2), the drop in dynamic pressure is

$$|\Delta\mathcal{P}| = 2(0.0791)\frac{\rho U^2 L}{h}\mathrm{Re}_H^{-1/4} = \frac{2(0.0791)(1.2)(2.31)^2(20)}{0.289}(4.17 \times 10^4)^{-1/4} = 4.91 \text{ Pa}.$$

Example 2.4-2 Material efficiency of square and circular ducts. The volume of wall material in a thin-walled duct with wetted perimeter C and length L is CL times the thickness. For a given thickness, which cross-section makes more efficient use of material, a square or a circular one? It will be assumed that C, L, and Q are each the same for the two shapes, and the measure of efficiency will be the pressure drop.[6] It will be assumed that the flow is turbulent and that Re is within the Blasius range.

With the fluid properties and the lengths the same, the two ducts are distinguished only by the differences in hydraulic diameter and mean velocity. Combining Eqs. (2.2-9) and (2.4-2) gives

$$\frac{|\Delta\mathcal{P}|}{2(0.0791)\rho L \nu^{1/4}} = \frac{U^{7/4}}{D_H^{5/4}}.$$

[6] Although the shape with the lower $|\Delta\mathcal{P}|$ will require less pumping power, to determine which is actually more economical we would also need the purchase and installation costs (Problem 2.8).

Accordingly, the ratio of pressure drops is

$$\frac{|\Delta\mathcal{P}|_c}{|\Delta\mathcal{P}|_s} = \left(\frac{U_c}{U_s}\right)^{7/4}\left(\frac{D_{H,s}}{D_{H,c}}\right)^{5/4}$$

where the subscripts c and s denote circular and square, respectively. The hydraulic diameters are a and D. With the same wetted perimeter in the two cases, $C_c = \pi D = C_s = 4a$ and $a/D = D_{H,s}/D_{H,c} = \pi/4$. Taking into account the differing cross-sectional areas, $U_c/U_s = A_s/A_c = 4a^2/(\pi D^2) = \pi/4$. Thus,

$$\frac{|\Delta\mathcal{P}|_c}{|\Delta\mathcal{P}|_s} = \left(\frac{\pi}{4}\right)^{7/4}\left(\frac{\pi}{4}\right)^{5/4} = \left(\frac{\pi}{4}\right)^{3} = 0.484.$$

It is seen that, for a given amount of material, the pressure drop in a circular duct is roughly half that in a square one.

2.5 WALL ROUGHNESS

No matter how carefully fluid conduits are manufactured, there are always at least microscopic irregularities on their surfaces. These are smallest for glass or drawn metals, intermediate for commercial steel, and relatively large for concrete. Irregularities can increase over time, as a result of corrosion, mineral deposits, or the growth of microorganisms. Wall roughness tends to increase the friction factor at a given Reynolds number, and f is found to be governed by roughness alone when Re is sufficiently large. The effects of roughness are noticeable only in turbulent flow. In that regime, where most of the velocity variation occurs very near the wall (Fig. 2.3), even small protrusions or indentations can affect the key part of the velocity profile. In laminar flow the profile is much less sensitive to what happens in the immediate vicinity of the wall.

What is needed to quantify the effects of roughness, at the very least, is the height of the irregularities. This new length scale, denoted as k, gives rise to an additional dimensionless group. For cylindrical tubes of diameter D we choose k/D as the new group. Thus, dimensional analysis indicates that $f = f(\text{Re}, k/D)$. Because roughness elements can differ in size or shape and have regular or random spacings, several other dimensionless groups might be relevant. However, to avoid having to characterize the microscopic details of each surface and having to measure their individual effects on f, it is customary to assign *effective* roughness heights to materials of engineering interest.

The scale used for k stems from experiments with artificially roughened tubes done by J. Nikuradse in Germany in the early 1930s. Sand that had been sieved into narrow size ranges was suspended in a lacquer, which was used to coat the walls of smooth tubes of varying diameter. After the coating dried, a densely packed monolayer of sand adhered to the wall (Nikuradse, 1933). At the highest flow rates examined, f became independent of Re. With k the diameter of a sand grain, it was found that

$$\frac{1}{\sqrt{f}} = -4\log\left(\frac{k}{3.7D}\right) \quad (\text{Re} \to \infty). \tag{2.5-1}$$

Subsequently, Eq. (2.5-1) has been used to convert the high-Re value of f for a naturally or artificially roughened tube to an effective sand-grain diameter. Roughness heights so obtained for several materials are shown in Table 2.2. It is customary to assume that k for a given material is independent of D, although there may be exceptions.

Table 2.2 Effective roughness of various pipe materials[a]

Material	k(mm)
Glass and drawn metals[b]	0.0015
Commercial steel or wrought iron	0.046
Asphalt-coated cast iron	0.12
Galvanized iron	0.15
Cast iron	0.26
Concrete	0.30–3.0

[a] Typical values, from Moody (1944). See Brater *et al.* (1996) for additional data.
[b] Includes copper, brass, and lead.

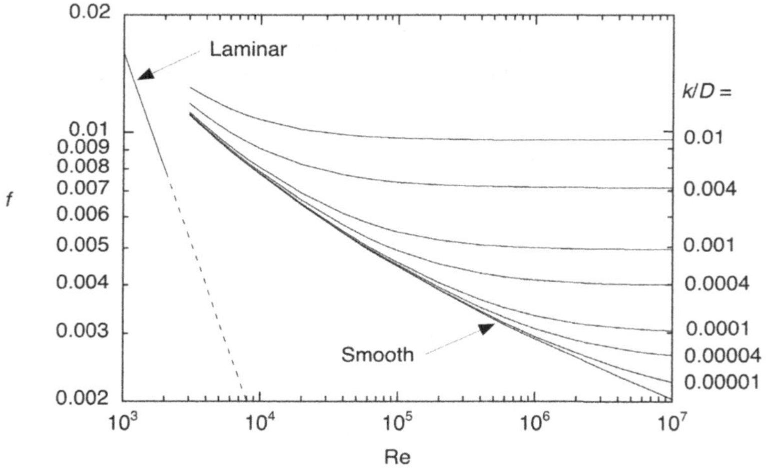

Figure 2.6 Friction factor for cylindrical tubes as a function of Reynolds number and relative roughness. The curves for Re > 3000 are based on Eq. (2.5-3).

The most widely accepted general formula for rough pipes is the *Colebrook–White equation* (Colebrook, 1939),

$$\frac{1}{\sqrt{f}} = -4\log\left(\frac{1.26}{\mathrm{Re}\sqrt{f}} + \frac{k}{3.7D}\right). \tag{2.5-2}$$

This reduces to Eq. (2.2-7) for $k/D \to 0$ and to Eq. (2.5-1) for $\mathrm{Re} \to \infty$. As with the Prandtl–Kármán equation, the presence of f on both sides makes this expression awkward for certain kinds of problems. An explicit relationship recommended by Haaland (1983), which differs from Eq. (2.5-2) by <2%, is

$$f = \left\{3.6\log\left[\frac{6.9}{\mathrm{Re}} + \left(\frac{k}{3.7D}\right)^{1.11}\right]\right\}^{-2}. \tag{2.5-3}$$

This reduces to Eq. (2.2-8) for $k/D \to 0$ and once again gives Eq. (2.5-1) for $\mathrm{Re} \to \infty$.

Friction factors calculated using Eq. (2.5-3) are shown in Fig. 2.6, a type of plot due to Moody (1944). In laminar flow f is unaffected by wall roughness, so there is only

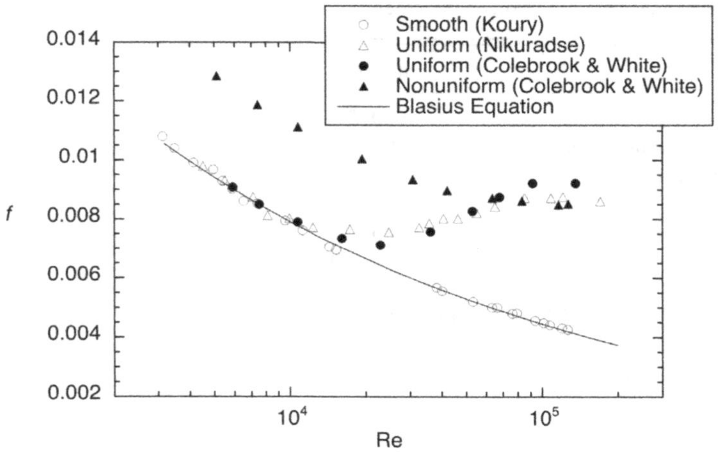

Figure 2.7 Friction factors in smooth and artificially roughened tubes. The data are from Koury (1995), Nikuradse (1933), and Colebrook and White (1937). "Uniform" and "nonuniform" are explained in the text. The curve is from Eq. (2.2-9).

one curve for Re < 2100. In turbulent flow a tube is functionally smooth only if Re is below a certain threshold, which depends on the relative roughness. In that case, f first declines monotonically with increasing Re, but eventually reaches a plateau determined by k/D. For $k/D \geq 0.001$, roughness becomes evident as soon as Re is large enough for turbulence. As k/D is reduced, the departure of f from that for smooth tubes is increasingly delayed. However, once Re is large enough, the "fully rough" behavior described by Eq. (2.5-1), in which f is independent of Re, is seen for all $k/D > 0$.

Although Eqs. (2.5-2) and (2.5-3) are satisfactory for typical commercial pipes, f can be influenced by factors in addition to Re and k/D. That is illustrated in Fig. 2.7, in which f measured for various kinds of surfaces is shown for a limited range of Re. Some of the data of Koury (1995) for functionally smooth tubes are repeated from Fig. 2.2. The other three data sets are for artificially roughened tubes, one from Nikuradse (1933) and two from Colebrook and White (1937). They are for tubes with approximately the same relative roughness ($k/D = 0.07$–0.09), so that the values of f at high Re are very similar. The similarities in k/D notwithstanding, there are noticeable differences in the transition from smooth to fully rough behavior. The Nikuradse experiments and one series from Colebrook and White involved particles that were closely packed on the wall, resulting in what is termed here "uniform" roughness. In these cases f from either laboratory declined gradually along the Blasius curve, reached a minimum, and then increased abruptly before leveling off. In the other series from Colebrook and White, the particles were larger but more widely spaced, yielding "nonuniform" roughness. In this case there was no local minimum and the effects of roughness were noticeable throughout the range of Re shown. As shown by these data and by other studies, f is affected by the spatial distribution of irregularities on the wall. Differences among varying kinds of roughness are discussed in Schlichting (1968, pp. 586–589), which summarizes results obtained using regularly shaped protrusions arrayed in various ways.

As discussed in Colebrook (1939), what is typically seen in commercial pipes with naturally occurring roughness is a monotonic decline in f, as in Fig. 2.6 and the "nonuniform" case in Fig. 2.7. That is the rationale for using the Colebrook–White equation (or the Haaland alternative) for design calculations.

Example 2.5-1 Effect of roughness on water flow in a process pipe. If the pipe in Example 2.3-1 were made of commercial steel, how much would the pressure drop be increased?

With $k = 0.046$ mm (Table 2.2) and $D = 0.1$ m, $k/D = 4.6 \times 10^{-5}/0.1 = 4.6 \times 10^{-4}$. As found before, $\mathrm{Re} = 1.02 \times 10^5$. From Eq. (2.5-3), the friction factor for the steel pipe is

$$f = \left\{3.6 \log\left[\frac{6.9}{1.02 \times 10^5} + \left(\frac{4.6 \times 10^{-4}}{3.7}\right)^{1.11}\right]\right\}^{-2} = 4.96 \times 10^{-3}.$$

This is 12% larger than f for a perfectly smooth pipe, as calculated using the Blasius equation ($f = 4.43 \times 10^{-3}$) or as found from Eq. (2.5-3) with $k/D = 0$ ($f = 4.44 \times 10^{-3}$). Thus, the pressure drop would also be 12% larger.

Example 2.5-2 Practical smoothness. How little wall roughness must there be for a tube to be functionally smooth? The answer depends, of course, on the level of precision desired. Also, as shown in Fig. 2.6, the value of k/D at which f for a real tube begins to deviate from that for a perfectly smooth one depends on Re.

Because it is difficult to quantify small deviations from smoothness in Fig. 2.6, we will rearrange Eq. (2.5-3) into a form that is convenient for this purpose. Denoting the friction factors for real and perfectly smooth tubes as f and f_0, respectively, the relative increase due to roughness is

$$\varepsilon = \frac{f - f_0}{f_0}. \tag{2.5-4}$$

What we seek is a simple way to calculate k/D for specified, small values of ε. Equation (2.5-3) may be rewritten as

$$f = \{3.6 \log [a(1 + b)]\}^{-2} \tag{2.5-5}$$

where $a = 6.9/\mathrm{Re}$ and $b = (\mathrm{Re}/6.9)(k/3.7D)^{1.11}$. Employing $(1 + x)^{1/2} \cong 1 + (x/2)$ and $a^x \cong 1 + x \ln a$ (each valid for $x \to 0$) to simplify intermediate results, it is found that

$$b = -\frac{\varepsilon}{2} \ln a. \tag{2.5-6}$$

Returning to the original variables gives

$$\frac{k}{D} = 3.7\left[-\frac{\varepsilon}{2}\left(\frac{6.9}{\mathrm{Re}}\right) \ln\left(\frac{6.9}{\mathrm{Re}}\right)\right]^{1/1.11} \quad (\varepsilon \ll 1). \tag{2.5-7}$$

Results from Eq. (2.5-7) are shown in Fig. 2.8. For the percentage increase in f to be no more than that on the abscissa, k/D must not exceed that on the ordinate. For example, limiting the effects of roughness to 1% at $\mathrm{Re} = 1 \times 10^5$ and 1×10^7 requires that k/D not exceed 4.3×10^{-5} and 9.6×10^{-7}, respectively. For Koury (1995), where $D = 10$ mm and $\mathrm{Re} \leq 1 \times 10^5$, conventional smoothness ($k = 0.0015$ mm, Table 2.2) would have sufficed to give the excellent agreement with the Blasius curves seen in Figs. 2.2 and 2.7. However, for Zagarola and Smits (1998), where $D = 129$ mm and $\mathrm{Re} \leq 3.5 \times 10^7$, extraordinary machining and polishing were needed to create a test section that was functionally smooth.

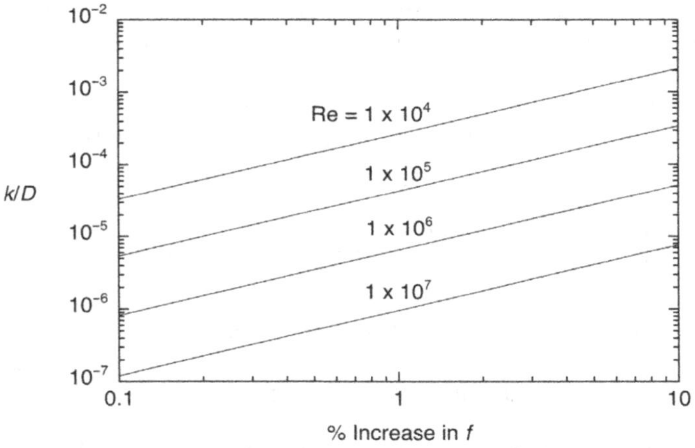

Figure 2.8 Relative roughnesses that give specified percentage increases in the friction factor at various Reynolds numbers, according to Eq. (2.5-7). The abscissa equals 100ε.

2.6 CONCLUSION

Relating the flow rate in a pipe to the pressure drop along it is essential for the design of any piping system. At the heart of such calculations is the friction factor (f), which is a dimensionless wall shear stress [Eq. (2.2-4)]. In a smooth pipe f depends only on the Reynolds number (Re), generally decreasing with increasing Re. However, there is an abrupt increase in f at Re $\cong 2100$, which corresponds to the transition from laminar to turbulent flow. The most notable formulas for $f(\text{Re})$ are a form of Poiseuille's law [Eq. (2.2-6), for laminar flow] and the Prandtl–Kármán equation [Eq. (2.2-7), for turbulent flow].

A force balance that relates the wall shear stress, pipe dimensions, and inlet and outlet pressures allows f to be expressed also as a dimensionless pressure drop [Eq. (2.3-6)], as needed for design calculations. Flow through horizontal, inclined, or vertical tubes can be described using a single set of relationships if the pressure drop is expressed in terms of the dynamic pressure ($\mathcal{P}$). This variable, also called the equivalent pressure or modified pressure, combines the effects of actual pressure differences and gravity.

Not all conduits are circular, and for laminar flow each cross-sectional shape must be examined individually. However, a fortuitous aspect of turbulent flow is that the results for cylindrical tubes can be applied to other shapes. That is done by replacing the tube diameter with the hydraulic diameter, which is calculated from the cross-sectional area and circumference of a conduit [Eq. (2.4-1)].

Any pipe wall has a certain amount of roughness. Depending on the material and the manufacturing method, the irregularities range from microscopic to plainly visible. Wall roughness does not affect f in laminar flow, but can greatly increase it in turbulent flow. Thus, more generally, f is a function not just of Re but also of k/D, where k is the effective roughness height and D is the tube diameter. The best-known relationship is the Colebrook–White equation [Eq. (2.5-2)]. The effects of k/D on f are evident only when Re exceeds a certain threshold; the smaller k/D, the larger the threshold for Re. As Re is increased, a "fully rough" regime is eventually reached in which f depends only on k/D.

REFERENCES

Brater, E. F., H. W. King, J. E. Lindell, and C. Y. Wei. *Handbook of Hydraulics*, 7th ed. McGraw-Hill, New York, 1996.
Colebrook, C. F. Turbulent flow in pipes, with particular reference to the transition region between the smooth and rough pipe laws. *J. Inst. Civ. Eng.* 11: 133–156, 1939.
Colebrook, C. F. and C. M. White. Experiments with fluid friction in roughened pipes. *Proc. Roy. Soc. Lond. A* 161: 367–381, 1937.
Haaland, S. E. Simple and explicit formulas for the friction factor in turbulent pipe flow. *J. Fluids Eng.* 105: 89–90, 1983.
Kármán, T. von. *Aerodynamics*. Cornell University Press, Ithaca, NY, 1954.
Koury, E. Drag reduction by polymer solutions in riblet pipes. Ph.D. thesis, Department of Chemical Engineering, Massachusetts Institute of Technology, 1995.
McComb, W. D. *The Physics of Fluid Turbulence*. Clarendon Press, Oxford, 1990.
Moody, L. F. Friction factors for pipe flow. *Trans. ASME* 66: 671–684, 1944.
Murray, C. D. The physiological principle of minimum work. I. The vascular system and the cost of blood volume. *Proc. Natl. Acad. Sci. (USA)* 12: 207–214, 1926.
Nikuradse, J. *Gesetzmäßigkeiten der turbulenten Strömung in glatten Rohren*. VDI Forschungsheft 356, 1932. [English translation in NASA Technical Report No. TT-F-10359, 1966.]
Nikuradse, J. *Strömungsgesetze in rauhen Rohren*. VDI Forschungsheft 361, 1933. [English translation in NACA Technical Memorandum No. 1292, 1950.]
Oh, K. W., K. Lee, B. Ahn, and E. P. Furlani. Design of pressure-driven microfluidic networks using electrical circuit analogy. *Lab Chip* 12: 515–545, 2012.
Peters, M. S., K. D. Timmerhaus, and R. E. West. *Plant Design and Economics for Chemical Engineers*, 5th ed. McGraw-Hill, New York, 2003.
Schlichting, H. *Boundary-Layer Theory*, 6th ed. McGraw-Hill, New York, 1968.
Sherman, T. F. On connecting large vessels to small. The meaning of Murray's law. *J. Gen. Physiol.* 78: 431–453, 1981.
Sturm, T. W. *Open Channel Hydraulics*, 2nd ed. McGraw-Hill, New York, 2010.
Sutera, S. P. and R. Skalak. The history of Poiseuille's law. *Annu. Rev. Fluid Mech.* 25: 1–19, 1993.
Tilton, J. N. Fluid and Particle Dynamics. In *Perry's Chemical Engineers' Handbook*, D. E. Green and R. H. Perry (Eds.), 8th ed. McGraw-Hill, New York, 2007.
Vincenti, W. G. Control-volume analysis: a difference in thinking between engineering and physics. *Technol. Culture* 23: 145–174, 1982.
Virk, P. S. Drag reduction fundamentals. *AIChE J.* 21: 625–656, 1975.
Zagarola, M. V. and A. J. Smits. Mean-flow scaling of turbulent pipe flow. *J. Fluid Mech.* 373: 33–79, 1998.

PROBLEMS

2.1. Cavitation Bubbles tend to form wherever the pressure in a liquid falls below its vapor pressure (P_v), a phenomenon called *cavitation*. Shock waves created by the rapid collapse of such bubbles can be very damaging to equipment. A system where this might be a concern is shown in Fig. P2.1. To empty a swimming pool for repairs, a pump with a flow rate $Q = 3.93 \times 10^{-3}$ m^3/s is to be connected to 50 m of rubber hose with a diameter of 5 cm. The hose is hydrodynamically smooth. The pump intake will be 3 m above the hose inlet, which will rest on the bottom, and the initial water depth will be 2 m. The water properties are as in Table 1.2.

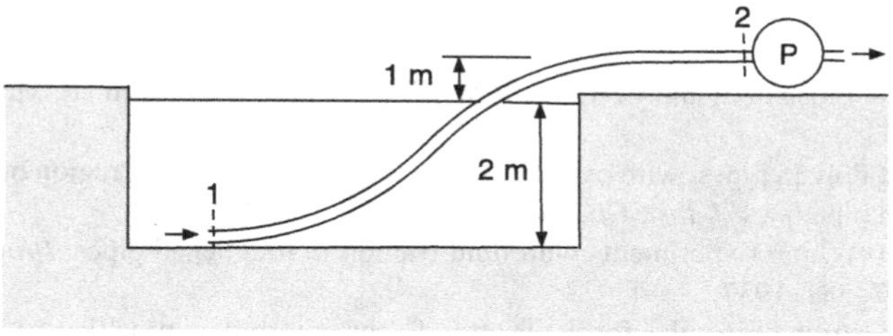

Figure P2.1 Pump and hose for emptying a swimming pool (not to scale).

(a) Calculate $|\Delta\mathcal{P}|$ from just inside the hose inlet (position 1) to the pump intake (position 2).
(b) Determine P_2 when the pool is nearly empty, at which time $P_1 = 1.01 \times 10^5$ Pa. For water at 20 °C, $P_V = 2339$ Pa. You should find that $P_V < P_2 < P_0$, where P_0 is atmospheric pressure.
(c) How much might Q be increased without risking cavitation? Why, at a given Q, does the risk increase as the water level in the pool goes down?

2.2. Bottling honey A food manufacturer is planning to bottle honey using gravity-driven flow from a tank. Fitting the equipment into the available space will require a pipe length of 4 m, with the inlet 2 m above the outlet. When the tank level is low, the pressure at the inlet will be close to atmospheric. The honey properties are as in Table 1.2.

What pipe diameter would be needed to obtain a production rate of at least one 17 oz. (0.482 kg) jar per minute? Is the proposed arrangement feasible?

2.3. Filling a boiler A particular steam boiler must be filled with 7.57 m^3 (2000 US gallons) of water before being used. It is connected to the water supply by copper tubing 15 m long that has a diameter of 2.60 cm. The fill point (tube outlet) is 3 m above the inlet and the supply pressure is such that $|\Delta P| = 4.0 \times 10^5$ Pa.

(a) How long would the filling take if water at 20 °C (Table 1.2) were used?
(b) An operator who knows that hot liquids flow more freely than cold ones decides to speed the filling by using water at 52 °C, for which $\mu = 5.29 \times 10^{-4}$ Pa · s and $\rho = 987$ kg m^{-3}. If the supply pressure is unchanged, how much time will be saved?

2.4. Syringe pump Suppose that a laboratory experiment requires that an aqueous solution be delivered to an open container at a constant flow rate of 100 μl/min for 30 min. A syringe pump is available, which has a motor and gear drive that will advance the plunger of any syringe at a set speed. As shown in Fig. P2.4, the syringe diameter is D_s and the pusher speed is U_s. The syringe will be connected to plastic tubing with a length $L = 50$ cm and a diameter $D = 0.86$ mm. There may be a height difference between the ends of the tubing. A syringe size must be chosen.

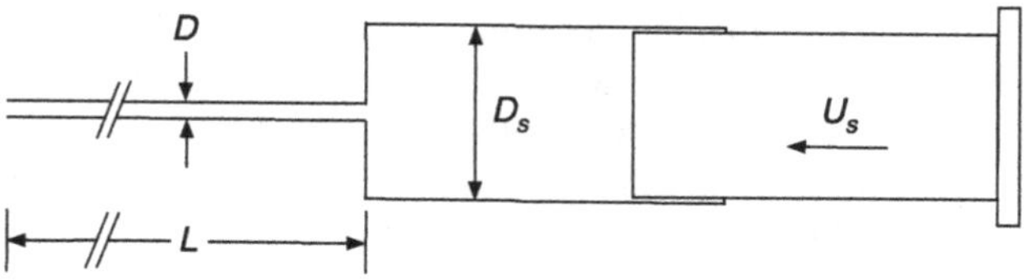

Figure P2.4 Syringe pump.

(a) According to the pump manufacturer, U_s can be controlled from 0.08 to 80 mm/min. Three syringe sizes available in the particular lab are 1 ml ($D_s = 4.8$ mm), 5 ml ($D_s = 12.1$ mm), and 50 ml ($D_s = 26.7$ mm). Which would be best?
(b) If the tubing inlet and outlet are at the same height, what will $|\Delta P|$ be within the tubing? Will it be affected by the size of the syringe?
(c) If the tubing outlet is raised by 10 cm, what will the new $|\Delta P|$ be?

2.5. Flue gases Suppose that the hot combustion products from a large furnace pass through horizontal steel pipes arranged in parallel, each 6.4 cm in diameter and 6.1 m in length. At the inlet temperature of 600 K, $\mu = 2.6 \times 10^{-5}$ Pa · s and $\rho = 0.63$ kg/m^3. The pressure drop is 63 Pa.

If the temperature remains constant, what is the mass flow rate per tube?

2.6. Hydraulic fracturing Hydraulic fracturing, or "fracking," is widely used to stimulate production from oil or gas deposits. A deep well is lined with steel pipe. A large volume of water is pumped into the well at a pressure sufficient to fracture the rock formation at the bottom. A slurry containing "proppants" (sand or ceramic particles) is then pumped into the formation, the particles acting to keep the cracks open when the applied pressure is released.

Suppose that the pipe diameter is 15 cm and the well depth is 2.3 km. The flow rate during the water-only phase is to be 0.21 m^3/s, which is predicted to give $P = 44$ MPa (6500 psi) at the well bottom, based on what is known about the geology. It is planned to use 60 °C water, for which $\mu = 0.467 \times 10^{-3}$ Pa · s and $\rho = 0.983 \times 10^3$ kg m^{-3} at atmospheric pressure. If these properties are constant throughout the pipe, what pressure P must be maintained at the top?

2.7. Drag reduction As mentioned in Section 2.2 and reviewed in Virk (1975), polymeric additives can greatly reduce the resistance in turbulent pipe flow. That is manifested by a decrease in f at a given Re. At the small concentrations used (typically tens to hundreds of ppm by weight), μ is very close to that of the solvent and non-Newtonian behavior such as shear thinning is not evident. Laminar flow is unaffected. In turbulent flow, decreases in f occur for $\tau_w > \tau_w^*$. The threshold shear stress τ_w^* decreases with increasing polymer molecular weight, but is independent of polymer concentration and solvent viscosity. Once τ_w^* is exceeded, increases in the concentration or molecular weight of the additive lead to greater reductions in f. However, the "maximum drag reduction" appears to be insensitive to the identity and concentration of the polymer and is described by (Virk, 1975)

$$\frac{1}{\sqrt{f}} = 19.0 \log(\text{Re}\sqrt{f}) - 32.4. \tag{P2.7-1}$$

This is compared with the laminar result and the Prandtl–Kármán correlation in Fig. P2.7.

For the oil pipeline in Example 2.3-2, how much might the pressure drop be reduced by use of a polymeric additive?[7]

[7] Because the incremental benefit of using more additive lessens as the additive concentration is increased, it tends to be most economical not to seek maximum drag reduction.

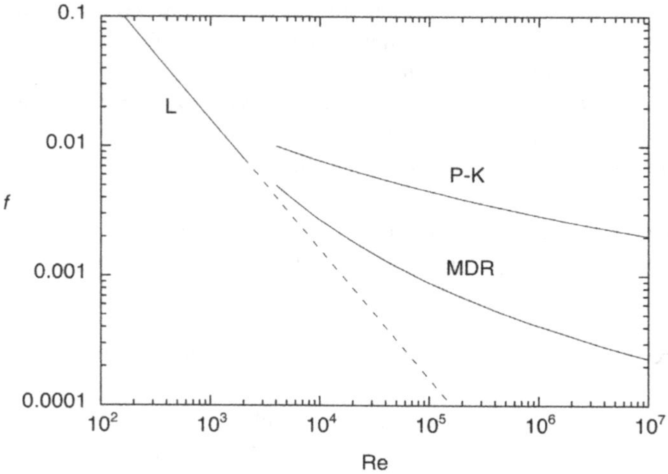

Figure P2.7 Comparison of friction factors for laminar flow [L, Eq. (2.2-6)], turbulent flow in smooth tubes [P–K, Eq. (2.2-7)], and turbulent flow with maximum drag reduction [MDR, Eq. (P2.7-1)].

2.8. Economic pipe diameter At numerous places in chemical plants there is a need for a pipe of length L to carry a given fluid at a specified flow rate Q. An appropriate pipe diameter D must be chosen by the designer. Larger-diameter pipes are more expensive to buy and install but cheaper to operate. More precisely, the total annual cost C of a pipe can be expressed as

$$C = K_1 D^n L + K_2 Q\,|\Delta \mathcal{P}|\,. \tag{P2.8-1}$$

Embedded in the dimensional constant K_1 are the purchase, installation, and maintenance costs of the pipe and associated fittings, each per unit length and each put on an annual basis by assuming a certain lifetime for the system (Peters *et al.*, 2003, pp. 401–406). For steel pipe, $n \cong 1.5$. The power that a pump must deliver for flow through a straight pipe is $Q|\Delta \mathcal{P}|$. The coefficient K_2 is the product of the electric rate, the operating time per year, and factors that account for pump efficiency and pressure drops in fittings. The fixed costs increase with D but the pumping costs decrease, such that there is an optimal value of D.

(a) In economic calculations the friction factor can be approximated as

$$f = a\,\mathrm{Re}^{-b} \tag{P2.8-2}$$

where a and b are constants. Peters *et al.* (2003) suggest $a = 0.04$ and $b = 0.16$ for turbulent flow in new steel pipes; Eqs. (2.2-6) or (2.2-9) also could be used, depending on the range of Re. Show that the diameter D^* that minimizes C is

$$D^* = \left[\frac{(5-b)cK_2}{nk_1}\right]^{1/(5+n-b)} \rho^{(1-b)/(5+n-b)}\mu^{b/(5+n-b)}Q^{(3-b)/(5+n-b)} \tag{P2.8-3}$$

and derive an expression for the dimensionless constant c.

(b) By examining the exponents in Eq. (P2.8-3), show that there is only a weak dependence of D^* on ρ and μ, for either laminar or turbulent flow. Thus, for a given Q there is one fairly narrow range of D^* for low-viscosity liquids and another for gases.

(c) Let U^* be the mean velocity when $D = D^*$. Show that

$$U^* \propto Q^{(n+b-1)/(5+n-b)} \cong Q^{0.1} \tag{P2.8-4}$$

for turbulent flow. Thus, unlike D^*, U^* is nearly independent of Q. For Schedule 40 steel pipe, U^* for liquids is typically between 1.8 and 2.4 m/s (Tilton, 2007).

2.9. Microfluidic device* Figure P2.9 shows the layout of a microfluidic device with both series and parallel channels. The first segment has a length $L_1 = 1$ mm and the relative lengths of the others are as indicated. Suppose that each channel has a square cross-section with side length 100 μm and that all flow is horizontal. The overall pressure drop for water is $P_A - P_D = 3000$ Pa. You may assume that the channels are long enough that the extra resistances associated with the bends and junctions are negligible.

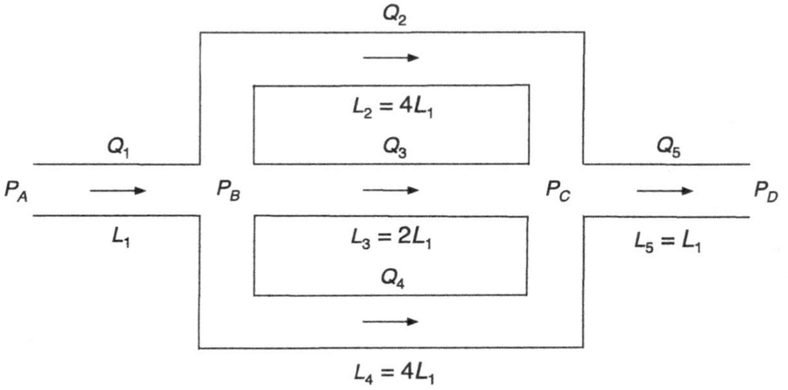

Figure P2.9 Microfluidic device, adapted from Oh *et al.* (2012). The channel lengths and widths are not to scale.

(a) Evaluate the flow-rate ratios Q_2/Q_1, Q_3/Q_1, Q_4/Q_1, and Q_5/Q_1.
(b) Calculate Q_1.

* This problem was suggested by P. S. Doyle.

2.10. Murray's law Figure P2.10 shows a blood vessel of radius R_1 branching into "daughter" vessels of radii R_2 and R_3. The respective flow rates are Q_1, Q_2, and Q_3. Anatomical studies of the mammalian circulation indicate that the radii in such a bifurcation tend to be related as

$$R_1^3 = R_2^3 + R_3^3 \quad \text{(P2.10-1)}$$

independently of the vessel lengths and branching angle. This is called *Murray's law*, after an American biologist who proposed the explanation that is outlined below.

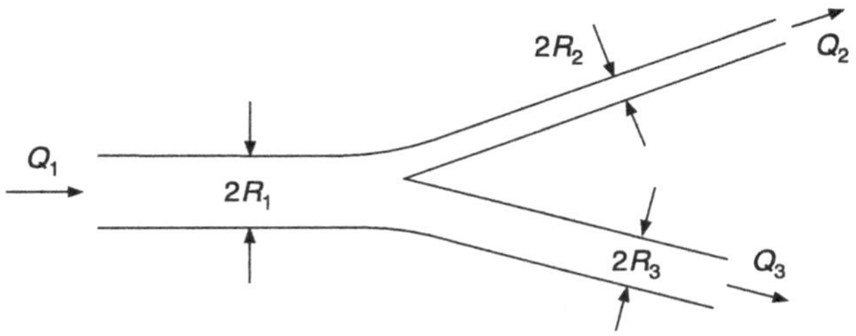

Figure P2.10 Branching of a blood vessel into two daughter vessels.

Murray (1926) postulated that the circulation evolved to minimize the metabolic energy needed for its operation and maintenance. The energy expenditure for a vessel of radius R and length L was viewed as having two parts: (1) the pumping power needed for a given blood flow rate Q, and (2) the energy per unit time required to renew the blood in the vessel and maintain the surrounding cells. The maintenance energy was assumed to be proportional to the blood volume. The total rate of energy usage E_t was expressed as

$$E_t = Q\,|\Delta\mathcal{P}| + b\pi R^2 L \qquad \text{(P2.10-2)}$$

where the proportionality constant b was assumed to be independent of vessel size. Blood flow in the circulation is almost entirely laminar. Thus, Poiseuille's law was used for Q, even though the flow is pulsatile (especially in large arteries) and not precisely Newtonian (especially in capillaries).

(a) Assuming that Q and L for a given vessel are fixed (by the need to supply oxygen and nutrients to a tissue), find the value of R that minimizes E_t for that vessel.
(b) Show that Murray's law is obtained by applying the result from part (a) to each vessel at a bifurcation.
(c) Each set of branchings in a network creates a new "generation" of vessels. Let R_{ij} be the radius of vessel i in generation j and let N_j be the total number of vessels in that generation. If the result from part (a) applies to all blood vessels, show that

$$\sum_{i=1}^{N_j} R_{ij}^3 = \sum_{i=1}^{N_k} R_{ik}^3 \qquad \text{(P2.10-3)}$$

where generations j and k are not necessarily adjacent. That is, the energy-minimization principle predicts that the sum of the cubes will remain constant as the network is traversed. That is approximately what is found (Sherman, 1981).

2.11. Open-channel flow Consider steady water flow in the open, rectangular channel in Fig. P2.11, which might be part of an irrigation or drainage system. The flow is due to gravity. The channel has a width W, is filled to a level H, and its sides and bottom each have a roughness height k. It is inclined at an angle θ that is small enough for the slope ($\tan\theta$) to be approximated as $S = \sin\theta$. Assuming these quantities are all known, it is desired to predict the mean velocity U.

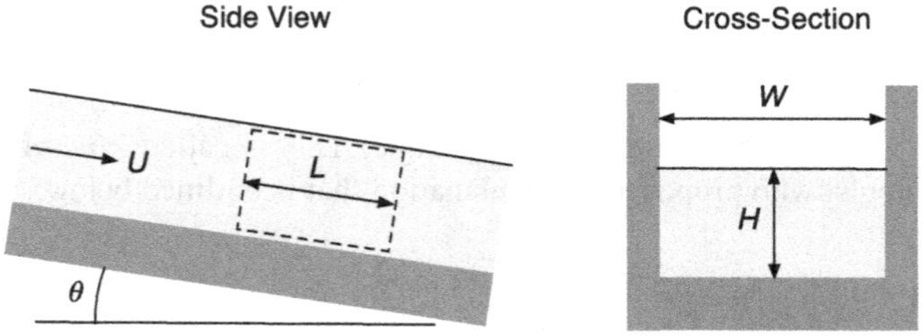

Figure P2.11 Flow down an inclined, open channel of rectangular cross-section.

(a) What can you conclude about U from dimensional analysis? (*Hint*: Include U, g, and H in the principal dimensionless group.)
(b) The shear stress exerted on a liquid by an adjacent gas is typically negligible. With that in mind, use a force balance for a length L of channel (Fig. P2.11) to relate the shear stress τ_w at the solid surfaces to ρ, g, S, W, and H.

(c) If an appropriate hydraulic diameter is used, the force balance of part (b) will apply to an open channel of any shape. To permit such a generalization, how should D_H be defined?

(d) In open-channel flow Re and k/D_H are each typically large enough for f to be independent of Re (i.e., to be in the fully rough regime). When this is so, show that

$$\frac{U}{(gD_H)^{1/2}} = \left(\frac{S}{2f}\right)^{1/2}. \tag{P2.11-1}$$

If, as recommended by the ASCE for open-channel flow, Eq. (2.5-1) is modified to

$$\frac{1}{\sqrt{f}} = 4.0 \log\left(\frac{3D_H}{k}\right) \tag{P2.11-2}$$

then Eq. (P2.11-1) is nearly equivalent to the *Gauckler–Manning formula* (Sturm, 2010, pp. 114–123). In that widely used equation f is approximated as

$$\frac{1}{\sqrt{f}} = C\left(\frac{D_H}{k}\right)^{1/6} \tag{P2.11-3}$$

where C is a constant. With $C = 4.45$, Eq. (P2.11-3) is within $\pm 5\%$ of Eq. (P2.11-2) for $16 < D_H/k < 2000$. It follows that $U \propto D_H^{2/3} S^{1/2}$ in such channels.

3

Drag, particles, and porous media

3.1 INTRODUCTION

The *drag* on an object is the fluid-dynamic force that resists its translational motion. The drag grows as the velocity difference between the object and the surrounding fluid increases. Evaluating drag is important in applications ranging from predicting diffusion rates of molecules to designing ships and aircraft. This chapter begins with a discussion of the mechanical origins of drag and how drag measurements are correlated using dimensionless groups. Results presented for spheres, disks, cylinders, and flat plates provide guidance on what to expect for various shapes. For objects propelled by a constant force, such as gravity, the relative velocity will increase or decrease until the drag balances the applied force, and be constant thereafter. The *terminal velocity* that is attained reflects a balance among drag, gravity, and buoyancy, as will be discussed. As it is the speed at which a dense solid particle or liquid drop will settle or a bubble will rise, the terminal velocity has numerous applications both in process design and in the analysis of natural phenomena.

In dilute suspensions of small objects such as macromolecules, cells, and fine particles, the motion of the individual entities is nearly independent and drag results for single objects can be used. Those results are less accurate for concentrated suspensions, and are inapplicable when particles or fibers join to form a porous material that consists of a rigid matrix with fluid-filled interstices. The attention shifts then from the drag on an individual object to the flow resistance or conductance of an array of obstacles. The *Darcy permeability* is a standard measure of the conductance. How it is related to the microstructure of a porous material will be discussed.

Packed beds are widely used in reactors, adsorbers, chromatography columns, and other devices requiring solid–fluid contact, because they provide a large area of solid per unit volume. A packed bed is constructed by partly or completely filling a vessel with particles, thereby forming a granular porous medium inside the vessel. A relationship that describes flow through such beds will be derived. A *fluidized bed* can be created by rapid upward flow through a packed bed. Typically, the fluid is a gas. As will be discussed, there is a range of velocities that will lift the particles, yet not carry them out the top of the vessel. The result is a dense particle–gas suspension with rapid mixing, which is ideal for certain kinds of reactors.

3.2 DRAG

Origins

The drag on an object is determined by its size and shape, the density and viscosity of the fluid, and the relative velocity of the object and fluid. Whether the object, bulk fluid,

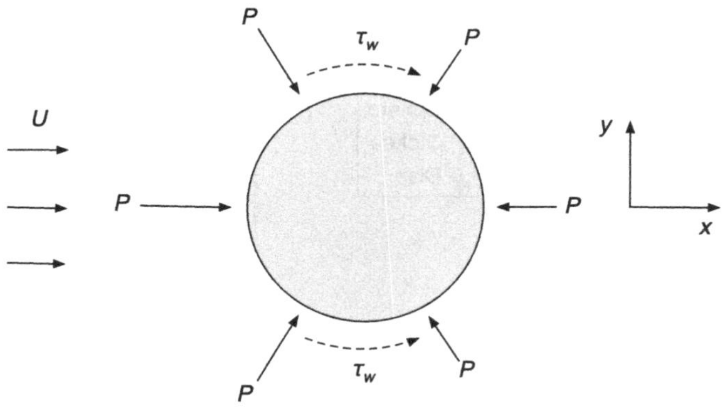

Figure 3.1 Fluid approaching a stationary object at velocity U. Pressure and viscous stresses due to the flow act locally in the directions shown by the solid and dashed arrows, respectively.

or both have a nonzero velocity in a particular reference frame is immaterial; what is important is the velocity difference. It is often convenient to regard the object as stationary and to view the distant fluid as approaching it at a uniform velocity U, as shown in Fig. 3.1. The flow near the object is complicated by the need for the fluid to pass around the obstacle; fluid paths must curve and a complex wake might develop on the downstream side. To avoid concern with such details, the approach velocity U is what is used in drag calculations.

With coordinates as in Fig. 3.1, there will be a force on the object acting in the x direction. This force, the drag, is the additive effect of pressure and viscous stresses. Locally, pressure acts perpendicular to the surface, as depicted by the solid arrows labeled "P." Because flow tends to elevate the pressure on the upstream side, the net effect is a force in the x direction. There is also a viscous stress acting tangent to the surface, as represented by the dashed arrows labeled "τ_w." Although the x and y components of this shear stress vary along the surface, the net effect again is a force in the x direction. The pressure and viscous contributions are customarily referred to as *form drag* and *friction drag*, respectively. Their relative importance depends on the shape of the object and the Reynolds number. The more bluff (or less "streamlined") the object and the larger the value of Re, the more prominent form drag becomes.[1]

In Part II we will show how to calculate forces such as drag from a knowledge of the pressure and velocity fields near an object. Our focus at present is the use of experimental results.

Drag coefficient

A quantity with the dimension of stress is obtained by dividing the drag force (F_D) by an area. In drag calculations it is customary to employ the *projected area* of the object ($A_\perp$). Imagine the shadow of an opaque object on a screen. If the screen is perpendicular to the direction of flow, $A_\perp$ is the area of the shadow. For a sphere of diameter D, the shadow would be a circle of area $A_\perp = \pi D^2/4$. A disk of diameter D whose surfaces are perpendicular to the flow, and a cylinder of diameter D with an end-on orientation, each

[1] For objects less symmetric than that in Fig. 3.1, the fluid-dynamic force might have additional components. The y component would be called a *lift* force. Although lift is crucial for certain applications, such as airfoil design, we are concerned here only with drag.

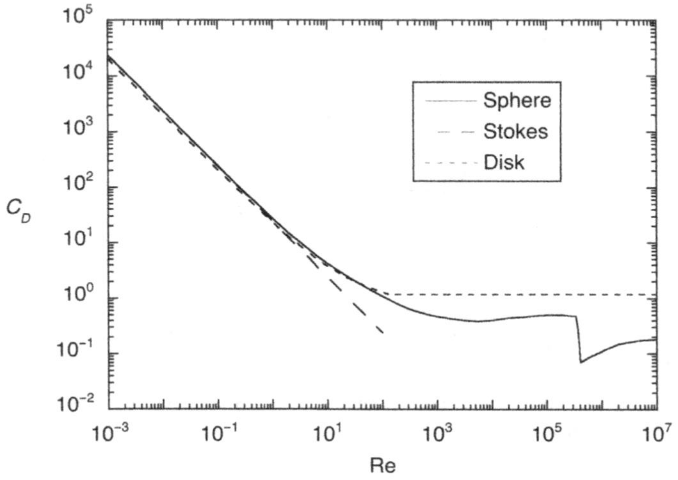

Figure 3.2 Drag coefficients for spheres and disks. The curves are fits to a large body of experimental data. The line labeled "Stokes" is the prediction for spheres given by Eq. (3.2-2).

have that same projected area. For a cylinder of length L whose axis is perpendicular to the flow, the shadow would be a rectangle of area $A_\perp = DL$.

A dimensionless group could be constructed by dividing $F_D/A_\perp$ by either the viscous or the inertial stress scale. As with the friction factor for pipe flow (Chapter 2), the conventional reference stress is $\rho U^2/2$. Dimensional analysis indicates that the dimensionless stress for an object of a given shape and orientation is a function only of the Reynolds number. Thus,

$$C_D(\mathrm{Re}) = \frac{2F_D}{\rho U^2 A_\perp} \tag{3.2-1}$$

where C_D is the *drag coefficient*. For spheres, disks, or cylinders the customary length scale in Re is the diameter. Surface roughness can also affect C_D and would introduce at least one more dimensionless group, as for pipe flow. However, we will not be considering roughness effects.

Spheres

Figure 3.2 summarizes C_D results for spheres from experiments spanning 10 orders of magnitude in Re. The solid curve is a best fit to data from many studies, as detailed in Clift *et al.* (1978, p. 111). In this log–log plot C_D first declines linearly with increasing Re and then levels off. Following a sudden drop at $\mathrm{Re} = 2 \times 10^5$, it finally increases gradually.

As shown by the line labeled "Stokes," the data for $\mathrm{Re} < 1$ are represented well by

$$C_D = \frac{24}{\mathrm{Re}} \quad (\mathrm{Re} < 1). \tag{3.2-2}$$

This result, which is exact in the limit $\mathrm{Re} \to 0$, is a dimensionless form of *Stokes' law*. As with the friction factor for laminar pipe flow (Section 2.2), the finding that $C_D \propto \mathrm{Re}^{-1}$

is evidence that inertia is negligible for small Re. Using dimensional quantities, Stokes' law is

$$F_D = 3\pi \mu U D. \tag{3.2-3}$$

This is derived from first principles in Chapter 8.[2]

Of the various expressions that have been proposed to describe the leveling off of C_D at intermediate Re, one that is both accurate and convenient is (Kürten *et al.*, 1966)

$$C_D = 0.28 + \frac{6}{\mathrm{Re}^{1/2}} + \frac{21}{\mathrm{Re}} \quad (0.1 < \mathrm{Re} < 4000). \tag{3.2-4}$$

This is reported to be good to within about ±7% (Clift *et al.*, 1978, p. 111). Within the indicated range of Re, it gives results that are visually indistinguishable from the solid curve in Fig. 3.2. Its range of applicability overlaps somewhat with that of Eq. (3.2-2).

The plateau region in the plot is approximated by

$$C_D = 0.445 \quad (750 < \mathrm{Re} < 2 \times 10^5) \tag{3.2-5}$$

to within ±13% (Clift *et al.*, 1978, p. 108). This constancy of C_D implies that viscous stresses are negligible and that the drag force is proportional to ρU^2. As it was hypothesized by Newton that the drag on any object should be proportional to the fluid density, the square of the velocity, and a characteristic area, this is called the *Newton's-law* regime. Again, there is some overlap with the preceding expression.

The sudden drop that follows the plateau, in which C_D falls from 0.5 at Re $= 2 \times 10^5$ to 0.07 at Re $= 4 \times 10^5$, is called the *Eiffel phenomenon*.[3] It is caused by the emergence of turbulence and a consequent reduction in the size of the wake. In this range of Re the scatter in the data is considerable and particular caution is needed in design calculations. For still larger Re, the recommended correlation (Clift *et al.*, 1978, p. 112) is

$$C_D = 0.19 - \frac{8 \times 10^4}{\mathrm{Re}} \quad (\mathrm{Re} > 10^6). \tag{3.2-6}$$

Disks

Also shown in Fig. 3.2 are results for supported disks that face the approaching fluid. The dashed curve was calculated from

$$C_D = \frac{64}{\pi \mathrm{Re}}(1 + 0.138\,\mathrm{Re}^{0.792}) \quad (\mathrm{Re} \le 133) \tag{3.2-7}$$

$$C_D = 1.17 \quad (\mathrm{Re} > 133). \tag{3.2-8}$$

[2] The Anglo-Irish physicist George Stokes (1819–1903) made numerous contributions to fluid mechanics, solid mechanics, optics, and mathematics during his career at Cambridge University. The 1851 paper in which he calculated the drag on a small sphere (cited in Chapter 8) is one of the most famous in fluid mechanics, in that it contains the first detailed analysis of a flow in which viscous stresses are dominant.

[3] Gustave Eiffel (1832–1923), a French civil engineer and architect, is best known for the tower in Paris that bears his name. After retirement from engineering, he did research in meteorology and aerodynamics. An exchange of data between him and Prandtl's group in Germany led to a small international incident (Kármán, 1954, p. 87). At one point, Prandtl's drag coefficient for Re near 2×10^5 was more than twice Eiffel's. A young engineer in Prandtl's laboratory suggested that Eiffel must have simply forgotten the 2 in the numerator of C_D. Angered by this comment, which had reached Paris from Göttingen, the elderly Eiffel extended his range of Re and thereby firmly established the sudden fall in C_D. The physical explanation for this finding was provided later by Prandtl.

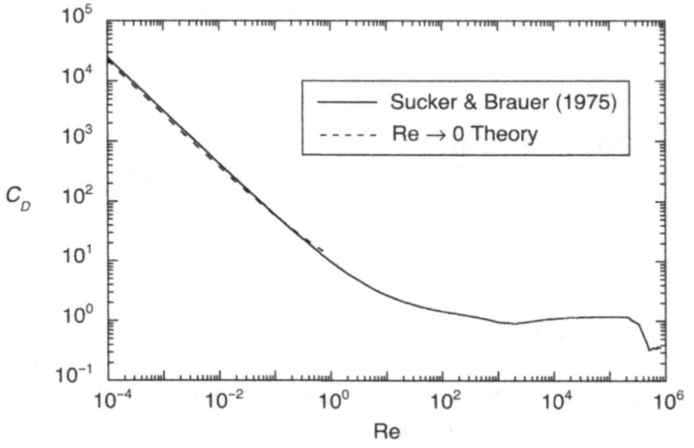

Figure 3.3 Drag coefficient for a long cylinder perpendicular to the flow. The solid curve is based on data in Sucker and Brauer (1975); the part up to $\mathrm{Re} = 2 \times 10^5$ is from Eq. (3.2-9). The dashed curve is the theoretical result given by Eq. (3.2-10).

Equation (3.2-7) gives values that are within 1% of those from other expressions for low or moderate Re, and Eq. (3.2-8) is said to fit most data at high Re to within 10% (Clift *et al.*, 1978, p. 145). For $\mathrm{Re} < 100$, C_D values for disks and spheres are very similar. For $\mathrm{Re} \to 0$, $\mathrm{Re}\, C_D \to 64/\pi = 20.4$ for disks, as compared with 24 for spheres, a difference that is hardly noticeable on the scale of Fig. 3.2. At high Re, disks have a more extended Newton's-law region than do spheres and the Eiffel phenomenon is absent. These results for fixed disks are applicable to freely settling or rising ones only for $\mathrm{Re} < 100$. At low-to-moderate Re the circular surface of a moving disk indeed remains perpendicular to the direction of motion, but disks tend to wobble, glide from side to side, or tumble when $\mathrm{Re} > 100$.

Cylinders

In general, the drag coefficient for a cylinder depends not only on Re, but also on its length-to-diameter ratio (L/D) and orientation. A cylinder is "long" if L/D is large enough to make the forces on its ends negligible. Of interest then is the drag per unit length, which is independent of L. Results will be presented only for long cylinders whose axes are perpendicular to the flow.

The drag coefficient for a cylinder is shown as a function of Re in Fig. 3.3. The solid curve is a fit to experimental data from various sources (Sucker and Brauer, 1975). The overall trends are again similar to those for spheres. That is, a nearly linear decline in a log–log plot is followed by a leveling off. The Eiffel phenomenon occurs, C_D declining suddenly from 1.2 to 0.3 at $\mathrm{Re} = 2 \times 10^5$. Thereafter, C_D increases gradually.

For Re below the Eiffel threshold, the solid curve is a plot of

$$C_D = 1.18 + \frac{6.8}{\mathrm{Re}^{0.89}} + \frac{1.96}{\mathrm{Re}^{1/2}} - \left(\frac{4 \times 10^{-4}\,\mathrm{Re}}{1 + 3.64 \times 10^{-7}\,\mathrm{Re}^2} \right) \quad (\mathrm{Re} < 2 \times 10^5). \tag{3.2-9}$$

A simpler but more limited expression, applicable for $\mathrm{Re} \to 0$ (Batchelor, 1967, p. 246), is

$$C_D = \frac{8\pi}{\mathrm{Re}\,\ln(7.4/\mathrm{Re})}. \tag{3.2-10}$$

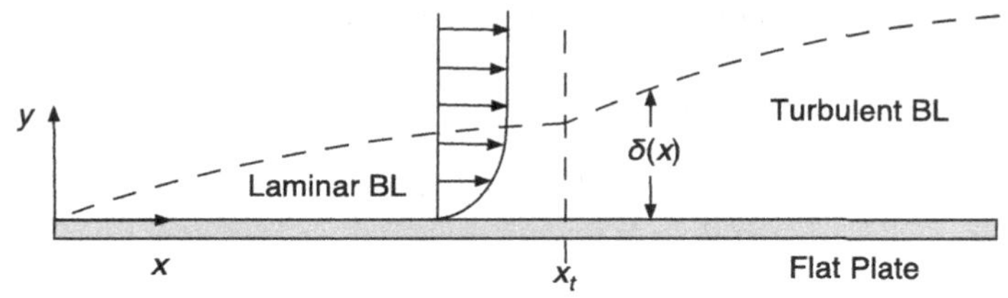

Figure 3.4 Development of the boundary layer (BL) along a flat plate. Velocity variations are confined to a fluid layer of thickness $\delta(x)$, where x is the distance from the leading edge. There is a laminar–turbulent transition at $x = x_t$.

As shown in the plot, Eqs. (3.2-9) and (3.2-10) are in good agreement for Re < 1. A peculiar aspect of indefinitely long cylinders is that C_D is never precisely proportional to Re^{-1}. That is, inertia is never entirely absent, as evidenced by the logarithmic term in Eq. (3.2-10).

Flat plates

Discussed thus far have been bluff objects with substantial projected areas. We focus now on thin, flat plates that are parallel to the approaching fluid. Such objects are quite different, in that $A_\perp$ is much smaller than the area $A_\parallel$ projected onto a plane parallel to the flow.[4] For a plate of length L and width W, $A_\parallel = LW$ and $A_\perp$ is ordinarily negligible. In this same category is a long cylinder aligned end-on, for which $A_\parallel = DL \gg A_\perp = \pi D^2/4$. If $A_\perp$ is negligible, pressure cannot act in the flow direction and form drag will be absent. Thus, a thin plate parallel to the flow experiences only friction drag. Conversely, $A_\parallel$ is negligible for a thin object facing the approaching fluid. In that case there will be no shear stress in the flow direction and friction drag will be absent.

The shear stress on a flat plate at large Re can be appreciable, for reasons illustrated in Fig. 3.4. The upstream edge of the stationary plate is at $x = 0$. As depicted by the arrows, the velocity in the x direction varies only within a thin region, called a *boundary layer* (the thickness of which has been greatly exaggerated for clarity in labeling). The velocity increases from zero at the solid surface ($y = 0$) to the approach velocity U at $y = \delta(x)$, which is the outer edge of the boundary layer. The boundary-layer thickness grows along the plate, but is always small relative to the plate length L if $\text{Re} = UL/\nu$ is large (see below). A velocity increase from 0 to U over a small distance δ can create a high shear rate and thus a large shear stress. Because δ increases with x, the local shear stress tends to decrease along the plate. Other objects, including spheres and cylinders, also have boundary layers when Re is large. What makes a flat plate special is that the shear stress is its only source of drag.

The boundary layer on a long plate is laminar for small x but turbulent for $x > x_t$. Where the transition occurs is governed by a Reynolds number based on x_t, namely $\text{Re}_t = Ux_t/\nu$. It has been found that Re_t ranges from 3×10^5 to 3×10^6. As shown qualitatively in Fig. 3.4, once the boundary layer becomes turbulent its thickness grows more rapidly. The turbulent velocity profile differs from the laminar one and, for a given δ, the shear stress greatly exceeds that for laminar flow. Turbulence immediately increases the shear

[4] There are any number of planes parallel to the flow, because the direction of view can be rotated about an axis aligned with the approaching fluid. If the object is symmetric about such an axis (such as a cylinder aligned with the flow), the area will be independent of the viewing angle. Otherwise, as used here $A_\parallel$ is the *maximum* such area.

stress and eventually causes it to fall more slowly with increasing x than if the flow had remained laminar.

Local boundary-layer thicknesses on flat plates may be estimated using

$$\frac{\delta(x)}{x} = 1.72\,\mathrm{Re}_x^{-1/2} \quad \text{(laminar)} \tag{3.2-11}$$

$$\frac{\delta(x)}{x} = 0.017\,\mathrm{Re}_x^{-0.14} \quad \text{(turbulent)} \tag{3.2-12}$$

where $\mathrm{Re}_x = Ux/\nu$ (Schlichting, 1968, pp. 130 and 605).[5] The second expression assumes that $x \gg x_t$. The maximum thickness is obtained by setting $x = L$ and $\mathrm{Re}_x = \mathrm{Re}$. For a plate just short enough to ensure entirely laminar flow ($\mathrm{Re} = 3 \times 10^5$), $\delta/L = 0.0031$; for one long enough to have predominantly turbulent flow ($\mathrm{Re} = 1 \times 10^7$), $\delta/L = 0.0018$. These values emphasize that the boundary layer is relatively thin.

Yet another Reynolds number is one based on the boundary-layer thickness, $\mathrm{Re}_\delta = U\delta/\nu$. It is analogous to Re for pipe flow in that it employs a length scale perpendicular to the main flow direction. It is related to the Reynolds number based on distance from the leading edge as $\mathrm{Re}_\delta = (\delta/x)\,\mathrm{Re}_x$. Evaluating δ/x using Eq. (3.2-11), at the laminar–turbulent transition $\mathrm{Re}_\delta = 1.72\,\mathrm{Re}_t^{1/2}$. For Re_t ranging from 3×10^5 to 3×10^6, Re_δ at the transition varies from 940 to 3000. These values are comparable to $\mathrm{Re} = 2100$ for the onset of turbulence in pipes, suggesting that the cross-stream length scale is what governs the transition to turbulence.

Once again choosing $\rho U^2/2$ as the reference stress, the dimensionless shear stress on a flat plate is defined as

$$C_f = \frac{2\langle\tau_w\rangle}{\rho U^2} \tag{3.2-13}$$

where C_f is analogous to the friction factor in pipe flow. Whereas in fully developed pipe flow the shear stress is uniform, on a plate it depends on x. The angle brackets indicate that τ_w is averaged over the plate length, from $x = 0$ to $x = L$. The drag on *one side* of a flat plate of width W and length L is

$$F_D = \langle\tau_w\rangle LW = \frac{C_f \rho U^2 LW}{2}. \tag{3.2-14}$$

If a plate is fully immersed, with boundary layers on both sides, the total area is $2LW$ and F_D is doubled.

As derived in Chapter 9, if the boundary layer is entirely laminar and $\mathrm{Re} > 100$, then

$$C_f = 1.328\,\mathrm{Re}^{-1/2} \quad (L < x_t). \tag{3.2-15}$$

If there are both laminar and turbulent regions, then

$$C_f = \frac{0.455}{(\log \mathrm{Re})^{2.58}} - \frac{B}{\mathrm{Re}} \quad (L > x_t). \tag{3.2-16}$$

The first term describes the shear stress for a fully turbulent boundary layer and the second corrects for the fact that the boundary layer is initially laminar. That correction, which reduces C_f, is negligible if Re is large enough. The value of B depends on where the

[5] Because the velocity in the boundary layer approaches that in the outer fluid asymptotically (the difference resembling a decaying exponential), there are different ways of arriving at a number for δ. All of the standard definitions give the same order of magnitude. Equations (3.2-11) and (3.2-12) provide the *displacement thickness* (Chapter 9).

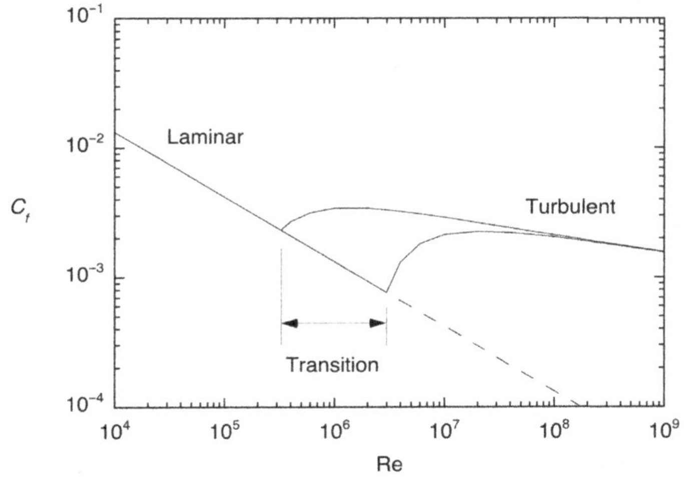

Figure 3.5 Drag coefficient for a flat plate of length L, from Eqs. (3.2-15) and (3.2-16).

transition occurs, with $B = 1050$ for $\text{Re}_t = 3 \times 10^5$ and $B = 8700$ for $\text{Re}_t = 3 \times 10^6$ (Schlichting, 1968, p. 601).

The results from Eqs. (3.2-15) and (3.2-16) are shown in Fig. 3.5. In this log–log plot C_f declines linearly with increasing Re when the entire boundary layer is laminar. Once Re is large enough for turbulence, C_f rises sharply before decreasing again. The two turbulent curves enclose a range of possible results, depending on x_t. A comparison with the laminar extrapolation indicates that, at $\text{Re} = 10^8$, turbulence increases the drag by an order of magnitude. Inertia is important over the entire range of Re shown, as evidenced by the fact that there is not a region where C_f varies as Re^{-1}. For purely laminar flow, $C_f \propto \text{Re}^{-1/2}$.

The two turbulent curves encompass the observed range of Re_t. The dashed line is an extrapolation of the results for purely laminar flow.

Example 3.2-1 Drag on a cylinder in water. It is desired to calculate the drag per unit length on a long cylinder due to water flow perpendicular to its axis. The diameter is $D = 0.10$ m, the relative velocity is $U = 1.0$ m/s, $\rho = 1.0 \times 10^3$ kg/m^3, and $\mu = 1.0 \times 10^{-3}$ Pa · s.

The Reynolds number is

$$\text{Re} = \frac{UD\rho}{\mu} = \frac{(1)(0.1)(1 \times 10^3)}{1 \times 10^{-3}} = 1.0 \times 10^5.$$

The projected area is $A_\perp = DL$ and, from Eq. (3.2-9), $C_D = 1.18$. The drag per unit length is then

$$\frac{F_D}{L} = \frac{\rho U^2 D}{2} C_D = \frac{(1 \times 10^3)(1)^2)(0.1)}{2}(1.18) = 59 \text{ N/m}.$$

Example 3.2-2 Comparative drag on a cylinder and a flat plate. The objective is to compare the drag on a long cylinder to that on a flat plate of equal surface area, for the conditions in Example 3.2-1. Imagine that the cylinder is flattened into a plate, giving a length in the flow direction that equals half the cylinder circumference ($L_p = \pi D/2$) and a width that equals the cylinder length ($W = L_c$). It is desired to compare F_D/W for the plate to F_D/L_c for the cylinder.

For $D = 0.1$ m, the length of the equivalent plate is $L_p = \pi(0.1)/2 = 0.157$ m and the plate Reynolds number is

$$\text{Re} = \frac{UL_p\rho}{\mu} = \frac{(1)(0.157)(1\times 10^3)}{1\times 10^{-3}} = 1.57\times 10^5.$$

Because this is less than 3×10^5, the boundary layer is entirely laminar. Accordingly,

$$C_f = 1.328\,\text{Re}^{-1/2} = 1.328(1.57\times 10^5)^{-1/2} = 3.35\times 10^{-3}.$$

Rearranging Eq. (3.2-14) and allowing for boundary layers on both sides of the plate, the drag per unit width is

$$\frac{F_D}{W} = \rho U^2 L_p C_f = (1\times 10^3)(1)^2(0.157)(3.35\times 10^{-3}) = 0.53 \text{ N/m}.$$

This is only about 1% of 59 N/m, the value of F_D/L_c from Example 3.2-1. Assuming that the flat-plate drag roughly equals the *friction* drag on the equivalent cylinder, this suggests that almost the entire drag on a cylinder at large Re is *form* drag. Pressure measurements and more precise calculations have confirmed that.

3.3 TERMINAL VELOCITY

Buoyancy and gravity

Any object immersed in a static fluid experiences an upward force caused by pressure variations. In a fluid at rest the pressure increases with depth (Chapter 4). This causes the pressure force acting upward on the bottom of an object to exceed that acting downward on the top. The resulting *buoyancy force* is given by *Archimedes' law*,

$$F_0 = \rho g V \tag{3.3-1}$$

where ρ is the density of the *fluid* and V is the volume of the object. The mass of fluid displaced by the object is ρV. Thus, the buoyancy force equals the weight of the displaced fluid.

Buoyancy is always opposed by gravity. For an object of density ρ_o, there is a downward force from gravity of magnitude $F_G = \rho_o g V$. Accordingly, the combined effect of buoyancy and gravity is an upward force of magnitude

$$F_B = F_0 - F_G = (\rho - \rho_o)gV \tag{3.3-2}$$

which we term the *net buoyancy force*. This is zero if $\rho = \rho_o$, in which case the object is said to be *neutrally buoyant*. For a sphere of diameter D, where $V = \pi D^3/6$, the net buoyancy force is

$$F_B = \frac{\pi}{6}(\rho - \rho_o)gD^3. \tag{3.3-3}$$

Terminal velocities for solid spheres

An object either falling or rising under the influence of gravity will eventually reach a speed at which the drag exactly balances the net buoyancy force. The forces in two such states are depicted in Fig. 3.6, in which heavy and light spheres are used as examples. In both cases a reference frame is chosen in which the sphere is stationary and the fluid

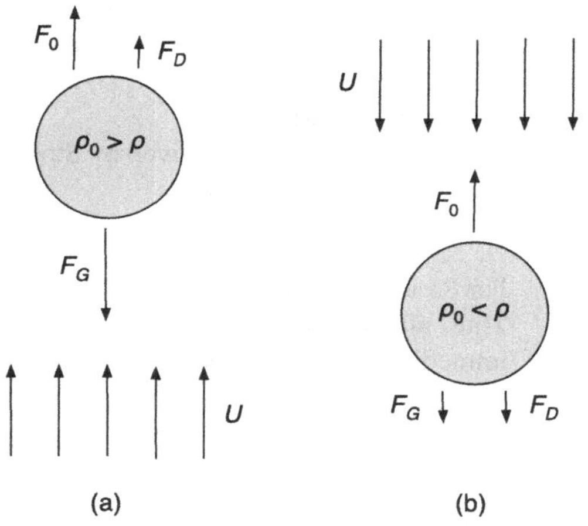

Figure 3.6 Forces on spheres that are either (a) falling in a fluid or (b) rising in a fluid. In both cases a reference frame in which the sphere is stationary is chosen.

velocity is U. For the heavy sphere in Fig. 3.6(a), the fluid moves upward in this reference frame and thus the drag force too is upward. For the light sphere in Fig. 3.6(b), the flow and drag are both downward. Buoyancy and gravity are each independent of the motion and always act upward and downward, respectively. The comparative lengths of the arrows in the two diagrams show the effects of a selective change in ρ_o, which will affect F_G but not F_0. For simplicity, it is supposed that F_B changes only in sign, so that the magnitudes of F_D and U are unchanged.

Using $A_\perp = \pi D^2/4$ in Eq. (3.2-1), the drag on any sphere is given by

$$F_D = \frac{\pi}{8}\rho U^2 D^2 C_D. \tag{3.3-4}$$

At constant velocity the total force must be zero. For a *falling* sphere ($\rho_o > \rho$), the net *upward* force at the terminal velocity is

$$0 = F_D + F_B = \frac{\pi}{8}\rho U^2 D^2 C_D + \frac{\pi}{6}(\rho - \rho_o)gD^3. \tag{3.3-5}$$

Solving for U^2 gives

$$U^2 = \frac{4}{3}\left(\frac{\Delta\rho}{\rho}\right)\frac{gD}{C_D} \tag{3.3-6}$$

where $\Delta\rho = \rho_o - \rho$. The same result is obtained for a *rising* sphere ($\rho_o < \rho$) if $\Delta\rho = \rho - \rho_o$. Hereafter we set $\Delta\rho = |\rho_o - \rho|$ to include both situations.

Equation (3.3-6) is of limited use as is because C_D is a function of Re and thus depends implicitly on U. It is helpful to rewrite the relationship in terms of Re and the *Archimedes number* (Ar), which is independent of U. Multiplying both sides by $(D\rho/\mu)^2 C_D$ gives

$$\mathrm{Re}^2 C_D(\mathrm{Re}) = \frac{4}{3}\mathrm{Ar} \tag{3.3-7}$$

where

$$\mathrm{Ar} = \left(\frac{\Delta\rho}{\rho}\right)\frac{gD^3}{\nu^2}. \tag{3.3-8}$$

The kinematic viscosity in Ar is based on the *fluid* density ($\nu = \mu/\rho$). For any given sphere and fluid, Ar can be calculated. The problem of finding U is reduced then mainly to solving Eq. (3.3-7) for Re ($= UD/\nu$). Once the terminal value of Re is known, calculating U is straightforward.

If Re < 1, then C_D is given by Stokes' law [Eq. (3.2-2)] and Eq. (3.3-7) becomes

$$\text{Re} = \frac{\text{Ar}}{18} \quad (\text{Ar} < 20). \tag{3.3-9}$$

For Re to be small enough to use Stokes' law, Ar must be less than about 20, as indicated. Thus, although Re is not known in advance, whether Eq. (3.3-9) is applicable is revealed immediately by the value of Ar. The Stokes terminal velocity in dimensional form is

$$U = \frac{gD^2 \Delta\rho}{18\mu} \quad (\text{Re} < 1). \tag{3.3-10}$$

In liquids, this expression applies for D as small as a few nm. For particles that small in gases, the continuum approximation breaks down and a correction factor involving the Knudsen number is needed.[6]

For the Newton's-law regime ($750 < \text{Re} < 2 \times 10^5$), combining Eqs. (3.2-5) and (3.3-7) gives

$$\text{Re} = (3\text{Ar})^{1/2} \quad (2 \times 10^5 < \text{Ar} < 1 \times 10^{10}). \tag{3.3-11}$$

Again, the applicability of this relationship can be judged immediately from Ar. Taking the square root of Eq. (3.3-6) and setting $C_D = 0.445$, the Newton terminal velocity is

$$U = \left(3\frac{\Delta\rho}{\rho} gD\right)^{1/2} \quad (750 < \text{Re} < 2 \times 10^5). \tag{3.3-12}$$

If Ar is between the Stokes and Newton ranges ($20 < \text{Ar} < 2 \times 10^5$) an iterative procedure involving Eq. (3.2-4) is needed, as illustrated in the next example. If Ar is above the Newton range, Eq. (3.2-6) can be employed for C_D.

Example 3.3-1 Sand grain falling in air. It is desired to find the terminal velocity of a coarse grain of sand in ambient air. The particle diameter is $D = 1$ mm and its density is $\rho_o = 3000$ kg/m^3. The air properties are $\rho = 1.2$ kg/m^3 and $\nu = 1.06 \times 10^{-5}$ m^2/s.

The first step is to identify the flow regime by evaluating Ar. Recognizing that ρ is negligible relative to ρ_o,

$$\text{Ar} = \left(\frac{3000}{1.2}\right)\left(\frac{9.81(0.001)^3}{(1.6 \times 10^{-5})^2}\right) = 9.58 \times 10^4.$$

This is below the Newton minimum of 2×10^5, but close enough to it that Eq. (3.3-11) should provide a good initial guess for Re. That starting value is

$$\text{Re} = [3(9.58 \times 10^4)]^{1/2} = 536.$$

[6] For fine particles in gases, U from Eq. (3.3-10) is multiplied by the *slip correction factor C*. From Jennings (1988),

$$C = 1 + \text{Kn}\,[1.252 + 0.399 \exp(-1.10/\text{Kn})]$$

where $\text{Kn} = 2\lambda/D$ is the Knudsen number based on the particle radius. In air at 20 °C and atmospheric pressure, where the mean free path is $\lambda = 0.0654$ μm, $C = 1.02$ for $D = 10$ μm, $C = 1.16$ for $D = 1$ μm, $C = 2.86$ for $D = 0.1$ μm, and $C = 22.2$ for $D = 0.01$ μm $= 10$ nm.

Equation (3.2-4) now provides a first estimate of the drag coefficient,

$$C_D = 0.28 + \frac{6}{\sqrt{536}} + \frac{21}{536} = 0.578.$$

Using this in Eq. (3.3-7) gives a refined value for Re, which is

$$\mathrm{Re} = \left(\frac{4}{3}\frac{\mathrm{Ar}}{C_D}\right)^{1/2} = \left[\frac{4}{3}\frac{(9.58 \times 10^4)}{0.578}\right]^{1/2} = 470.$$

Repeating the last two steps gives $C_D = 0.601$ and $\mathrm{Re} = 461$. A final iteration, giving $C_D = 0.605$ and $\mathrm{Re} = 459$, indicates that both quantities have converged to within 1% after three iterations. There is little reason to continue, because the correlation for C_D at intermediate Re [Eq. (3.2-4)] itself is only good to within about 7%. With $\mathrm{Re} = 459$,

$$U = \frac{\nu}{D}\mathrm{Re} = \left(\frac{1.6 \times 10^{-5}}{1 \times 10^{-3}}\right) 459 = 7.4 \text{ m/s}.$$

Example 3.3-2 Microfluidic cell separation. It is proposed to isolate certain rare cells from mixtures by flowing dilute cell suspensions through a horizontal microfluidic channel. Antibodies specific to the target cells will be immobilized on the bottom of the channel. For a cell to bind to the antibodies, it must reach the bottom by sedimenting. For a given channel height and mean velocity, it is desired to know the minimum channel length (L) that would permit any target cell to be captured. The cells are approximately spherical, with $D = 10$ μm and $\rho_o = 1.11 \times 10^3$ kg/m^3. The liquid is a dilute aqueous solution at body temperature (37 °C), for which $\rho = 1.01 \times 10^3$ kg/m^3 and $\nu = 6.85 \times 10^{-7}$ m^2/s. The channel height is $h = 100$ μm and it is intended to operate the device at a mean velocity of $u = 1$ mm/s.

The Archimedes number for these conditions is

$$\mathrm{Ar} = \left(\frac{100}{1010}\right)\left(\frac{9.81(1 \times 10^{-5})^3}{(6.85 \times 10^{-7})^2}\right) = 2.09 \times 10^{-3}$$

which is well within the Stokes range. Accordingly, from Eq. (3.3-9), $\mathrm{Re} = \mathrm{Ar}/18 = 1.16 \times 10^{-4}$. The terminal velocity of a cell is then

$$U = \frac{\nu}{D}\mathrm{Re} = \frac{(6.85 \times 10^{-7})}{(1.0 \times 10^{-5})}(1.16 \times 10^{-4}) = 8.0 \times 10^{-6} \text{ m/s}$$

or 8.0 μm/s. A cell that enters at the top of the channel will take longest to settle. To go from contact at the top to contact at the bottom, the center of such a cell must fall 90 μm at velocity U, requiring a time $t = (90\ \mu\text{m})/(8\ \mu\text{m/s}) = 11.3$ s. The transit time through the channel must exceed this. The transit time based on the mean velocity is L/u, suggesting that what is needed is $L > ut = (0.1\ \text{cm/s})(11.3\ \text{s}) = 1.13$ cm.

Terminal velocities for fluid spheres

A difficulty in characterizing the mechanics of drops or bubbles is that their shape is determined by a combination of gravitational, pressure, viscous, and surface tension forces, and can vary from moment to moment. However, surface tension tends to make small drops or bubbles nearly spherical, and the absence of inertia permits an extension of

Stokes' theoretical result. The drag on a small fluid sphere was derived independently by J. S. Hadamard and W. Rybczyński in 1911, leading to

$$C_D = \frac{8}{\text{Re}}\left(\frac{2+3\kappa}{1+\kappa}\right) \quad (\text{Re} < 1) \tag{3.3-13}$$

where $\kappa = \mu_o/\mu$, the viscosity of the object divided by that of the surrounding fluid. The result for a solid sphere [Eq. (3.2-2)] is recovered for $\kappa \to \infty$ and a prediction for bubbles is obtained by setting $\kappa = 0$. The corresponding expression for the terminal velocity is

$$U = \frac{1}{6}\left(\frac{1+\kappa}{2+3\kappa}\right)\frac{gD^2\Delta\rho}{\mu} \quad (\text{Re} < 1). \tag{3.3-14}$$

This predicts that the terminal velocity of a bubble ($\kappa \to 0$) will be 50% larger than that of a solid particle ($\kappa \to \infty$) if the other parameters are the same.

In practice, another consequence of surface tension is that small drops and bubbles commonly obey Eq. (3.3-10), rather than Eq. (3.3-14) (Clift *et al.*, 1978, pp. 35–41). Unless great care is taken, impurities cause the surface tension to vary with position. That occurs in such a way that circulation within the drop or bubble is suppressed. In the absence of that internal motion, the drag is like that on a solid particle. Water is especially prone to have such trace contaminants. Accordingly, in systems of ordinary purity, Eq. (3.3-10) is usually the better choice for predicting the terminal velocities of small drops or bubbles. Equation (3.3-14) provides an upper bound for U.

Drops and bubbles greater than about 1 mm in diameter are nonspherical enough to make these results inapplicable. Clift *et al.* (1978) reviews the findings for nonspherical as well as spherical fluid objects. Large drops or bubbles often closely resemble "spherical caps" (truncated spheres). The terminal velocity of a spherical-cap bubble is discussed in Problem 9.13.

Approach to terminal velocity

An object that is initially at rest needs a certain time to reach its terminal velocity. For any sphere with solid-like drag, the time scale for the acceleration is

$$t_o = \frac{(\alpha + 1/2)}{|\alpha - 1|}\frac{U}{g} \tag{3.3-15}$$

where $\alpha = \rho_o/\rho$. After a time that is a small multiple of t_o, the velocity of a solid sphere will closely approach U. The exact multiple depends on the flow regime and desired precision. Large spheres are examined in the next example and small ones are considered in Problem 3.6.

The value of t_o decreases rapidly as objects become smaller. For a 1 mm sand grain in air (Example 3.3-1), $t_o = 0.75$ s, and for a 10 μm cell in water (Example 3.3-2), $t_o = 13$ μs. Whether the initial transient in the velocity is important can be judged either from t_o or the distance traveled during the acceleration, which is roughly Ut_o. For the sand grain, $U = 7.4$ m/s and $Ut_o = 5.6$ m. The transient could be ignored in calculating a settling time if the sand were falling from a height much greater than that. Implicit in Example 3.3-2 was the assumption that the cells immediately settle at their terminal velocity. That is justified by the fact that the time constant of 13 μs is much smaller than the estimated settling time of 12.5 s.

Example 3.3-3 Approach to terminal velocity for large spheres. The objective is to show that Eq. (3.3-15) applies to spheres that follow Newton's law. A diameter of even a few mm can make a sphere "large" in this sense, as will be seen. The time constant for acceleration will be identified by considering the instantaneous velocity of a sphere that is settling ($\alpha > 1$). A reference frame will be chosen in which the bulk fluid is stationary and $u(t)$ is the downward velocity of the sphere at time t. The terminal velocity is $u(\infty) = U$.

A complication in applying Newton's second law is that, as the neighboring fluid is displaced by an accelerating sphere, it too accelerates. The changing velocity of the fluid retards the acceleration of the sphere, just as if the sphere mass were larger than its actual value. This can be accounted for by replacing the sphere mass with an effective or virtual mass m^*. For a sphere of volume V the *added mass* is $\rho V/2$, or half the mass of the displaced fluid.[7] Accordingly, $m^* = (\rho_o + \rho/2)V = (\alpha + 1/2)\rho V$.

Equating virtual mass times acceleration to the sum of the downward forces gives

$$m^* \frac{du}{dt} = (\rho_o - \rho)gV - \frac{\pi}{8}\rho u^2 D^2 C_D, \quad u(0) = 0 \tag{3.3-16}$$

for a sphere that is initially at rest. Dividing by ρV and setting $C_D = 0.445$,

$$(\alpha + 1/2)\frac{du}{dt} = (\alpha - 1)g - \frac{u^2}{3D}, \quad u(0) = 0. \tag{3.3-17}$$

The Newton terminal velocity [Eq. (3.3-12)] may be obtained by setting $du/dt = 0$. Equation (3.3-17) is simplified by adopting the dimensionless variables $\theta = u/U$ and $\tau = t/t_o$, where t_o is to be determined. With those definitions,

$$(\alpha + 1/2)\frac{U}{t_o}\frac{d\theta}{d\tau} = (\alpha - 1)g - \frac{U^2}{3D}\theta^2, \quad \theta(0) = 0. \tag{3.3-18}$$

Multiplying both sides by $3D/U^2$ collects all the parameters into the coefficient of the time derivative:

$$\frac{3D(\alpha + 1/2)}{Ut_o}\frac{d\theta}{d\tau} = 1 - \theta^2, \quad \theta(0) = 0. \tag{3.3-19}$$

If t_o is chosen as in Eq. (3.3-15), this simplifies to

$$\frac{d\theta}{d\tau} = 1 - \theta^2, \quad \theta(0) = 0. \tag{3.3-20}$$

This first-order equation is nonlinear but separable. Given that $\int (1 - x^2)^{-1}\, dx = \tanh^{-1} x$, the solution is

$$\theta(\tau) = \tanh \tau. \tag{3.3-21}$$

Noting that $\tanh^{-1}(0.95) = 1.83$, this indicates that the acceleration is 95% complete at $\tau = 1.83$ or $t = 1.83 t_o$. This confirms that Eq. (3.3-15) provides the time scale.

In that Re for a sphere that is initially at rest does not immediately exceed 750 (the lower Newton limit), the assumed constancy of C_D will create a certain amount of error in calculating $u(t)$. A more fundamental issue is that $C_D = 0.445$, as with all other correlations in Section 3.2, is based on steady-state data, whereas the flow around an accelerating body is time-dependent. Figure 3.7 compares the prediction of Eq. (3.3-21) with data obtained for a steel sphere of diameter 3.18 mm settling in water (Owen and Ryu, 2005). Although Re based on the measured terminal velocity was only 3250, the model described above is seen to be remarkably accurate.

[7] A derivation of this result is outlined in Problem 9.5.

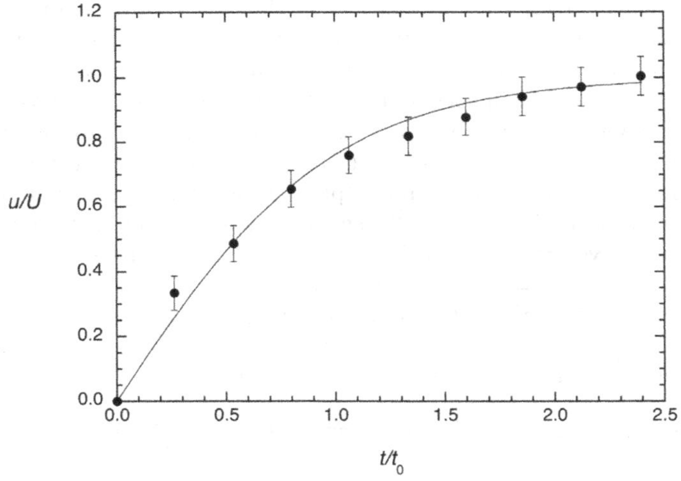

Figure 3.7 Predicted and observed velocity for a steel sphere settling in water. The solid curve is from Eq. (3.3-21) and the data are from Owen and Ryu (2005). The reported velocities have been normalized using the measured terminal value, $U = 1.003$ m/s, and the times have been expressed relative to t_o. Using $\alpha = 8.02$ and $U = 1.003$ m/s in Eq. (3.3-15), $t_o = 0.124$ s.

3.4 POROUS MEDIA

Stated simply, *porous media* are solids with holes. Innumerable things fall into this category, including materials such as hydrogels, paper, soil, and rocks, and devices such as packed beds and synthetic membranes. Connective tissues, walls of blood capillaries, and cell membranes are also porous. Hard or soft, natural or synthetic, what these materials and objects have in common are fluid-filled spaces embedded within some type of structure. The sizes of the spaces may range from molecular to readily visible, and the fraction of the volume occupied by fluid may range from nearly zero to almost unity, but for classification as a porous material the fluid spaces must be much smaller than the system under consideration. Effective pore diameters are typically orders of magnitude smaller than sample dimensions. Accordingly, flow through porous media can be analyzed on either of two length scales: a *macroscopic* scale corresponding to the sample or system dimensions, and a *microscopic* one related to the dimensions of the fluid spaces. At the macroscopic level the flow is governed by the *Darcy permeability*, which will be defined shortly. Microscopic analyses seek to relate the Darcy permeability to the internal structure of the material, as will be illustrated by a few examples.

Darcy permeability

The macroscopic viewpoint is depicted in Fig. 3.8. A sample of a porous material is assumed to have a thickness L and a uniform cross-sectional area A. It is assumed that an applied pressure creates a volume flow rate Q in the x direction. If the sample volume and fluid density are each constant, then Q must be independent of x, as for flow through a pipe. Although downward in the diagram, the x direction is not necessarily the direction of gravity.

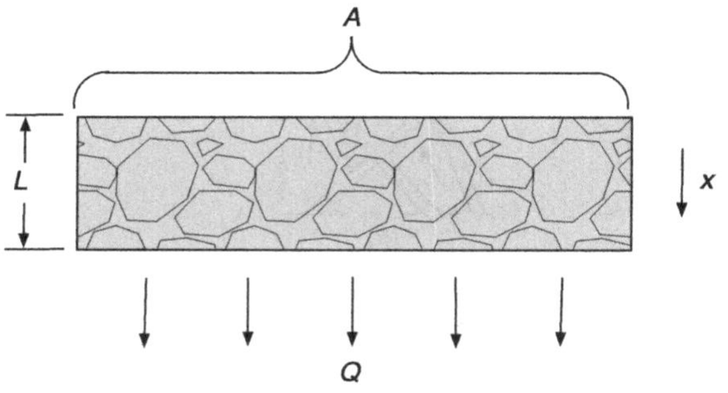

Figure 3.8 Flow through a porous material of possibly unknown structure and composition.

The sizes and shapes of the fluid spaces within the material might be entirely unknown or at least difficult to characterize. To avoid having to describe flow at the level of a single channel, we define a *superficial velocity* as

$$v_s = \frac{Q}{A}. \tag{3.4-1}$$

In the absence of solids, v_s would equal the mean velocity inside the sample, similar to U in pipe flow (Chapter 2). However, by restricting the area for flow, the particles, fibers, or other obstacles cause the velocity in the pores to exceed v_s. Although it underestimates the internal velocities, the advantage of v_s is that it is determined unambiguously by measuring only Q and A.

Flow in porous materials is customarily described using *Darcy's law*, which assumes that the superficial velocity is proportional to the pressure gradient. To allow for gravitational effects and thus include the possibility of vertical flow, the dynamic pressure $\mathcal{P}$ is used [Eq. (2.3-5)]. As in pipes, the flow is in the direction of decreasing $\mathcal{P}$. The superficial velocity in the x direction is

$$v_s = -\frac{k}{\mu}\frac{d\mathcal{P}}{dx} \tag{3.4-2}$$

where k is the *Darcy permeability*. In that v_s and $d\mathcal{P}/dx$ are analogous to a current density and a voltage gradient, respectively, k/μ is a conductivity and μ/k is a resistivity. The viscosity can be factored out, as shown, because at the low Reynolds numbers typical of porous media, inertia is negligible and the flow resistance is proportional to μ. With the effects of viscosity accounted for in this manner, k is independent of the fluid properties and is a function only of the pore geometry. It has the dimension of length squared.

If the sample is homogeneous, then k will be independent of x. With k and v_s each constant, the pressure gradient also must be constant. If $|\Delta\mathcal{P}|$ is the drop in dynamic pressure over the thickness L, then Eq. (3.4-2) may be rearranged as

$$k = \frac{\mu v_s L}{|\Delta\mathcal{P}|}. \tag{3.4-3}$$

Thus, measuring v_s as a function of $|\Delta\mathcal{P}|$ for a given fluid and known sample thickness provides a way to evaluate k. In principle, k for a given porous material should be the same for all fluids. However, for a gas in which the mean free path approaches or exceeds the pore width, the apparent value of μ will be affected, which in turn will influence what is inferred for k. Also, if a liquid cannot fill the entire pore volume, due to its surface

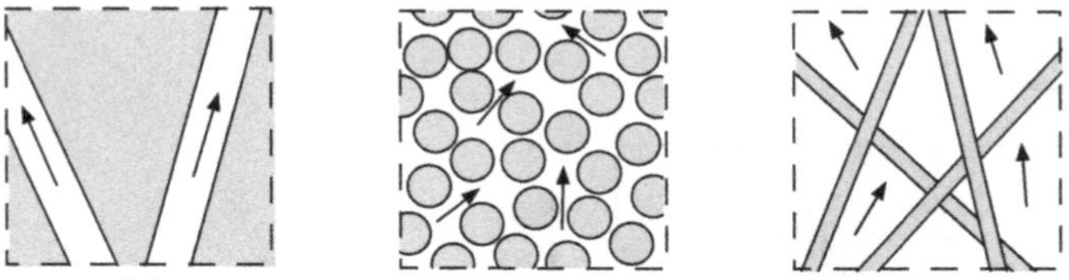

Figure 3.9 Some microstructures of porous media: (a) discrete pores; (b) packed particles; and (c) array of fibers. The arrows in (b) and (c) show only some of the possible flow paths.

tension and the limited wettability of the solids, that too would affect the measured value of k.

Microstructural models

Models that relate the Darcy permeability to the microstructure can be used either to predict k for materials of known pore geometry, or to make structural inferences from measurements of k in materials that are not well characterized. Three kinds of microstructures are depicted in Fig. 3.9: (a) discrete pores passing through a solid; (b) a packed bed of particles; and (c) a randomly oriented array of fibers. In each case there is an identifiable length scale, namely, the pore, particle, or fiber diameter, which will be denoted as d. Each material is characterized also by the amount of fluid or solid present. The fraction of the total volume occupied by fluid, which is called either the *porosity* or the *void fraction*, is denoted as ε. Dimensional analysis suggests that in each case

$$\frac{k}{d^2} = f(\varepsilon). \tag{3.4-4}$$

The function $f(\varepsilon)$ will depend on the type of structure, but should be independent of the fluid properties. Dimensional analysis provides no guidance on the form of $f(\varepsilon)$. The Reynolds number does not appear, because the assumption that $\mathrm{Re} \to 0$ is implicit in Darcy's law. Three examples of $f(\varepsilon)$ will be derived.

The simplest variation of the kind of structure in Fig. 3.9(a) is an array of pores of uniform diameter, all perpendicular to the surfaces of the sample. With this arrangement, the pore length equals the sample thickness (L) and the fraction of the cross-sectional area occupied by pores is the same as the fractional pore volume (ε). The fractional pore area is the number of pores per unit area (n) times the cross-sectional area per pore, or $\varepsilon = n\pi d^2/4$. The superficial velocity is the volume flow rate per unit of total area, or n times the volume flow rate per pore. Using Poiseuille's equation [Eq. (2.3-7)] to calculate the flow rate per pore gives

$$v_s = \left(\frac{4\varepsilon}{\pi d^2}\right)\left(\frac{\pi d^4|\Delta\mathcal{P}|}{128\,\mu L}\right) = \frac{\varepsilon d^2|\Delta\mathcal{P}|}{32\,\mu L} = \varepsilon v_p \tag{3.4-5}$$

where v_p is the mean velocity within a pore. The fact that $\varepsilon < 1$ makes it evident that $v_s < v_p$. A comparison of Eqs. (3.4-5) and (3.4-3) indicates that

$$k = \frac{\varepsilon d^2}{32}. \tag{3.4-6}$$

Thus, $f(\varepsilon) = \varepsilon/32$ for this idealized material. This approach can be extended to materials having distributions of pore radii and/or orientation angles.

It should be emphasized that Eq. (3.4-6) applies only to flow *perpendicular* to the sample surfaces. The apparent value of k for flow *parallel* to those surfaces would be zero, because no pores run in those directions. Any material with preferred flow directions, as occurs with aligned pores, is *anisotropic*. The remaining two structures to be considered are both *isotropic*, in that k is the same for all directions.

The geometric complexity of the fluid spaces in a packed bed like that in Fig. 3.9(b) is forbidding. A way around this difficulty is to replace the tortuous paths between particles by an equivalent set of straight channels of length L and hydraulic diameter d_H. As introduced in Section 2.4, for a channel of constant but possibly noncircular cross-section, $d_H/4$ equals the cross-sectional area divided by the wetted perimeter or, equivalently, the fluid volume divided by the wetted surface area. Using the latter ratio,

$$d_H = \frac{4\varepsilon AL}{S} \tag{3.4-7}$$

where AL is the total bed volume, εAL the fluid volume, and S the total particle surface area. If there are N spheres of diameter d in the bed, then $S = N\pi d^2$. The number of spheres is obtained from the total sphere volume and volume per sphere as $N = (1 - \varepsilon)AL/(\pi d^3/6)$. Accordingly, the total sphere area is $S = 6(1-\varepsilon)AL/d$ and the hydraulic diameter is

$$d_H = 4\varepsilon AL\left(\frac{d}{6(1-\varepsilon)AL}\right) = \frac{2}{3}\left(\frac{\varepsilon}{1-\varepsilon}\right)d. \tag{3.4-8}$$

Setting $d = d_H$ in Eq. (3.4-5) and introducing a dimensionless constant c_0 in the denominator,

$$v_s = \frac{\varepsilon d_H^2 |\Delta\mathcal{P}|}{32 c_0 \mu L}. \tag{3.4-9}$$

For cylindrical channels $c_0 = 1$, but for packed beds we expect that $c_0 \neq 1$. Evaluating the hydraulic diameter as in Eq. (3.4-8) gives

$$v_s = \left[\left(\frac{2}{3}\right)^2 \frac{1}{32c_0}\right] \frac{\varepsilon^3}{(1-\varepsilon)^2} \frac{d^2\,|\Delta\mathcal{P}|}{\mu L} \equiv \frac{\varepsilon^3}{c(1-\varepsilon)^2} \frac{d^2\,|\Delta\mathcal{P}|}{\mu L}. \tag{3.4-10}$$

The last expression was simplified by replacing c_0 with another constant, $c = 72c_0$. The corresponding Darcy permeability is

$$k = \frac{d^2}{c}\frac{\varepsilon^3}{(1-\varepsilon)^2} = \frac{d^2}{150}\frac{\varepsilon^3}{(1-\varepsilon)^2}. \tag{3.4-11}$$

This is the *Kozeny–Carman equation*.[8] The first equality shows that the porosity function for this granular structure is $f(\varepsilon) = \varepsilon^3/[c(1-\varepsilon)^2]$. This particular dependence of k on ε, which would have been difficult to anticipate without a model, is supported well by experimental results. The best fit to data for packed beds of spheres is obtained with $c = 150$ (Ergun, 1952), as shown by the second equality. This implies that $c_0 = 150/72 = 2.1$. This is of the same order of magnitude as the value for cylindrical tubes ($c_0 = 1$), as it should be. Various correction factors have been used to extend Eq. (3.4-11) to mixtures of spheres and to more complex particle shapes (Carman, 1956, p. 18).

[8] As reviewed in Carman (1956), application of the hydraulic diameter concept to packed beds was pioneered by F. C. Blake in 1922. In 1927, J. Kozeny independently reported a derivation similar to that of Blake, and his results were refined by P. C. Carman in 1937. Also called the Kozeny–Carman equation is the expression for v_s obtained by using Eq. (3.4-11) to evaluate k in Eq. (3.4-3).

It is worth noting that the resistance to flow in the spaces between particles will differ from that for the spaces between particles and container walls. Neglecting the effects of walls, as was done in deriving Eq. (3.4-11), requires that the container dimensions be at least an order of magnitude larger than d.

In the kinds of porous materials considered thus far, much of the volume is occupied by solids. For example, $\varepsilon \cong 0.36$ in a bed of "randomly close packed" spheres. In contrast, in fibrous media as in Fig. 3.9(c), $\varepsilon > 0.9$ is not unusual. With so little solid present, there is no longer any resemblance to a set of discrete channels. In calculations for fibrous media it is customary to replace ε by the volume fraction of fibers, $\phi = 1 - \varepsilon$. For randomly oriented cylindrical fibers of diameter d, the data from numerous studies are represented well by (Jackson and James, 1986)

$$\frac{k}{d^2} = \frac{3}{80\phi}(-\ln\phi - 0.931). \tag{3.4-12}$$

The theoretical origins of this expression and others for estimating k for either random or ordered arrays of fibers are discussed in Clague and Phillips (1997); see also Problem 7.6. Notice that $k \to \infty$ as $\phi \to 0$. The conductance becomes infinite upon removal of the fibers because nothing remains to resist the flow. In reality, a fiber array would be inside something, and the wall resistance eventually would exceed that of the fibers as $\phi \to 0$. As with Eq. (3.4-11), wall effects were not considered in deriving Eq. (3.4-12).

Example 3.4-1 Air flow through a packed bed of spheres. It is desired to predict the value of $|\Delta P|$ needed to give air flow at a given superficial velocity through a closely packed bed of spherical particles, at ambient temperature. A question that arises is whether it matters whether the flow is horizontal or vertical. The particle diameter is $d = 1$ mm, the void fraction is $\varepsilon = 0.36$, the bed length is $L = 1$ m, the superficial velocity is $v_s = 1$ cm/s, and the air density and viscosity are $\rho = 1.2$ kg/m^3 and $\mu = 1.8 \times 10^{-5}$ Pa · s, respectively.

The Darcy permeability of the bed will be calculated first, followed by $|\Delta\mathcal{P}|$ and then $|\Delta P|$. From Eq. (3.4-11),

$$k = \frac{d^2}{150}\frac{\varepsilon^3}{(1-\varepsilon)^2} = \frac{(1\times 10^{-3})^2(0.36)^3}{150(0.64)^2} = 7.59 \times 10^{-10}\ \text{m}^2.$$

Assuming that the Reynolds number is small enough for us to use Darcy's law, Eq. (3.4-3) gives

$$|\Delta\mathcal{P}| = \frac{\mu v_s L}{k} = \frac{(1.8\times 10^{-5})(1\times 10^{-2})(1)}{7.59\times 10^{-10}} = 237\ \text{Pa}.$$

If the flow is horizontal, then $|\Delta P| = |\Delta\mathcal{P}|$. If it is vertical, then L will be the height difference between the ends and $|\Delta P|$ and $|\Delta\mathcal{P}|$ will differ by $\pm\rho g L = \pm(1.2)(9.81)(1) = \pm 12$ Pa. For upward flow, which is opposed by gravity, $|\Delta P| = |\Delta\mathcal{P}| + \rho g L = 249$ Pa; for downward flow, which is assisted by gravity, $|\Delta P| = |\Delta\mathcal{P}| - \rho g L = 225$ Pa. Thus, the low density of air and moderate height difference notwithstanding, the effects of orientation would be noticeable.

As derived in Section 3.5, the Reynolds number for packed beds is Re_p, which has as scales the sphere diameter d and the velocity $v_s/(1-\varepsilon)$. The Kozeny–Carman equation is found to be accurate when $\mathrm{Re}_p < 10$ (Ergun, 1952). For the present problem,

$$\mathrm{Re}_p = \frac{v_s d \rho}{(1-\varepsilon)\mu} = \frac{(1\times10^{-2})(1\times10^{-3})(1.2)}{(0.64)(1.8\times10^{-5})} = 1.0$$

which supports the use of Eqs. (3.4-3) and (3.4-11).

As mentioned in Section 1.2, the mean free path in ambient air is about 0.1 μm. A typical channel width in a packed bed of spheres will be no smaller than about $0.1d$, or 100 μm in this example. This is roughly 10^3 times the mean free path, which justifies the use of μ for bulk air.

Example 3.4-2 Comparative properties of granular and fibrous media. In catalytic reactors, adsorbers, and other units in continuous-flow processes it is desirable to maximize the surface area per unit volume (s_v), while also keeping pressure drops small. For a given s_v, which will offer less flow resistance, a packed bed of spheres or a random array of fibers? Suppose that the solids are relatively concentrated in each case, such that $\varepsilon = 0.36$ for the bed of spheres and $\phi = 0.10$ for the array of fibers.

For a bed of spheres, s_v equals the surface area per volume of a sphere times the sphere volume per total bed volume, or

$$s_v = \left(\frac{\pi d^2}{\pi d^3/6}\right)(1-\varepsilon) = \frac{6(1-\varepsilon)}{d}.$$

Using this to eliminate d from Eq. (3.4-11) in favor of s_v gives

$$k = \frac{6}{25}\frac{\varepsilon^3}{s_v^2} = \frac{0.0112}{s_v^2}$$

for $\varepsilon = 0.36$. Likewise, for an array of long fibers of diameter d and length L,

$$s_v = \left(\frac{\pi d L}{\pi d^2 L/4}\right)\phi = \frac{4\phi}{d}$$

where the fiber ends have been neglected. Eliminating d from Eq. (3.4-12) in favor of s_v gives

$$k = \frac{3}{5}\frac{\phi}{s_v^2}(-\ln\phi - 0.931) = \frac{0.0823}{s_v^2}$$

for $\phi = 0.10$. In both cases, k varies as s_v^{-2}. Thus, $k(\text{fibers})/k(\text{spheres}) = 0.0823/0.0112 = 7.35$ for any s_v. In other words, for the volume fractions assumed here, the flow resistance of a fiber array will be about 1/7 that of a bed of spheres.

3.5 PACKED BEDS AND FLUIDIZED BEDS

Packed beds

At low Reynolds number the flow through a packed bed is governed by the Darcy permeability. However, packed beds may be operated also at flow rates high enough that inertia makes Darcy's law inapplicable. Ergun (1952) employed a hydraulic diameter approach like that used for packed spheres in Section 3.4, but recast the result in terms of a bed friction factor (f_p) that is a function solely of a pore Reynolds number (Re_p). He was able to find a simple functional form that fit the available data for a wide range of operating conditions, as will be described.

By analogy with Eq. (2.3-6) for tubes, the friction factor for a packed bed is

$$f_p = c_1 \frac{d_H}{\rho v_p^2} \frac{|\Delta \mathcal{P}|}{L} \tag{3.5-1}$$

where $|\Delta \mathcal{P}|$ is the dynamic pressure drop over a bed of length L, and v_p and d_H are the effective pore velocity and the hydraulic diameter, respectively. The dimensionless constant c_1 will be chosen so as to simplify the final expression. As discussed in Section 3.4, a measure of the pore velocity is $v_p = v_s/\varepsilon$, where v_s is the superficial velocity and ε is the void fraction. Evaluating v_p in this manner, using Eq. (3.4-8) to relate d_H to the diameter of the spherical particles (d), and choosing $c_1 = 3/2$, Eq. (3.5-1) becomes

$$f_p = \frac{d}{\rho v_s^2} \frac{|\Delta \mathcal{P}|}{L} \left(\frac{\varepsilon^3}{1 - \varepsilon} \right). \tag{3.5-2}$$

Likewise, the pore Reynolds number is written as

$$\mathrm{Re}_p = c_2 \frac{v_p d_H \rho}{\mu}. \tag{3.5-3}$$

Evaluating v_p and d_H as before and setting $c_2 = 3/2$, the Reynolds number becomes

$$\mathrm{Re}_p = \frac{v_s d \rho}{(1 - \varepsilon) \mu}. \tag{3.5-4}$$

Equations (3.5-2) and (3.5-4) define the dimensionless groups used to correlate experimental results for packed beds.

The empirical relationship between f_p and Re_p for packed beds of spheres is

$$f_p = \frac{150}{\mathrm{Re}_p} + 1.75 \tag{3.5-5}$$

which is called the *Ergun equation*. As seen in Sections 2.2 and 3.2, an inverse dependence of a friction factor or drag coefficient on a Reynolds number is a hallmark of flow where inertia is negligible. Thus, when Re_p is small enough to make the second term in Eq. (3.5-5) negligible (roughly, for $\mathrm{Re}_p < 10$), the friction-factor relationship is equivalent to the Kozeny–Carman equation [Eq. (3.4-11)]. When the first term is negligible (roughly, for $\mathrm{Re}_p > 1000$), there is a Newton-like regime in which inertia is dominant and f_p is nearly constant.

Sometimes one needs the total drag on all particles in a packed bed, as in the analysis of fluidization that concludes this section. Just as the flow resistance is the same in a horizontal or vertical pipe, the drag in a packed bed is the same for horizontal or vertical flow. The drag calculation for the horizontal case is slightly simpler. The control volume is defined as all fluid within the bed (excluding the particles). For steady flow, the net pressure force on the fluid at the ends of the bed, plus the force exerted on the fluid by the particles, must be zero. (The viscous force on the fluid due to the vessel walls is assumed to be negligible.) The particle force on the fluid is equal and opposite to the total drag on

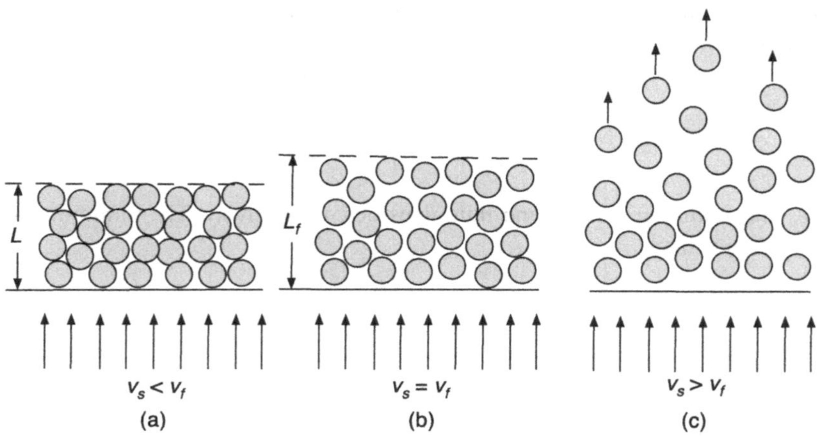

Figure 3.10 Stages of fluidization: (a) a packed bed with a superficial velocity (v_s) that is less than the minimum for fluidization (v_f); (b) incipient fluidization, in which the particles no longer rest on the bottom and the bed height expands from L to L_f; (c) entrainment and loss of particles.

the particles. Thus, $F_D = |\Delta P|A$ for a horizontal bed of cross-sectional area A. The high- and low-Re_p limits of Eq. (3.5-5) then give

$$F_D = 150\frac{(1-\varepsilon)^2}{\varepsilon^3}\frac{\mu v_s AL}{d^2} \quad (\text{Re}_p < 10) \tag{3.5-6}$$

$$F_D = 1.75\frac{(1-\varepsilon)}{\varepsilon^3}\frac{\rho v_s^2 AL}{d} \quad (\text{Re}_p > 1000). \tag{3.5-7}$$

For a packed bed with $\varepsilon = 0.4$, either expression for F_D leads to a drag per sphere that is roughly 80 times that obtained from the value of C_D for an isolated sphere (Section 3.2). This emphasizes that there are strong hydrodynamic interactions among the closely spaced particles in a packed bed.

Fluidized beds

The process of fluidization is illustrated in Fig. 3.10. Three conditions are shown, each with upward flow of a gas through a collection of spherical particles. At low velocities the upward-acting drag is insufficient to displace the particles, which form a packed bed that rests on a screen or perforated plate at the bottom, as in Fig. 3.10(a). However, as the velocity is increased, the drag on the particles will grow until it is just large enough to overcome the net effect of gravity and buoyancy. The superficial velocity at which that occurs is the *minimum fluidization velocity*, v_f. The particles then seem weightless, as they no longer rest on one another or on the bottom. As this occurs the bed will expand somewhat, increasing its height from L to L_f and increasing its void fraction from ε to ε_f. This is the situation in Fig. 3.10(b). Continuing to raise the velocity will expand the bed further and make the particles increasingly mobile, but this will eventually start to carry them out the top, as in Fig. 3.10(c). Of interest is the range of velocities for which fluidization without particle loss is possible. Fluidization may occur at widely varying Reynolds numbers, $0.001 \leq \text{Re}_p \leq 4000$ (Kunii and Levenspiel, 1977, p. 73). In the example that follows, Re_p is assumed to be small. Large values of Re_p are considered in Problem 3.11.

Example 3.5-1 Fluidization at low Reynolds number. It is desired to determine the velocity range for fluidization when $\text{Re}_p < 10$. Using Eq. (3.5-6) for the drag, the sum of the upward forces at the onset of fluidization ($v_s = v_f$) is

$$0 = F_D + F_B = 150\frac{(1-\varepsilon_f)^2}{\varepsilon_f^3}\frac{\mu v_f A L_f}{d^2} + (\rho - \rho_o)g(1-\varepsilon_f)AL_f \qquad (3.5\text{-}8)$$

where ρ_o and ρ are the particle and fluid densities, respectively. The minimum velocity for fluidization is then

$$v_f = \frac{(\rho_o - \rho)gd^2}{150\mu}\frac{\varepsilon_f^3}{(1-\varepsilon_f)}. \qquad (3.5\text{-}9)$$

At the maximum operating velocity (v_m), we suppose that the bed has expanded enough that single-particle results can be used. Particles will be entrained by the upward flow when the fluid velocity exceeds their terminal settling velocity (v_t). If $\text{Re} < 1$ for a sphere, the terminal velocity is given by Eq. (3.3-10). Combining the results for v_f and v_m,

$$\frac{(\rho_o - \rho)gd^2}{150\mu}\frac{\varepsilon_f^3}{(1-\varepsilon_f)} < v_s < \frac{(\rho_o - \rho)gd^2}{18\mu} \qquad (3.5\text{-}10)$$

for fluidization. Both inequalities will be satisfied when $v_f < v_m$, or

$$\frac{v_f}{v_m} = \frac{3}{25}\frac{\varepsilon_f^3}{(1-\varepsilon_f)} < 1. \qquad (3.5\text{-}11)$$

For beds of spheres at small Re_p it has been found that $(1-\varepsilon_f)/\varepsilon_f^3 \cong 11$ (Kunii and Levenspiel, 1977, p. 73), which corresponds to $\varepsilon_f \cong 0.38$ and gives $v_f/v_m \cong 0.011$. This indicates that beds of small spheres can be fluidized over nearly a 100-fold range of superficial velocities.

3.6 CONCLUSION

The drag on an object is the fluid-dynamic force that resists its translational motion. Its magnitude depends on the size and shape of the object, the fluid density and viscosity, and the relative velocity. In general, the drag on a solid is due to a combination of pressure (form drag) and viscous shear (friction drag). The drag coefficient C_D [Eq. (3.2-1)] is the force divided by the product of the projected area and modified inertial stress scale. That dimensionless force applies to most kinds of objects, such as spheres, disks, and cylinders. For flow parallel to a thin plate, where only friction drag is present, the analogous drag coefficient is C_f [Eq. (3.2-13)]. For any family of objects, C_D (or C_f) depends only on the Reynolds number if the surface roughness is negligible.

When particles, drops, or bubbles are formed or released, they tend to accelerate or decelerate until the drag, gravity, and buoyancy forces are in balance. Thereafter, they fall or rise at a constant speed, the terminal velocity. The Archimedes number [Ar, Eq. (3.3-8)], which involves the object size, object density, and fluid properties, provides an efficient way to identify the flow regime in terminal-velocity calculations. The time required for an object to reach its terminal velocity U is typically a few times U/g [Eq. (3.3-15)].

The fluid spaces in porous media can be discrete pores or the interstices in granular or fibrous materials. When the pores are small enough for inertia to be negligible and the system size is large enough for the material to appear homogeneous, the flow resistance is described by the Darcy permeability k [Eq. (3.4-2)]. Experimental and theoretical results for tube flow and for flow past arrays of objects can be used to relate k to various microstructures. For porous media consisting of packed spheres, k is related to the sphere diameter and void fraction by the Kozeny–Carman equation [Eq. (3.4-11)], a result derived using the hydraulic-diameter concept.

Packed beds, as used in equipment such as catalytic reactors, are a type of granular porous medium in which the particles are macroscopic, rather than microscopic. Flow of gases or liquids through packed beds of spheres is described by the Ergun equation [Eq. (3.5-5)]. When the particle Reynolds number is small, the Ergun equation is equivalent to combining Darcy's law with the Kozeny–Carman equation. However, the Ergun equation applies also to beds in which the Reynolds number is large enough for inertia to be important. Fluidization can be achieved by flowing a gas upward through a packed bed, within a certain range of superficial velocities [Eqs. (3.5-10) and (P3.11-1)]. As the velocity in such a packed bed is increased, a point is reached at which the particles are suspended and become mobile. The minimum fluidization velocity is a kind of collective terminal velocity, in which the total drag on a large number of closely spaced particles just balances the gravitational force on them. The upper limit on the velocity in a fluidized bed, set by the requirement that particles not be carried out the top, is the terminal velocity of an individual particle.

REFERENCES

Batchelor, G. K. *An Introduction to Fluid Dynamics*. Cambridge University Press, Cambridge, 1967.

Carman, P. C. *Flow of Gases Through Porous Media*. Butterworths, London, 1956.

Chisnell, R. F. The unsteady motion of a drop moving vertically under gravity. *J. Fluid Mech.* 176: 443–464, 1987.

Clague, D. S. and R. J. Phillips. A numerical calculation of the hydraulic permeability of three-dimensional disordered fibrous media. *Phys. Fluids* 9: 1562–1572, 1997.

Clift, R., J. R. Grace, and M. E. Weber. *Bubbles, Drops, and Particles*. Academic Press, New York, 1978.

Ergun, S. Fluid flow through packed columns. *Chem. Eng. Progr.* 48: 89–94, 1952.

Hagerman, F. C. Applied physiology of rowing. *Sports Medicine* 1: 303–326, 1984.

Jackson, G. W. and D. F. James. The permeability of fibrous porous media. *Can. J. Chem. Eng.* 64: 364–374, 1986.

Jennings, S. G. The mean free path in air. *J. Aerosol Sci.* 19: 159–166, 1988.

Kármán, T. von. *Aerodynamics*. Cornell University Press, Ithaca, NY, 1954.

Kunii, D. and O. Levenspiel. *Fluidization Engineering*. Krieger, Huntington, NY, 1977.

Kürten, H., J. Raasch, and H. Rumpf. Beschleunigung eines kugelförmigen Feststoffteilchens im Strömungsfeld konstanter Geschwindigkeit. *Chem. Ing. Tech.* 38: 941–948, 1966.

McClements, D. J. *Food Emulsions: Principles, Practices, and Techniques*, 2nd ed. CRC Press, Boca Raton, FL, 2005.

Owen, J. P. and W. S. Ryu. The effect of linear and quadratic drag on falling spheres: an undergraduate laboratory. *Eur. J. Phys.* 26: 1085–1091, 2005.

Schlichting, H. *Boundary-Layer Theory*, 6th ed. McGraw-Hill, New York, 1968.

Sucker, D. and H. Brauer. Fluiddynamik bei quer angeströmten Zylindern. *Wärme- und Stoffübertragung* 8: 149–158, 1975.

Wang, C.-S. *Inhaled Particles*. Elsevier, Amsterdam, 2005.

PROBLEMS

3.1. Chain-link fence To predict the ability of a chain-link fence to withstand high winds, an estimate is needed for the horizontal force that might be exerted on it. The 1.8 m-high fence is to be supported by steel posts 6.0 cm in diameter that are placed every 3.0 m. Between the posts will be steel wire 3.2 mm in diameter that is woven into 5.0 cm squares.

For ambient air with $U = 25$ m/s, calculate the drag per post and the drag on one 5.0 cm length of wire. For the complete fence, how will the total drag on the wire mesh compare with that on the posts?

3.2. Rowing power Racing shells are very narrow, causing the water drag to be predominantly frictional. A typical eight-rower (plus coxswain) shell has a length at the waterline of 16.9 m and a wetted area of 9.41 m^2. A competitive speed for a men's eight in international competition is 6 m/s. You are asked to estimate the power each rower must generate to maintain that speed, and to check your estimate by comparing it with the power well-conditioned athletes have been found to generate.

(a) Confirm that, if the water boundary layer is similar to that on a flat plate, it will be very thin compared with the hull dimensions. This supports the idea of "unwrapping" the wetted surface into an equivalent flat plate.
(b) Estimate the water drag. Assume that the results for flat plates are applicable, based on the gradual taper of the hull and the thinness of the boundary layer.
(c) Estimate the air drag. You may assume that the lead rower experiences most of it, the others "drafting" behind him. A rower's sitting height and width are roughly 1.0 and 0.5 m, respectively. Based on results for other bluff objects (spheres, disks, cylinders), what is a reasonable value for C_D?
(d) Use the total drag to estimate the power (in W) each rower must provide. Exercise physiologists have found that elite oarsmen can average 390 W over 6 min (Hagerman, 1984), the approximate time for a 2000 m race. However, not all of the energy expended propels the boat. There are losses associated with the motion of the oar, the rower's body, and the sliding seat in the shell. The mechanical efficiency, defined as the propulsion power relative to the rower's total output, is thought to be about 60%. In light of that, is your power estimate reasonable?

3.3. Dispersion of pollen Particles of plant pollen are often nearly spherical, with diameters ranging from about 10 to 100 μm and densities of about 1000 kg/m^3. Suppose that a sudden gust of wind dislodges such particles and lifts them to a height of 5 m, followed by a steady breeze at 4 m/s.

(a) Calculate the terminal velocities for two sizes of pollen, $D = 10$ and 100 μm.
(b) Assuming that the terminal velocity is reached immediately, how far will each be carried before reaching the ground?
(c) Confirm that for both particle sizes the time needed to attain the terminal velocity is negligible.

3.4. Downhill ski racing World-class downhill ski racers reach speeds up to 150 km/h. The objective of this problem is to estimate the forces on a skier at that speed. For simplicity, the skier will be modeled as two cylinders (legs) attached to a sphere (the rest of the body, including the trunk, arms, and head). Some very approximate dimensions are

$D = 14$ cm and $L = 70$ cm for each leg and $D = 70$ cm for the rest of the body. For air at 0 °C, $\rho = 1.29$ kg/m^3 and $\nu = 1.32 \times 10^{-5}$ m^2/s.

(a) Calculate the air drag on the skier.
(b) Also resisting the motion is the force on the skis, which is a combination of air drag on them and friction where they contact the snow. Suppose that a skier with a mass of 90 kg is moving at a constant speed down a 45° slope. What fraction of the total resistance is due to the force on the skis?
(c) Modeling each ski as a flat plate of width $W = 7$ cm and length $L = 220$ cm that has one side exposed to the air, estimate the air drag. How important is that relative to the friction on the snow?

3.5. Homogenized milk Milk is an emulsion in which fat globules are dispersed within an aqueous solution that contains lactose, proteins, and minerals. Because the density of the fat is about 80 kg/m^3 less than that of water, it tends to rise to the top and create a layer of cream. The mechanical breakup of the fat globules by *homogenization* stabilizes the milk against cream formation. The average diameter of the globules is reduced from about 2 μm in raw bovine milk to about 0.2 μm in homogenized milk (McClements, 2005, pp. 519–520).

Compare the time required for all fat globules to rise to the top of a 20 cm-tall carton of homogenized milk with that for raw milk. Can it be assumed that each size of globule reaches its terminal velocity immediately?

3.6. Approach to terminal velocity for small fluid spheres The objective is to predict how long it takes a small fluid sphere, initially at rest, to reach its terminal velocity. To derive the most general results for Re < 1, assume that the Hadamard–Rybczyński drag coefficient [Eq. (3.3-13)] is applicable. Results for solid spheres (or fluid spheres behaving as solids) may be obtained then by letting $\kappa \to \infty$.

(a) Starting with the differential equation that governs the particle velocity, show that the time constant is

$$t_o = \frac{1}{6}\left(\frac{1+\kappa}{2+3\kappa}\right)\left(\alpha + \frac{1}{2}\right)\frac{D^2}{\nu}. \qquad \text{(P3.6-1)}$$

This is the same as that derived in Chisnell (1987) from detailed solutions for the pressure and velocity fields inside and outside a small fluid sphere. Show that it is also equivalent to Eq. (3.3-15).
(b) Using the notation of Example 3.3-3, show that the dimensionless velocity is

$$\theta(\tau) = 1 - e^{-\tau}. \qquad \text{(P3.6-2)}$$

Given that $e^{-3} = 0.05$, this indicates that the velocity will reach 95% of its terminal value at $\tau = 3$ or $t = 3t_o$.

3.7. Inhaled particles Two mechanisms for the deposition of inhaled particles in the airways are gravitational settling, discussed in Section 3.3, and inertial impaction, depicted in Fig. P3.7. If fluid paths suddenly curve, as when an airway bifurcates, relatively large particles tend to continue along straight lines and may collide with a wall. It is desired to compare the rates of gravitational settling and inertial impaction at various locations within the respiratory tract.

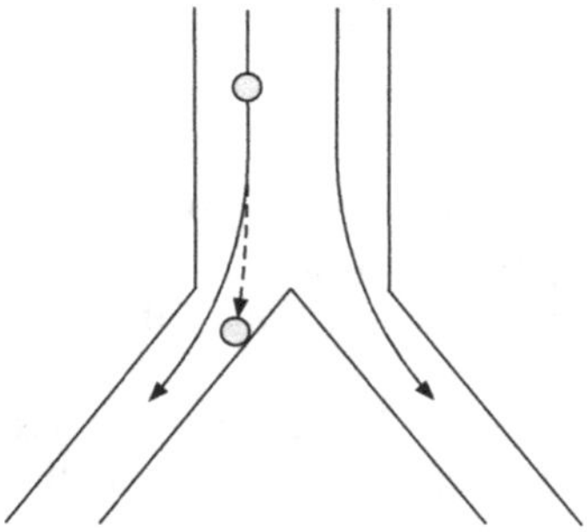

Figure P3.7 Particle deposition at an airway bifurcation. The inertia of the particle causes its trajectory (dashed curve) to depart from the fluid path it had been following (solid curve).

(a) The measure of particle inertia is the stopping distance D_s, which is how far a particle drifts when its propelling force is removed. For simplicity, suppose that a particle is being carried initially at a velocity $u(0) = u_o$ in the x direction, and then the fluid suddenly makes a 90° turn. Because there is no longer an x component of the fluid velocity, drag gradually reduces $u(t)$ to zero. This can be modeled by setting $g = 0$ in Eq. (3.3-16). Show that, for a particle with a diameter d and $u_o d/\nu < 1$,

$$D_s = \frac{\rho_o d^2 u_o}{18\mu}. \tag{P3.7-1}$$

(b) There are millions of airway segments, from the trachea to tiny alveoli deep in the lungs. According to a standard anatomical model, human airways bifurcate 23 times, ultimately generating 8.39×10^6 branches. The diameter (D), length (L), and air velocity (U) at selected locations are given in Table P3.7. An order-of-magnitude estimate of the probability of inertial impaction within a given segment is $p_i \sim D_s/D$. Calculate D_s/D as a function of d and location. If a therapeutic aerosol needs to reach far into the lungs, explain why one with $d = 1$ μm or 10 μm will be much more effective than one with $d = 100$ μm. A typical particle density is $\rho_o = 1000$ kg/m^3, and $\mu = 1.90 \times 10^{-5}$ Pa s and $\rho = 1.11$ kg/ m^3 for humid air at body temperature (37 °C).

Table P3.7 Characteristics of selected human airways[a]

Generation	No. of airways	D(cm)	L(cm)	U(cm/s)
1	2	1.220	4.760	215
4	16	0.450	1.270	202
7	128	0.230	0.760	98
10	1024	0.130	0.460	37
13	8192	0.082	0.270	11
16	65,536	0.049	0.112	2.2
19	524,288	0.036	0.070	0.51
22	4,194,304	0.029	0.067	0.099

[a] From a more complete tabulation in Wang (2005, p. 40).

(c) The probability of gravitational deposition is roughly the settling distance divided by D, or $p_s \sim u_s L/(UD)$, where u_s is the Stokes terminal velocity. Because D_s and u_s both vary as d^2, the relative importance of the two mechanisms is independent of particle size (within the Stokes range). However, it depends very much on location.

Show that $p_i/p_s \sim U^2/(gL) = \text{Fr}$ (a Froude number) and plot Fr as a function of the generation number. Where in the respiratory network is inertial impaction more important and where is gravitational settling more important?

3.8. Flocculation Particles may aggregate in water as a result of their hydrophobicity or Van der Waals attractions. Such aggregates are called *flocs* and the process is termed *flocculation*. (Aerosols also can flocculate.) The objective is to explore how this affects gravitational settling. Suppose that an individual particle is a sphere with diameter $d = 10\ \mu\text{m}$ and density $\rho_o = 2000\ \text{kg/m}^3$.

(a) Evaluate the terminal velocity u of an individual particle.
(b) Suppose that 1000 particles form a nearly spherical floc of diameter D. If the floc has a void fraction of $\varepsilon = 0.50$, what will D be?
(c) If the floc of part (b) behaves as an impermeable sphere, what will its terminal velocity U be? How does U compare with u?
(d) Whether water flow through the floc is actually negligible can be checked by viewing the floc as a miniature packed bed and estimating the superficial velocity. Predict k for the floc. To obtain v_s from Eq. (3.4-3), let $L = D/2$ and (because $\text{Re}_p < 1$) use the viscous pressure scale as an estimate for $|\Delta\mathcal{P}|$. Is $v_s \ll U$, as assumed in part (c)?

3.9. Hydrogel disks Suppose that it is desired to measure the Darcy permeability k of certain gels that consist of crosslinked polymers with water-filled interstices. These hydrogels are routinely fabricated as disks of diameter $D = 5$ mm, thickness $h = 0.5$ mm, and density $\rho_o = 1030\ \text{kg/m}^3$. Someone suggests that k be inferred from the terminal velocity U of such a disk in water. The presumption is that there will be enough flow through the gel to make U noticeably different from that of an equivalent solid disk. You are asked to evaluate this idea before any equipment is purchased.

(a) For a solid disk that stays horizontal as it settles, relate Re to Ar and C_D. How should Ar be defined for this object?
(b) Determine U for a solid disk with the given properties.
(c) Assuming there is only form drag, calculate the pressure difference ΔP across the disk. This is the average pressure at the bottom (leading) surface minus that at the top.
(d) Suppose that the polymer chains are thought to bundle into fibers of approximately 2 nm diameter and that the volume fraction of solids (polymer) ranges from 0.03 to 0.10. Approximately how large might k and the superficial velocity v_s be? Are the suggested experiments promising?

3.10. Bypassing a packed bed Suppose that a reactor used to pretreat a hot gas is a cylinder of diameter $D = 20$ cm and length $L = 2$ m that is filled with spherical catalyst particles of diameter $d = 5$ mm. The void fraction is $\varepsilon = 0.35$, the gas properties are $\rho = 0.705\ \text{kg/m}^3$ and $\mu = 2.71 \times 10^{-5}\ \text{Pa} \cdot \text{s}$, and the superficial velocity is $v_s = 0.5$ m/s.

(a) Calculate $|\Delta\mathcal{P}|$ for the packed bed.
(b) Suppose now that a change in the composition of the feed gas has made the pretreatment unnecessary and that it has been decided to bypass the packed bed with a pipe of the same length. If it were desired to keep $|\Delta\mathcal{P}|$ the same, so as not to affect other units in the plant, what pipe diameter D_p would be needed? (*Hint*: Re for the equivalent pipe is in the Blasius range, which makes it possible to derive an explicit expression for $x = D_p/D$.)

3.11. Fluidization at high Reynolds number Show that, when $\mathrm{Re}_p > 1000$ and C_D is governed by Eq. (3.2-5), the range of feasible velocities for fluidizing spheres of diameter d and density ρ_o is

$$\left[\frac{\varepsilon_f^3}{1.75}\left(\frac{\rho_o-\rho}{\rho}\right)gd\right]^{1/2} < v_s < \left[3\left(\frac{\rho_o-\rho}{\rho}\right)gd\right]^{1/2}. \qquad \text{(P3.11-1)}$$

Given that $\varepsilon_f \cong 0.41$ under these conditions (Kunii and Levenspiel, 1977, p. 73), $v_f/v_m = 0.11$. That is, there is approximately a 10-fold range of feasible velocities at high Re_p, in contrast to the 100-fold range at low Re_p.

Part II

Fundamentals of fluid dynamics

4

Fluid statics: pressure, gravity, and surface tension

4.1 INTRODUCTION

In a static fluid, gravity causes the pressure to increase with depth. Pressures at fluid–fluid interfaces may be influenced also by surface tension. Both kinds of pressure variation have numerous practical consequences, as will be discussed in this chapter. An advantage of beginning a more detailed analysis of fluid mechanics with statics is that viscous stresses, which are the most difficult conceptually and mathematically, are absent in fluids at rest. Thus, we can concentrate first on pressure, gravity, and surface tension, and return later to a more general and precise description of viscous stresses. Quantifying momentum changes within fluids also can be deferred. Viscous stresses and inertia are considered in detail in Chapter 6.

This chapter begins the use of vector notation, and it is suggested that the reader review certain parts of the appendix before proceeding. Needed particularly is familiarity with vector representation (Section A.2), vector dot products (Section A.3), the gradient operator (Section A.4), and cylindrical and spherical coordinates (last part of Section A.5).

4.2 PRESSURE IN STATIC FLUIDS

Properties of pressure

Pressure is a force per unit area, making it a type of stress. It has three properties, whether or not there is flow.

(i) Pressure forces only act normal to (perpendicular to) surfaces.
(ii) Positive pressures are *compressive* (rather than *tensile*). In other words, pressure pushes on (rather than pulls on) a surface.
(iii) Pressure is isotropic. That is, the pressure P has a single value at any point in a fluid, and tends to act equally in all directions.

Properties (i) and (ii) provide a mechanical definition of pressure and establish the direction of pressure forces when $P > 0$. Property (iii) is a consequence of (i), as shown at the end of this section. If (i) and (iii) seem contradictory, keep in mind that the pressure itself is a scalar, whereas the pressure *force* acting on a surface is a vector. The orientation of the surface determines the direction of the force vector, but not the value of P.

Static pressure equation

A differential equation that describes pressure variations in a static fluid will be derived using a force balance on the small fluid cube in Fig. 4.1. The edges of this tiny control

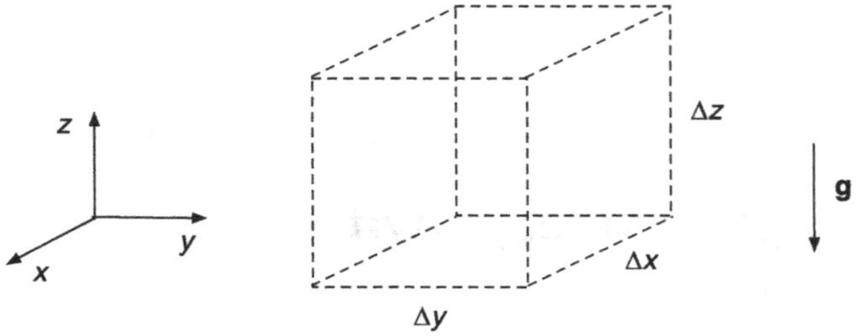

Figure 4.1 Cubic control volume of differential size.

volume parallel the Cartesian coordinate axes and its dimensions are Δx by Δy by Δz. The gravitational acceleration vector **g** is assumed (for now) to point in the $-z$ direction.

For the fluid inside to be stationary, the net force exerted on the control volume by its surroundings must be zero. We begin with the z component. If density variations within the control volume are negligible, the mass of fluid it contains is the density times its volume, or $\rho\Delta x\,\Delta y\,\Delta z$. The z component of the gravitational force is $-\rho g\Delta x\,\Delta y\,\Delta z$, where g is the magnitude of the vector **g**. The minus sign is needed because the z axis points upward and gravity acts downward. The pressure force that the surrounding fluid exerts on each face has a magnitude equal to P times the surface area. Its direction is inward and normal to the surface. Denoting the heights of the bottom and top surfaces as z and $z + \Delta z$, respectively, the corresponding pressure forces in the z direction are $+P(x, y, z)\Delta x\,\Delta y$ and $-P(x, y, z + \Delta z)\Delta x\,\Delta y$. The z component of the force balance is then

$$P(x, y, z)\Delta x\,\Delta y - P(x, y, z + \Delta z)\Delta x\,\Delta y - \rho g\Delta x\,\Delta y\,\Delta z = 0. \tag{4.2-1}$$

Dividing by $\Delta x\,\Delta y\,\Delta z$ and taking the limit $\Delta z \to 0$ gives

$$\frac{\partial P}{\partial z} = -\rho g. \tag{4.2-2}$$

This strategy for deriving a differential equation, which consists of writing a balance for a small control volume and then reducing that volume to a point, is something we will use numerous times. Equation (4.2-1), which involves a region of differential dimensions, is an example of what is called a *shell balance.*

The procedure for the x and y components is the same, but in those directions gravity is absent. Accordingly,

$$\frac{\partial P}{\partial x} = 0 = \frac{\partial P}{\partial y}. \tag{4.2-3}$$

Taken together, Eqs. (4.2-2) and (4.2-3) indicate that P decreases with height (or increases with depth), but does not vary within any horizontal plane.

A more general version of Eqs. (4.2-2) and (4.2-3) is needed to calculate pressure forces and permit the use of other coordinate systems. It is obtained by first writing the gravitational vector **g** in component form as

$$\mathbf{g} = \mathbf{e_x}g_x + \mathbf{e_y}g_y + \mathbf{e_z}g_z. \tag{4.2-4}$$

For the special case in Fig. 4.1, $g_x = g_y = 0$ and $g_z = -g$. Replacing $-g$ by g_z in Eq. (4.2-2) gives

$$\frac{\partial P}{\partial z} = \rho g_z. \tag{4.2-5}$$

Similarly, for the other components,

$$\frac{\partial P}{\partial x} = \rho g_x \tag{4.2-6}$$

$$\frac{\partial P}{\partial y} = \rho g_y. \tag{4.2-7}$$

Collecting these three results yields a single vector equation. Multiplying each by the corresponding unit vector and adding them gives

$$\nabla P = \rho \mathbf{g} \tag{4.2-8}$$

where

$$\nabla = \mathbf{e_x}\frac{\partial}{\partial x} + \mathbf{e_y}\frac{\partial}{\partial y} + \mathbf{e_z}\frac{\partial}{\partial z} \tag{4.2-9}$$

is the Cartesian *gradient operator*. Equation (4.2-8) is the most general form of *static pressure equation*. It indicates once again that the pressure in a static fluid increases with depth, because that is the direction in which **g** points. Although derived by adding Cartesian components, Eq. (4.2-8) is applicable to any coordinate system. The only change needed is replacement of Eq. (4.2-9). The cylindrical and spherical gradient operators are given in Tables A.3 and A.4, respectively.

Pressure distributions

The pressure distribution in a static fluid is found by integrating whichever form of the static pressure equation is most convenient. For a Cartesian system in which the z axis points upward the starting point is Eq. (4.2-2) and

$$P(z) = -g\int \rho\, dz + C. \tag{4.2-10}$$

To evaluate the constant C we need to know P at some height. If $P(0) = P_0$, then

$$P(z) = P_0 - g\int_0^z \rho\, dZ \tag{4.2-11}$$

where Z is a dummy variable (i.e., one that disappears when the definite integral is evaluated). In most situations ρ is nearly constant. That is true even for gases if height differences are moderate. In such cases ρ is taken out of the integral in Eq. (4.2-10) to give

$$P(z) = -\rho g z + C. \tag{4.2-12}$$

Example 4.2-1 Manometer. A simple application of Eq. (4.2-12) involves the U-tube manometer in Fig. 4.2, which contains a liquid of density ρ_m. It is desired to determine the pressure difference between the two interfaces, $P_1 - P_2$, in terms of the heights shown.

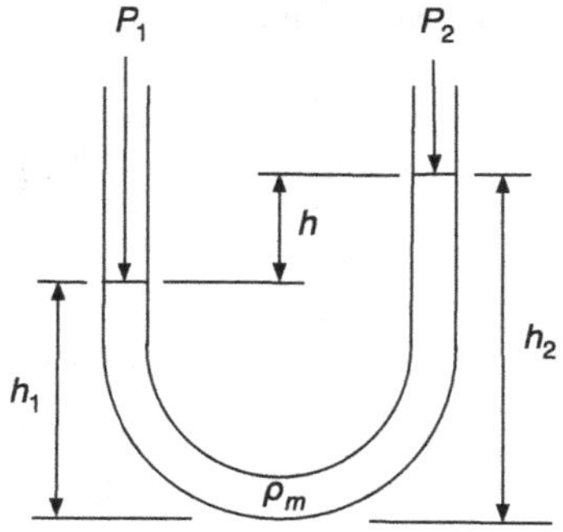

Figure 4.2 Manometer.

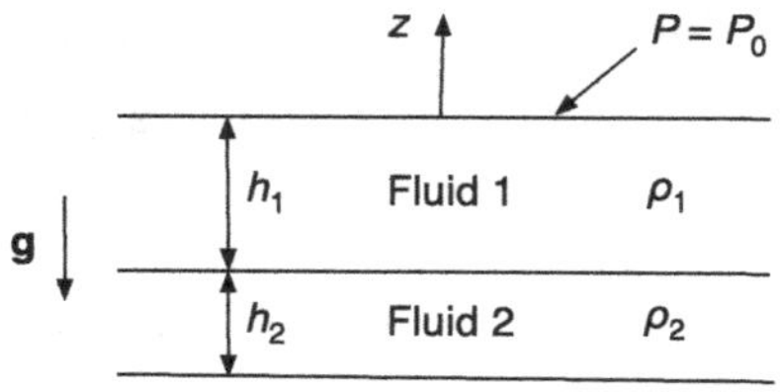

Figure 4.3 Layered fluids of different densities.

Beginning at the left, the pressure at the bottom of the tube is found to be $P_1 + \rho_m g h_1$. Starting at the right, the pressure at the bottom is $P_2 + \rho_m g h_2$. As there is just one value of pressure at any height, those two expressions must be equal. Accordingly,

$$P_1 - P_2 = \rho_m g(h_2 - h_1) = \rho_m g h. \tag{4.2-13}$$

Example 4.2-2 Layered fluids. If two or more fluids of differing density form layers, as with oil on water or immiscible solvents in a separatory funnel, Eq. (4.2-12) can be applied piecewise within each layer. At each planar interface the pressure on one side equals that on the other, providing a starting value for the next layer. This will be illustrated using the system in Fig. 4.3. The objective is to find $P(z)$ in both fluids, assuming that there is a known pressure P_0 at $z = 0$, the top of fluid 1.

Working downward from where the pressure is known, the pressure within fluid 1 is

$$P(z) = P_0 - \rho_1 g z \quad (-h_1 \leq z \leq 0). \tag{4.2-14}$$

At the interface between fluids 1 and 2,

$$P(-h_1) = P_0 + \rho_1 g h_1. \tag{4.2-15}$$

To continue downward we must change the density from ρ_1 to ρ_2. The pressure distribution within fluid 2 is

$$\begin{aligned} P(z) &= P(-h_1) - \rho_2 g(z + h_1) \\ &= P_0 + \rho_1 g h_1 - \rho_2 g(z + h_1) \quad (-h_1 - h_2 \leq z \leq -h_1). \end{aligned} \tag{4.2-16}$$

The pressure at the bottom is then

$$P(-h_1 - h_2) = P_0 + g(\rho_1 h_1 + \rho_2 h_2). \tag{4.2-17}$$

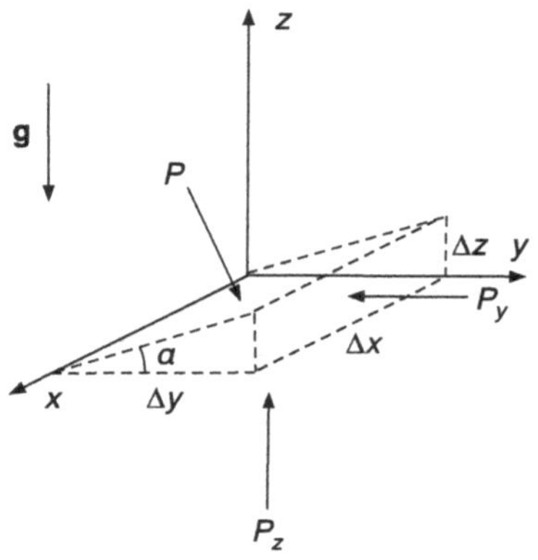

Figure 4.4 Control volume for proof that pressure is isotropic.

Additional note: Pascal's law

It will be shown now that pressure is an *isotropic* stress [property (iii)], a result known as *Pascal's law*. A static force balance will be performed on the small, wedge-shaped fluid volume in Fig. 4.4. We will start by assuming that the pressure has values P_x, P_y, and P_z when acting on surfaces perpendicular to the x, y, and z axes, respectively. The pressure on the top of the wedge (inclined at angle α) is P. The objective is to prove that all these pressures must be identical.

Taking into account that pressure acts only normal to surfaces, the z component of the force on the wedge is

$$(-P\cos\alpha)\frac{\Delta x\,\Delta y}{\cos\alpha} + P_z\Delta x\,\Delta y - \rho g\frac{\Delta x\,\Delta y\,\Delta z}{2} = 0. \qquad (4.2\text{-}18)$$

In the first term, $-P\cos\alpha$ is the z component of the normal stress on the top surface and $\Delta x\,\Delta y/\cos\alpha$ is the area of that surface. (The same total force is obtained by multiplying P by the projected area, $\Delta x\,\Delta y$.) Dividing by $\Delta x\,\Delta y$. and rearranging,

$$-P + P_z = \frac{\rho g}{2}\,\Delta z. \qquad (4.2\text{-}19)$$

For a vanishingly small control volume, such that $\Delta z \to 0$, we conclude that $P = P_z$.

The y component of the force balance involves the pressures on the top and right-hand surfaces, but not gravity. It is

$$(P\sin\alpha)\frac{\Delta x\,\Delta y}{\cos\alpha} - P_y\Delta x\,\Delta z = 0. \qquad (4.2\text{-}20)$$

Dividing by Δx and rearranging,

$$\frac{P_y}{P} = \frac{\sin\alpha}{\cos\alpha}\frac{\Delta y}{\Delta z} = \tan\alpha\frac{\Delta y}{\Delta z} = \frac{\Delta z}{\Delta y}\frac{\Delta y}{\Delta z} = 1. \qquad (4.2\text{-}21)$$

Rotating the wedge by 90° in the x–y plane shows that $P_x/P = 1$. Accordingly, $P = P_x = P_y = P_z$.

4.3 PRESSURE FORCES

Stress and force vectors

The defining properties of pressure listed at the beginning of Section 4.2 are encapsulated by stating that the corresponding *stress vector* is $-\mathbf{n}P$, where $\mathbf{n}$ is a unit vector that

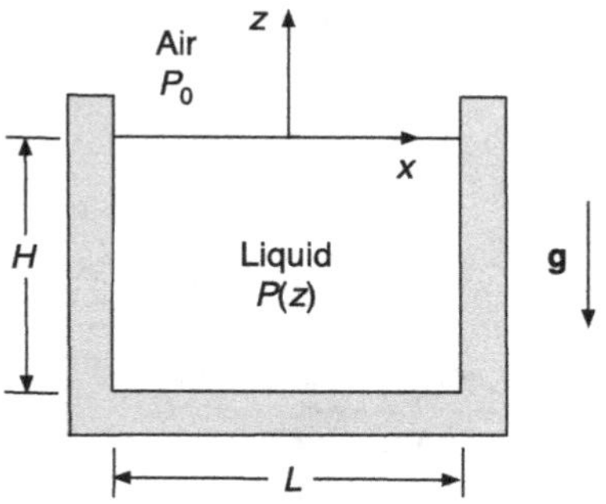

Figure 4.5 A tank partly filled with a static liquid. The top of the liquid and sides of the tank are exposed to air at constant pressure P_0. The dimension in the y direction (not shown) is W.

is normal to the surface and points toward the fluid that is exerting the force. Multiplying P by $-\mathbf{n}$ creates a vector with the proper direction (perpendicular to the surface and compressive for $P > 0$). This expression for the pressure stress applies to any fluid, static or flowing. The *pressure force* on a differential area dS is $d\mathbf{F_P} = -\mathbf{n}P\,dS$. Accordingly, the pressure force $\mathbf{F_P}$ on a surface S is

$$\mathbf{F_P} = -\int_S \mathbf{n}P\,dS \tag{4.3-1}$$

where $\int_S dS$ denotes integration over the surface. (Likewise, $\int_V dV$ will denote integration over a volume V.) In general, the direction of $\mathbf{n}$ will vary from point to point on S; only for a planar surface will $\mathbf{n}$ be independent of position.

Boundaries

A pressure force might need to be calculated in either of two types of situations. In one, only one side of a container wall or other object is in contact with the fluid of interest. In that case the surface in question may be just part of what bounds the fluid. In the other, the object is completely immersed and the pressure acts on a closed surface. Pressure forces at partial boundaries will be discussed first.

Example 4.3-1 Rectangular tank. Suppose that an open rectangular tank that occupies an area L by W is filled to a height H with a static liquid, as shown in Fig. 4.5. The tank might be a laboratory container or a large vessel in a plant. It is desired to calculate the fluid forces on it.

We begin with the liquid force on the right-hand wall. The unit normal pointing into the liquid is $\mathbf{n} = -\mathbf{e_x}$, indicating that the only nonzero component of the pressure force vector is F_{Px}. The area element is $dS = dy\,dz$. With the coordinate origin at the top (as shown), $0 \le y \le W$ and $-H \le z \le 0$. Accordingly, the force is

$$F_{Px} = \int_S P\,ds = \int_{-H}^{0}\int_{0}^{W} P(z)\,dy\,dz = W\int_{-H}^{0} P(z)\,dz. \tag{4.3-2}$$

For an air pressure P_0 and liquid density ρ, the pressure in the liquid is

$$P(z) = P_0 - \rho g z. \tag{4.3-3}$$

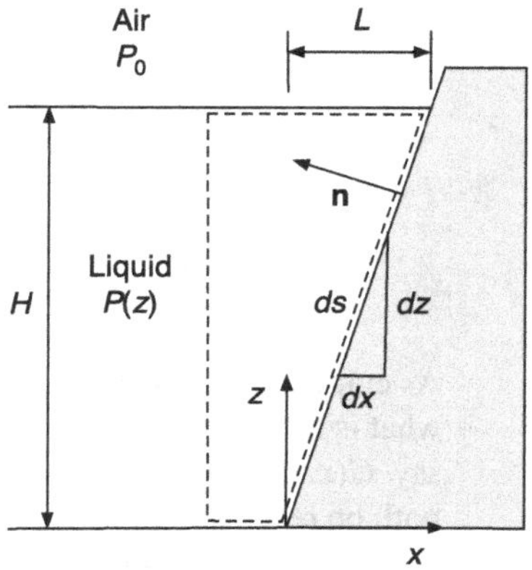

Figure 4.6 Pressure force on an inclined wall.

Substituting this into Eq. (4.3-2) and evaluating the last integral gives

$$F_{Px} = WH\left(P_0 + \frac{\rho g H}{2}\right). \tag{4.3-4}$$

The calculation for the left-hand wall is the same, except that $\mathbf{n} = \mathbf{e_x}$ instead of $-\mathbf{e_x}$. This simply changes the sign of F_{Px}.

At the bottom, $\mathbf{n} = \mathbf{e_z}$ and the pressure force is in the $-z$ direction. Here $dS = dx\, dy$, $-L/2 \le x \le L/2$, and $0 \le y \le W$. Because P is independent of x and y, the force may be written immediately as

$$F_{Pz} = -WL(P_0 + \rho g H). \tag{4.3-5}$$

Although calculated as the pressure–area product at the bottom, notice that this is also the force of the air on the liquid surface ($P_0 WL$) plus the weight of the liquid ($\rho g HWL$). Indeed, the forces on the entire volume of static fluid must sum to zero, including gravity and the pressure forces on the top and bottom.

Acting on the *outside* of the tank is a pressure force from the ambient air. For either the left- or right-hand walls its magnitude up to the liquid level is WHP_0. Above that, the pressure force on the inside balances that on the outside. The ambient pressure cancels the P_0 term in Eq. (4.3-4), revealing that the *net* horizontal force on both walls is $\rho g W H^2/2$. That is the force to consider when judging the strength of construction.

Example 4.3-2 Inclined planar surface. Suppose that a wall of an open tank is inclined. As shown in Fig. 4.6, the surface rises to a height H over a horizontal distance L. It is desired to determine the pressure force that the liquid exerts on the inclined wall.

The new feature is that the wall of interest is not a *coordinate surface*; that is, it does not correspond to a constant value of a coordinate such as x. Its inclination relative to the z axis will cause $\mathbf{n}$ to have both x and z components, although each will be constant because the surface is planar. Unit normal vectors are evaluated generally as

$$\mathbf{n} = \frac{\nabla G}{|\nabla G|} \tag{4.3-6}$$

where $G(x, y, z) = 0$ defines the location of the surface. The function G is chosen so that $\mathbf{n}$ points in the desired direction. For the inclined surface in Fig. 4.6, let

$G(x, z) = -Hx + Lz$. Then

$$\nabla G = \mathbf{e_x}\frac{\partial G}{\partial x} + \mathbf{e_y}\frac{\partial G}{\partial y} + \mathbf{e_z}\frac{\partial G}{\partial z} = -H\mathbf{e_x} + L\mathbf{e_z} \tag{4.3-7}$$

$$|\nabla G| = (H^2 + L^2)^{1/2} \tag{4.3-8}$$

$$\mathbf{n} = \frac{-H\mathbf{e_x} + L\mathbf{e_z}}{(H^2 + L^2)^{1/2}}. \tag{4.3-9}$$

As evidenced by the fact that $n_x < 0$ and $n_z > 0$, this vector points into the liquid, which is what is needed to calculate the pressure force that the liquid exerts on the wall. Choosing, say, $G(x, z) = Hx - Lz$ would have given the wrong signs for n_x and n_z and required that both be reversed.

With the coordinate origin at the bottom of the liquid layer of depth H, the pressure is

$$P(z) = P_0 + \rho g(H - z). \tag{4.3-10}$$

Letting s be the arc length along the surface, the differential area is $dS = dy\,ds$. Referring to Fig. 4.6, $(ds)^2 = (dx)^2 + (dz)^2$ and

$$ds = \left[\left(\frac{dx}{dz}\right)^2 + 1\right]^{1/2} dz = \left[\left(\frac{L}{H}\right)^2 + 1\right]^{1/2} dz. \tag{4.3-11}$$

Substitution of these expressions into Eq. (4.3-1) and integration over y gives

$$\mathbf{F_P} = W\left(\mathbf{e_x} - \frac{L}{H}\mathbf{e_z}\right)\int_0^H [P_0 + \rho g(H - z)]\,dz. \tag{4.3-12}$$

The integral is evaluated as

$$\int_0^H [P_0 + \rho g(H - z)]\,dz = H\left(P_0 + \frac{\rho g H}{2}\right). \tag{4.3-13}$$

The horizontal and vertical components of the pressure force are then

$$F_{Px} = WH\left(P_0 + \frac{\rho g H}{2}\right) \tag{4.3-14}$$

$$F_{Pz} = -WL\left(P_0 + \frac{\rho g H}{2}\right). \tag{4.3-15}$$

Projected areas

Notice that F_{Px} for the inclined wall [Eq. (4.3-14)] equals that for the corresponding vertical wall [Eq. (4.3-4)]. That is not coincidental. Rather, for static fluids such an equality is guaranteed for any pair of surfaces that have the same height and width. This provides a very helpful shortcut for force calculations that is described now.

What is special about a static fluid is that the net force exerted on it by its surroundings is zero. That holds for the entire fluid volume or any part of it. That concept was applied to *vertical* forces in Eq. (4.2-1). The *horizontal* forces in a static fluid arise from pressure alone. Consider the control volume shown by the dashed lines in Fig. 4.6. For there to

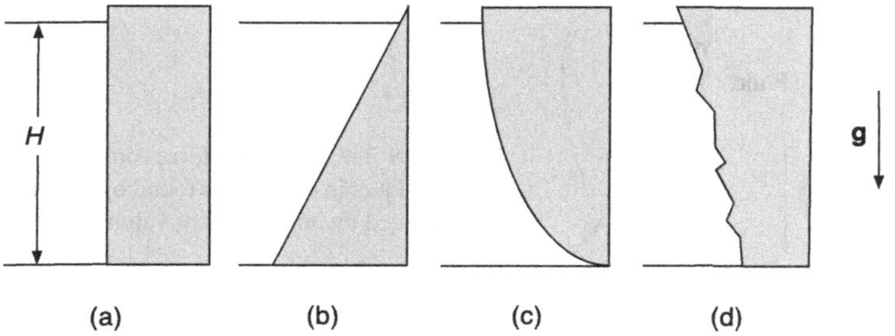

Figure 4.7 Surfaces of varying shape, but identical width W (not shown), in contact with a liquid of height H. Each surface has the same area projected onto a vertical plane.

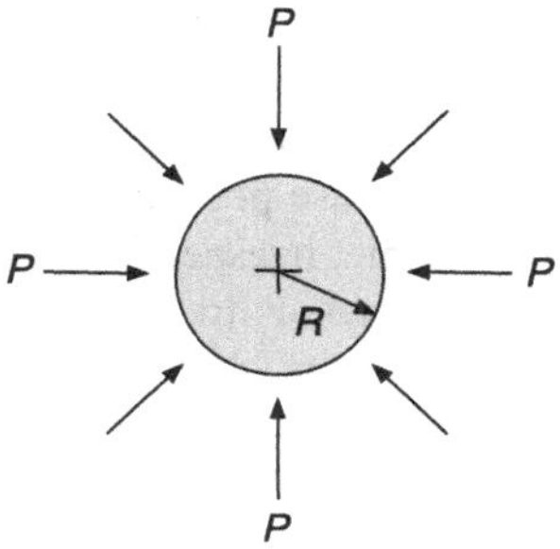

Figure 4.8 Force on a cylinder of radius R surrounded by a fluid at constant pressure P.

be no net force in the x direction, $|F_{Px}|$ on the vertical surface at the left must equal that on the inclined one at the right. The advantage of knowing this is that the force at the left is much easier to calculate. It is simply the area of the vertical surface (WH) times its average pressure, which is the quantity in parentheses in Eq. (4.3-14). The area of the vertical surface is the *projected area* of the inclined one. Thus, the horizontal force on the inclined surface is its projected area times the pressure averaged over that projection.

Surfaces can be much more complicated, as illustrated in Fig. 4.7. What we have seen already is that the horizontal forces on those in Figs. 4.7(a) and (b) are identical. Because the projected areas are the same, also equal to the others are the forces on the gently curved surface in Fig. 4.7(c) and on the irregular one in Fig. 4.7(d). Evaluating $\mathbf{n}$ and integrating $\mathbf{n}P$ over the irregular surface would be laborious at best, and perhaps even impractical. Thus, the projected-area approach can save a great deal of effort in force calculations involving static fluids.

Immersed objects at constant pressure

We turn now to objects that are fully immersed. If P is constant, then $\mathbf{F_P} = \mathbf{0}$ for any such object. For a symmetric body, such as the cylinder in Fig. 4.8, it is fairly obvious that if the pressure is uniform, the differential force at any point on the curved surface will be balanced by that on the opposite side, as depicted by the arrows. Likewise, the forces on the ends will balance, resulting in no net pressure force.

That $\mathbf{F_P} = \mathbf{0}$ for an object of *any* shape at constant pressure may be shown using Eq. (A.4-30), an identity that is related to the divergence theorem of calculus. If a volume V is enclosed by a surface S, then

$$\int_S \mathbf{n} f \, dS = \int_V \nabla f \, dV \tag{4.3-16}$$

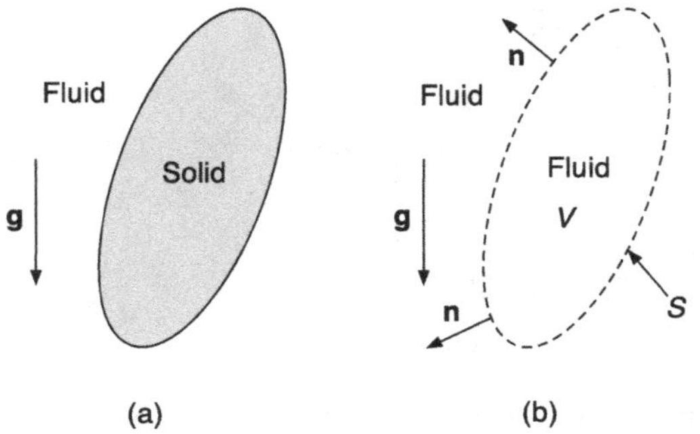

Figure 4.9 The pressure force on the fully submerged object in (a) may be found by imagining that it is replaced by an equivalent volume of fluid, as in (b).

for any differentiable scalar function f. Imagine now that the arbitrary object in Fig. 4.9(a) is replaced by a fluid volume of the same size and shape, as in Fig. 4.9(b). In a static fluid, that would not change P anywhere outside the control volume shown by the dashed curve, because pressure depends only on height. That is, P at any point on the surface S is the same, irrespective of whether S encloses a solid or some of the fluid. Replacing the solid allows us to evaluate P (which we have defined only for fluids) within V. If $f = P =$ constant, then $\nabla P = \mathbf{0}$ everywhere and the volume integral vanishes. Accordingly,

$$\mathbf{F_P} = -\int_S \mathbf{n} P\, dS = \mathbf{0} \quad \text{(closed surface at constant } P\text{)} \tag{4.3-17}$$

for an object of any shape.

Equation (4.3-17) implies that adding any constant to P will leave the pressure force on a closed surface unchanged. Accordingly, that force is independent of the ambient pressure. As a consequence, calculations can be done using either absolute or gauge pressures, whichever is more convenient. Pressure gauges measure the difference between the absolute and ambient pressure, and subtracting a constant ambient value from P does not affect the net force.

Buoyancy

The vector $\mathbf{F_0}$ will denote the force on an object caused by *static* pressure variations. As discovered by Archimedes and used already in Chapter 3, an object of volume V that is surrounded by a fluid of constant density ρ experiences an upward force of magnitude

$$F_0 = \rho g V. \tag{4.3-18}$$

Because pressure increases with depth, the upward force on the bottom of any object exceeds the downward one on the top. Thus, there is a buoyancy force that acts upward, and its magnitude happens to equal the weight of the displaced fluid. Equation (4.3-18) applies equally to objects within liquids or gases, although for the latter ρ is so small that F_0 is ordinarily negligible.

The derivation of Archimedes' law is straightforward. Replacing the object by an equivalent volume of fluid (as in Fig. 4.9) and recalling that $\nabla P = \rho \mathbf{g}$ throughout a static fluid,

$$\mathbf{F_P} = -\int_S \mathbf{n} P\, dS = -\int_V \nabla P\, dV = -\int_V \rho \mathbf{g}\, dV = -\rho \mathbf{g} V = \mathbf{F_0} \tag{4.3-19}$$

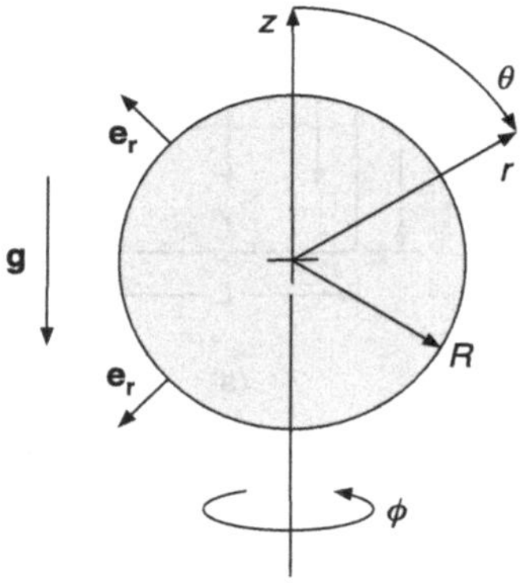

Figure 4.10 Cartesian and spherical coordinates for a stationary sphere of radius R.

for constant ρ. The last equality is the vector form of Eq. (4.3-18). The minus sign confirms that the net pressure force is upward (opposite the direction of $\mathbf{g}$). It is seen now also that there is no *horizontal* pressure force on any object immersed in a static fluid.

Archimedes' law is readily extended to an object that resides at the boundary between two fluids, such as an air–water interface. Suppose that something rests with a volume V_1 within fluid 1, which has a density ρ_1, and a volume V_2 within fluid 2, which has a density ρ_2. The total volume of the object is $V_1 + V_2$. By analogy with what was done to obtain Eq. (4.3-19), the object can be replaced conceptually by two fluid volumes and the forces on them added. Thus, its total buoyancy force is $\rho_1 g V_1 + \rho_2 g V_2$. However many layers of fluid surround the object, the buoyancy force always equals the total weight of the fluid it displaces.

Example 4.3-3 Buoyancy of a sphere. The buoyancy force can be found also by directly evaluating the surface integral in Eq. (4.3-1), as will be shown now for a sphere of radius R immersed in a static fluid, as in Fig. 4.10. In addition to verifying Archimedes' law for spheres, this will illustrate the general procedure for calculating pressure forces. Evaluation of surface integrals like this will be needed later to predict the form drag on objects.

The Cartesian z in Fig. 4.10 will be convenient for describing the pressure variations, and the spherical coordinates will facilitate integration over the surface. What is to be evaluated is

$$F_0 = \mathbf{e_z} \cdot \mathbf{F_P} = -\int_S (\mathbf{e_z} \cdot \mathbf{n}) P \, dS. \qquad (4.3\text{-}20)$$

The unit normal pointing into the fluid is the radial vector $\mathbf{e_r}$. While of constant magnitude, its direction depends on the spherical angles θ and ϕ. Using Eq. (A.5-42a),

$$\mathbf{e_z} \cdot \mathbf{n} = \mathbf{e_z} \cdot \mathbf{e_r} = \mathbf{e_z} \cdot (\sin\theta \cos\phi \, \mathbf{e_x} + \sin\theta \sin\phi \, \mathbf{e_y} + \cos\theta \, \mathbf{e_z}) = \cos\theta. \qquad (4.3\text{-}21)$$

Setting $P = P_0$ (an arbitrary constant) at $z = 0$, the fluid pressure is $P(z) = P_0 - \rho g z$. This must be evaluated at $r = R$ or [using Eq. (A.5-40)] $z = R\cos\theta$, which gives $P(\theta) =$

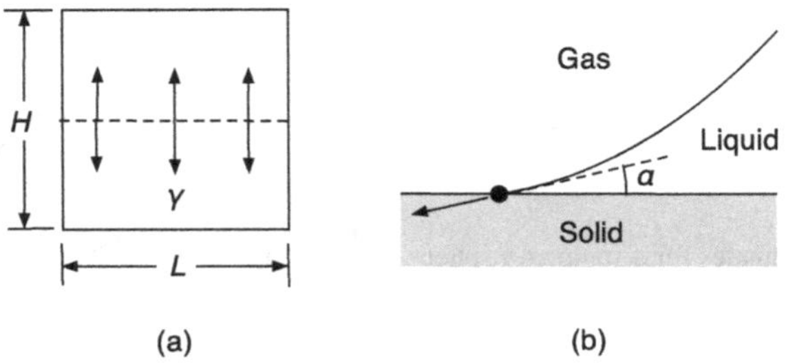

Figure 4.11 Surface tension acting (a) within a planar interface, and (b) at a three-phase contact line.

$P_0 - \rho g R \cos\theta$ at the sphere surface. From Eq. (A.5-46), the differential area is $dS = R^2 \sin\theta\, d\theta\, d\phi$. Substituting these expressions into Eq. (4.3-1) results in

$$F_0 = -\int_0^{2\pi}\int_0^{\pi} (P_0 - \rho g R \cos\theta) R^2 \cos\theta \sin\theta\, d\theta\, d\phi. \tag{4.3-22}$$

Whereas ϕ goes from 0 to 2π, covering the surface requires only that θ vary from 0 to π. In that everything is independent of rotations about the z axis (i.e., independent of ϕ), this is an *axisymmetric* problem. Integrating over ϕ and factoring out the constants gives

$$F_0 = -2\pi R^2 \left(P_0 \int_0^{\pi} \cos\theta \sin\theta\, d\theta - \rho g R \int_0^{\pi} \cos^2\theta \sin\theta\, d\theta \right). \tag{4.3-23}$$

The trigonometric integrals are

$$\int_0^{\pi} \cos\theta \sin\theta\, d\theta = 0, \qquad \int_0^{\pi} \cos^2\theta \sin\theta\, d\theta = \frac{2}{3}. \tag{4.3-24}$$

Thus, the force on the closed surface is independent of the additive constant P_0 (as discussed earlier) and

$$F_0 = \rho g \left(\frac{4}{3}\pi R^3 \right) = \rho g V \tag{4.3-25}$$

in accord with Eq. (4.3-18).

4.4 SURFACE TENSION

Tensile forces and contact lines

As discussed in Section 1.2, the surface tension (γ) at a fluid–fluid interface may be viewed either as a force per unit length or as an energy per unit area. We will favor the force interpretation, although the energy one also has uses in fluid mechanics (e.g., Problem 4.12). The surface tension creates a force that is tangent to the interface. That is illustrated in Fig. 4.11(a), which is a downward-looking view at a flat liquid–air interface. At an imaginary cut through the interface (dashed line) there are tensile forces of magnitude $F_\gamma = \gamma L$ acting on each side of the line of length L. If γ is constant and the interface is planar, the forces will balance and the surface tension will not be evident. However, at an interface with curvature, the surface tension generally creates a pressure difference

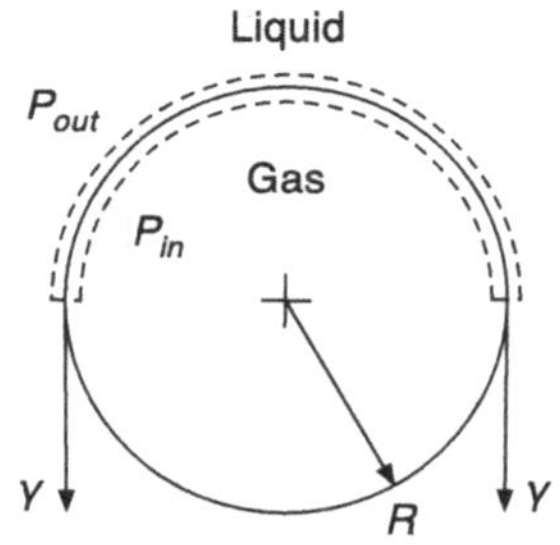

Figure 4.12 Pressure and surface tension acting on a spherical bubble of radius R. The control volume is the hemispherical shell indicated by the dashed curves.

between the phases. For example, the pressure inside a small drop or bubble exceeds that in the surrounding fluid (Example 4.4-1).

A special situation exists wherever a fluid–fluid interface meets a solid, creating a three-phase *contact line*. A gas–liquid–solid contact line is shown "end-on" in Fig. 4.11(b). At such locations the gas–liquid interface meets the solid at an angle α, called the *contact angle*, which is a characteristic of the materials.[1] If $\alpha < 90°$ (as shown), the liquid film will tend to spread and the liquid is said to *wet* the solid; if $\alpha > 90°$, the liquid will tend to contract into drops and is said to be *nonwetting*. The spreading or contraction may be attributed to an unbalanced surface tension force at the contact line. As shown by the arrow in Fig. 4.11(b), the surroundings can be viewed as exerting a force that is tangent to the gas–liquid interface and away from the liquid.

In addition to creating pressure differences across curved interfaces, surface tension causes the spontaneous uptake of liquids into small channels within wettable materials, such as capillary tubes or pores, and can make small objects float, even when they are denser than the liquid. These phenomena are illustrated by the following examples and some of the end-of-chapter problems.

Example 4.4-1 Young–Laplace equation. It is desired to relate the pressure difference across the surface of a stationary bubble to the surface tension and bubble size. A stationary bubble can be created, for example, by using a syringe to gently expel air upward through a capillary tube that is immersed in water. A small enough bubble will adhere to the tip of the tube. Consider a spherical bubble of radius R, as shown in Fig. 4.12. We will assume that the bubble is small enough that the internal and external pressures are each nearly uniform at P_{in} and P_{out}, respectively. After the pressure difference has been found, the limitations of that assumption will be examined.

A convenient control volume is a hemispherical shell of infinitesimal thickness that encloses half of the gas–liquid interface. With this choice, the surface tension force that the surroundings exert on the control volume is downward in the diagram. The length of the "cut" formed by the intersection of the control surface with the interface is the circumference of a circle of radius R. Thus, the surface tension force is $2\pi R\gamma$. The calculation of the pressure force is simplified by the projected-area concept introduced in Section 4.3. The uniform pressures each act on a projected area πR^2. Gravity need not be considered, because the control volume is so thin as to contain negligible mass. The net force acting downward is then

$$\pi R^2 P_{out} - \pi R^2 P_{in} + 2\pi R\gamma = 0. \tag{4.4-1}$$

[1] Although under static conditions the contact angle is a material property, it may be affected by the motion of the contact line.

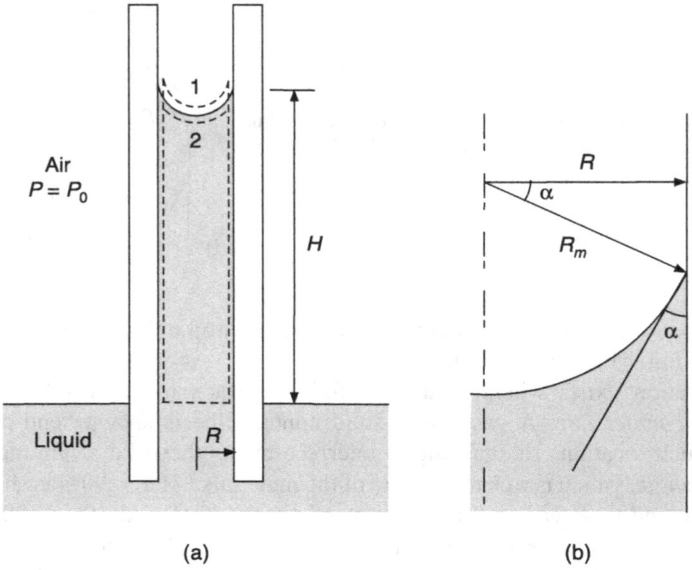

Figure 4.13 Rise of a liquid inside a small, wettable tube: (a) overall view, with the tops of control volumes CV1 and CV2 labeled 1 and 2, respectively; and (b) enlargement of part of the meniscus.

Solving for the pressure difference gives

$$\Delta P = P_{in} - P_{out} = \frac{2\gamma}{R} \tag{4.4-2}$$

which is called the *Young–Laplace equation*. The same result holds for a spherical droplet in a gas. Again, the higher pressure is that inside the fluid sphere. The analogous result for a cylindrical interface of radius R is $\Delta P = \gamma / R$.

The assumption of uniform internal and external pressures requires the bubble (or droplet) to be small. More precisely, the pressure variations within each fluid must be negligible relative to ΔP. Of more concern is the liquid, because of its much greater density (ρ_L). The gravity-induced pressure variation over a height $2R$ of liquid is $2\rho_L gR$. Accordingly, the Young–Laplace equation requires that $2\rho_L gR \ll 2\gamma / R$. Stated another way, what is needed is that $\text{Bo} = \rho_L gR^2 / \gamma \ll 1$, where Bo is the Bond number (Table 1.6). For an air–water interface at room temperature, the radius for which $\text{Bo} = 1$ (the capillary length) is 3 mm. Radii smaller than this are needed for a static bubble or drop to be spherical. If P_{in} or P_{out} varies noticeably with height, then a sphere is no longer an equilibrium shape.

Example 4.4-2 Capillary rise. When one end of a narrow tube is immersed in a wetting liquid (e.g., a glass capillary held upright in water), the liquid rises inside to a certain final height H, as depicted in Fig. 4.13(a). The objective is to show how H depends on the tube radius (R), surface tension, and contact angle (α).

The analysis will be done using force balances on either of two control volumes, denoted as CV1 and CV2. Both encompass the liquid inside the tube, with a bottom at the level of the undisturbed liquid and a top at the meniscus. The difference is that the top of CV1 is on the air side of the meniscus and that of CV2 is on the liquid side

[Fig. 4.13(a)]. If R is small enough, the meniscus will be part of a spherical surface of radius $R_m = R/\cos\alpha$ [Fig. 4.13(b)].

With CV1 the control surface cuts through the air–liquid interface. The surroundings therefore exert a surface tension force on the liquid, the upward component of which is $2\pi R\gamma \cos\alpha$. Opposing that is the gravitational force on the liquid. There is no net pressure force, because the top and bottom of CV1 are at the same pressure (the ambient value, P_0) and have the same projected area (πR^2). Requiring that the upward forces in this static system sum to zero, we find that

$$2\pi R\gamma \cos\alpha - \rho g \pi R^2 H = 0 \tag{4.4-3}$$

where ρ is the liquid density. Solving for H gives

$$H = \frac{2\gamma \cos\alpha}{\rho g R}. \tag{4.4-4}$$

Thus, the smaller the tube radius and the smaller the contact angle, the higher the liquid will rise. If γ, ρ, and R are known, measuring H provides a way to determine α for a particular liquid–solid combination. Implicit in this derivation was the assumption that $H \gg R$, which makes the elevated liquid volume very nearly that of a circular cylinder ($\pi R^2 H$).

With CV2 the control surface does not cut through the air–liquid interface, and therefore there is no surface-tension force from the surroundings. However, there is a net pressure force, because surface tension within the curved interface lowers the pressure at the top of the liquid column. If $\text{Bo} = \rho g R^2/\gamma \ll 1$ the gas–liquid interface will resemble part of a spherical bubble and the Young–Laplace equation will apply. From Eq. (4.4-2), the pressure on the liquid side is then $P_0 - 2\gamma/R_m$. The vertical force balance for CV2 is

$$\pi R^2 P_0 - \pi R^2 \left(P_0 - \frac{2\gamma \cos\alpha}{R}\right) - \rho g \pi R^2 H = 0. \tag{4.4-5}$$

Solving for H leads again to Eq. (4.4-4). Although the final results are the same, the verbal explanation depends on the choice of control volume. It is equally valid to attribute capillary rise to (1) surface tension pulling upward on the liquid, or (2) surface tension creating a partial vacuum at the top of the liquid column.

Interfaces with variable curvature

The preceding examples involved interfaces where the pressure difference caused by surface tension was independent of position. In general, the pressure difference will vary from point to point in proportion to the curvature of the interface. The local pressure difference between static fluids A and B is

$$\Delta P = P_A - P_B = -2\mathcal{H}\gamma \tag{4.4-6}$$

where $\mathcal{H}$, the *mean curvature*, is a function of position and the unit normal $\mathbf{n}$ points from A to B. The more curved a surface, the more rapidly the direction of $\mathbf{n}$ will vary with position, and $\mathcal{H}$ is a measure of that rate. The value of $\mathcal{H}$ at a given point on an interface positioned at $z = f(x, y)$ can be calculated in either of two ways: by averaging two curvatures (hence the name) or by computing the divergence of $\mathbf{n}$. The first approach is adopted here. The second, which yields an expression involving all first and second partial derivatives of f (Deen, 2012, p. 636), tends to be preferable for complex shapes.

Imagine two perpendicular planes that intersect along $\mathbf{n}$ and cut through the interface. Any convenient pair of such planes can be used, rotating them about $\mathbf{n}$ as desired. The

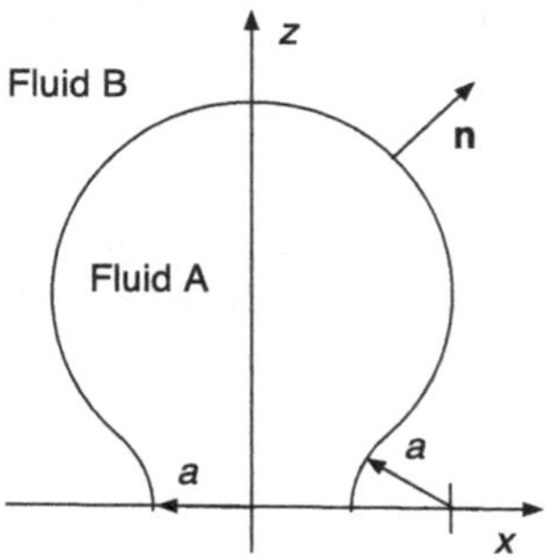

Figure 4.14 Fluid–fluid interface with variable curvature.

intersection of each with the interface defines a curve that passes though the point in question. For simplicity, assume that both curves are segments of circles. The *curvature* of a circle is the inverse of its radius (e.g., Thomas and Finney, 1984, p. 772). If the radii are R_1 and R_2, then

$$\mathcal{H} = \frac{1}{2}\left(\pm\frac{1}{R_1} \pm \frac{1}{R_2}\right). \tag{4.4-7}$$

The choice of signs depends on whether the curve is concave or convex when looking in the direction of **n** (i.e., from A toward B), the sign being positive for concave and negative for convex.

Spheres and the sides of cylinders are special, in that $\mathcal{H}$ is independent of position. For a sphere of radius R with **n** pointing outward, $R_1 = R_2 = R$, both curves are convex, and $\mathcal{H} = -1/R$ anywhere on the surface. Substitution of this into Eq. (4.4-6) gives Eq. (4.4-2). For a cylinder of radius R with **n** again pointing outward, choosing one plane perpendicular to the axis and the other parallel to it gives $R_1 = R$, $R_2 = \infty$, $\mathcal{H} = -1/(2R)$, and $\Delta P = \gamma/R$. That result is readily confirmed by using a force balance as in Example 4.4-1.

An insight from Eqs. (4.4-6) and (4.4-7) is that it is possible to have $\Delta P = 0$ at certain locations on an interface, even if the surface tension is sizable and the interface is not flat anywhere. One way this might occur is shown in Fig. 4.14, in which the interface resembles a light bulb. This shape is qualitatively correct for, say, a bubble emerging from a tube immersed in a liquid, although a true equilibrium shape would be somewhat more complex. Suppose that the base forms a circle of radius a in the x–y plane, and that in any vertical plane, including the x–z plane that is shown, it curves outward with the same radius. In that one of those curves is convex and the other concave, all the points at $z = 0$ are *saddle points*. Because the x–y and x–z planes intersect along the x axis, they are suitable for calculating the mean curvature at $(x, y, z) = (a, 0, 0)$, where $\mathbf{n} = \mathbf{e_x}$. Although the radii are equal, the signs in Eq. (4.4-7) are opposite, and $\mathcal{H} = 0$ and $\Delta P = 0$ at $z = 0$. Moving upward, at first $\mathcal{H} > 0$ and $\Delta P < 0$, because the surface is predominantly concave when looking outward, but eventually $\mathcal{H} < 0$ and $\Delta P > 0$, as it becomes more spherical and convex.

Surface tension tends to smooth ripples on the surface of a liquid. It does so by increasing the liquid pressure at a wave peak, where the surface is convex looking upward, and decreasing it at a wave trough, where the surface is concave. Such pressure variations promote peak-to-trough flow. However, this effect is significant only for small disturbances. If $\mathrm{Bo} \gg 1$, then gravity is the more important leveling force (Example 1.3-1).

For more information on surface tension and contact angles, see Adamson and Gast (1997). A broad and engaging survey of capillarity, wetting, and other phenomena associated with surface tension may be found in De Gennes *et al.* (2003).

4.5 CONCLUSION

From a mechanical viewpoint, pressure is a stress that acts normal to surfaces and is compressive when $P > 0$. In a static fluid, where viscous stresses are absent, the effects of gravity and surface tension are mediated by pressure variations. The effect of gravity is embodied in the static pressure equation [Eq. (4.2-8)], which indicates that in a motionless fluid P decreases with height (in proportion to ρg) but does not vary within any horizontal plane. In gases, ρ is small enough that vertical variations in P are ordinarily negligible. In a static fluid the horizontal component of the pressure force on any surface, however irregular, is its area projected onto a vertical plane times the average pressure on that projection. The downward component of the pressure force exerted by a liquid may be calculated similarly, but including now the weight of the liquid. There is no net pressure force on any object that is completely immersed in a fluid in which P is uniform. The buoyancy force is the upward component of the pressure force on a partially or completely submerged object. As discovered by Archimedes, it equals the weight of the displaced fluid [Eq. (4.3-18)].

In mechanics, surface tension is interpreted as a force per unit length acting on an imaginary contour within a fluid–fluid interface. It pulls away from the contour in a direction that is tangent to the interface. At a nonplanar interface, the surface tension ordinarily creates a pressure difference between the fluids. As described by the Young–Laplace equation [Eq. (4.4-2)], the pressure inside a spherical bubble or drop exceeds that outside. A three-phase contact line is a contour at which the surroundings pull in a direction determined by the contact angle. The resulting force imbalance is responsible for a variety of phenomena, including drop spreading or contraction on surfaces and fluid uptake into narrow channels (capillary flow).

REFERENCES

Adamson, A. W. and A. P. Gast. *Physical Chemistry of Surfaces*, 6th ed. Wiley, New York, 1997.

Blanchard, D. C. and L. D. Syzdek. Production of air bubbles of a specified size. *Chem. Eng. Sci.* 32: 1109–1112, 1977.

Deen, W. M. *Analysis of Transport Phenomena*, 2nd ed. Oxford University Press, New York, 2012.

De Gennes, P.-G., F. Brochard-Wyart, and D. Quéré. *Capillarity and Wetting Phenomena*. Springer, New York, 2003.

Hu, D. L., B. Chan, and J. W. M. Bush. The hydrodynamics of water strider locomotion. *Nature* 424: 663–666, 2003.

Thomas, G. B., Jr. and R. L. Finney. *Calculus and Analytic Geometry*, 6th ed. Addison-Wesley, Reading, MA, 1984.

PROBLEMS

4.1. Manometry for liquid pipe flow One way to monitor the flow rate in a pipe is to use a manometer to measure the pressure drop $P_1 - P_2$ between taps spaced a known distance apart. Two arrangements for liquid pipe flow are shown in Fig. P4.1. The U-tube is placed below the pipe if the manometer fluid is relatively dense and above it if the manometer fluid is less dense. The distance of point i (= A, B, or C) from the nearest pipe wall is h_i. Flow through the U-tube itself can be made negligible by using a small diameter.

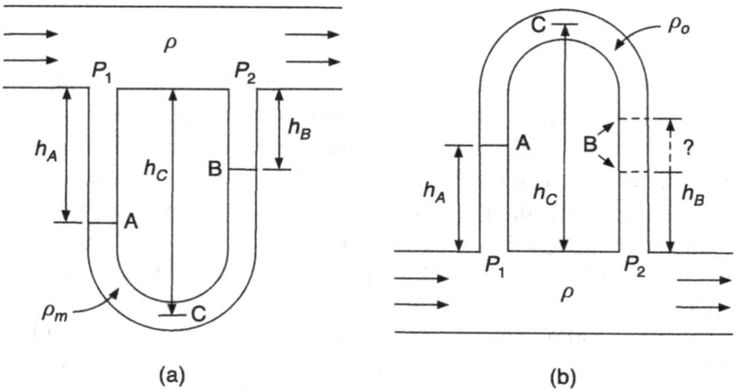

Figure P4.1 Liquid-filled pipes with manometers: (a) manometer liquid denser than that in the pipe ($\rho_m > \rho$) and (2) manometer liquid less dense ($\rho_o < \rho$).

(a) Relate $P_1 - P_2$ to $h = |h_A - h_B|$ for the manometer in Fig. P4.1(a).
(b) For the manometer in Fig. P4.1(b), will h_B be greater than or less than h_A? Again relate $P_1 - P_2$ to h.
(c) If the same pressure drop for water is measured by manometers containing mercury ($\rho_m/\rho = 13$) or an oil ($\rho_o/\rho = 0.80$), how will the values of h compare?

4.2. Hydraulic lift In a hydraulic lift an extendable chamber of diameter D is connected to a pipe of diameter $d \ll D$, as shown in Fig. P4.2. Both are filled with an oil of density ρ. During lifting a pump delivers additional oil until the platform reaches its final height h_2. The pump is then shut off and a pressure P_1 is maintained at height h_1 in the pipe.

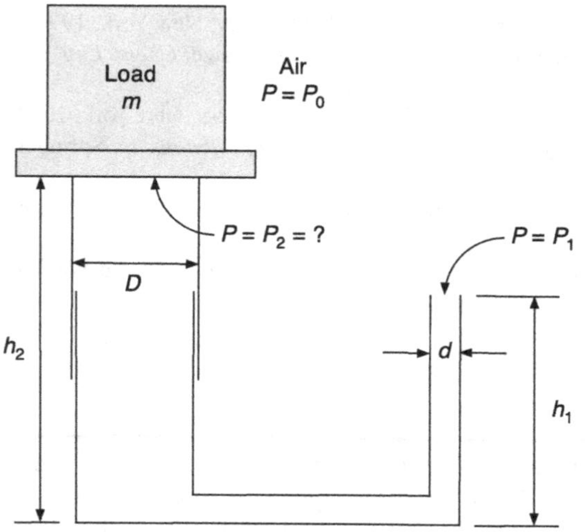

Figure P4.2 Hydraulic lift in a raised position.

(a) Relate the pressure P_2 at the base of the stationary platform to P_1 and the dimensions.
(b) Derive an expression for the total mass m (load plus platform) that can be supported in the position shown.
(c) A friend who knows that force equals pressure times area is puzzled that a large load can be supported, even though with $d \ll D$ the pressure force at position 1 is much smaller than that at position 2. What is his or her misconception?

4.3. Static pressure variations in air In force calculations it is usually assumed that the pressure in still air is constant. The objective is to examine the limitations of that assumption.

(a) Derive an expression for $P(z)$ in static air, with the z axis pointing upward and $P(0) = P_0$. Use the ideal-gas law to find $\rho(z)$, assuming that the temperature is constant at T_0.
(b) As an application of the result from part (a), consider the pressure force on a vertical plane of height H and width W. If $T_0 = 15$ °C, how large must H be for the pressure variation to affect the *force* by 1%?
(c) Repeat part (a), assuming now that $T(z) = T_0 - Gz$, where G is a positive constant.
(d) On average, the atmospheric temperature on Earth declines linearly from about 15 °C at $z = 0$ to –25 °C at $z = 10$ km. (The height of the troposphere, the lower part of the atmosphere, is 17 km in the middle latitudes.) Plot $P(z)/P_0$ for z up to 10 km. Where does P first deviate from P_0 by 1%?

4.4. Force on Hoover Dam When viewed from downstream, Hoover Dam in the US is roughly trapezoidal (see www.usbr.gov/dataweb/dams/nv10122.htm), as in Fig. P4.4. The width at the top is $W_H = 1244$ ft; the width at the bottom is about 2/3 of that, or $W_0 = 830$ ft; and the height of the reservoir, assumed here to reach almost to the top, is $H = 576$ ft. The dam contains 4,400,000 cubic yards of concrete, which has a specific gravity of 2.4.

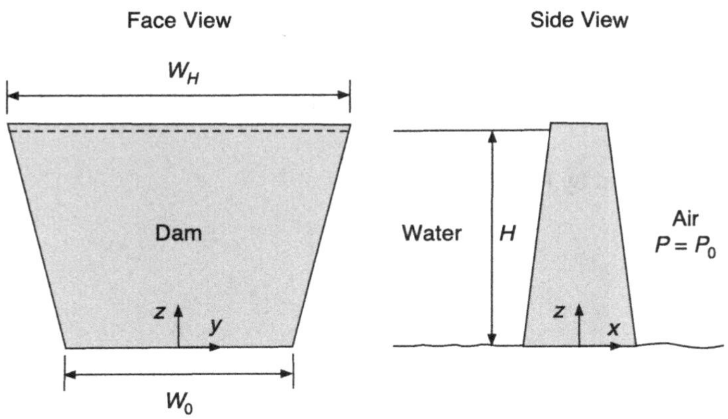

Figure P4.4 Hoover Dam (not to scale).

(a) Derive an expression for the net pressure force acting horizontally on the dam. With coordinates as in Fig. P4.4, that is F_{Px}.
(b) How does the horizontal force (in N) compare with the weight of the concrete?

4.5. Floating cup Suppose that a ceramic cup floats in water as in Fig. P4.5. The height of the cylindrical cup equals its diameter ($2R$) and its walls are thin ($W \ll R$). Its rim rests at a height H above the water surface. The dimensions are large enough that surface tension is negligible.

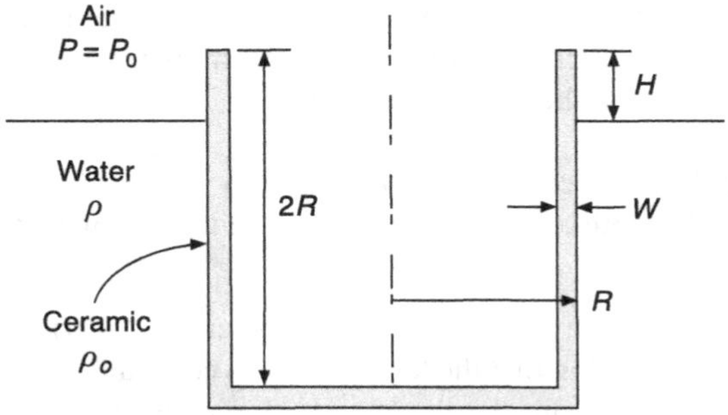

Figure P4.5 Ceramic cup floating in water.

(a) Show how to calculate the specific gravity of the ceramic (ρ_o/ρ) from R, H, and W. If $W/R = 0.1$ and the cup is three-quarters submerged $(H = R/2)$, what is the value of ρ_o/ρ?
(b) Suppose that the cup in part (a) is now partly filled with water. Beyond what water depth D would it sink?
(c) For what values of ρ_o/ρ would such a cup never float, even if empty?

4.6. Sedimentation in a sucrose gradient* Centrifugation in liquids with spatially varying densities is used to purify cell components and viruses. This problem concerns simple gravitational settling in such a system. One way to vary the liquid density is with a gradient in sucrose concentration. Suppose that a rod-like virus or other particle of radius R, length L, and density ρ_p is submerged in a tube of height H, as shown in Fig. P4.6. (The particle dimensions are greatly exaggerated.) A sucrose gradient causes the solution density to vary as

$$\rho(z) = \rho_0\left(1.23 - 0.23\frac{z}{H}\right) \qquad \text{(P4.6-1)}$$

where $z = 0$ at the tube bottom and $\rho = \rho_0$ at the top. If the rod settles end-on, what will be its final height h?

* This problem was suggested by P. S. Doyle.

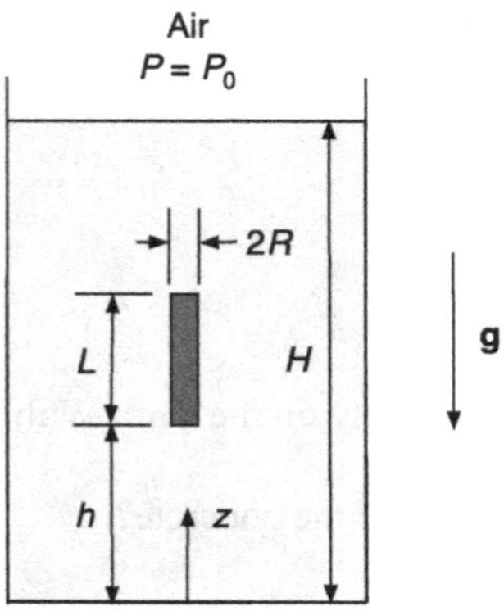

Figure P4.6 Rod-like particle settling in a sucrose gradient.

4.7. Half-submerged cylinder Suppose that a long cylinder of radius a is exactly half submerged in water, as shown in Fig. P4.7. The cylinder density is ρ_o, the water density is ρ, and the air density is negligible.

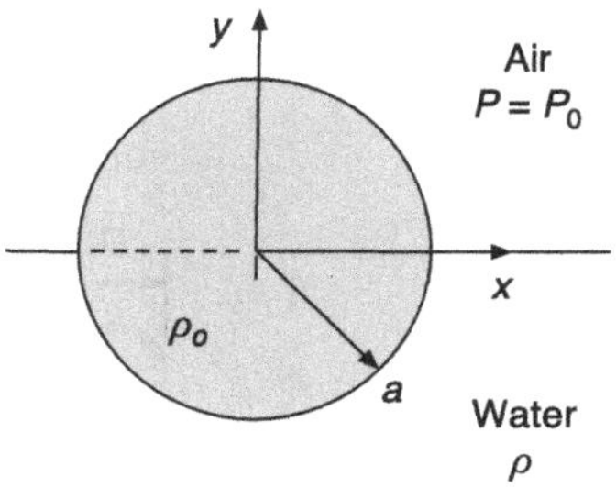

Figure P4.7 Half-submerged cylinder.

(a) Verify Archimedes' law for this case by directly calculating F_{Py} using Eq. (4.3-1).
(b) If the surface-tension force is negligible, what must be the value of ρ_o/ρ?

4.8. Buoyancy of a cone Verify Archimedes' law for an upright cone of radius R and height H by direct calculation using Eq. (4.3-1). It is convenient to use a combination of Cartesian and spherical coordinates, as shown in Fig. P4.8, and to set $P = 0$ at $z = 0$. Note that the side of the cone corresponds to $\theta = \pi - \beta$, where β is the cone angle.

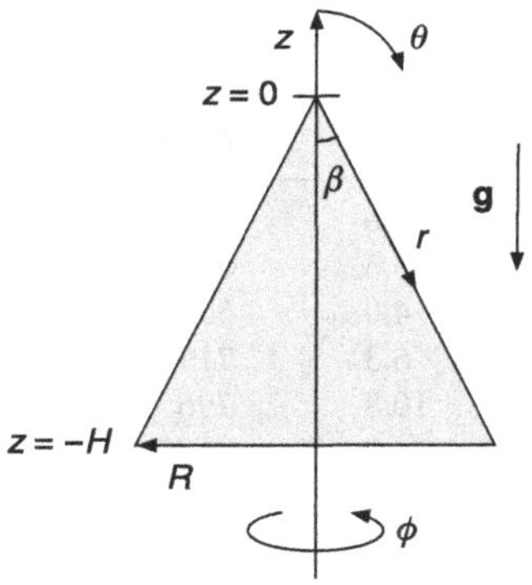

Figure P4.8 Coordinates for calculating the buoyancy force on a cone.

4.9. Formation of small bubbles Blanchard and Syzdek (1977) formed individual air bubbles in water under near-static conditions using a method shown in Fig. P4.9. Air was pushed through an immersed glass capillary to form a slowly growing bubble at its tip. After reaching a critical size and being released, the bubble rose until it rested on an inverted cup attached to a sensitive balance. The balance measured the upward force F that the bubble exerted on the cup. We will denote the inner diameter of the capillary tip and critical bubble diameter as d and D, respectively. Values of D (calculated from F) were obtained as a function of d at 20–22 °C.

(a) What is the relationship between F and D?
(b) As it nears release from a vertical tube, the actual shape of a bubble will resemble that in Fig. 4.14. The volume V of the "light bulb" equals that of the final sphere of

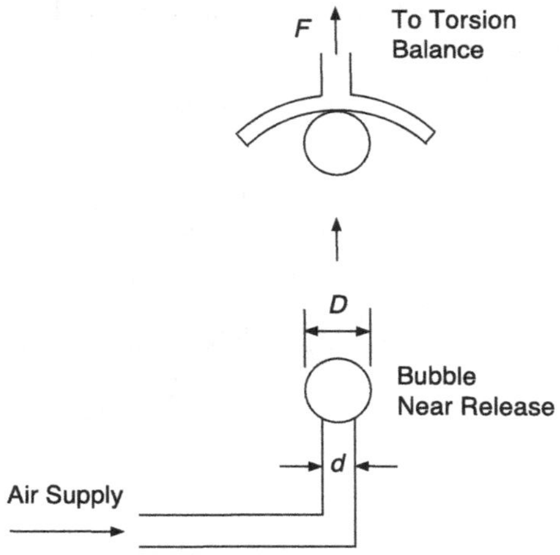

Figure P4.9 Method for forming individual bubbles and measuring their size.

diameter D. Use a force balance on a carefully defined control volume to relate D to d and γ.

(c) Create a log–log plot to compare your predicted curve for $D(d)$ with the experimental results in Table P4.9. How accurate is your expression?

Table P4.9 Capillary and bubble sizes from Blanchard and Syzdek (1977)[a]

d (μm)	D (μm)	d (μm)	D (μm)
1.6	409	56.9	1360
3.6	601	117	1720
4.0	557	320	2330
6.3	719	674	2940
10.8	796	1220	3510
14.2	860	2700	4770
15.5	860		

[a] Values read from a graph with an estimated uncertainty of ±2%–3%.

4.10. Capillary adhesion Suppose that a drop of water is placed between two glass plates, which are then pressed together to create a film of thickness $2h$ and radius R, as shown in Fig. P4.10. You may assume that the contact angle for water on the glass is nearly zero.

(a) Derive a general expression for P within the water film, which you may then simplify by assuming that $h \ll R$. (*Hint*: Although the air–water interface is *concave* outward in the horizontal view that is shown, when viewed vertically it is *convex* outward.)

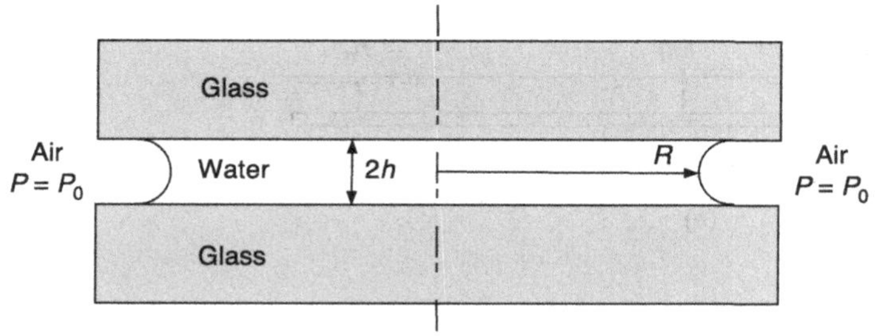

Figure P4.10 Capillary adhesion due to a thin film of water between glass plates.

(b) How large a load (mass of lower plate plus anything attached to it) could be supported if $h = 5$ μm and $R = 1$ cm?

4.11. Capillary flotation An object far denser than water can float if it is small enough and has the right surface properties. A result of surface tension, this is called *capillary flotation*. In what follows, suppose that each object is a cylinder of diameter D and length L, where $L \gg D$.

Figure P4.11 Cylinders floating in water: (a) poorly wettable surface ($\alpha > 90°$) and (b) readily wettable surface ($\alpha < 90°$).

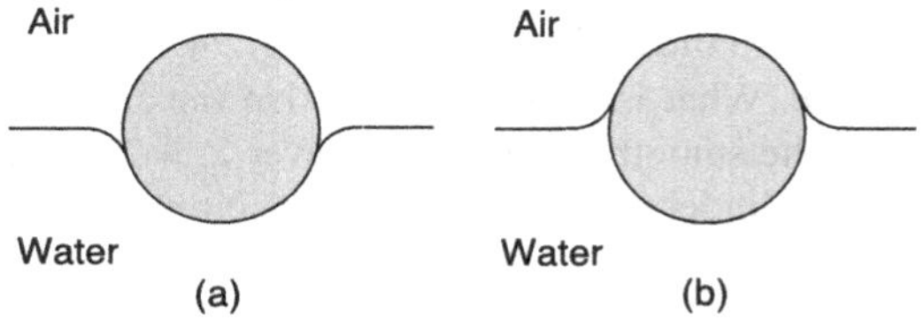

(a) In which of the two situations in Fig. P4.11 might a dense cylinder ($\rho_o > \rho$) be able to float? Explain qualitatively.
(b) Derive an expression for the *maximum* upward force due to surface tension. When can end effects be neglected?
(c) *Water striders* are insects that routinely walk on water (Hu *et al.*, 2003). A representative member of this family has six legs of 80 μm diameter, and when at rest on the surface the average length per leg that is in contact with water is 4 mm. For pond water, $\gamma = 67$ mN/m. Show that buoyancy is negligible relative to surface tension. What is the maximum mass m that this insect could have without sinking? Its actual mass is about 0.01 g.
(d) Although wood ordinarily floats, occasionally a log stored in a pond is denser than water and sinks to the bottom. Explain why surface tension could do little to help avoid such an inconvenience, even if the surface properties of the wood (contact angle) were optimal.

4.12. Plateau–Rayleigh instability Surface tension stabilizes planar gas–liquid interfaces by smoothing ripples. It may be surprising, then, that at cylindrical interfaces it causes the liquid to break up. This is called the *Plateau–Rayleigh instability*. In Fig. P4.12(a) a wettable fiber of radius b is coated with a liquid film of outer radius a, where a is small enough that Bo $\ll 1$ and gravity is negligible. In response to small disturbances (e.g., vibrations), such a film ordinarily rearranges into a necklace-like set of droplets, as shown in Fig. P4.12(b).

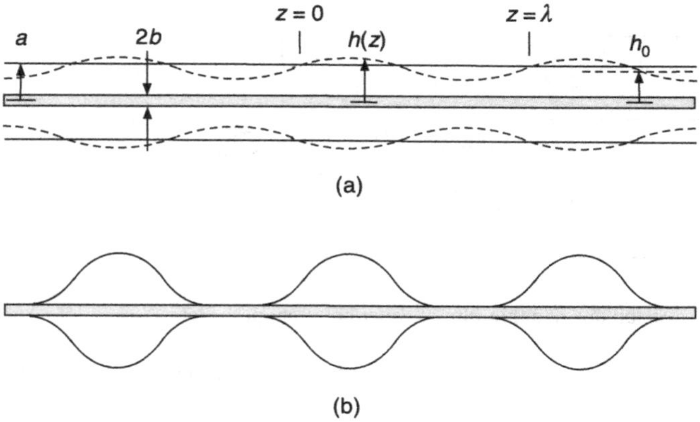

Figure P4.12 Plateau–Rayleigh instability. A liquid film that coats a small wettable fiber, as in (a), rearranges spontaneously into a string of droplets, as in (b).

The instability of the cylindrical film is a consequence of the system seeking a lower energy. With gravity negligible, the change in total energy equals that in the surface energy, $\gamma \Delta S$. Suppose that something causes a sinusoidal deformation of the interface, as in the dashed curves in Fig. P4.12(a). Such a deformation might have any wavelength λ. What is to be shown is that the area of the wavy surface (S_w) is smaller than that of the smooth one (S_0) whenever $\lambda > 2\pi a$. A practical consequence of this is that a liquid cylinder that is longer than its circumference will break up spontaneously.

(a) The local radius of a slightly wavy cylinder with average radius h_0 is

$$h(z) = h_0 \left[1 + \varepsilon \sin\left(\frac{2\pi z}{\lambda}\right)\right] \tag{P4.12-1}$$

where $\varepsilon \ll 1$. [For clarity of labeling, the wave amplitude εh_0 in Fig. P4.12(a) is greatly exaggerated.] For the wavy and smooth films to have the same volume, show that

$$\frac{h_0}{a} = \left(1 + \frac{\varepsilon^2}{2}\right)^{-1/2} \cong 1 - \frac{\varepsilon^2}{4}. \tag{P4.12-2}$$

Use is made here of $(1+x)^{1/2} = 1 + x/2 + \cdots$ and $(1+x)^{-1} = 1 - x + \cdots$, where the terms involving higher powers of x are neglected. Thus, the average radius of the wavy cylinder is slightly less than the unperturbed radius.

(b) The change in interfacial area per wavelength is[2]

$$\Delta S = 2\pi \int_0^\lambda h \left[1 + \frac{1}{2}\left(\frac{dh}{dz}\right)^2\right] dz - 2\pi a \lambda. \tag{P4.12-3}$$

[2] The differential area of the wavy surface is the circumference ($2\pi h$) times the differential change in arc length (ds). For small $|dh/dz|$,

$$ds = \left[1 + \left(\frac{dh}{dz}\right)^2\right]^{1/2} dz \cong \left[1 + \frac{1}{2}\left(\frac{dh}{dz}\right)^2\right] dz.$$

Show that $\Delta S < 0$ for $\lambda > 2\pi a$ and any small value of ε. Accordingly, the cylindrical surface is unstable with respect to any long-wavelength disturbance, no matter how small its initial amplitude. The amplitude grows until the film forms the string of "pearls" in Fig. P4.12(b). How to calculate the final droplet shape is discussed in De Gennes *et al.* (2003, p. 11).

Notice that the critical wavelength is independent of the fiber radius b. Indeed, the key parts of the derivation are unchanged if there is no inner cylinder at all. A critical wavelength $\lambda = \pi d$ (where d is the diameter of a liquid cylinder) was discovered experimentally by the Belgian physicist Joseph Plateau about 1870 and later confirmed theoretically by Lord Rayleigh. As described in De Gennes *et al.* (2003, pp. 119–120), Plateau found that oil droplets immersed in water could be made to coalesce into cylinders by pushing them together. However, the cylinders broke up when λ/d was greater than 3.13 to 3.18. Of importance for technologies such as inkjet printing is that a cylindrical jet of liquid also starts to break up as soon as its length exceeds its circumference (Problem 1.7).

5

Fluid kinematics

5.1 INTRODUCTION

Kinematics is the branch of mechanics that focuses on the description of motion, without concern for the associated forces. This chapter provides background that is needed for deriving the basic equations of fluid dynamics in Chapter 6 and analyzing the various types of flow in Part III. Any velocity field must satisfy conservation of mass at each point in the fluid. That requirement is embodied in the continuity equation, which is derived first. Considered then are rates of change as perceived by moving observers. Of particular interest in fluid mechanics are the rates of change of density and velocity seen by a hypothetical observer moving with the fluid. Addressed next is how to quantify rates of fluid deformation as rates of strain, which provides a foundation for describing viscous stresses. Concluding the chapter are introductions to two kinematic functions: the vorticity, which is a local measure of rotation, and the stream function, which is valuable for visualizing flow direction and speed.

In addition to the appendix material mentioned in Section 4.1, familiarity is needed now with vector cross products (Section A.3), the divergence and curl of a vector (Section A.4), and the Laplacian operator (Section A.4).

5.2 CONTINUITY

Until now we have made only incidental use of the concept of conservation of mass. Throughout Chapter 2 we used the fact that, for steady pipe flow, the mass flow rate at the inlet equals that at the outlet. While sufficient for some purposes, overall balances like that fall short of what is required to predict the details of a flow. What is needed is a differential equation that expresses conservation of mass at any point in a fluid.

A local form of conservation of mass will be derived by employing again the small cubic volume in Fig. 4.1. It is desired to equate the rate of accumulation of mass with the net rate of mass inflow. The total mass in the control volume is $\rho \Delta x\, \Delta y\, \Delta z$ and the rate at which it increases with time is $\partial(\rho \Delta x\, \Delta y\, \Delta z)/\partial t$. Mass can enter or leave the cube through any of its six surfaces. At a surface perpendicular to the x axis (an "x surface") the volume flow rate is $v_x \Delta y\, \Delta z$ and the mass flow rate is $\rho v_x \Delta y\, \Delta z$. The expressions for the y and z surfaces are analogous. Noting that positive velocities at the surfaces positioned at x, y, and z give mass inflows, whereas those at $x + \Delta x$, $y + \Delta y$, and $z + \Delta z$ give outflows, the mass balance is

$$\begin{aligned} \frac{\partial}{\partial t}(\rho \Delta x\, \Delta y\, \Delta z) &= (\rho v_x)|_x\, \Delta y\, \Delta z - (\rho v_x)|_{x+\Delta x}\, \Delta y\, \Delta z \\ &\quad + (\rho v_y)|_y\, \Delta x\, \Delta z - (\rho v_y)|_{y+\Delta y}\, \Delta x\, \Delta z \\ &\quad + (\rho v_z)|_z\, \Delta x\, \Delta y - (\rho v_z)|_{z+\Delta z}\, \Delta x\, \Delta y. \end{aligned} \tag{5.2-1}$$

Table 5.1 Continuity equation in Cartesian, cylindrical, and spherical coordinates

Cartesian (x, y, z):

$$\frac{\partial \rho}{\partial t} + \frac{\partial}{\partial x}(\rho v_x) + \frac{\partial}{\partial y}(\rho v_y) + \frac{\partial}{\partial z}(\rho v_z) = 0$$

Cylindrical (r, ϕ, z):

$$\frac{\partial \rho}{\partial t} + \frac{1}{r}\frac{\partial}{\partial r}(r\rho v_r) + \frac{1}{r}\frac{\partial}{\partial \theta}(\rho v_\theta) + \frac{\partial}{\partial z}(\rho v_z) = 0$$

Spherical (r, θ, ϕ):

$$\frac{\partial \rho}{\partial t} + \frac{1}{r^2}\frac{\partial}{\partial r}(r^2 \rho v_r) + \frac{1}{r \sin\theta}\frac{\partial}{\partial \theta}(\rho v_\theta \sin\theta) + \frac{1}{r \sin\theta}\frac{\partial}{\partial \phi}(\rho v_\phi) = 0$$

Dividing by $\Delta x \, \Delta y \, \Delta z$ and reducing the control volume to a point gives

$$\frac{\partial \rho}{\partial t} = -\frac{\partial}{\partial x}(\rho v_x) - \frac{\partial}{\partial y}(\rho v_y) - \frac{\partial}{\partial z}(\rho v_z). \tag{5.2-2}$$

This shows that spatial variations in the velocity components are needed to compensate for any change in the density.

A more compact and general form of Eq. (5.2-2) is obtained by noticing that the right-hand side is minus the divergence of the vector $\rho\mathbf{v}$. Moving that term to the left gives

$$\frac{\partial \rho}{\partial t} + \nabla \cdot (\rho \mathbf{v}) = 0. \tag{5.2-3}$$

This is the most general form of the *continuity equation*, which expresses conservation of mass at any point in a fluid. Written in this manner using vectors, it is valid for any coordinate system. Table 5.1 gives its forms for Cartesian, cylindrical, and spherical coordinates.

In the flows encountered in chemical engineering, fluid densities are usually almost constant. That is true for gases as well as liquids.[1] Equation (5.2-3) simplifies then to

$$\nabla \cdot \mathbf{v} = 0 \quad (\text{constant } \rho). \tag{5.2-4}$$

This relationship for incompressible fluids is what we will be using nearly always. Its Cartesian, cylindrical, and spherical forms are obtained from Table 5.1 by setting $\partial\rho/\partial t = 0$ and factoring ρ out of the other terms; $\nabla \cdot \mathbf{v}$ for the three coordinate systems is given also in Tables A.2, A.3, and A.4, respectively. Although incompressibility is our default assumption in examples and problems, in key derivations we will continue to allow for variations in ρ.

Example 5.2-1 Unknown velocity component. Equation (5.2-4) can be used to find an unknown velocity component when the others are given. To be considered is a steady flow in which v_x and v_y each depend at most on x and y and $v_z = 0$. Because only two coordinates are involved, such a problem is *two-dimensional*; with two nonzero velocity components, the flow is *bidirectional*. In particular, suppose that the fluid occupies the space $y > 0$ and that $v_y = -Cy$, where $C > 0$. Notice that $v_y = 0$ at $y = 0$, as if that plane

[1] Exceptions include gas flows in very long conduits or at velocities near the speed of sound, as discussed in Sections 12.4 and 12.5, respectively.

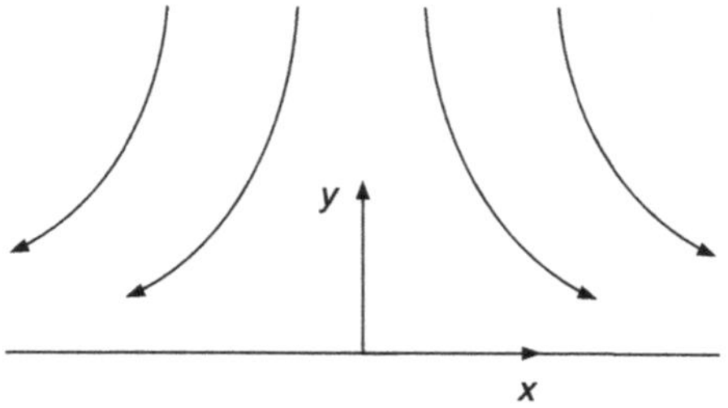

Figure 5.1 Qualitative depiction of planar stagnation flow.

were an impermeable surface. Suppose also that there is *reflective symmetry* about the plane $x = 0$ (i.e., the velocity field for $x < 0$ is the mirror image of that for $x > 0$). Such a flow is shown qualitatively in Fig. 5.1. A flow with reflective symmetry that is directed toward a flat surface is called a *planar stagnation flow*. The objective is to determine $v_x(x, y)$, assuming that C is known.

For $v_z = 0$ and constant ρ, the Cartesian continuity equation reduces to

$$\frac{\partial v_x}{\partial x} + \frac{\partial v_y}{\partial y} = 0. \tag{5.2-5}$$

Solving for $\partial v_x/\partial x$ and integrating over x gives

$$v_x(x, y) = -\int \frac{\partial v_y}{\partial y}\,dx = -\int (-C)\,dx = Cx + D \tag{5.2-6}$$

where D is a constant. Because flow in either direction across the plane $x = 0$ would violate the assumed symmetry, it is necessary that $v_x = 0$ at $x = 0$. Thus, $D = 0$ and

$$v_x(x) = Cx. \tag{5.2-7}$$

It is seen that v_x in this flow actually depends only on x.

Example 5.2-2 Expansion of the Universe. The Universe is a system in which ρ depends on time, but not position. The American astronomer Edwin Hubble (1889–1953) is famous for observing that each object recedes from the Earth at a velocity that is proportional to its distance. This relationship is known as *Hubble's law*. The proportionality factor between velocity and distance, called *Hubble's constant* and denoted here as H, is actually believed to be a function of time. It has also been found that, on the megaparsec length scale needed to justify continuum modeling (1 parsec $= 3 \times 10^{13}$ km), the density of the Universe is spatially uniform. The objective is to see what these observations imply about $\rho(t)$.

Because Hubble's law holds equally in all directions, this is a problem with *spherical symmetry*. That is, positions can be described using just a spherical radial coordinate (r), with the origin at the Earth. The motion is purely radial and does not depend on either spherical angle. In this coordinate system Hubble's law is simply $v_r = Hr$. The spherical continuity equation simplifies to

$$\frac{d\rho}{dt} = -\frac{\rho}{r^2}\frac{\partial}{\partial r}(r^2 v_r). \tag{5.2-8}$$

Setting $v_r = Hr$ and grouping the density terms together gives

$$\frac{1}{\rho}\frac{d\rho}{dt} = -3H(t). \tag{5.2-9}$$

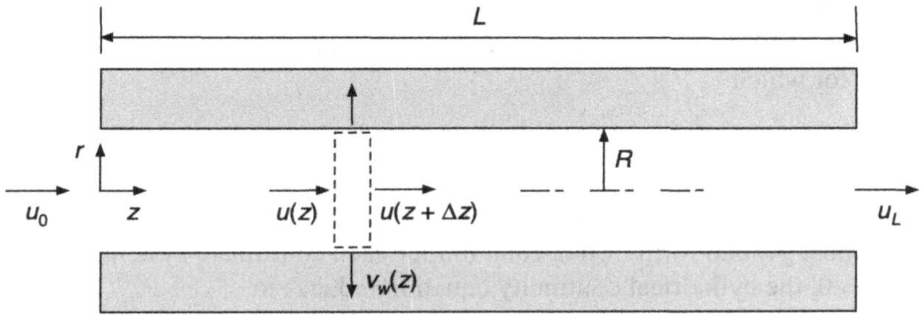

Figure 5.2 Filtration in a hollow fiber.

To solve for $\rho(t)$ we would need the function $H(t)$. Assuming that $\rho(0) = \rho_0$ and that H is constant would give $\rho(t) = \rho_0 \exp(-3Ht)$.

Example 5.2-3 Filtration in a hollow fiber. A common type of membrane filtration unit consists of many hollow fibers arranged in parallel. The wall of each fiber has fine pores, and a pressure difference causes outward flow of the solvent, thereby concentrating macromolecules or particles that are retained inside the fiber. Figure 5.2 shows a single fiber of radius R and length L. Predicting the performance of the device requires knowledge of $\mathbf{v}$ inside the fiber. The cylindrical velocity components of interest are $v_z(r, z)$ and $v_r(r, z)$. The flow is assumed to be axisymmetric (independent of θ) and without swirl ($v_\theta = 0$). At any position along the fiber the mean axial velocity is $u(z)$ and the radial velocity at the wall is $v_w(z) = v_r(R, z)$. The mean velocities at the inlet and outlet are u_0 and u_L, respectively. To be illustrated are three levels of detail at which conservation of mass can be applied.

An overall relationship is obtained by choosing as the control volume all liquid inside the fiber. With ρ constant, a mass balance is the same as a volume balance. As the volume inside the fiber is constant, the inflow and outflow must be equal and

$$\pi R^2 u_0 = \pi R^2 u_L + 2\pi R \int_0^L v_w \, dz. \tag{5.2-10}$$

This may be rearranged as

$$u_L = u_0 - \frac{2}{R} \int_0^L v_w \, dz \tag{5.2-11}$$

which shows how much the mean velocity at the outlet is reduced by the filtration.

An intermediate level of detail comes from selecting a control volume of radius R and differential length Δz that is located at some position z, as shown by the dashed lines in Fig. 5.2. Because one of the dimensions is differential, what is to be constructed is a kind of shell balance. Equating volume inflow and outflow for this region gives

$$\pi R^2 u(z) = \pi R^2 u\,(z + \Delta z) + (2\pi R \, \Delta z)\, v_w(z). \tag{5.2-12}$$

Dividing by Δz and letting $\Delta z \to 0$ provides a differential equation that governs the mean velocity,

$$\frac{du}{dz} = -\frac{2v_w}{R}, \quad u(0) = u_0. \tag{5.2-13}$$

If $v_w(z)$ is known, this can be solved for $u(z)$. Thus, this approach provides the mean velocity at intermediate locations along the fiber. The simplest possibility is a constant wall velocity, for which

$$u(z) = u_0 - \frac{2v_w z}{R} \quad (\text{constant } v_w). \tag{5.2-14}$$

The most detailed information on **v** is contained in Eq. (5.2-4), and Table 5.1 already provides the most general form of that equation for each coordinate system. With ρ constant and $v_\theta = 0$, the cylindrical continuity equation reduces to

$$\frac{1}{r}\frac{\partial}{\partial r}(rv_r) + \frac{\partial v_z}{\partial z} = 0. \tag{5.2-15}$$

Equations (5.2-11) and (5.2-13) each could have been obtained also from a more detailed form of conservation of mass. Equation (5.2-11) could have been found by integrating Eq. (5.2-13) over the channel length. Likewise, Eq. (5.2-13) could have been derived by integrating Eq. (5.2-15) over the cross-section, as follows. The differential area for the circular cross-section is $dS = r\,dr\,d\theta$. Integrating each term in Eq. (5.2-15) gives

$$\int_0^{2\pi}\int_0^R \left[\frac{1}{r}\frac{\partial}{\partial r}(rv_r)\right] r\,dr\,d\theta = 2\pi(rv_r)\Big|_{r=0}^{r=R} = 2\pi R v_w \tag{5.2-16}$$

$$\int_0^{2\pi}\int_0^R \frac{\partial v_z}{\partial z} r\,dr\,d\theta = 2\pi \frac{d}{dz}\int_0^R v_z r\,dr = \pi R^2 \frac{du}{dz} \tag{5.2-17}$$

and assembling these results leads to Eq. (5.2-13). Although either approach gives the same result, a shell balance such as Eq. (5.2-12) is generally the simpler way to determine how a mean velocity varies along a conduit.

Of the three ways to express conservation of mass, only the continuity equation permits evaluation of an unknown velocity component. It was assumed at the outset that $v_\theta = 0$, and it is shown in Example 8.2-2 that

$$v_z(r, z) = 2u(z)\left[1 - \left(\frac{r}{R}\right)^2\right] \tag{5.2-18}$$

if v_w is not too large. What is unknown is v_r. That component is found by using Eqs. (5.2-18) and (5.2-13) in Eq. (5.2-15) to obtain

$$\frac{1}{r}\frac{\partial}{\partial r}(rv_r) = -\frac{\partial v_z}{\partial z} = -2\frac{du}{dz}\left[1 - \left(\frac{r}{R}\right)^2\right] = \frac{4v_w(z)}{R}\left[1 - \left(\frac{r}{R}\right)^2\right]. \tag{5.2-19}$$

Multiplying by r, integrating, and then dividing by r gives

$$v_r(r, z) = \frac{2v_w(z)}{R}\left(r - \frac{r^3}{2R^2}\right) + \frac{C}{r}. \tag{5.2-20}$$

The constant C is evaluated by requiring that v_r be finite everywhere, including $r = 0$. Accordingly, $C = 0$ and

$$\frac{v_r(r, z)}{v_w(z)} = 2\left(\frac{r}{R}\right) - \left(\frac{r}{R}\right)^3. \tag{5.2-21}$$

Table 5.2 Material derivative in Cartesian, cylindrical, and spherical coordinates

Cartesian (x, y, z):

$$\frac{D}{Dt} = \frac{\partial}{\partial t} + v_x \frac{\partial}{\partial x} + v_y \frac{\partial}{\partial y} + v_z \frac{\partial}{\partial z}$$

Cylindrical (r, θ, z):

$$\frac{D}{Dt} = \frac{\partial}{\partial t} + v_r \frac{\partial}{\partial r} + \frac{v_\theta}{r} \frac{\partial}{\partial \theta} + v_z \frac{\partial}{\partial z}$$

Spherical (r, θ, ϕ):

$$\frac{D}{Dt} = \frac{\partial}{\partial t} + v_r \frac{\partial}{\partial r} + \frac{v_\theta}{r} \frac{\partial}{\partial \theta} + \frac{v_\phi}{r \sin\theta} \frac{\partial}{\partial \phi}$$

5.3 RATES OF CHANGE FOR MOVING OBSERVERS

Often needed when analyzing fluid flow or other transport processes is a rate of change that would be sensed by a moving observer. Consider a scalar function f, such as density or temperature, which might depend on both time and position. For an observer moving at velocity $\mathbf{u}$ we denote the apparent rate of change of f as $(df/dt)_\mathbf{u}$. At a fixed location, the change in f over a small time interval Δt is $(\partial f/\partial t)\Delta t$. If f varies with position, the motion of the observer will cause an additional change. For example, movement in the x direction by a small amount Δx will produce a change $(\partial f/\partial x)\Delta x$. Noting that $\Delta x = u_x \, \Delta t$ for small time increments, this variation in f equals $u_x(\partial f/\partial x)\,\Delta t$. Similar contributions arise from observer motion in the y and z directions. The total change in f over the time interval is then

$$\Delta f = \Delta t \left(\frac{\partial f}{\partial t} + u_x \frac{\partial f}{\partial x} + u_y \frac{\partial f}{\partial y} + u_z \frac{\partial f}{\partial z} \right) = \Delta t \left(\frac{\partial f}{\partial t} + \mathbf{u} \cdot \nabla f \right). \quad (5.3\text{-}1)$$

In the last expression it is recognized that the combination of velocity components and spatial derivatives equals a dot product. For $\Delta t \to 0$,

$$\left(\frac{df}{dt} \right)_\mathbf{u} = \frac{\partial f}{\partial t} + \mathbf{u} \cdot \nabla f. \quad (5.3\text{-}2)$$

Thus, the perceived rate of change combines variations over time at fixed locations with those stemming from the combination of observer motion and spatial gradients.

Of particular interest is an observer moving with the fluid. The rate of change for $\mathbf{u} = \mathbf{v}$ is

$$\left(\frac{df}{dt} \right)_\mathbf{v} = \frac{\partial f}{\partial t} + \mathbf{v} \cdot \nabla f \equiv \frac{Df}{Dt}. \quad (5.3\text{-}3)$$

The differential operator D/Dt is called the *material derivative* (or *substantial derivative*). This combination of a time derivative, fluid velocity, and gradient arises often enough to merit its own symbol. A *streamline* is the path that a small element of fluid follows (Section 5.6). Thus, the rate of change for an observer moving with the fluid is the rate of change along a streamline. The D/Dt operator is given for the three common coordinate systems in Table 5.2.

Although presented here for a scalar function f, the material derivative can operate also on a vector. Indeed, as shown in Chapter 6, $D\mathbf{v}/Dt$ describes fluid acceleration in the differential equation for conservation of linear momentum. How to evaluate $\mathbf{v} \cdot \nabla \mathbf{v}$ in various coordinate systems is explained in Section A.5.

Example 5.3-1 Temperature changes sensed by a weather balloon. It is desired to calculate the rate of temperature change that would be measured by an instrument on a balloon that is rising due to its buoyancy and also being carried by the wind. The vertical and horizontal velocity components of the balloon are assumed to be constant at 2 m/s and 5 m/s, respectively. Suppose that the atmospheric temperature is steady and varies with height above the Earth's surface (z) according to

$$T(z) = T_0 - Gz \tag{5.3-4}$$

where $T_0 = 15$ °C and $G = 6.5$ °C/km.

With the temperature a function of z only, Eq. (5.3-2) reduces to

$$\left(\frac{dT}{dt}\right)_{\mathbf{u}} = u_z \frac{\partial T}{\partial z} = -u_z G. \tag{5.3-5}$$

Notice that the values of u_x and u_y do not matter, because $\partial T/\partial x = 0 = \partial T/\partial y$. The observed rate of temperature change would be (2 m/s)(10^{-3} km/m)(–6.5 °C/km) = –0.013 °C/s.

The material derivative allows the continuity equation to be written in another form. From Table A.1, $\nabla \cdot (\rho \mathbf{v}) = \mathbf{v} \cdot \nabla \rho + \rho \nabla \cdot \mathbf{v}$. Substituting this into Eq. (5.2-3) gives

$$\frac{1}{\rho}\frac{D\rho}{Dt} = -\nabla \cdot \mathbf{v}. \tag{5.3-6}$$

This shows that an observer moving at velocity $\mathbf{v}$ would experience compression of the fluid (an increase in ρ) if $\nabla \cdot \mathbf{v} < 0$ and expansion of the fluid (a decrease in ρ) if $\nabla \cdot \mathbf{v} > 0$.

5.4 RATE OF STRAIN

The defining property of a fluid is that it deforms continuously when subjected to shearing forces. In solid mechanics a deformation is generally expressed as a *strain*, which is a ratio of lengths. For a sample of initial length L that is elongated by an amount ΔL, the strain is $\Delta L/L$. Because it is the *rate* of deformation that determines viscous stresses, in fluid mechanics we are concerned with a *rate of strain*, which is defined as $L^{-1}\, dL/dt$ and has the dimension of inverse time.

Before starting to quantify rates of strain in a fluid, we should be clear about when deformation occurs and when it does not. Four basic types of two-dimensional fluid motion are depicted in Fig. 5.3. In each case the solid squares represent the positions of small elements of fluid at time t and the dashed shapes are their locations at time $t + \Delta t$. Such elements, which move at the fluid velocity and retain their identity, are called *material points*.

No deformation results from uniform translation [Fig. 5.3(a)] because the velocity is the same everywhere. In the figure, any two points on the square remain the same distance from one another as they move to the right. Deformation is absent also in *rigid-body rotation* [Fig. 5.3(b)], despite the fact that the linear velocity varies with position. Although it is not associated with deformation, rigid-body rotation is related to vorticity and is

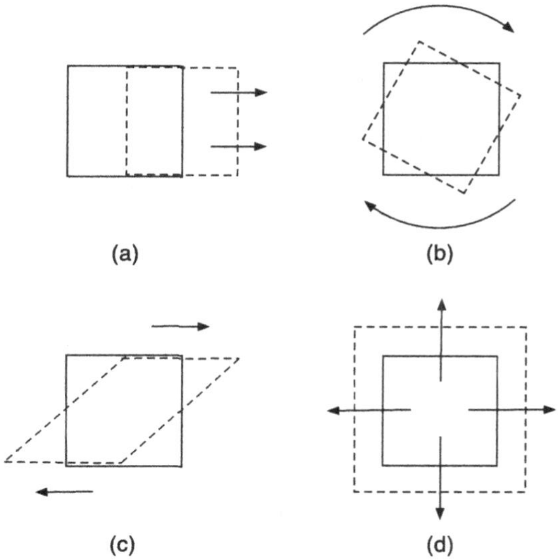

Figure 5.3 Basic types of fluid motion in two dimensions: (a) uniform translation; (b) rigid-body rotation; (c) simple shear; and (d) dilatation. The positions of fluid elements at times t and $t + \Delta t$ are shown by solid and dashed lines, respectively.

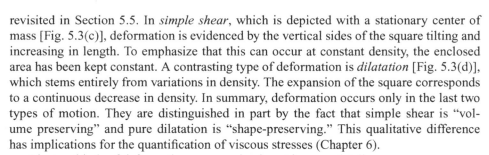

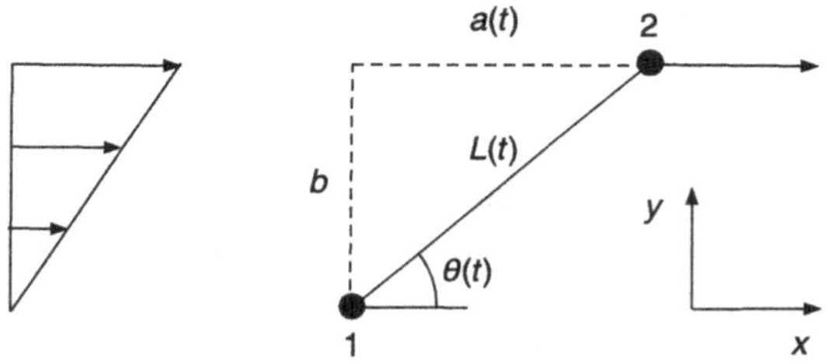

Figure 5.4 Two material points in a simple shear flow. With point 1 stationary and point 2 moving to the right, the distance $L(t)$ increases with time and the angle $\theta(t)$ decreases.

revisited in Section 5.5. In *simple shear*, which is depicted with a stationary center of mass [Fig. 5.3(c)], deformation is evidenced by the vertical sides of the square tilting and increasing in length. To emphasize that this can occur at constant density, the enclosed area has been kept constant. A contrasting type of deformation is *dilatation* [Fig. 5.3(d)], which stems entirely from variations in density. The expansion of the square corresponds to a continuous decrease in density. In summary, deformation occurs only in the last two types of motion. They are distinguished in part by the fact that simple shear is "volume preserving" and pure dilatation is "shape-preserving." This qualitative difference has implications for the quantification of viscous stresses (Chapter 6).

The two kinds of deformation are examined next in more detail.

Example 5.4-1 Rate of strain in simple shear flow. It is desired to calculate the rate of strain for a pair of material points in a simple shear flow. As shown in Fig. 5.4, a reference frame is chosen in which point 1 is stationary and point 2 moves parallel to the x axis. At a given instant the x and y coordinates of point 2 differ from those of point 1 by $a(t)$ and b, respectively. The points are separated by a distance $L(t)$, and a line between them meets the x axis at an angle $\theta(t)$. Placing point 1 at $y = 0$, the fluid velocity is given by

$$v_x(y) = cy, \quad v_y = 0 = v_z \tag{5.4-1}$$

where c is a constant that corresponds to the shear rate defined in Section 1.2.

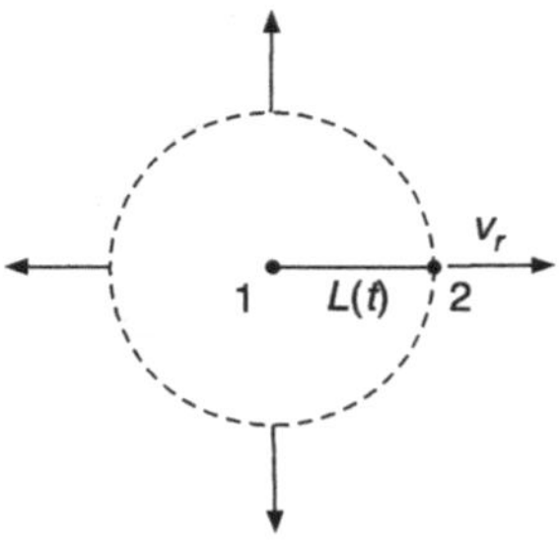

Figure 5.5 Schematic for calculating the rate of strain in dilatation. Point 1 is at the origin, point 2 moves radially at velocity v_r, and $L(t)$ is the distance between the points.

In that $L^2 = a^2 + b^2$ and b is constant,

$$\frac{d(L^2)}{dt} = 2L\frac{dL}{dt} = 2a\frac{da}{dt}. \tag{5.4-2}$$

The rate of change in a is just the value of v_x at $y = b$, or $da/dt = cb$. Substituting this into Eq. (5.4-2) and dividing by L^2 gives

$$\frac{1}{L}\frac{dL}{dt} = \frac{abc}{L^2}. \tag{5.4-3}$$

Using $a = L\cos\theta$ and $b = L\sin\theta$, the rate of strain becomes

$$\frac{1}{L}\frac{dL}{dt} = c\,\sin\theta\cos\theta. \tag{5.4-4}$$

It is seen that the rate of strain in simple shear is proportional to, but not the same as, the shear rate c. In this special type of flow the rate of strain is spatially uniform; that is, it is independent of the location of point 1. However, it depends on the direction of the line that connects the points. It is positive if that line is inclined to the right (as in Fig. 5.4) and negative if it is inclined to the left. It vanishes whenever the points are aligned with either the x or the y axis (θ an integer multiple of $\pi/2$), and its absolute value is greatest when θ is a multiple of $\pi/4$.

Example 5.4-2 Rate of strain in pure dilatation. The objective is to calculate the rate of strain in a fluid in which the density is spatially uniform but varies with time in a known manner.

If the density is known, then so is the volume per unit mass, $\hat{V}(= 1/\rho)$. With ρ and $\hat{V}$ independent of position, Eq. (5.3-6) becomes

$$\frac{1}{\hat{V}}\frac{d\hat{V}}{dt} = \nabla\cdot\mathbf{v} = f(t). \tag{5.4-5}$$

The rate of change of $\hat{V}$, relative to its instantaneous value, is the *rate of dilatation*. Thus, in a fluid of uniform density, the rate of dilatation equals the divergence of the velocity. The rate of dilatation could be computed from the given $\rho(t)$, so $f(t)$ is considered to be known.

This is a spherically symmetric problem. The material points for calculating the rate of strain are chosen such that point 1 is at the origin and point 2 is at radial position $r = L(t)$, as shown in Fig. 5.5. Although the outward arrows imply a decreasing density (or increasing specific volume), the derivation that follows applies equally to an increasing density. The only difference is that $f > 0$ for decreasing ρ and $f < 0$ for increasing ρ.

Table 5.3 Vorticity in Cartesian, cylindrical, and spherical coordinates

Cartesian (x, y, z):

$$\mathbf{w} = \left(\frac{\partial v_z}{\partial y} - \frac{\partial v_y}{\partial z}\right)\mathbf{e_x} + \left(\frac{\partial v_x}{\partial z} - \frac{\partial v_z}{\partial x}\right)\mathbf{e_y} + \left(\frac{\partial v_y}{\partial x} - \frac{\partial v_x}{\partial y}\right)\mathbf{e_z}$$

Cylindrical (r, θ, z):

$$\mathbf{w} = \left(\frac{1}{r}\frac{\partial v_z}{\partial \theta} - \frac{\partial v_\theta}{\partial z}\right)\mathbf{e_r} + \left(\frac{\partial v_r}{\partial z} - \frac{\partial v_z}{\partial r}\right)\mathbf{e_\theta} + \left(\frac{1}{r}\frac{\partial}{\partial r}(r v_\theta) - \frac{1}{r}\frac{\partial v_r}{\partial \theta}\right)\mathbf{e_z}$$

Spherical (r, θ, ϕ):

$$\mathbf{w} = \frac{1}{r\sin\theta}\left(\frac{\partial}{\partial \theta}(v_\phi \sin\theta) - \frac{\partial v_\theta}{\partial \phi}\right)\mathbf{e_r} + \left(\frac{1}{r\sin\theta}\frac{\partial v_r}{\partial \phi} - \frac{1}{r}\frac{\partial}{\partial r}(r v_\phi)\right)\mathbf{e_\theta}$$
$$+\left(\frac{1}{r}\frac{\partial}{\partial r}(r v_\theta) - \frac{1}{r}\frac{\partial v_r}{\partial \theta}\right)\mathbf{e_\phi}$$

Recognizing that dL/dt equals the radial velocity at point 2, the rate of strain is

$$\frac{1}{L}\frac{dL}{dt} = \frac{1}{L}v_r(L, t). \tag{5.4-6}$$

To find v_r, we first write the divergence of the velocity as

$$\nabla \cdot \mathbf{v} = \frac{1}{r^2}\frac{\partial}{\partial r}(r^2 v_r) = f(t). \tag{5.4-7}$$

Multiplying by r^2, integrating, and then dividing by r^2 gives

$$v_r(r, t) = \frac{r f(t)}{3} + \frac{a(t)}{r^2} \tag{5.4-8}$$

where $a(t)$ is an integration "constant" (actually, what might be a function of time). For the velocity at $r = 0$ to be finite we must set $a(t) = 0$. Accordingly,

$$v_r(r, t) = \frac{r f(t)}{3}. \tag{5.4-9}$$

Evaluating this at $r = L$ and substituting the result into Eq. (5.4-6) gives

$$\frac{1}{L}\frac{dL}{dt} = \frac{f(t)}{3} = \frac{1}{3}\nabla \cdot \mathbf{v}. \tag{5.4-10}$$

Thus, in pure dilatation the rate of strain is one-third the divergence of the velocity.

5.5 VORTICITY

Definition

The *vorticity vector* **w** is defined as the *curl* of the velocity,

$$\mathbf{w} = \nabla \times \mathbf{v}. \tag{5.5-1}$$

Thus, evaluating certain spatial derivatives allows **w** to be calculated as a function of position and time for any known velocity field. General expressions for the three common coordinate systems are given in Table 5.3.

As the derivation of the word "vorticity" from "vortex" (a swirling body of fluid) suggests, **w** is a measure of rotational motion. That is clearest for a fluid undergoing

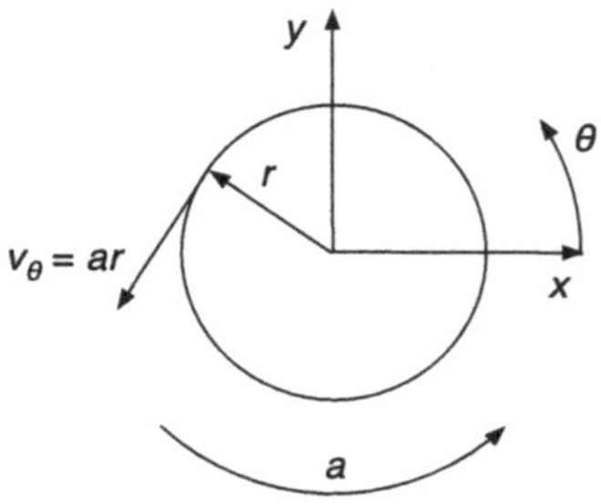

Figure 5.6 Circular path of a material point during rigid-body rotation about the z axis at angular velocity a. Cartesian and cylindrical coordinates are shown.

rigid-body rotation. Suppose there is rotation about the z axis at an angular velocity a, as in Fig. 5.6. Using cylindrical coordinates, the linear velocity in the θ direction is $v_\theta = ar$. There is no radial or axial flow ($v_r = 0 = v_z$). Accordingly, the one nonzero component of $\mathbf{w}$ is

$$w_z = \frac{1}{r}\frac{\partial}{\partial r}(rv_\theta) = \frac{1}{r}\frac{\partial}{\partial r}(ar^2) = 2a. \tag{5.5-2}$$

Thus, for rigid-body rotation the vorticity is spatially uniform and its magnitude is twice the angular velocity.

In simple shear flow, as with rigid-body rotation, there is a uniform, nonzero vorticity. For $v_x = cy$ and $v_y = 0 = v_z$ (as in Example 5.4-1),

$$w_z = -\frac{\partial v_x}{\partial y} = -c \tag{5.5-3}$$

and the other components of $\mathbf{w}$ are zero. Here and in Eq. (5.5-2) the sign of w_z indicates the direction of rotation. When viewed from positive z, it is counterclockwise for $w_z > 0$ (Fig. 5.6) and clockwise for $w_z < 0$ (Fig. 5.4). In more complex flows, $\mathbf{w}$ depends on position and there are not readily identifiable axes of rotation.

Irrotational flow

If for any reason $\mathbf{w} = \nabla \times \mathbf{v} = \mathbf{0}$ everywhere, then a flow is said to be *irrotational*. Planar stagnation flow (Example 5.2-1) is an example. Irrotational flows have special mathematical properties. In that $\nabla \times \nabla f = \mathbf{0}$ for any scalar function f (Table A.1), if $\mathbf{w} = \mathbf{0}$ there evidently exists a *velocity potential* ϕ such that

$$\mathbf{v} = \nabla\phi. \tag{5.5-4}$$

Continuity for an incompressible fluid ($\nabla \cdot \mathbf{v} = 0$) requires that such a function satisfy *Laplace's equation*,

$$\nabla^2\phi = 0 \tag{5.5-5}$$

where $\nabla^2 = \nabla \cdot \nabla$ is the Laplacian operator. Expressions for ∇^2 in the common coordinate systems are given in Section A.5. When Eq. (5.5-5) applies, there exists a *potential flow*. Thus, irrotational flow and potential flow are synonymous.

The existence of a velocity potential when $\mathbf{w} = \mathbf{0}$ indicates that the velocity field could be evaluated by solving Eq. (5.5-5) with suitable boundary conditions. For Laplace's equation it is sufficient to specify $\mathbf{n} \cdot \nabla\phi$ at each boundary ($\mathbf{n}$ being the unit normal), and this information is available if $\mathbf{v}\,(= \nabla\phi)$ is known at each boundary. What makes this attractive is that the linearity of Laplace's equation and the absence of a time derivative make it one of the easier partial differential equations to solve. What makes it physically meaningful is that any irrotational velocity field is an exact solution of the *Navier–Stokes equation*, which expresses conservation of momentum for an incompressible Newtonian fluid. The Navier–Stokes equation is derived in Chapter 6 and why it is satisfied when $\mathbf{w} = \mathbf{0}$ is explained in Chapter 9. The ability of the potential-flow formulation to satisfy conservation of both mass and momentum, together with the many methods that are available for solving Laplace's equation analytically, have made such problems a favorite of generations of applied mathematicians.

With $\mathbf{n} \cdot \nabla\phi$ specified at all boundaries, the solution of Eq. (5.5-5) will be unique only to within an additive constant. As with $\mathcal{P}$ or an electrical potential, only the gradient in ϕ is physically meaningful and the constant may be chosen at will. Actually, the "constant" can be a function of time, in that adding a function $f(t)$ to ϕ affects neither Eq. (5.5-4) nor Eq. (5.5-5). The additive constant or function of time aside, if the boundary velocities depend on t, then ϕ will depend on t as well as position. Even so, ϕ will still be governed by Eq. (5.5-5).

In summary, if a flow is known to be irrotational ($\mathbf{w} = \mathbf{0}$), then a velocity potential exists and determining the velocity field becomes relatively straightforward. Otherwise ($\mathbf{w} \neq \mathbf{0}$), there is no function ϕ that satisfies Eq. (5.5-4) and some other approach is needed. Circumstances in which real flows are nearly irrotational are discussed in Chapter 9.

5.6 STREAM FUNCTION

Definitions

The *stream function* (ψ) can be used to visualize a velocity field that has been found either analytically or numerically.[2] It is applicable to incompressible flows that are bidirectional. There is little need for ψ in a unidirectional flow, and ψ is undefined if there are three velocity components or if ρ varies.

Many flows of technological or scientific interest are unidirectional or bidirectional, provided that the coordinate system is chosen properly. Bidirectional flows fall into two broad categories, according to the type of symmetry that causes the third velocity component to be absent. In *planar* flow, $v_z = 0$ and the other velocity components are independent of z. Planar stagnation flow (Example 5.2-1) is an example. Depending on what is being modeled, a planar problem might be analyzed using either Cartesian (x, y) or cylindrical (r, θ) coordinates. In *axisymmetric* flows there is rotational symmetry about the z axis, such that all quantities are independent of a cylindrical or spherical angle. In cylindrical coordinates the nonzero velocity components are v_r and v_z and neither depends on θ. In such situations it is assumed that there is no swirling motion about the z axis, such that $v_\theta = 0$. An example is flow in a permeable tube (Example 5.2-3). In spherical coordinates the angle describing rotation about the z axis is ϕ, and, if the flow is axisymmetric, then $v_\phi = 0$ and nothing depends on ϕ. The nonzero velocity components in this case are the spherical v_r and v_θ. In this category are flows past bodies of revolution, such as spheroids and cones.

For planar flows in Cartesian coordinates, $\psi(x, y, t)$ is defined as

$$v_x = \frac{\partial \psi}{\partial y}, \quad v_y = -\frac{\partial \psi}{\partial x}. \tag{5.6-1}$$

Because only its spatial derivatives are specified, ψ contains an additive constant that can be chosen freely, as with $\mathcal{P}$, electrical potentials, or velocity potentials. A key aspect of

[2] As illustrated, for example, in Deen (2012), ψ can be helpful also in solving the differential equations that govern the velocity.

Table 5.4 Definitions of the stream function for planar or axisymmetric flows

Planar, Cartesian (x, y):

$$v_x = \frac{\partial \psi}{\partial y}, \quad v_y = -\frac{\partial \psi}{\partial x}$$

Planar, cylindrical (r, θ):

$$v_r = \frac{1}{r}\frac{\partial \psi}{\partial \theta}, \quad v_\theta = -\frac{\partial \psi}{\partial r}$$

Axisymmetric, cylindrical (z, r):

$$v_z = \frac{1}{r}\frac{\partial \psi}{\partial r}, \quad v_r = -\frac{1}{r}\frac{\partial \psi}{\partial z}$$

Axisymmetric, spherical (r, θ):

$$v_r = \frac{1}{r^2 \sin\theta}\frac{\partial \psi}{\partial \theta}, \quad v_\theta = -\frac{1}{r \sin\theta}\frac{\partial \psi}{\partial r}$$

Eq. (5.6-1) is its consistency with the continuity equation. That is,

$$\nabla \cdot \mathbf{v} = \frac{\partial v_x}{\partial x} + \frac{\partial v_y}{\partial y} = \frac{\partial}{\partial x}\left(\frac{\partial \psi}{\partial y}\right) + \frac{\partial}{\partial y}\left(-\frac{\partial \psi}{\partial x}\right) = 0. \quad (5.6\text{-}2)$$

Because the continuity equation for constant ρ does not involve time, the definition in Eq. (5.6-1) works equally well for steady and unsteady flows.

Stream functions can be constructed for any orthogonal, curvilinear coordinate system [Deen (2012, pp. 255–256)]. Defining equations for the two planar and two axisymmetric cases are given in Table 5.4.

Streamlines and streaklines

What makes the stream function valuable for flow visualization is that constant values of ψ correspond to *streamlines*. A streamline is a curve that is tangent everywhere to the velocity vector. The circle in Fig. 5.6 is a particularly simple streamline. That curves of constant ψ correspond to streamlines can be seen by recalling that, in a steady flow, the rate of change of f seen by an observer moving at the local fluid velocity is $\mathbf{v} \cdot \nabla f$ (Section 5.3). Using Eq. (5.6-1),

$$\mathbf{v} \cdot \nabla \psi = v_x \frac{\partial \psi}{\partial x} + v_y \frac{\partial \psi}{\partial y} = \left(\frac{\partial \psi}{\partial y}\right)\frac{\partial \psi}{\partial x} + \left(-\frac{\partial \psi}{\partial x}\right)\left(\frac{\partial \psi}{\partial y}\right) = 0. \quad (5.6\text{-}3)$$

Thus, ψ is constant along a given streamline, as stated. If a set of streamlines is calculated, each corresponding to a different value of ψ, the flow pattern can be visualized using a contour plot, as will be illustrated shortly.

By definition, there is no flow normal to (crossing) any streamline. Accordingly, an impermeable surface always coincides with a streamline and corresponds to some constant value of ψ. The same is true for a symmetry plane.

In general, the path that a material point follows is a *streakline*. Whereas in steady flows streaklines are the same as streamlines, in time-dependent flows they are not. The distinction stems from the fact that streamlines provide only an instantaneous "snapshot" of the velocity field. A velocity that is changing with time as well as position will alter

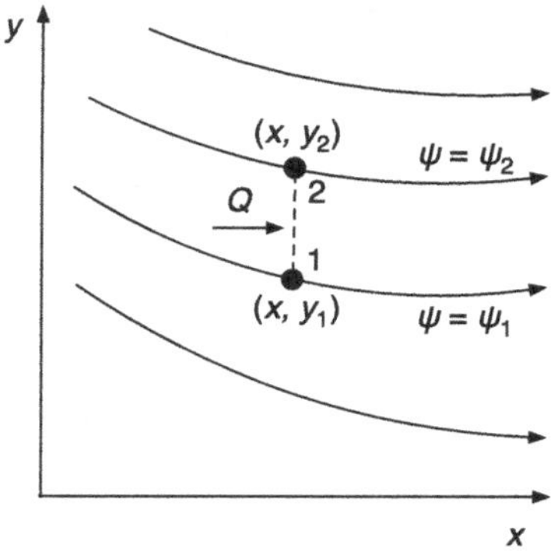

Figure 5.7 Hypothetical streamline plot, showing the coordinates of points 1 and 2 on adjacent streamlines.

the streamline pattern from moment to moment. Accordingly, in an unsteady flow a material point does not have time to trace out the streamline it resides on at a given instant. For this reason, time-exposure photographs showing the paths of bubbles or reflective particles are easiest to interpret if the flow is steady, in which case the streaks in the photos correspond to streamlines that might be calculated theoretically. Many attractive and informative images have been generated in this manner (Van Dyke, 1982).

A streamline plot indicates relative flow speed as well as direction. If the interval in ψ between each pair of streamlines is the same, the distance between streamlines is inversely proportional to the local velocity. This can be seen by considering points on adjacent streamlines, as shown in Fig. 5.7. The coordinates of points 1 and 2 are (x, y_1) and (x, y_2), respectively. If the dimension in the z direction is W, the total flow rate across the plane shown by the dashed line is

$$Q = W \int_{y_1}^{y_2} v_x \, dy = W \int_{y_1}^{y_2} \frac{\partial \psi}{\partial y} \, dy = W(\psi_2 - \psi_1) \tag{5.6-4}$$

where ψ_1 and ψ_2 are the values of ψ on the streamlines. Thus, if the increment in ψ is constant, the flow rate between any pair of streamlines is the same.[3] Closer spacings (smaller areas for flow) must then correspond to larger velocities. Incidentally, the arrows on the streamlines in Fig. 5.7, which indicate that $v_x = \partial\psi/\partial y > 0$, correspond to $\psi_2 > \psi_1$.

Example 5.6-1 Streamlines from the stream function. It is desired to plot streamlines for the planar stagnation flow in Example 5.2-1, where it was shown that $v_x = Cx$ and $v_y = -Cy$ for a fluid in the space $y > 0$, with C a constant. Equation (5.6-1) indicates that $\partial\psi/\partial y = Cx$. Integrating gives

$$\psi(x, y) = Cxy + f(x). \tag{5.6-5}$$

Because we began with a partial derivative, the integration "constant" might depend on x and is denoted as $f(x)$. Similarly, Eq. (5.6-1) indicates that $\partial\psi/\partial x = Cy$ and integration leads to

$$\psi(x, y) = Cxy + g(y) \tag{5.6-6}$$

[3] Constant values of ψ actually correspond to surfaces rather than curves. The curves in a streamline plot are the intersections of those surfaces with a plane, such as the x–y plane. Although "stream surface" is more precise and is sometimes used, it is common to call the result of holding ψ constant a streamline.

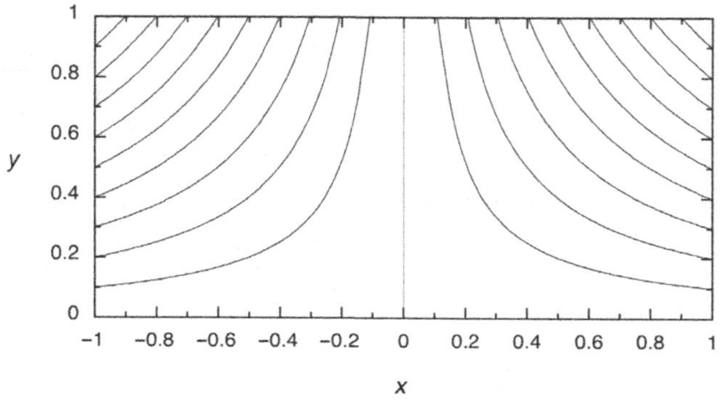

Figure 5.8 Streamlines for planar stagnation flow.

where $g(y)$ is another function. The two expressions for $\psi(x, y)$ can be reconciled only if $f(x) = g(y)$. Because they are functions of different variables, f and g can be equal only if both are constant. As mentioned earlier, there is always an additive constant in ψ that can be chosen freely. Choosing $f = g = 0$, the stream function is

$$\psi(x, y) = Cxy. \tag{5.6-7}$$

Notice that the planes $x = 0$ and $y = 0$ each correspond to the same constant value of ψ, which we have chosen as zero. In other words, both are part of a streamline for which $\psi = 0$. As mentioned earlier, ψ is constant on any symmetry plane or impermeable surface. Because ψ is a continuous function of position and the surfaces $x = 0$ (symmetry plane) and $y = 0$ (impermeable surface) intersect at the origin, ψ must equal the same constant on both surfaces.

To plot streamlines we choose a set of values ψ_n such that $\Delta\psi = \psi_{n+1} - \psi_n$ is constant. For $\Delta\psi = 0.1C$, the equation for the first streamline to the right of the y axis is $y = 0.1/x$, that for the second one is $y = 0.2/x$, and so on. The first streamline to the left of the y axis is described by $y = -0.1/x$. Figure 5.8 shows such streamlines. For $C > 0$, the flow direction is downward and out toward the sides; for $C < 0$, it is in from the sides and upward. The origin, where both velocity components vanish, is a *stagnation point*. The central streamline bifurcates there into branches for negative and positive x.

Trajectories

A streakline, the path followed by a material point, is a kind of trajectory. The correspondence between streaklines and streamlines in steady flows provides another way to calculate the latter. If we follow the movement of a material point for a certain period, then time is a variable even though the flow is steady. Consider a planar flow and suppose that at $t = 0$ the coordinates of the point are (x_0, y_0). Its coordinates at later times will be governed by

$$\frac{dx}{dt} = v_x(x, y), \quad x(0) = x_0 \tag{5.6-8}$$

$$\frac{dy}{dt} = v_y(x, y), \quad y(0) = y_0. \tag{5.6-9}$$

Solving these equations for $x(t)$ and $y(t)$ allows construction of $y(x)$, the function that describes the streamline that passes through (x_0, y_0).

In that the flows under consideration are steady, introducing t as a variable is artificial. Indeed, the final function $y(x)$ is independent of how fast or slow the material point traces out its streamline. Thus, arc length or anything proportional to arc length would serve as well as t in following the movement of the point through space.

Example 5.6-2 Streamlines from trajectories. The objective is to use Eqs. (5.6-8) and (5.6-9) to define the streamlines for a planar stagnation flow.

With $v_x = Cx$ and $v_y = -Cy$, the trajectory of a point that starts at (x_0, y_0) is governed by

$$\frac{dx}{dt} = Cx, \quad x(0) = x_0 \tag{5.6-10}$$

$$\frac{dy}{dt} = -Cy, \quad y(0) = y_0. \tag{5.6-11}$$

These differential equations happen to be uncoupled and are solved independently as

$$\frac{x(t)}{x_0} = e^{Ct}, \quad \frac{y(t)}{y_0} = e^{-Ct}. \tag{5.6-12}$$

Multiplying the results eliminates the variable t and gives

$$xy = x_0 y_0 \tag{5.6-13}$$

as the equation for the streamline. As with Eq. (5.6-7), this shows that the product xy is constant along any streamline.

5.7 CONCLUSION

The most important kinematic consideration is that the velocity field be consistent with conservation of mass. Conservation of mass at any point in a fluid is embodied in the continuity equation [Eq. (5.2-3) and Table 5.1], which describes how spatial variations in the velocity vector ($\mathbf{v}$) are related to changes in density (ρ). For an incompressible fluid (constant ρ), an idealization that is usually applicable in chemical engineering, the continuity equation simplifies to $\nabla \cdot \mathbf{v} = 0$. Among its uses, the incompressible continuity equation permits calculation of a missing velocity component when the others are known. A material point is a small element of fluid that retains its identity as it moves with the flow. Rates of change of various quantities along the path of a material point are needed often in the analysis of fluid flow and other transport processes. The material derivative [D/Dt, Eq. (5.3-3) and Table 5.2] permits such rates of change to be calculated. It combines in one operator the effects of temporal variations at fixed locations with the effects of spatial variations sampled by a moving observer.

Fluids are materials that deform continuously when subjected to shearing forces. Deformation is evidenced by changes in the distances between pairs of material points. Uniform translation and rigid-body rotation do not involve deformation, whereas simple shear and pure dilatation (changes everywhere in ρ) are two types of motion that do. Rates of strain are the key to quantifying how rapidly a body of fluid is being deformed and, ultimately, describing how the fluid resists deformation. In a simple shear flow, the rate of strain is proportional to (but not equal to) the shear rate introduced in Section 1.2. In pure dilatation, the rate of strain is $(\nabla \cdot \mathbf{v})/3$.

The vorticity ($\mathbf{w} = \nabla \times \mathbf{v}$, Table 5.3) and stream function (ψ, Table 5.4) can each be calculated whenever $\mathbf{v}$ is known. Vorticity is a measure of rotational motion. In rigid-body rotation, **w** is constant and its magnitude is twice the angular velocity. In general, **w** varies from point to point. If for any reason $\mathbf{w} = \mathbf{0}$ everywhere, the flow is said to be irrotational. Irrotational flow permits definition of a velocity potential [Eq. (5.5-4)], which can simplify the prediction of **v**. The stream function, which is used with bidirectional, incompressible flows, provides a way to visualize flow patterns. Constant values of ψ correspond to streamlines, which are the paths of material points in steady flows. Contour plots constructed using equally spaced values of ψ indicate relative velocity as well as flow direction.

REFERENCES

Deen, W. M. *Analysis of Transport Phenomena*, 2nd ed. Oxford University Press, New York, 2012.
Van Dyke, M. *An Album of Fluid Motion*. Parabolic Press, Stanford, CA, 1982.

PROBLEMS

5.1. Flow past a bubble Consider steady flow at velocity U relative to a spherical bubble of radius R, as in Fig. P5.1. The flow is axisymmetric (independent of ϕ and with $v_\phi = 0$) and $UR/\nu \to 0$ in both fluids. The Hadamard–Rybczyński analysis predicts that

$$v_\theta(r, \theta) = -U \sin\theta \left[\left(\frac{r}{R}\right)^2 - \frac{1}{2} \right] \qquad \text{(P5.1-1)}$$

inside the bubble. Determine $v_r(r, \theta)$ inside $(0 \le r \le R)$. Note that $v_r(R, \theta) = 0$ for a bubble of fixed size.

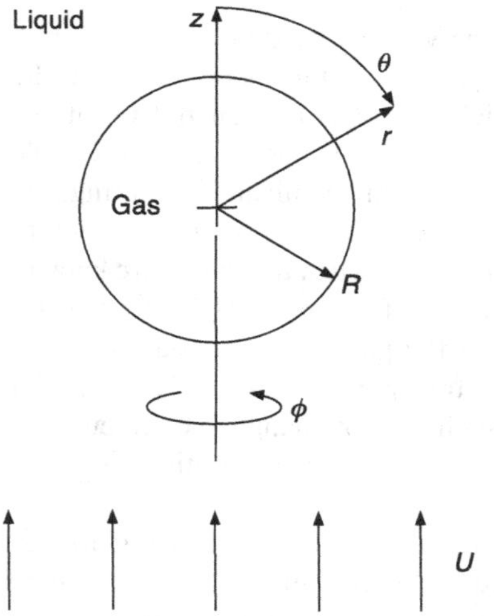

Figure P5.1 Flow past a spherical bubble.

5.2. Channel with wavy walls Consider steady flow in the two-dimensional channel in Fig. P5.2, in which the half-height varies as

$$h(x) = h_0 \left(1 + \varepsilon \sin \frac{2\pi x}{L} \right). \qquad \text{(P5.2-1)}$$

Assume that $q = Q/W$ is known, where Q is the volume flow rate and W is the channel width.

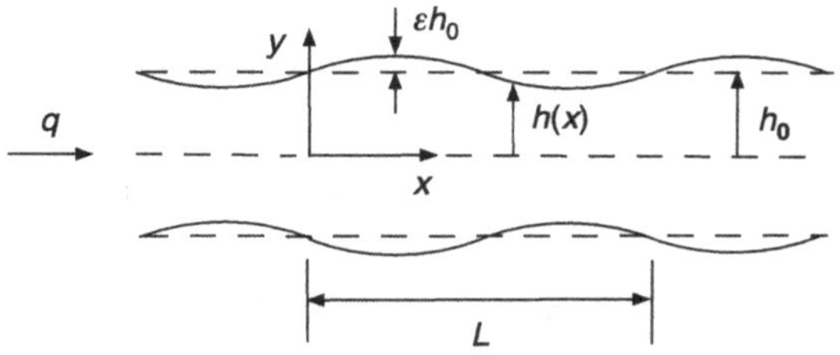

Figure P5.2 Channel with wavy walls.

(a) Relate $u(x)$, the mean velocity in the x direction, to q and $h(x)$.
(b) If the height variations are gradual enough that $2\pi\varepsilon h_0/L \ll 1$, it can be shown using the methods of Section 8.2 that

$$v_x(x, y) = \frac{3}{2}u(x)\left[1 - \left(\frac{y}{h(x)}\right)^2\right]. \qquad \text{(P5.2-2)}$$

Derive an expression for $v_y(x, y)$ in terms of q and $h(x)$. Note that there is reflective symmetry about the plane $y = 0$.

5.3. Condensation on a vertical wall Suppose that a vapor condenses at a constant rate on a cold vertical wall, as shown in Fig. P5.3. As the liquid runs down the wall, the film thickness $\delta(x)$ and mean velocity $u(x)$ both increase. Assume that the condensation begins at $x = 0$, such that $u(0) = 0 = \delta(0)$. The increase in volume flow from position x to $x + \Delta x$ is $v_c W \Delta x$, where the condensation velocity v_c is given and W is the width of the wall.

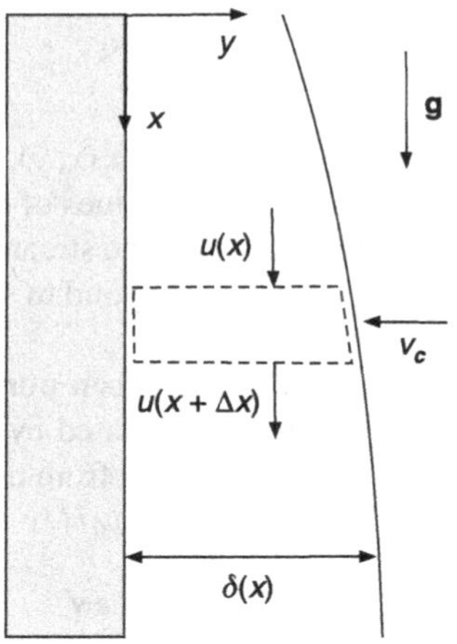

Figure P5.3 Condensation on a vertical wall.

(a) Use a shell balance to relate $\delta(x)$ and $u(x)$ to v_c.
(b) If $d\delta/dx \ll 1$, it is found using the methods in Section 8.2 that the downward velocity is

$$v_x(x, y) = \frac{g\delta^2(x)}{2\nu}\left[2\left(\frac{y}{\delta(x)}\right) - \left(\frac{y}{\delta(x)}\right)^2\right]. \qquad \text{(P5.3-1)}$$

Use this and the result from part (a) to find $u(x)$ and $\delta(x)$.
(c) Determine $v_y(x, y)$.

5.4. Flow past a solid sphere For flow past a solid sphere with coordinates as in Fig. P5.1 and with $UR/\nu \to 0$,

$$v_r(r,\theta) = U\cos\theta\left[1 - \frac{3}{2}\left(\frac{R}{r}\right) + \frac{1}{2}\left(\frac{R}{r}\right)^3\right] \tag{P5.4-1}$$

$$v_\theta(r,\theta) = -U\sin\theta\left[1 - \frac{3}{4}\left(\frac{R}{r}\right) - \frac{1}{4}\left(\frac{R}{r}\right)^3\right] \tag{P5.4-2}$$

as shown in Example 8.3-2. Show that this is *not* an irrotational flow. Where is the vorticity largest and where is it smallest?

5.5. Wedge flow Consider two-dimensional flow near a solid wedge with angle $\beta\pi$, as in Fig. P5.5. The flow along the top of this indefinitely long object can be described using the (x, y) coordinates shown. If the fluid has negligible viscosity, then for $x \geq 0$ it is found that

$$v_x(x) = Cx^m \tag{P5.5-1}$$

where C is a constant and $m = \beta/(2-\beta)$.

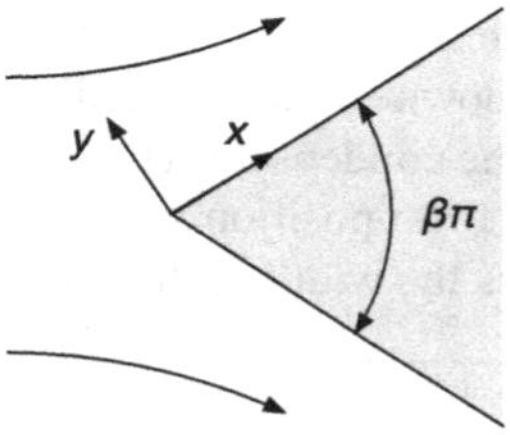

Figure P5.5 Flow past a wedge.

(a) Determine $v_y(x, y)$.
(b) For what values of m and β (if any) is such a flow irrotational?
(c) Calculate the stream function $\psi(x, y)$ for $m = 1/3$ (a 90° wedge) and plot streamlines that correspond to several values of ψ/C.

5.6. Flow between porous and solid disks Suppose that two disks of indefinitely large radius are separated by a distance H, as in Fig. P5.6. One is porous and the other solid. Fluid is injected from the porous disk into the intervening space at a constant velocity v_0, such that $\mathrm{Re} = v_0 H/\nu \ll 1$. The radial velocity averaged over the height H is $u(r)$.

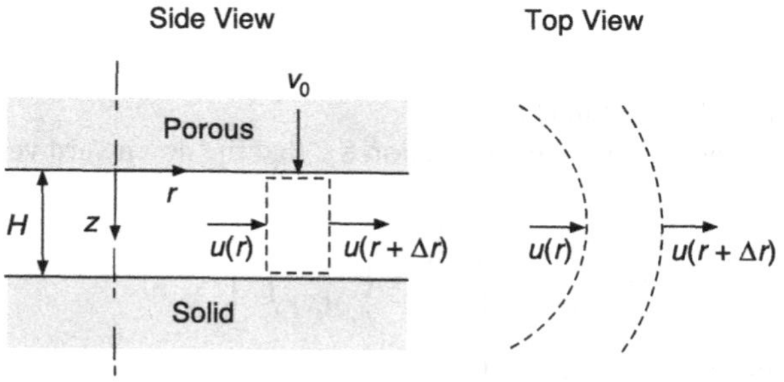

Figure P5.6 Flow between porous and solid disks.

(a) Use a shell balance to determine $u(r)$.
(b) It is shown in Example 8.3-1 that

$$v_r(r,z) = \frac{3v_0 r}{H}\left[\left(\frac{z}{H}\right) - \left(\frac{z}{H}\right)^2\right] \quad \text{(P5.6-1)}$$

$$v_z(z) = v_0\left[1 - 3\left(\frac{z}{H}\right)^2 + 2\left(\frac{z}{H}\right)^3\right]. \quad \text{(P5.6-2)}$$

Determine the stream function and plot streamlines corresponding to several values of $\psi/(H^2 v_0)$. The expression for ψ is simplest if the dimensionless coordinates $\eta = r/H$ and $\zeta = z/H$ are used.

5.7. Trajectories of sedimenting particles Suppose there is steady, laminar flow of a suspension of spherical particles through a parallel-plate channel, as in Fig. P5.7. The plate spacing is h, the channel length is L, the mean fluid velocity along the channel is u, and each particle has a constant downward velocity U. As can be shown using the methods of Chapter 7, the local fluid velocity is

$$v_x(y) = 6u\left[\left(\frac{y}{h}\right) - \left(\frac{y}{h}\right)^2\right], \quad v_y = 0 = v_z. \quad \text{(P5.7-1)}$$

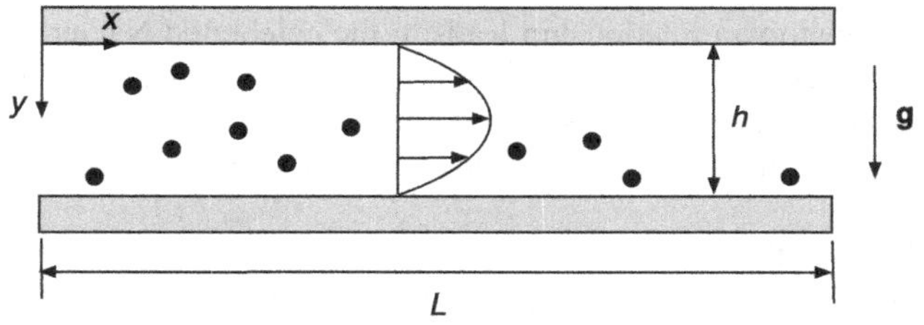

Figure P5.7 Particles settling in a parallel-plate channel. The particle size is greatly exaggerated.

(a) Confirm that u is indeed the average of v_x for $0 \leq y \leq h$.
(b) Calculate the trajectory of a point-size particle that enters the channel at a distance y_0 from the top. That is, determine $x(t)$ and $y(t)$ when $x(0) = 0$ and $y(0) = y_0$.
(c) Suppose that L is just large enough to allow all particles to reach the bottom before the suspension exits the channel. Plot the trajectories for two entering heights, $y_0/h = 0$ and $y_0/h = 1/2$.
(d) A simpler but less accurate approach is to assume that all particles move horizontally at the mean velocity u (as in Example 3.3-2). Compare the resulting trajectories with the ones from part (c). How would the simpler approach affect the calculated minimum value of L needed for all particles to settle?

6

Stress and momentum

6.1 INTRODUCTION

In Chapter 5 the continuity equation was derived and several other aspects of fluid motion were discussed, including the concept of rate of strain. The main task in completing the derivation of the basic equations of fluid dynamics is to adapt Newton's second law to fluids. That requires a much more general description of viscous stresses than in Chapter 1. A more precise notation for stresses is introduced now and an expression is derived for the force per unit volume at any point in a fluid. That is combined with a description of inertia to express conservation of linear momentum at any point. Constitutive equations to evaluate viscous stresses are presented. When combined with conservation of momentum, the Newtonian relationship leads to the celebrated Navier–Stokes equation. The chapter concludes with general discussions of the mechanical conditions at interfaces (which provide boundary conditions for the momentum and continuity equations) and force calculations. Applications of the basic equations to specific kinds of flow begin in Chapter 7.

Familiarity is needed now with tensor representation (Section A.2), vector–tensor dot products (Section A.3), and differential operations involving tensors (Section A.4).

6.2 STRESS VECTOR AND STRESS TENSOR

The stress, or force per unit area, is a vector that varies from point to point on a surface. Imagine dividing a force vector by the area of the surface on which the force acts and letting the area approach zero. Stresses are due to the additive effects of pressure and viscosity. Although when comparing forces in Section 1.3 it was useful to infer stress scales for gravity, surface tension, and inertia, those are not surface forces. Gravity acts on a mass or volume of fluid (making it a *body force*), surface tension acts on a contour within an interface, and inertia is not a force at all, but rather an equivalent rate of change of momentum.

Beyond its properties discussed in Section 4.2, little more need be said about pressure. However, the challenge provided by viscosity is that each of its contributions to the total stress is associated with two directions. There is, of course, the direction of the resultant force. But, because a stress is a force per unit area, implicit also is the choice of an imaginary *test surface* within the fluid. The test-surface concept provides a way to describe how stresses are transmitted within fluids and how they are felt at actual interfaces. A stress is undefined until the orientation of the test surface is stated, which is done by specifying a unit vector **n** that is normal to the surface. This need to keep track of two directions at once (those of the resultant force and of the unit normal **n**) is what motivates the use of tensors.

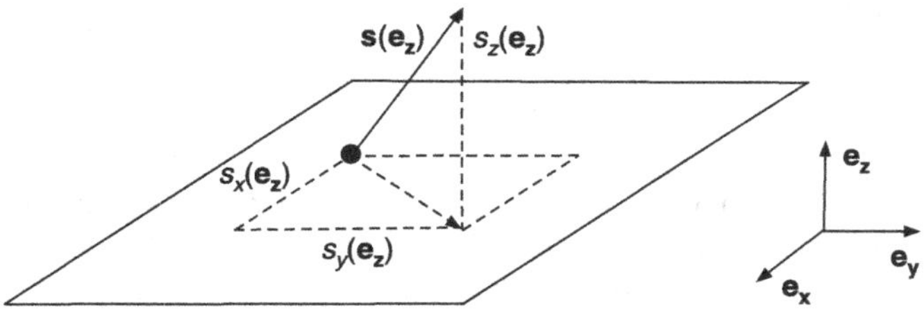

Figure 6.1 Stress notation for a test surface that is normal to the z axis.

That stresses depend on surface orientation is evident already from the properties of pressure. If **n** points toward the fluid exerting the force, the pressure force acts in the $-\mathbf{n}$ direction. Stated another way, the stress vector arising from the pressure P is $-\mathbf{n}P$. The *magnitude* of this vector is the same for any surface orientation, but its *direction* is unknown until **n** is specified. Because all directional information for pressure happens to be provided by **n**, the dual-directional nature of stresses did not arise in Chapter 4. What is needed now is an efficient way to acknowledge that the stress at an arbitrary test surface can act in any direction.

Stress notation

Consider a point located on a test surface that is normal to $\mathbf{e_z}$, as shown in Fig. 6.1. The total stress exerted by the fluid at larger z is denoted as $\mathbf{s}(\mathbf{e_z})$; this is the *stress vector*. The functional argument $\mathbf{e_z}$ is a reminder that we are concerned with a z surface (one where z is constant). The stress vector is shown by the solid arrow and its projection onto the test surface is the dashed arrow. The stress vector has Cartesian components $s_x(\mathbf{e_z})$, $s_y(\mathbf{e_z})$, and $s_z(\mathbf{e_z})$, as shown by the dashed lines. The components of this particular stress vector are all positive. That is, the fluid above the surface is pulling upward on it (in the $+z$ direction), as well as in the $+x$ and $+y$ directions. The z component is a *normal stress* and the others are *shear stresses*.

A test surface could be perpendicular to any coordinate axis. In general, stress vectors for such coordinate surfaces are defined as

$$\mathbf{s}(\mathbf{e_i}) = \text{force/area on an } i \text{ surface that is exerted by the fluid on the side toward which } \mathbf{e_i} \text{ points.} \tag{6.2-1}$$

As emphasized already, the surface orientation is a crucial part of the definition of a stress. That is, even if evaluated at the same point in a fluid, the stress vectors for test surfaces facing different directions are generally unequal. In other words, $\mathbf{s}(\mathbf{e_x}) \neq \mathbf{s}(\mathbf{e_y}) \neq \mathbf{s}(\mathbf{e_z})$. Also important is the side of the test surface on which the stress is computed. Choosing a side gives each stress component a definite algebraic sign, as will be seen.

While general and precise, the $\mathbf{s}(\mathbf{e_i})$ notation soon becomes awkward. To describe fluid stresses more conveniently, we define a 3×3 array with elements $\sigma_{ij} = s_j(\mathbf{e_i})$. Notice that in σ_{ij} the first subscript refers to the orientation of the test surface and the second refers to the direction of the resultant force. We can speak of σ_{ij} as the j component of the stress on an i surface. The stresses for Cartesian surfaces are then written in component form as

$$\mathbf{s}(\mathbf{e_x}) = \sigma_{xx}\mathbf{e_x} + \sigma_{xy}\mathbf{e_y} + \sigma_{xz}\mathbf{e_z} \tag{6.2-2}$$

$$\mathbf{s}(\mathbf{e_y}) = \sigma_{yx}\mathbf{e_x} + \sigma_{yy}\mathbf{e_y} + \sigma_{yz}\mathbf{e_z} \tag{6.2-3}$$

$$\mathbf{s}(\mathbf{e_z}) = \sigma_{zx}\mathbf{e_x} + \sigma_{zy}\mathbf{e_y} + \sigma_{zz}\mathbf{e_z}. \tag{6.2-4}$$

These three relationships may be summarized as

$$\mathbf{s}(\mathbf{e_i}) = \sum_j \sigma_{ij}\mathbf{e_j} \tag{6.2-5}$$

where i and $j = x, y$, or z. Thus, knowing the nine quantities σ_{ij} permits us to evaluate the stresses on all three coordinate surfaces. It will be shown that the same set of nine scalars can be used to calculate stresses on arbitrarily tilted surfaces. In other words, the set of numbers σ_{ij} completely describes the state of stress at a given point in a fluid, at a given instant. These nine scalars are the components of the *total stress tensor*, $\boldsymbol{\sigma}$.

Stress at an arbitrary surface

One of the dot products of a vector $\mathbf{a}$ and tensor $\boldsymbol{\tau}$ is

$$\mathbf{a} \cdot \boldsymbol{\tau} = \sum_i \sum_j a_i \tau_{ij} \mathbf{e_j}. \tag{6.2-6}$$

If $\mathbf{a} = \mathbf{e_i}$ and $\boldsymbol{\tau} = \boldsymbol{\sigma}$, this is the same as Eq. (6.2-5). (The component of $\mathbf{e_i}$ in the i direction is unity and the others are zero, thereby eliminating the sum over i.) Accordingly, the stress vector for any coordinate surface is related to the stress tensor as

$$\mathbf{s}(\mathbf{e_i}) = \mathbf{e_i} \cdot \boldsymbol{\sigma}. \tag{6.2-7}$$

This is not only a much more concise way to write Eqs. (6.2-2)–(6.2-4), it can be generalized to a surface having any orientation. If $\mathbf{n}$ is a unit normal directed *toward* the fluid that is exerting the force, then

$$\mathbf{s}(\mathbf{n}) = \mathbf{n} \cdot \boldsymbol{\sigma}. \tag{6.2-8}$$

The proof of this key relationship relies on the concept of local stress equilibrium, which follows in turn from conservation of linear momentum. It is given at the end of Section 6.4.

Another general result is that the stress vectors at the front and back of a test surface are related as

$$\mathbf{s}(-\mathbf{n}) = -\mathbf{s}(\mathbf{n}). \tag{6.2-9}$$

Thus, each component of the stress exerted on one side (unit normal $-\mathbf{n}$ pointing into the fluid) is equal in magnitude but opposite in sign to that exerted on the other side (unit normal $\mathbf{n}$ pointing into the fluid). For the example in Fig. 6.1, where all components of $\mathbf{s}(\mathbf{e_z})$ are positive, all components of $\mathbf{s}(-\mathbf{e_z})$ (not shown) are negative. The proof of Eq. (6.2-9) again relies on local stress equilibrium and is given at the end of Section 6.4.

6.3 FORCE AT A POINT

Describing conservation of momentum at any point in a fluid requires an expression for the total force per unit volume at a point. As seen already in Section 4.2, the gravitational force per unit volume is $\rho\mathbf{g}$. The contribution of pressure and viscous stresses can be derived by returning again to the small fluid cube in Fig. 4.1. The total force exerted by the surrounding fluid on the control surface is denoted as $\Delta\mathbf{F_S}$. This will be expressed in terms of the stress vector and then the stress tensor. The x component of the force on the six surfaces is

$$\begin{aligned}\Delta F_{Sx} &= s_x(\mathbf{e_x})|_{x+\Delta x}\,\Delta y\,\Delta z + s_x(-\mathbf{e}_x)|_x\,\Delta y\,\Delta z \\ &+ s_x(\mathbf{e_y})|_{y+\Delta y}\,\Delta x\,\Delta z + s_x(-\mathbf{e_y})|_y\,\Delta x\,\Delta z \\ &+ s_x(\mathbf{e_z})|_{z+\Delta z}\,\Delta x\,\Delta y + s_x(-\mathbf{e_z})|_z\,\Delta x\,\Delta y.\end{aligned} \tag{6.3-1}$$

Notice that in each term the unit vector describing the orientation of the surface points *outward*, as needed to compute the force that the surrounding fluid exerts on the control surface.

It follows from Eqs. (6.2-7) and (6.2-9) that

$$s_x(\mathbf{e_i}) = \mathbf{e_x}\cdot(\mathbf{e_i}\cdot\boldsymbol{\sigma}) = \sigma_{ix} \tag{6.3-2}$$

$$s_x(-\mathbf{e_i}) = -s_x(\mathbf{e_i}) = -\sigma_{ix}. \tag{6.3-3}$$

Equation (6.3-1) can then be rewritten in terms of the stress tensor as

$$\begin{aligned}\Delta F_{Sx} &= \sigma_{xx}|_{x+\Delta x}\,\Delta y\,\Delta z - \sigma_{xx}|_x\,\Delta y\,\Delta z \\ &+ \sigma_{yx}|_{y+\Delta y}\,\Delta x\,\Delta z - \sigma_{yx}|_y\,\Delta x\,\Delta z \\ &+ \sigma_{zx}|_{z+\Delta z}\,\Delta x\,\Delta y - \sigma_{zx}|_z\,\Delta x\,\Delta y.\end{aligned} \tag{6.3-4}$$

The x component of the *surface force per unit volume* is $\Delta F_{Sx}/\Delta V$, where $\Delta V = \Delta x\,\Delta y\,\Delta z$. Dividing by the volume and letting all lengths approach zero gives

$$\lim_{\Delta V\to 0}\frac{\Delta F_{Sx}}{\Delta V} = \frac{\partial\sigma_{xx}}{\partial x} + \frac{\partial\sigma_{yx}}{\partial y} + \frac{\partial\sigma_{zx}}{\partial z}. \tag{6.3-5}$$

Doing the same for the y and z components, the complete surface force per unit volume is

$$\begin{aligned}\lim_{\Delta V\to 0}\frac{\Delta \mathbf{F_S}}{\Delta V} &= \left(\frac{\partial\sigma_{xx}}{\partial x} + \frac{\partial\sigma_{yx}}{\partial y} + \frac{\partial\sigma_{zx}}{\partial z}\right)\mathbf{e_x} \\ &+ \left(\frac{\partial\sigma_{xy}}{\partial x} + \frac{\partial\sigma_{yy}}{\partial y} + \frac{\partial\sigma_{zy}}{\partial z}\right)\mathbf{e_y} \\ &+ \left(\frac{\partial\sigma_{xz}}{\partial x} + \frac{\partial\sigma_{yz}}{\partial y} + \frac{\partial\sigma_{zz}}{\partial z}\right)\mathbf{e_z}.\end{aligned} \tag{6.3-6}$$

The right-hand side of Eq. (6.3-6) is the divergence of the stress tensor. This allows us to write the surface force per unit volume much more simply as

$$\lim_{\Delta V\to 0}\frac{\Delta \mathbf{F_S}}{\Delta V} = \nabla\cdot\boldsymbol{\sigma}. \tag{6.3-7}$$

As already mentioned, the gravitational force per unit volume is $\rho\mathbf{g}$. Thus, if $\mathbf{\Delta F}$ is the *total* force on the differential control volume, the force per unit volume at a point is

$$\lim_{\Delta V\to 0}\frac{\Delta \mathbf{F}}{\Delta V} = \rho\mathbf{g} + \nabla\cdot\boldsymbol{\sigma}. \tag{6.3-8}$$

This is the main result we were seeking, although we will go a bit further to separate the total stress into its pressure and viscous parts.

The static pressure equation derived in Section 4.2 helps identify the contribution of pressure to $\nabla\cdot\boldsymbol{\sigma}$. From Eq. (4.2-8), for a static fluid,

$$\mathbf{0} = \rho\mathbf{g} - \nabla P. \tag{6.3-9}$$

The right-hand side is the net force per unit volume due to pressure and gravity, which in a static fluid is zero. A comparison of Eqs. (6.3-8) and (6.3-9) reveals that in a static fluid $\nabla \cdot \boldsymbol{\sigma} = -\nabla P$. When there is flow, the form of the pressure force is unchanged, but there will be an additional contribution to $\nabla \cdot \boldsymbol{\sigma}$ from viscous stresses. Denoting the viscous part as $\nabla \cdot \boldsymbol{\tau}$, where $\boldsymbol{\tau}$ is the *viscous stress tensor*, Eq. (6.3-8) is rewritten as

$$\lim_{\Delta V \to 0} \frac{\Delta \mathbf{F}}{\Delta V} = \rho \mathbf{g} - \nabla P + \nabla \cdot \boldsymbol{\tau}. \tag{6.3-10}$$

The pressure and viscous contributions have now been distinguished.

To relate $\boldsymbol{\sigma}$ itself to P and $\boldsymbol{\tau}$, recall that in a static fluid the stress vector for a surface with unit normal $\mathbf{n}$ is $-\mathbf{n}P$. Thus,

$$\mathbf{s}(\mathbf{n}) = \mathbf{n} \cdot \boldsymbol{\sigma} = -\mathbf{n}P. \tag{6.3-11}$$

The second equality will hold if $\boldsymbol{\sigma} = -P\boldsymbol{\delta}$. That is, because the identity tensor $\boldsymbol{\delta}$ has the property $\mathbf{a} \cdot \boldsymbol{\delta} = \mathbf{a}$ (Section A.3), $\mathbf{n} \cdot (P\boldsymbol{\delta}) = \mathbf{n}P$ and Eq. (6.3-11) is satisfied. Thus, equating the pressure part of the stress tensor with $-P\boldsymbol{\delta}$ neatly encapsulates the mechanical properties of pressure, which are the same for static or flowing fluids. When viscous stresses are present as well, the total stress tensor is

$$\boldsymbol{\sigma} = -P\boldsymbol{\delta} + \boldsymbol{\tau}. \tag{6.3-12}$$

All we have asserted thus far about the viscous stress tensor is that $\boldsymbol{\tau} = \mathbf{0}$ in a static fluid. How to relate $\boldsymbol{\tau}$ to rates of strain and material properties is the subject of Section 6.5. However, the next step (in Section 6.4) is to obtain a momentum equation that is valid for any $\boldsymbol{\sigma}$ or $\boldsymbol{\tau}$. Before proceeding, it is worth noting that Eq. (6.3-12) actually defines $\boldsymbol{\tau}$. That is, in general, $\boldsymbol{\tau}$ is the part of the total stress tensor not due to pressure. In polymeric fluids with a partly elastic character, this excess stress is not purely viscous. Because viscoelastic phenomena are beyond the scope of this book, we will continue to call $\boldsymbol{\tau}$ the *viscous* stress tensor.

6.4 CONSERVATION OF MOMENTUM

The linear momentum of a rigid body is its mass times its translational velocity. In a deforming fluid the velocity varies with position and $\rho\mathbf{v}$ is the linear momentum per unit volume at a given point. The total momentum of a volume of fluid is the integral of $\rho\mathbf{v}$ over that volume.

For a control volume that is fixed in size and position, conservation of linear momentum requires that

$$\text{force} = (\text{rate of change of momentum}) + (\text{net rate of momentum outflow}). \tag{6.4-1}$$

If there were no flow across the control surface, this would be Newton's second law in its most basic form, namely, force equals mass times acceleration. However, in an open system, momentum can be carried in or out of the control volume by flow across its boundaries. A rocket is such a system. Even in a force-free environment (i.e., without gravity or drag), it will accelerate according to the momentum outflow in its exhaust.

Transport of momentum by flow is analogous to transport of mass. The concentration of mass at any point (mass per unit volume) is the density. The rate at which mass crosses an imaginary surface within a fluid equals ρ times the volume flow rate across the surface. Similarly, the concentration of momentum at any point (momentum per unit volume) is $\rho\mathbf{v}$, and the rate of momentum flow across a surface is $\rho\mathbf{v}$ times the volume flow rate.

The force term in Eq. (6.4-1) was derived in Section 6.3. To describe momentum accumulation and outflow, we return yet again to the differential control volume in Fig. 4.1. As with the force derivation, we will consider one component of the momentum vector at a time. The concentration of x momentum is ρv_x and the total amount of x momentum within the differential volume is therefore $\rho v_x \Delta x\, \Delta y\, \Delta z$. The rate of change of x momentum is the derivative of this with respect to time. Including also the flow terms, Eq. (6.4-1) becomes

$$\begin{aligned} \textit{force} = {} & \frac{\partial(\rho v_x)}{\partial t}\,\Delta x\,\Delta y\,\Delta z \\ & + (\rho v_x v_x)|_{x+\Delta x}\,\Delta y\,\Delta z - (\rho v_x v_x)|_x\,\Delta y\,\Delta z \\ & + (\rho v_x v_y)|_{y+\Delta y}\,\Delta x\,\Delta z - (\rho v_x v_y)|_y\,\Delta x\,\Delta z \\ & + (\rho v_x v_z)|_{z+\Delta z}\,\Delta x\,\Delta y - (\rho v_x v_z)|_z\,\Delta x\,\Delta y. \end{aligned} \tag{6.4-2}$$

At positions x, y, and z the flow is inward; hence, the minus signs in the outflow terms. Dividing each term by $\Delta x\,\Delta y\,\Delta z$, letting the volume go to zero, and using Eq. (6.3-8) to evaluate the force per unit volume, the x-momentum equation becomes

$$\rho g_x + \mathbf{e}_x \cdot \nabla \cdot \boldsymbol{\sigma} = \frac{\partial}{\partial t}(\rho v_x) + \frac{\partial}{\partial x}(\rho v_x v_x) + \frac{\partial}{\partial y}(\rho v_x v_y) + \frac{\partial}{\partial z}(\rho v_x v_z). \tag{6.4-3}$$

The dot product with $\mathbf{e}_x$ is needed to obtain the x component of the vector $\nabla \cdot \boldsymbol{\sigma}$.

The right-hand side of Eq. (6.4-3) can be simplified considerably. This is seen by first expanding the various derivatives of products as

$$\begin{aligned} & \frac{\partial}{\partial t}(\rho v_x) + \frac{\partial}{\partial x}(\rho v_x v_x) + \frac{\partial}{\partial y}(\rho v_x v_y) + \frac{\partial}{\partial z}(\rho v_x v_z) \\ & = v_x\left[\frac{\partial \rho}{\partial t} + \frac{\partial}{\partial x}(\rho v_x) + \frac{\partial}{\partial y}(\rho v_y) + \frac{\partial}{\partial z}(\rho v_z)\right] \\ & \quad + \rho\left[\frac{\partial v_x}{\partial t} + v_x\frac{\partial v_x}{\partial x} + v_y\frac{\partial v_x}{\partial y} + v_z\frac{\partial v_x}{\partial z}\right]. \end{aligned} \tag{6.4-4}$$

The Cartesian form of the continuity equation is (from Table 5.1)

$$\frac{\partial \rho}{\partial t} + \frac{\partial}{\partial x}(\rho v_x) + \frac{\partial}{\partial y}(\rho v_y) + \frac{\partial}{\partial z}(\rho v_z) = 0. \tag{6.4-5}$$

This indicates that the first term in square brackets in Eq. (6.4-4) is identically zero. The second term in square brackets is just the material derivative of v_x (Section 5.3). Thus, Eq. (6.4-3) becomes

$$\rho\frac{Dv_x}{Dt} = \rho g_x + \mathbf{e}_x \cdot \nabla \cdot \boldsymbol{\sigma}. \tag{6.4-6}$$

Separating the contributions of pressure and viscous stresses as in Eq. (6.3-10), an equivalent form of the x-momentum equation is

$$\rho\frac{Dv_x}{Dt} = \rho g_x - \frac{\partial P}{\partial x} + \mathbf{e}_x \cdot \nabla \cdot \boldsymbol{\tau}. \tag{6.4-7}$$

The other components of the momentum equation are found simply by replacing x in Eq. (6.4-6) or Eq. (6.4-7) by y or z. A single vector equation is obtained by multiplying

Table 6.1 Cauchy momentum equation in Cartesian coordinates

x component:	$\rho\left[\frac{\partial v_x}{\partial t}+v_x\frac{\partial v_x}{\partial x}+v_y\frac{\partial v_x}{\partial y}+v_z\frac{\partial v_x}{\partial z}\right]=\rho g_x-\frac{\partial P}{\partial x}+\left[\frac{\partial \tau_{xx}}{\partial x}+\frac{\partial \tau_{yx}}{\partial y}+_z\frac{\partial \tau_{zx}}{\partial z}\right]$
y component:	$\rho\left[\frac{\partial v_y}{\partial t}+v_x\frac{\partial v_y}{\partial x}+v_y\frac{\partial v_y}{\partial y}+v_z\frac{\partial v_y}{\partial z}\right]=\rho g_y-\frac{\partial P}{\partial y}+\left[\frac{\partial \tau_{xy}}{\partial x}+\frac{\partial \tau_{yy}}{\partial y}+\frac{\partial \tau_{zy}}{\partial z}\right]$
z component:	$\rho\left[\frac{\partial v_z}{\partial t}+v_x\frac{\partial v_z}{\partial x}+v_y\frac{\partial v_z}{\partial y}+v_z\frac{\partial v_z}{\partial z}\right]=\rho g_z-\frac{\partial P}{\partial z}+\left[\frac{\partial \tau_{xz}}{\partial x}+\frac{\partial \tau_{yz}}{\partial y}+\frac{\partial \tau_{zz}}{\partial z}\right]$

Table 6.2 Cauchy momentum equation in cylindrical coordinates

r component:	$\rho\left[\frac{\partial v_r}{\partial t}+v_r\frac{\partial v_r}{\partial r}+\frac{v_\theta}{r}\frac{\partial v_r}{\partial \theta}-\frac{v_\theta^2}{r}+v_z\frac{\partial v_r}{\partial z}\right]$
	$=\rho g_r-\frac{\partial P}{\partial r}+\left[\frac{1}{r}\frac{\partial}{\partial r}(r\tau_{rr})+\frac{1}{r}\frac{\partial \tau_{\theta r}}{\partial \theta}-\frac{\tau_{\theta\theta}}{r}+\frac{\partial \tau_{zr}}{\partial z}\right]$
θ component:	$\rho\left[\frac{\partial v_\theta}{\partial t}+v_r\frac{\partial v_\theta}{\partial r}+\frac{v_\theta}{r}\frac{\partial v_\theta}{\partial \theta}+\frac{v_r v_\theta}{r}+v_z\frac{\partial v_\theta}{\partial z}\right]$
	$=\rho g_\theta-\frac{1}{r}\frac{\partial P}{\partial \theta}+\left[\frac{1}{r^2}\frac{\partial}{\partial r}(r^2\tau_{r\theta})+\frac{1}{r}\frac{\partial \tau_{\theta\theta}}{\partial \theta}+\frac{\partial \tau_{z\theta}}{\partial z}\right]$
z component:	$\rho\left[\frac{\partial v_z}{\partial t}+v_r\frac{\partial v_z}{\partial r}+\frac{v_\theta}{r}\frac{\partial v_z}{\partial \theta}+v_z\frac{\partial v_z}{\partial z}\right]$
	$=\rho g_z-\frac{\partial P}{\partial z}+\left[\frac{1}{r}\frac{\partial}{\partial r}(r\tau_{rz})+\frac{1}{r}\frac{\partial \tau_{\theta z}}{\partial \theta}+\frac{\partial \tau_{zz}}{\partial z}\right]$

the component equations by the corresponding unit vectors and adding the results. This gives

$$\rho\frac{D\mathbf{v}}{Dt}=\rho\mathbf{g}+\nabla\cdot\boldsymbol{\sigma}=\rho\mathbf{g}-\nabla P+\nabla\cdot\boldsymbol{\tau}. \tag{6.4-8}$$

Either form (with $\boldsymbol{\sigma}$ or with $\boldsymbol{\tau}$) is called the *Cauchy momentum equation*. This fundamental expression of conservation of linear momentum applies to any fluid that can be modeled as a continuum.

The components of the P–$\boldsymbol{\tau}$ form of Eq. (6.4-8) in Cartesian, cylindrical, and spherical coordinates are given in Tables 6.1, 6.2, and 6.3, respectively. The bracketed terms on the left and right sides of each entry are the components of $D\mathbf{v}/Dt$ and $\nabla\cdot\boldsymbol{\tau}$, respectively.

Additional note: stress equilibrium

A stress balance that applies at any point in a fluid is derived now, and the key properties of the stress vector [Eqs. (6.2-8) and (6.2-9)] are shown to be consequences of that balance. Consider a small control volume, fixed in space, which has a linear dimension L and surface area S. Its volume is $V=\alpha L^3$, where α is a dimensionless constant that depends only on its shape. For example, if the control volume is spherical and L is the radius, then $\alpha=4\pi/3$. Conservation of momentum for the control volume as a whole can be expressed by integrating the $\boldsymbol{\sigma}$ form of Eq. (6.4-8) over V, using Eq. (A.4-32) to

Table 6.3 Cauchy momentum equation in spherical coordinates

r component:

$$\rho\left[\frac{\partial v_r}{\partial t}+v_r\frac{\partial v_r}{\partial r}+\frac{v_\theta}{r}\frac{\partial v_r}{\partial \theta}+\frac{v_\phi}{r\sin\theta}\frac{\partial v_r}{\partial \phi}-\frac{v_\theta^2+v_\phi^2}{r}\right]$$

$$=\rho g_r-\frac{\partial P}{\partial r}+\left[\frac{1}{r^2}\frac{\partial}{\partial r}(r^2\tau_{rr})+\frac{1}{r\sin\theta}\frac{\partial}{\partial\theta}(\tau_{\theta r}\sin\theta)+\frac{1}{r\sin\theta}\frac{\partial\tau_{\phi r}}{\partial\phi}-\frac{\tau_{\theta\theta}+\tau_{\phi\phi}}{r}\right]$$

θ component:

$$\rho\left[\frac{\partial v_\theta}{\partial t}+v_r\frac{\partial v_\theta}{\partial r}+\frac{v_\theta}{r}\frac{\partial v_\theta}{\partial \theta}+\frac{v_\phi}{r\sin\theta}\frac{\partial v_\theta}{\partial \phi}+\frac{v_r v_\theta}{r}-\frac{v_\phi^2\cot\theta}{r}\right]$$

$$=\rho g_\theta-\frac{1}{r}\frac{\partial P}{\partial \theta}+\left[\frac{1}{r^2}\frac{\partial}{\partial r}(r^2\tau_{r\theta})+\frac{1}{r\sin\theta}\frac{\partial}{\partial\theta}(\tau_{\theta\theta}\sin\theta)+\frac{1}{r\sin\theta}\frac{\partial\tau_{\phi\theta}}{\partial\phi}+\frac{\tau_{r\theta}}{r}-\frac{\cot\theta}{r}\tau_{\phi\phi}\right]$$

ϕ component:

$$\rho\left[\frac{\partial v_\phi}{\partial t}+v_r\frac{\partial v_\phi}{\partial r}+\frac{v_\theta}{r}\frac{\partial v_\phi}{\partial \theta}+\frac{v_\phi}{r\sin\theta}\frac{\partial v_\phi}{\partial \phi}+\frac{v_\phi v_r}{r}+\frac{v_\theta v_\phi\cot\theta}{r}\right]$$

$$=\rho g_\phi-\frac{1}{r\sin\theta}\frac{\partial P}{\partial \phi}+\left[\frac{1}{r^2}\frac{\partial}{\partial r}(r^2\tau_{r\phi})+\frac{1}{r}\frac{\partial\tau_{\theta\phi}}{\partial\theta}+\frac{1}{r\sin\theta}\frac{\partial\tau_{\phi\phi}}{\partial\phi}+\frac{\tau_{r\phi}}{r}+\frac{2\cot\theta}{r}\tau_{\theta\phi}\right]$$

convert the volume integral of $\nabla\cdot\boldsymbol{\sigma}$ to the surface integral of $\mathbf{n}\cdot\boldsymbol{\sigma}$, and rewriting the surface integral in terms of $\mathbf{s}(\mathbf{n})$. The result is

$$\int_V \rho\frac{D\mathbf{v}}{Dt}\,dV=\int_V \rho\mathbf{g}\,dV+\int_S \mathbf{s}(\mathbf{n})\,dS. \tag{6.4-9}$$

The volume integrals are combined and evaluated as

$$\int_V \rho\left(\frac{D\mathbf{v}}{Dt}-\mathbf{g}\right)dV=\alpha L^3\left\langle\rho\left(\frac{D\mathbf{v}}{Dt}-\mathbf{g}\right)\right\rangle_V \tag{6.4-10}$$

where the angle brackets denote an average over V and use has been made of the fact that any volume integral equals the volume times the mean value of the integrand. Combining these results and dividing by L^2 gives

$$\alpha L\left\langle\rho\left(\frac{D\mathbf{v}}{Dt}-\mathbf{g}\right)\right\rangle_V=\frac{1}{L^2}\int_S \mathbf{s}(\mathbf{n})\,dS. \tag{6.4-11}$$

In the limit $L\to 0$, the left-hand side vanishes. That is because α is constant and the bracketed quantity approaches its value at a point and becomes independent of L. Thus, the left-hand side vanishes in proportion to L. Consequently,

$$\lim_{L\to 0}\frac{1}{L^2}\int_S \mathbf{s}(\mathbf{n})\,dS=0. \tag{6.4-12}$$

This equation represents the *stress equilibrium* at a point. It shows how the stresses must balance if the control volume is small enough.

To prove that the stresses on the two sides of a test surface are equal and opposite [Eq. (6.2-9)], we apply Eq. (6.4-12) to the control volume in Fig. 6.2, in which h/L is arbitrarily small. The top and bottom faces of area L^2 have outward normals $\mathbf{n}$ and $-\mathbf{n}$, respectively. The other four surfaces (the edges) are each of area hL. The integral of

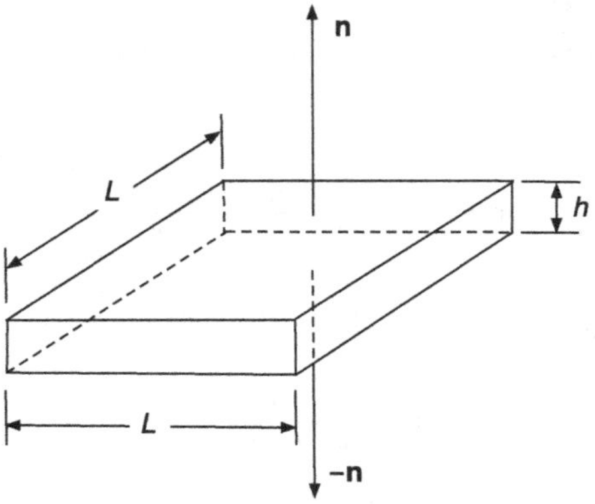

Figure 6.2 Control volume for proving that $\mathbf{s}(\mathbf{n}) = -\mathbf{s}(-\mathbf{n})$.

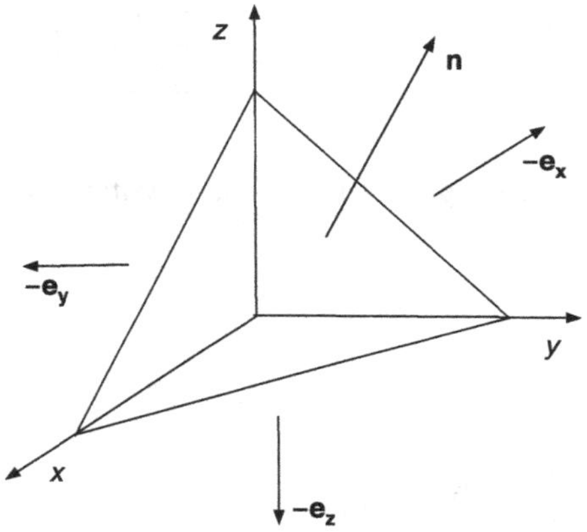

Figure 6.3 Control volume for proving that $\mathbf{s}(\mathbf{n}) = \mathbf{n} \cdot \boldsymbol{\sigma}$.

the stress vector over the complete surface S is evaluated by adding the face and edge contributions:

$$\int_S \mathbf{s}(\mathbf{n})\, dS = \langle \mathbf{s}(\mathbf{n}) \rangle_{S_1} L^2 + \langle \mathbf{s}(-\mathbf{n}) \rangle_{S_2} L^2 + \text{edge contributions} \tag{6.4-13}$$

where S_1 and S_2 are the top and bottom, respectively, and angle brackets now denote area averages. Because $hL \ll L^2$, the edge contributions are negligible. Accordingly,

$$\frac{1}{L^2} \int_S \mathbf{s}(\mathbf{n})\, dS = \langle \mathbf{s}(\mathbf{n}) \rangle_{S_1} + \langle \mathbf{s}(-\mathbf{n}) \rangle_{S_2}. \tag{6.4-14}$$

In the limit $L \to 0$, the stresses become point quantities and angle brackets are no longer needed. Moreover, S_1 and S_2 become the same. Referring to Eq. (6.4-12), we conclude that $\mathbf{s}(\mathbf{n}) = -\mathbf{s}(-\mathbf{n})$.

The tetrahedral control volume in Fig. 6.3 will be used now to derive the general relationship between $\mathbf{s}(\mathbf{n})$ and $\boldsymbol{\sigma}$ [Eq. (6.2-8)]. Three of the surfaces are perpendicular to the coordinate axes and the fourth (tilted) surface has a unit outward normal $\mathbf{n}$ and area ΔA. The length scale L is defined by setting $L^2 = \Delta A$. The area of each coordinate surface can be evaluated as a projection of ΔA onto the plane of interest. For example, the area of the face that is normal to $\mathbf{e_x}$ is $(\mathbf{e_x} \cdot \mathbf{n})L^2$. Evaluating the integral in Eq. (6.4-12) piecewise and omitting the angle brackets for simplicity gives

$$\lim_{L\to 0}\frac{1}{L^2}[\mathbf{s}(\mathbf{n})L^2+\mathbf{s}(-\mathbf{e_x})(\mathbf{e_x}\cdot\mathbf{n})L^2+\mathbf{s}(-\mathbf{e_y})(\mathbf{e_y}\cdot\mathbf{n})L^2+\mathbf{s}(-\mathbf{e_z})(\mathbf{e_z}\cdot\mathbf{n})L^2]=0. \tag{6.4-15}$$

Canceling the L^2 terms and solving for $\mathbf{s}(\mathbf{n})$ yields

$$\mathbf{s}(\mathbf{n})=-(\mathbf{e_x}\cdot\mathbf{n})\mathbf{s}(-\mathbf{e_x})-(\mathbf{e_y}\cdot\mathbf{n})\mathbf{s}(-\mathbf{e_y})-(\mathbf{e_z}\cdot\mathbf{n})\mathbf{s}(-\mathbf{e_z}). \tag{6.4-16}$$

Using Eq. (6.2-9) to eliminate the minus signs, this becomes

$$\mathbf{s}(\mathbf{n})=(\mathbf{e_x}\cdot\mathbf{n})\mathbf{s}(\mathbf{e_x})+(\mathbf{e_y}\cdot\mathbf{n})\mathbf{s}(\mathbf{e_y})+(\mathbf{e_z}\cdot\mathbf{n})\mathbf{s}(\mathbf{e_z}). \tag{6.4-17}$$

Relating $\mathbf{s}(\mathbf{e_i})$ to $\boldsymbol{\sigma}$ using Eq. (6.2-7) and noting that $\mathbf{e_i}\cdot\mathbf{n}=n_i$, we find that

$$\begin{aligned}\mathbf{s}(\mathbf{n})&=(\mathbf{e_x}\cdot\mathbf{n})(\mathbf{e_x}\cdot\boldsymbol{\sigma})+(\mathbf{e_y}\cdot\mathbf{n})(\mathbf{e_y}\cdot\boldsymbol{\sigma})+(\mathbf{e_z}\cdot\mathbf{n})(\mathbf{e_z}\cdot\boldsymbol{\sigma})\\&=\sum_i(\mathbf{e_i}\cdot\mathbf{n})(\mathbf{e_i}\cdot\boldsymbol{\sigma})=\left(\sum_i n_i\mathbf{e_i}\right)\cdot\boldsymbol{\sigma}=\mathbf{n}\cdot\boldsymbol{\sigma}.\end{aligned} \tag{6.4-18}$$

6.5 VISCOUS STRESS

Rate-of-strain tensor

The concepts in Section 5.4 are extended now to obtain a general description of rates of strain in a fluid. Central to this discussion is the *rate-of-strain tensor* ($\boldsymbol{\Gamma}$),

$$\boldsymbol{\Gamma}=\frac{1}{2}[\nabla\mathbf{v}+(\nabla\mathbf{v})^t]. \tag{6.5-1}$$

Displayed as a 3×3 array, the Cartesian form of $\boldsymbol{\Gamma}$ is

$$\boldsymbol{\Gamma}=\begin{bmatrix}\dfrac{\partial v_x}{\partial x} & \dfrac{1}{2}\left(\dfrac{\partial v_y}{\partial x}+\dfrac{\partial v_x}{\partial y}\right) & \dfrac{1}{2}\left(\dfrac{\partial v_z}{\partial x}+\dfrac{\partial v_x}{\partial z}\right)\\ \dfrac{1}{2}\left(\dfrac{\partial v_x}{\partial y}+\dfrac{\partial v_y}{\partial x}\right) & \dfrac{\partial v_y}{\partial y} & \dfrac{1}{2}\left(\dfrac{\partial v_z}{\partial y}+\dfrac{\partial v_y}{\partial z}\right)\\ \dfrac{1}{2}\left(\dfrac{\partial v_x}{\partial z}+\dfrac{\partial v_z}{\partial x}\right) & \dfrac{1}{2}\left(\dfrac{\partial v_y}{\partial z}+\dfrac{\partial v_z}{\partial y}\right) & \dfrac{\partial v_z}{\partial z}\end{bmatrix}. \tag{6.5-2}$$

Notice that this tensor is *symmetric* (i.e., $\Gamma_{ij}=\Gamma_{ji}$). A complementary construct is the *vorticity tensor* ($\boldsymbol{\Omega}$),

$$\boldsymbol{\Omega}=\frac{1}{2}[\nabla\mathbf{v}-(\nabla\mathbf{v})^t]. \tag{6.5-3}$$

Displayed as an array, the Cartesian vorticity tensor is

$$\boldsymbol{\Omega}=\begin{bmatrix}0 & \dfrac{1}{2}\left(\dfrac{\partial v_y}{\partial x}-\dfrac{\partial v_x}{\partial y}\right) & \dfrac{1}{2}\left(\dfrac{\partial v_z}{\partial x}-\dfrac{\partial v_x}{\partial z}\right)\\ \dfrac{1}{2}\left(\dfrac{\partial v_x}{\partial y}-\dfrac{\partial v_y}{\partial x}\right) & 0 & \dfrac{1}{2}\left(\dfrac{\partial v_z}{\partial y}-\dfrac{\partial v_y}{\partial z}\right)\\ \dfrac{1}{2}\left(\dfrac{\partial v_x}{\partial z}-\dfrac{\partial v_z}{\partial x}\right) & \dfrac{1}{2}\left(\dfrac{\partial v_y}{\partial z}-\dfrac{\partial v_z}{\partial y}\right) & 0\end{bmatrix}. \tag{6.5-4}$$

This tensor is *antisymmetric* (i.e., $\Omega_{ij} = -\Omega_{ji}$). Whereas $\mathbf{\Gamma}$ has six independent components, $\mathbf{\Omega}$ has only three, the same as the vorticity vector (Section 5.5).[1]

The rate-of-strain and vorticity tensors have been constructed so that, when added, they give the velocity gradient tensor:

$$\nabla \mathbf{v} = \mathbf{\Gamma} + \mathbf{\Omega}. \tag{6.5-5}$$

This decomposition of $\nabla \mathbf{v}$ into symmetric and antisymmetric parts is unremarkable in itself, in that it can be done with any tensor. However, the piece of $\nabla \mathbf{v}$ represented by $\mathbf{\Gamma}$ turns out to be just what is needed to evaluate $\boldsymbol{\tau}$. Conversely, $\mathbf{\Omega}$ has no bearing on viscous stresses.

The relationship between $\mathbf{\Gamma}$ and rates of strain may be seen by considering a pair of material points in a fluid. Imagine a vector $\mathbf{L}$ that is directed from point 1 to point 2, such that $L(t)$ is the instantaneous distance between the points. A unit vector $\mathbf{p}$ that is parallel to $\mathbf{L}$ is $\mathbf{p} = \mathbf{L}/|\mathbf{L}| = \mathbf{L}/L$. What will be shown is that

$$\frac{1}{L}\frac{dL}{dt} = \mathbf{p} \cdot \mathbf{\Gamma} \cdot \mathbf{p} \tag{6.5-6}$$

for closely neighboring points. In other words, the local rate of strain in any flow can be computed from $\mathbf{\Gamma}$.

The proof of Eq. (6.5-6) begins by recognizing that the velocity difference between two nearby points can be approximated as

$$\Delta \mathbf{v} = \mathbf{v}_2 - \mathbf{v}_1 = \mathbf{L} \cdot \nabla \mathbf{v}|_1 \tag{6.5-7}$$

where the velocity gradient is evaluated at point 1. This comes from a Taylor expansion of the velocity, in which terms involving higher powers of L and higher derivatives of $\mathbf{v}$ have been neglected. As an example of how such a multidimensional expansion works, suppose that at some instant points 1 and 2 are positioned on the z axis at $z = H$ and $z = H + L$, respectively, and that the fluid velocity is given by $\mathbf{v} = az^2\mathbf{e}_x + bz^2\mathbf{e}_y$, where a and b are constants. The exact velocity difference is

$$\mathbf{\Delta v} = [(H + L)^2 - H^2](a\mathbf{e}_x + b\mathbf{e}_y) = (2HL + L^2)(a\mathbf{e}_x + b\mathbf{e}_y). \tag{6.5-8}$$

To obtain $\mathbf{\Delta v}$ from Eq. (6.5-7), $\mathbf{L} = L\mathbf{e}_z$ and $\nabla \mathbf{v}|_1 = 2z(a\mathbf{e}_z\mathbf{e}_x + b\mathbf{e}_z\mathbf{e}_y)|_{z=H} = 2H(a\mathbf{e}_z\mathbf{e}_x + b\mathbf{e}_z\mathbf{e}_y)$. All other components of $\nabla \mathbf{v}$ are zero. The velocity difference from the truncated expansion is

$$\mathbf{\Delta v} = L\mathbf{e}_z \cdot [2H(a\mathbf{e}_z\mathbf{e}_x + b\mathbf{e}_z\mathbf{e}_y)] = 2HL(a\mathbf{e}_x + b\mathbf{e}_y) \tag{6.5-9}$$

which approaches the exact result in Eq. (6.5-8) as $L \to 0$.

Expressing the velocity difference as in Eq. (6.5-7),

$$\mathbf{\Delta v} \cdot \mathbf{L} = \mathbf{L} \cdot \nabla \mathbf{v} \cdot \mathbf{L}. \tag{6.5-10}$$

However, $\mathbf{\Delta v}$ is also the rate of change in $\mathbf{L}$, such that $\mathbf{\Delta v} = d\mathbf{L}/dt$. This gives

$$\mathbf{\Delta v} \cdot \mathbf{L} = \frac{d\mathbf{L}}{dt} \cdot \mathbf{L} = L\frac{dL}{dt}. \tag{6.5-11}$$

[1] Comparing Eq. (6.5-4) with $\mathbf{w} = \nabla \times \mathbf{v}$ reveals that $\Omega_{xy} = w_z/2 = -\Omega_{yx}$, $\Omega_{xz} = -w_y/2 = -\Omega_{zx}$, and $\Omega_{yz} = w_x/2 = -\Omega_{zy}$. Thus, $\mathbf{\Omega}$ and $\mathbf{w}$ contain the same three independent quantities.

Equating the two expressions for $\Delta\mathbf{v} \cdot \mathbf{L}$, dividing by L^2, and using the definitions of $\mathbf{p}$, $\boldsymbol{\Gamma}$, and $\boldsymbol{\Omega}$ leads to

$$\frac{1}{L}\frac{dL}{dt} = \mathbf{p} \cdot \nabla\mathbf{v} \cdot \mathbf{p} = (\mathbf{p} \cdot \boldsymbol{\Gamma} \cdot \mathbf{p}) + (\mathbf{p} \cdot \boldsymbol{\Omega} \cdot \mathbf{p}). \qquad (6.5\text{-}12)$$

All that remains to complete the proof of Eq. (6.5-6) is to realize that $\mathbf{p} \cdot \boldsymbol{\Omega} \cdot \mathbf{p} = 0$, because $\boldsymbol{\Omega}$ is antisymmetric (Example A.3-1). Recalling from Chapter 5 that vorticity is related to rotation and that rotation can occur without deformation, it is not too surprising that the rate of strain in a fluid is determined only by $\boldsymbol{\Gamma}$ and is unaffected by $\boldsymbol{\Omega}$.

Example 6.5-1 Rate of strain in simple shear flow. The objective is to show that the general prescription for calculating rates of strain in Eq. (6.5-6) reduces to what was found in Example 5.4-1.

With reference to Fig. 5.4, $\mathbf{p} = \cos\theta\mathbf{e_x} + \sin\theta\mathbf{e_y}$, $\boldsymbol{\Gamma} = c\mathbf{e_y}\mathbf{e_x}$, and

$$\begin{aligned}\frac{1}{L}\frac{dL}{dt} &= (\cos\theta\mathbf{e_x} + \sin\theta\mathbf{e_y}) \cdot (c\,\mathbf{e_y}\mathbf{e_x}) \cdot (\cos\theta\mathbf{e_x} + \sin\theta\mathbf{e_y}) \\ &= c\cos^2\theta(\mathbf{e_x} \cdot \mathbf{e_y}\mathbf{e_x} \cdot \mathbf{e_x}) + c\sin^2\theta(\mathbf{e_y} \cdot \mathbf{e_y}\mathbf{e_x} \cdot \mathbf{e_y}) \\ &\quad + c\sin\theta\cos\theta[(\mathbf{e_y} \cdot \mathbf{e_y}\mathbf{e_x} \cdot \mathbf{e_x}) + (\mathbf{e_x} \cdot \mathbf{e_y}\mathbf{e_x} \cdot \mathbf{e_y})] \\ &= c\sin\theta\cos\theta\end{aligned}$$

in agreement with Eq. (5.4-4).

We are prepared now to relate $\boldsymbol{\tau}$ to the material properties and motion of various kinds of fluids. The constitutive equations to be presented have two features in common. First, when a fluid is not being deformed, as in uniform translation, rigid-body rotation, or static conditions, $\boldsymbol{\tau} = \mathbf{0}$. Second, $\boldsymbol{\tau}$ is symmetric. As shown at the end of this section, symmetry of $\boldsymbol{\sigma}$ and of $\boldsymbol{\tau}$ is ordinarily implied by conservation of angular momentum.

Newtonian fluids

In Section 1.2 a Newtonian fluid was defined as one in which the shear stress is proportional to the shear rate. The context was a hypothetical experiment involving a simple shear flow. More generally, each component of $\boldsymbol{\tau}$ for a Newtonian fluid is proportional to a rate of strain. Because $\boldsymbol{\Gamma}$ describes rates of strain and, like $\boldsymbol{\tau}$, is symmetric, we might suppose that $\boldsymbol{\tau}$ is simply proportional to $\boldsymbol{\Gamma}$. For a Newtonian fluid of constant density this presumption is correct. Choosing 2μ as the proportionality constant gives

$$\boldsymbol{\tau} = 2\mu\boldsymbol{\Gamma} = \mu[\nabla\mathbf{v} + (\nabla\mathbf{v})^t] \quad (\text{Newtonian, } \rho = \text{constant}). \qquad (6.5\text{-}13)$$

This is the starting point for nearly all viscous-stress calculations in subsequent chapters.

The expression we used originally to define viscosity, $\tau_w = \mu s$ [Eq. (1.2-1)], is obtained by applying Eq. (6.5-13) to simple shear flow. In the idealized experiment in Fig. 1.1, $v_x = sy$ and $v_y = v_z = 0$, where $s = U/H$ is the shear rate, a constant. Evaluating each of the velocity derivatives in Eq. (6.5-2), we find that $\Gamma_{xy} = \Gamma_{yx} = s/2$ and $\Gamma_{ij} = 0$ otherwise. Thus, Eq. (6.5-13) indicates that the nonzero components of $\boldsymbol{\tau}$ are $\tau_{xy} = \tau_{yx} = \mu s$. To see that $\tau_w = \mu s$ is a special case of Eq. (6.5-13), it is necessary then only to recognize that $\tau_w = \tau_{yx}$ at $y = H$.

Equation (6.5-13) must be modified if the Newtonian fluid is not incompressible. As discussed in Section 5.4, deformations due to variations in ρ are qualitatively different from those due to shear. In pure dilatation, the fluid expands or contracts equally in all directions; in simple shear, layers of fluid slide past one another, like a deck of cards being spread on a table. This suggests the possibility of differing amounts of molecular-level

Table 6.4 Viscous stress components for Newtonian fluids in Cartesian coordinates

$$\tau_{xx} = 2\mu\left[\frac{\partial v_x}{\partial x} - \frac{1}{3}(\nabla \cdot \mathbf{v})\right]$$

$$\tau_{yy} = 2\mu\left[\frac{\partial v_y}{\partial y} - \frac{1}{3}(\nabla \cdot \mathbf{v})\right]$$

$$\tau_{zz} = 2\mu\left[\frac{\partial v_z}{\partial z} - \frac{1}{3}(\nabla \cdot \mathbf{v})\right]$$

$$\tau_{xy} = \tau_{yx} = \mu\left[\frac{\partial v_x}{\partial y} + \frac{\partial v_y}{\partial x}\right]$$

$$\tau_{yz} = \tau_{zy} = \mu\left[\frac{\partial v_y}{\partial z} + \frac{\partial v_z}{\partial y}\right]$$

$$\tau_{zx} = \tau_{xz} = \mu\left[\frac{\partial v_z}{\partial x} + \frac{\partial v_x}{\partial z}\right]$$

$$\nabla \cdot \mathbf{v} = \frac{\partial v_x}{\partial x} + \frac{\partial v_y}{\partial y} + \frac{\partial v_z}{\partial z}$$

friction, and thus different proportionality constants, when relating $\boldsymbol{\tau}$ to the corresponding rates of strain. We have seen that $\boldsymbol{\Gamma}$ provides the total rate of strain. As shown in Example 5.4-2, the rate of strain in pure dilatation is one-third the rate of dilatation, or $(1/3)(\nabla \cdot \mathbf{v})$; the corresponding isotropic tensor is $(1/3)(\nabla \cdot \mathbf{v})\boldsymbol{\delta}$. Separating the two kinds of deformation and calling the proportionality constants 2μ and 3κ (by convention), the more general constitutive equation for a Newtonian fluid is

$$\boldsymbol{\tau} = 2\mu\left(\boldsymbol{\Gamma} - \frac{1}{3}(\nabla \cdot \mathbf{v})\boldsymbol{\delta}\right) + 3\kappa\left(\frac{1}{3}(\nabla \cdot \mathbf{v})\boldsymbol{\delta}\right) = 2\mu\boldsymbol{\Gamma} + \left(\kappa - \frac{2}{3}\mu\right)(\nabla \cdot \mathbf{v})\boldsymbol{\delta} \tag{6.5-14}$$

where μ is the usual *shear viscosity* and κ is the *bulk viscosity* or *dilatational viscosity*. In the first equality, in the term multiplied by 2μ, the dilatational rate of strain has been subtracted from the total to avoid double-counting of dilatation.

The effects of κ on fluid dynamics turn out to be difficult to detect. It is found from kinetic theory that $\kappa = 0$ for ideal monatomic gases, and the evidence is that $\kappa \ll \mu$ for other fluids. Further minimizing observable effects of κ is that $\nabla \cdot \mathbf{v}$ is ordinarily quite small. Accordingly, it is customary to set $\kappa = 0$ and simplify the constitutive equation for variable density to

$$\boldsymbol{\tau} = 2\mu\left(\boldsymbol{\Gamma} - \frac{1}{3}(\nabla \cdot \mathbf{v})\boldsymbol{\delta}\right) \quad \text{(Newtonian, } \rho \neq \text{constant).} \tag{6.5-15}$$

If ρ is constant, our default assumption in examples and problems, $\nabla \cdot \mathbf{v} = 0$ and Eq. (6.5-15) reduces to Eq. (6.5-13).

The components of τ from Eq. (6.5-15) in Cartesian, cylindrical, and spherical coordinates are shown in Tables 6.4, 6.5, and 6.6, respectively. Except for τ_{ii}, the components of Eq. (6.5-13) are the same. Setting $\nabla \cdot \mathbf{v} = 0$ in the table entries yields τ_{ii} from Eq. (6.5-13).

Table 6.5 Viscous stress components for Newtonian fluids in cylindrical coordinates

$$\tau_{rr} = 2\mu\left[\frac{\partial v_r}{\partial r} - \frac{1}{3}(\nabla\cdot\mathbf{v})\right]$$

$$\tau_{\theta\theta} = 2\mu\left[\frac{1}{r}\frac{\partial v_\theta}{\partial\theta} + \frac{v_r}{r} - \frac{1}{3}(\nabla\cdot\mathbf{v})\right]$$

$$\tau_{zz} = 2\mu\left[\frac{\partial v_z}{\partial z} - \frac{1}{3}(\nabla\cdot\mathbf{v})\right]$$

$$\tau_{r\theta} = \tau_{\theta r} = \mu\left[r\frac{\partial}{\partial r}\left(\frac{v_\theta}{r}\right) + \frac{1}{r}\frac{\partial v_r}{\partial\theta}\right]$$

$$\tau_{\theta z} = \tau_{z\theta} = \mu\left[\frac{\partial v_\theta}{\partial z} + \frac{1}{r}\frac{\partial v_z}{\partial\theta}\right]$$

$$\tau_{zr} = \tau_{rz} = \mu\left[\frac{\partial v_z}{\partial r} + \frac{\partial v_r}{\partial z}\right]$$

$$\nabla\cdot\mathbf{v} = \frac{1}{r}\frac{\partial}{\partial r}(rv_r) + \frac{1}{r}\frac{\partial v_\theta}{\partial\theta} + \frac{\partial v_z}{\partial z}$$

Table 6.6 Viscous stress components for Newtonian fluids in spherical coordinates

$$\tau_{rr} = 2\mu\left[\frac{\partial v_r}{\partial r} - \frac{1}{3}(\nabla\cdot\mathbf{v})\right]$$

$$\tau_{\theta\theta} = 2\mu\left[\frac{1}{r}\frac{\partial v_\theta}{\partial\theta} + \frac{v_r}{r} - \frac{1}{3}(\nabla\cdot\mathbf{v})\right]$$

$$\tau_{\phi\phi} = 2\mu\left[\frac{1}{r\sin\theta}\frac{\partial v_\phi}{\partial\phi} + \frac{v_r}{r} + \frac{v_\theta\cot\theta}{r} - \frac{1}{3}(\nabla\cdot\mathbf{v})\right]$$

$$\tau_{r\theta} = \tau_{\theta r} = \mu\left[r\frac{\partial}{\partial r}\left(\frac{v_\theta}{r}\right) + \frac{1}{r}\frac{\partial v_r}{\partial\theta}\right]$$

$$\tau_{\theta\phi} = \tau_{\phi\theta} = \mu\left[\frac{\sin\theta}{r}\frac{\partial}{\partial\theta}\left(\frac{v_\phi}{\sin\theta}\right) + \frac{1}{r\sin\theta}\frac{\partial v_\theta}{\partial\phi}\right]$$

$$\tau_{\phi r} = \tau_{r\phi} = \mu\left[\frac{1}{r\sin\theta}\frac{\partial v_r}{\partial\phi} + r\frac{\partial}{\partial r}\left(\frac{v_\phi}{r}\right)\right]$$

$$\nabla\cdot\mathbf{v} = \frac{1}{r^2}\frac{\partial}{\partial r}(r^2 v_r) + \frac{1}{r\sin\theta}\frac{\partial}{\partial\theta}(v_\theta\sin\theta) + \frac{1}{r\sin\theta}\frac{\partial v_\phi}{\partial\phi}$$

Non-Newtonian fluids

Any incompressible fluid that does not follow Eq. (6.5-13) is non-Newtonian. Power-law and Bingham fluids, introduced in Section 1.2, are examples of *generalized Newtonian fluids*. Such liquids obey the stress equations in Tables 6.4 through 6.6 with $\nabla\cdot\mathbf{v} = 0$.

Table 6.7 Magnitude of the rate of strain

Rectangular:	$(2\Gamma)^2 = 2\left[\left(\frac{\partial v_x}{\partial x}\right)^2 + \left(\frac{\partial v_y}{\partial y}\right)^2 + \left(\frac{\partial v_z}{\partial z}\right)^2\right] + \left[\frac{\partial v_y}{\partial x} + \frac{\partial v_x}{\partial y}\right]^2 + \left[\frac{\partial v_z}{\partial y} + \frac{\partial v_y}{\partial z}\right]^2 + \left[\frac{\partial v_x}{\partial z} + \frac{\partial v_z}{\partial x}\right]^2$
Cylindrical:	$(2\Gamma)^2 = 2\left[\left(\frac{\partial v_r}{\partial r}\right)^2 + \left(\frac{1}{r}\frac{\partial v_\theta}{\partial \theta} + \frac{v_r}{r}\right)^2 + \left(\frac{\partial v_z}{\partial z}\right)^2\right] + \left[r\frac{\partial}{\partial r}\left(\frac{v_\theta}{r}\right) + \frac{1}{r}\frac{\partial v_r}{\partial \theta}\right]^2 + \left[\frac{1}{r}\frac{\partial v_z}{\partial \theta} + \frac{\partial v_\theta}{\partial z}\right]^2 + \left[\frac{\partial v_r}{\partial z} + \frac{\partial v_z}{\partial r}\right]^2$
Spherical:	$(2\Gamma)^2 = 2\left[\left(\frac{\partial v_r}{\partial r}\right)^2 + \left(\frac{1}{r}\frac{\partial v_\theta}{\partial \theta} + \frac{v_r}{r}\right)^2 + \left(\frac{1}{r\sin\theta}\frac{\partial v_\phi}{\partial \phi} + \frac{v_r}{r} + \frac{v_\theta \cot\theta}{r}\right)^2\right] + \left[r\frac{\partial}{\partial r}\left(\frac{v_\theta}{r}\right) + \frac{1}{r}\frac{\partial v_r}{\partial \theta}\right]^2 + \left[\frac{\sin\theta}{r}\frac{\partial}{\partial \theta}\left(\frac{v_\phi}{\sin\theta}\right) + \frac{1}{r\sin\theta}\frac{\partial v_\theta}{\partial \phi}\right]^2 + \left[\frac{1}{r\sin\theta}\frac{\partial v_r}{\partial \phi} + r\frac{\partial}{\partial r}\left(\frac{v_\phi}{r}\right)\right]^2$

What makes them non-Newtonian is that μ depends on the rate of strain. That is, $\mu = \mu(\Gamma)$, where Γ is the magnitude of the rate-of-strain tensor. From Eq. (A.3-27),

$$\Gamma = \left(\frac{1}{2}\sum_i \sum_j \Gamma_{ij}^2\right)^{1/2}. \qquad (6.5\text{-}16)$$

Like the magnitude of a vector, the value of Γ is independent of the coordinate system. Rates of strain, expressed as $(2\Gamma)^2$, are shown in Table 6.7 for the three common coordinate systems.

In the simple shear flow in Fig. 1.1, $\Gamma = |\Gamma_{xy}| = |\Gamma_{yx}| = s/2$, where s is the shear rate. Thus, the shear rate is twice the rate of strain. Replacing s by 2Γ in Eqs. (1.2-4) and (1.2-6), the general forms of the constitutive equations for power-law and Bingham fluids are

$$\mu = m(2\Gamma)^{n-1} \quad \text{(power-law)} \qquad (6.5\text{-}17)$$

$$\mu = \begin{cases} \infty & \text{for } \tau < \tau_0 \\ \mu_0 + \dfrac{\tau_0}{2\Gamma} & \text{for } \tau \geq \tau_0 \end{cases} \quad \text{(Bingham).} \qquad (6.5\text{-}18)$$

Unlike what was presented in Chapter 1, these expressions are not limited to simple shear flows. The magnitude of the viscous stress tensor (τ), which is needed for the Bingham model, is calculated as in Eq. (6.5-16). The Cartesian result is

$$\tau = \left\{\frac{1}{2}\left[\left(\tau_{xx}^2 + \tau_{yy}^2 + \tau_{zz}^2\right) + \left(\tau_{xy}^2 + \tau_{yx}^2\right) + \left(\tau_{yz}^2 + \tau_{zy}^2\right) + \left(\tau_{xz}^2 + \tau_{zx}^2\right)\right]\right\}^{1/2}. \qquad (6.5\text{-}19)$$

For the flow in Fig. 1.1, in which τ_{yx} and τ_{xy} are the only nonzero components of $\boldsymbol{\tau}$, this reduces to $\tau = |\tau_{xy}| = |\tau_{yx}|$.

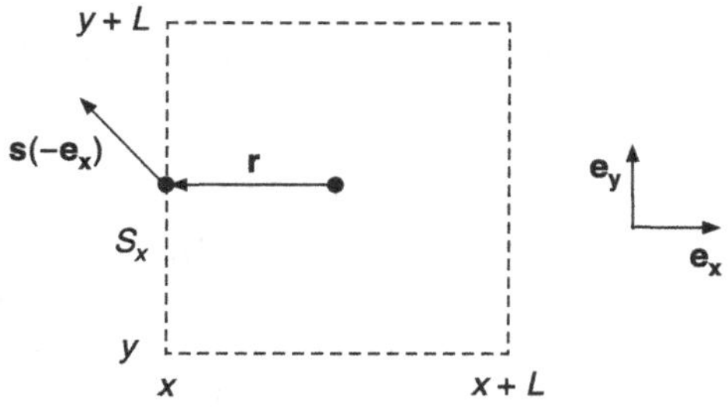

Figure 6.4 Control volume for the proof that σ is symmetric.

Additional note: stress symmetry

Reasoning like that used to derive Eq. (6.4-12) indicates that, for a fixed control volume of linear dimension L,

$$\lim_{L \to 0} \frac{1}{L^3} \int_S \mathbf{r} \times \mathbf{s}(\mathbf{n})\, dS = 0 \tag{6.5-20}$$

where $\mathbf{r}$ is a vector that extends from an origin to a point on the control surface (Whitaker, 1968, p. 120). Thus, just as the stresses on a vanishingly small control volume are in balance, so are the surface torques.

To show that the stress tensor is symmetric, Eq. (6.5-20) will be applied to a cubic control volume of edge length L, with the origin for $\mathbf{r}$ placed at the center of the cube. A view along the z axis is shown in Fig. 6.4. At the left-hand side (position x) is a surface S_x. The position vector at its center is $\mathbf{r} = -(L/2)\mathbf{e_x}$, the outward normal is $-\mathbf{e_x}$, and the stress vector is

$$\mathbf{s}(-\mathbf{e_x}) = -(\mathbf{e_x} \cdot \boldsymbol{\sigma})|_x = -\sum_j \sigma_{xj}|_x \mathbf{e_j}. \tag{6.5-21}$$

The torque on S_x is[2]

$$\begin{aligned} \int_{S_x} \mathbf{r} \times \mathbf{s}(-\mathbf{e_x})\, dS &= \left(-\frac{L}{2}\mathbf{e_x}\right) \times (-\sigma_{xx}\mathbf{e_x} - \sigma_{xy}\mathbf{e_y} - \sigma_{xz}\mathbf{e_z})\Big|_x L^2 \\ &= \frac{L^3}{2}(\sigma_{xy}\mathbf{e_z} - \sigma_{xz}\mathbf{e_y})\Big|_x . \end{aligned} \tag{6.5-22}$$

The same expression applies at the right-hand edge (position $x + L$), because the signs of $\mathbf{r}$ and $\mathbf{s}(\mathbf{n})$ are both reversed. Adding the contributions of the six surfaces and canceling the $L^3/2$ factors, Eq. (6.5-20) becomes

$$\begin{aligned} 0 = \lim_{L \to 0} [&(\sigma_{xy}|_x + \sigma_{xy}|_{x+L})\mathbf{e_z} - (\sigma_{xz}|_x + \sigma_{xz}|_{x+L})\mathbf{e_y} \\ &+ (\sigma_{yz}|_y + \sigma_{yz}|_{y+L})\mathbf{e_x} - (\sigma_{yx}|_y + \sigma_{yx}|_{y+L})\mathbf{e_z} \\ &+ (\sigma_{zx}|_z + \sigma_{zx}|_{z+L})\mathbf{e_y} - (\sigma_{zy}|_z + \sigma_{zy}|_{z+L})\mathbf{e_x}]. \end{aligned} \tag{6.5-23}$$

[2] In the first equality in Eq. (6.5-22), the surface average of $\mathbf{r} \times \mathbf{s}(\mathbf{n})$ is approximated by evaluating $\mathbf{r}$ at the center of the surface and treating $\mathbf{s}(\mathbf{n})$ as if it were uniform. This is permissible because the control volume is later reduced to a point.

As $L \to 0$, $\sigma_{xy}|_{x+L} \to \sigma_{xy}|_x$, $\sigma_{yz}|_{y+L} \to \sigma_{yz}|_y$, and so on. Taking that limit, canceling the resulting factors of 2, and grouping the vector components, we obtain

$$\mathbf{0} = (\sigma_{yz} - \sigma_{zy})\mathbf{e_x} + (\sigma_{zx} - \sigma_{xz})\mathbf{e_y} + (\sigma_{xy} - \sigma_{yx})\mathbf{e_z}. \tag{6.5-24}$$

For each component of this vector equation to vanish, it is necessary that $\sigma_{ij} = \sigma_{ji}$. That is, $\boldsymbol{\sigma}$ must be symmetric. Because $\boldsymbol{\sigma}$ and $\boldsymbol{\tau}$ have the same off-diagonal components (i.e., $\sigma_{ij} = \tau_{ij}$ for $i \neq j$), symmetry of $\boldsymbol{\sigma}$ implies symmetry of $\boldsymbol{\tau}$.

A caveat is that only surface torques were considered in this derivation. In a suspension of particles that can align in response to a body force, there may be an additional "microscopic" or "internal" torque that makes $\boldsymbol{\sigma}$ and $\boldsymbol{\tau}$ asymmetric (Dahler and Scriven, 1963). Microscopic torques arise in suspensions of colloidal magnetite or other particles with permanent or induced magnetic dipoles. Such suspensions, called *ferrofluids*, are important in certain technologies (Rosensweig, 1985). However, in common fluids $\boldsymbol{\sigma}$ and $\boldsymbol{\tau}$ are indeed symmetric.

6.6 GOVERNING EQUATIONS

In *incompressible* flows, which are our main focus, there may be as many as four primary unknowns: pressure and the three components of the velocity vector. Conservation of mass and the three components of conservation of linear momentum provide the requisite number of differential equations. The derivation of the differential equations is now largely complete; what remains is to substitute the appropriate forms of the viscous stress tensor from Section 6.5 into the momentum equation from Section 6.4. Also requiring discussion are the velocity and stress components at interfaces, which provide the boundary conditions for the differential equations. The final topic in this chapter is a discussion of force calculations.

Newtonian fluids with constant properties

Our default assumption is that the fluid is Newtonian and nearly isothermal, such that μ and ρ are both known constants. The divergence of the stress tensor is evaluated then as

$$\nabla \cdot \boldsymbol{\tau} = \mu \nabla \cdot [\nabla \mathbf{v} + (\nabla \mathbf{v})^t] = \mu[\nabla^2 \mathbf{v} + \nabla(\nabla \cdot \mathbf{v})] = \mu \nabla^2 \mathbf{v}. \tag{6.6-1}$$

The identity used to obtain the second equality is proven in Example A.4-2. Substituting Eq. (6.6-1) into the Cauchy momentum equation [Eq. (6.4-8)] gives

$$\rho \frac{D\mathbf{v}}{Dt} = \rho \mathbf{g} - \nabla P + \mu \nabla^2 \mathbf{v} \tag{6.6-2}$$

which is one form of the *Navier–Stokes equation*.[3] This vector equation provides three of the four differential equations needed to find P and the three components of $\mathbf{v}$. The fourth is the continuity equation for an incompressible fluid,

$$\nabla \cdot \mathbf{v} = 0. \tag{6.6-3}$$

It was shown in Section 2.3 that the same formulas can be used to calculate flow rates in horizontal, inclined, and vertical pipes, if P is replaced by $\mathcal{P}$. The dynamic pressure $\mathcal{P}$ arose also when discussing flow through porous media and packed beds in Sections 3.4 and 3.5. In that it conveniently combines the effects of pressure and gravity, $\mathcal{P}$ is useful quite generally in the analysis of incompressible flows. Previously, we defined $\mathcal{P}$ in terms

[3] This incorporation of Newtonian viscous stresses into the momentum equation was reported independently by the French physicist L. Navier (1785–1836) in 1822 and by G. G. Stokes (see Chapter 3) in 1845.

Table 6.8 Navier–Stokes equation in Cartesian coordinates

x component:	$\rho\left[\frac{\partial v_x}{\partial t}+v_x\frac{\partial v_x}{\partial x}+v_y\frac{\partial v_x}{\partial y}+v_z\frac{\partial v_x}{\partial z}\right]=-\frac{\partial \mathcal{P}}{\partial x}+\mu\left[\frac{\partial^2 v_x}{\partial x^2}+\frac{\partial^2 v_x}{\partial y^2}+\frac{\partial^2 v_x}{\partial z^2}\right]$
y component:	$\rho\left[\frac{\partial v_y}{\partial t}+v_x\frac{\partial v_y}{\partial x}+v_y\frac{\partial v_y}{\partial y}+v_z\frac{\partial v_y}{\partial z}\right]=-\frac{\partial \mathcal{P}}{\partial y}+\mu\left[\frac{\partial^2 v_y}{\partial x^2}+\frac{\partial^2 v_y}{\partial y^2}+\frac{\partial^2 v_y}{\partial z^2}\right]$
z component:	$\rho\left[\frac{\partial v_z}{\partial t}+v_x\frac{\partial v_z}{\partial x}+v_y\frac{\partial v_z}{\partial y}+v_z\frac{\partial v_z}{\partial z}\right]=-\frac{\partial \mathcal{P}}{\partial z}+\mu\left[\frac{\partial^2 v_z}{\partial x^2}+\frac{\partial^2 v_z}{\partial y^2}+\frac{\partial^2 v_z}{\partial z^2}\right]$

of the height h above a reference plane ($\mathcal{P} \equiv P + \rho g h$). An equivalent definition using the gradient operator is

$$\nabla\mathcal{P} \equiv \nabla P - \rho\mathbf{g}. \tag{6.6-4}$$

An alternative form of the Navier–Stokes equation is then

$$\rho\frac{D\mathbf{v}}{Dt} = -\nabla\mathcal{P} + \mu\nabla^2\mathbf{v}. \tag{6.6-5}$$

Setting $\mathbf{v} = \mathbf{0}$ in Eq. (6.6-5) reveals that $\nabla\mathcal{P} = \mathbf{0}$ in a static fluid. Thus, spatial variations in $\mathcal{P}$ are either the cause or result of fluid motion. Because $\mathcal{P}$ is defined in terms of either its gradient or the height above a reference plane that is chosen at will, its constant value in a static fluid is arbitrary. It is often convenient to set $\mathcal{P} = 0$ in a static fluid.

Knowledge of the absolute pressure (P) in incompressible fluids tends to be important only if there are fluid–fluid interfaces. When such interfaces are absent, as in liquid-filled pipes, Eq. (6.6-5) is the preferred form of the Navier–Stokes equation. As seen already, using $\mathcal{P}$ often leads to more universal results. Moreover, calculating pressure forces using $\mathcal{P}$ automatically separates the effects of fluid motion from those of buoyancy, as shown later in this section.

The components of Eq. (6.6-5) in Cartesian, cylindrical, and spherical coordinates are given in Tables 6.8, 6.9, and 6.10, respectively. The bracketed terms on the left and right side of each entry are the components of $D\mathbf{v}/Dt$ and $\nabla^2\mathbf{v}$, respectively. The components of Eq. (6.6-2) can be found by replacing the $\mathcal{P}$ terms in these tables with the P and g_i terms from Tables 6.1, 6.2, and 6.3. Likewise, using the present tables as guidance, the components of the Cauchy momentum equation in the earlier tables can be rewritten in terms of $\mathcal{P}$. The incompressible continuity equation [Eq. (6.6-3)] may be obtained for each coordinate system by holding ρ constant in Table 5.1. Expressions for $\nabla \cdot \mathbf{v}$ are given also at the bottom of Tables 6.4, 6.5, and 6.6.

Example 6.6-1 Pressure in planar stagnation flow. If the velocity is known, the Navier–Stokes equation can be used to find the pressure. This will be illustrated for planar stagnation flow, where $v_x = Cx$, $v_y = -Cy$, and $v_z = 0$ (Example 5.2-1).

Most relevant for this two-dimensional flow are the x and y components of Eq. (6.6-5), as shown in Table 6.8. For the given velocity field, all viscous terms and most of the inertial terms are zero. Simplifying accordingly and solving for the partial derivatives of

Table 6.9 Navier–Stokes equation in cylindrical coordinates

r component:

$$\rho\left[\frac{\partial v_r}{\partial t}+v_r\frac{\partial v_r}{\partial r}+\frac{v_\theta}{r}\frac{\partial v_r}{\partial \theta}-\frac{v_\theta^2}{r}+v_z\frac{\partial v_r}{\partial z}\right]$$
$$=-\frac{\partial \mathcal{P}}{\partial r}+\mu\left[\frac{\partial}{\partial r}\left(\frac{1}{r}\frac{\partial}{\partial r}(rv_r)\right)+\frac{1}{r^2}\frac{\partial^2 v_r}{\partial \theta^2}-\frac{2}{r^2}\frac{\partial v_\theta}{\partial \theta}+\frac{\partial^2 v_r}{\partial z^2}\right]$$

θ component:

$$\rho\left[\frac{\partial v_\theta}{\partial t}+v_r\frac{\partial v_\theta}{\partial r}+\frac{v_\theta}{r}\frac{\partial v_\theta}{\partial \theta}+\frac{v_r v_\theta}{r}+v_z\frac{\partial v_\theta}{\partial z}\right]$$
$$=-\frac{1}{r}\frac{\partial \mathcal{P}}{\partial \theta}+\mu\left[\frac{\partial}{\partial r}\left(\frac{1}{r}\frac{\partial}{\partial r}(rv_\theta)\right)+\frac{1}{r^2}\frac{\partial^2 v_\theta}{\partial \theta^2}+\frac{2}{r^2}\frac{\partial v_r}{\partial \theta}+\frac{\partial^2 v_\theta}{\partial z^2}\right]$$

z component:

$$\rho\left[\frac{\partial v_z}{\partial t}+v_r\frac{\partial v_z}{\partial r}+\frac{v_\theta}{r}\frac{\partial v_z}{\partial \theta}+v_z\frac{\partial v_z}{\partial z}\right]$$
$$=-\frac{\partial \mathcal{P}}{\partial z}+\mu\left[\frac{1}{r}\frac{\partial}{\partial r}\left(r\frac{\partial v_z}{\partial r}\right)+\frac{1}{r^2}\frac{\partial^2 v_z}{\partial \theta^2}+\frac{\partial^2 v_z}{\partial z^2}\right]$$

Table 6.10 Navier–Stokes equation in spherical coordinates

r component:

$$\rho\left[\frac{\partial v_r}{\partial t}+v_r\frac{\partial v_r}{\partial r}+\frac{v_\theta}{r}\frac{\partial v_r}{\partial \theta}+\frac{v_\phi}{r\sin\theta}\frac{\partial v_r}{\partial \phi}-\frac{v_\theta^2+v_\phi^2}{r}\right]$$
$$=-\frac{\partial \mathcal{P}}{\partial r}+\mu\left[\nabla^2 v_r-\frac{2}{r^2}v_r-\frac{2}{r^2}\frac{\partial v_\theta}{\partial \theta}-\frac{2}{r^2}v_\theta\cot\theta-\frac{2}{r^2\sin\theta}\frac{\partial v_\phi}{\partial \phi}\right]$$

θ component:

$$\rho\left[\frac{\partial v_\theta}{\partial t}+v_r\frac{\partial v_\theta}{\partial r}+\frac{v_\theta}{r}\frac{\partial v_\theta}{\partial \theta}+\frac{v_\phi}{r\sin\theta}\frac{\partial v_\theta}{\partial \phi}+\frac{v_r v_\theta}{r}-\frac{v_\phi^2\cot\theta}{r}\right]$$
$$=-\frac{1}{r}\frac{\partial \mathcal{P}}{\partial \theta}+\mu\left[\nabla^2 v_\theta+\frac{2}{r^2}\frac{\partial v_r}{\partial \theta}-\frac{v_\theta}{r^2\sin^2\theta}-\frac{2\cos\theta}{r^2\sin^2\theta}\frac{\partial v_\phi}{\partial \phi}\right]$$

ϕ component:

$$\rho\left[\frac{\partial v_\phi}{\partial t}+v_r\frac{\partial v_\phi}{\partial r}+\frac{v_\theta}{r}\frac{\partial v_\phi}{\partial \theta}+\frac{v_\phi}{r\sin\theta}\frac{\partial v_\phi}{\partial \phi}+\frac{v_\phi v_r}{r}+\frac{v_\theta v_\phi\cot\theta}{r}\right]$$
$$=-\frac{1}{r\sin\theta}\frac{\partial \mathcal{P}}{\partial \phi}+\mu\left[\nabla^2 v_\phi-\frac{v_\phi}{r^2\sin^2\theta}+\frac{2}{r^2\sin\theta}\frac{\partial v_r}{\partial \phi}+\frac{2\cos\theta}{r^2\sin^2\theta}\frac{\partial v_\theta}{\partial \phi}\right]$$

Laplacian:

$$\nabla^2=\frac{1}{r^2}\frac{\partial}{\partial r}\left(r^2\frac{\partial}{\partial r}\right)+\frac{1}{r^2\sin\theta}\frac{\partial}{\partial \theta}\left(\sin\theta\frac{\partial}{\partial \theta}\right)+\frac{1}{r^2\sin^2\theta}\left(\frac{\partial^2}{\partial \phi^2}\right)$$

the pressure,

$$\frac{\partial \mathcal{P}}{\partial x}=-\rho v_x\frac{\partial v_x}{\partial x}=-\rho C^2 x \qquad (6.6\text{-}6)$$

$$\frac{\partial \mathcal{P}}{\partial y}=-\rho v_y\frac{\partial v_y}{\partial y}=-\rho C^2 y. \qquad (6.6\text{-}7)$$

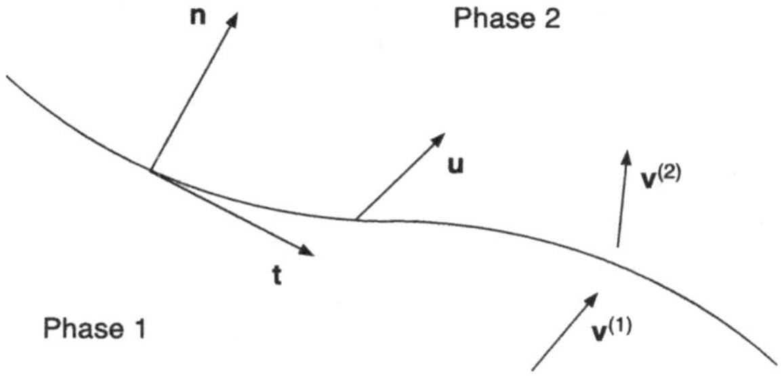

Figure 6.5 Unit vectors and velocities at an interface. The normal and tangent vectors are **n** and **t**, respectively. The velocity in phase 1 is $\mathbf{v}^{(1)}$, that in phase 2 is $\mathbf{v}^{(2)}$, and that of the interface itself is **u**. The unit normal is directed toward phase 2.

The z component is uninformative, because it is known already that $v_z = 0$ and that nothing depends on z. Integration of Eqs. (6.6-6) and (6.6-7) gives

$$\mathcal{P}(x,y) = -\frac{\rho C^2 x^2}{2} + f(y) + a \qquad (6.6\text{-}8)$$

$$\mathcal{P}(x,y) = -\frac{\rho C^2 y^2}{2} + g(x) + b \qquad (6.6\text{-}9)$$

where $f(y)$ and $g(x)$ are unknown functions and a and b are unknown constants. A comparison of these results indicates that $f(y) = -\rho C^2 y^2/2$, $g(x) = -\rho C^2 x^2/2$, and $a = b$. Choosing the additive constant as zero, for convenience, the final result is

$$\mathcal{P}(x,y) = -\frac{\rho C^2}{2}(x^2 + y^2). \qquad (6.6\text{-}10)$$

Fluids with varying viscosity

In generalized Newtonian fluids, μ depends on the local rate of strain. Returning to the Cauchy equation [Eq. (6.4-8)] and keeping in mind that μ might vary with position, the momentum equation for such a fluid is

$$\rho \frac{D\mathbf{v}}{Dt} = \rho \mathbf{g} - \nabla P + \nabla \cdot \{\mu(\Gamma)[\nabla \mathbf{v} + (\nabla \mathbf{v})^t]\}. \qquad (6.6\text{-}11)$$

The function $\mu(\Gamma)$ is determined by the specific type of fluid, as exemplified by Eqs. (6.5-17) and (6.5-18). Because ρ is constant, Eq. (6.6-3) applies again.

Velocities at phase boundaries

The mechanical conditions at an interface will be described using the notation in Fig. 6.5. Phase 1 and phase 2 might both be fluids, or one might be a solid. Although interfacial regions typically extend a few molecular diameters into either phase, the length scales of interest in continuum modeling allow interfaces to be viewed as mathematical surfaces. The unit normal vector directed from phase 1 to phase 2 is **n** and a unit tangent is **t**. What follows applies to any of the tangent vectors that could be created by rotations of **t** about **n**. In general, the directions of **n** and **t** will vary with position on the interface, and each relationship applies at any given point. The velocities in the two phases are $\mathbf{v}^{(1)}$ and $\mathbf{v}^{(2)}$. Even when evaluated at the same point on the interface, these velocities may differ from that of the interface itself, which is denoted as **u**. For example, a melting ice cube whose center is stationary has zero velocity throughout the solid, an interfacial velocity corresponding to the rate of shrinkage, and another (small) velocity on the liquid side of the interface.

The velocity components normal to or tangent to the interface will be identified using the subscripts n or t. For example, the component of $\mathbf{u}$ that is normal to the interface is $u_n = \mathbf{n} \cdot \mathbf{u}$ and a tangential component is $u_t = \mathbf{t} \cdot \mathbf{u}$.

Normal velocity components. A relationship among the velocities normal to an interface is provided by conservation of mass. Ordinarily, mass does not accumulate at an interface.[4] Equating the rate at which mass arrives via phase 1 with the rate at which it leaves via phase 2 gives

$$\rho_1\left(v_n^{(1)} - u_n\right) = \rho_2\left(v_n^{(2)} - u_n\right). \tag{6.6-12}$$

Within each phase, the mass flux (mass per unit time per unit area) *relative to the interface* is the density times the relative velocity, as shown. Equation (6.6-12) is more general than what is usually needed. If phase 1 is a stationary solid that is not undergoing a phase change such as melting or solidification, then $v_n^{(1)} = u_n = 0$ and

$$v_n^{(2)} = 0. \tag{6.6-13}$$

This vanishing of the normal component of the fluid velocity is called a *no-penetration condition*. It applies to any solid surface that is stationary and impermeable.

Tangential velocity components. The tangential velocities at an interface are usually assumed to be equal, such that

$$v_t^{(1)} = v_t^{(2)}. \tag{6.6-14}$$

This absence of relative tangential motion is called a *no-slip condition*. If phase 1 is a stationary solid, then

$$v_t^{(2)} = 0 \tag{6.6-15}$$

which is the form of no-slip condition usually encountered.

Unlike the normal-velocity condition and the stress conditions to be presented shortly, the no-slip condition cannot be derived from continuum principles. Although supported by molecular-dynamics simulations, it is primarily an empirical observation. Because continuum mechanics is silent on this issue, the validity of the no-slip condition was questioned, especially for high-speed flows, until about 1900. By then there was enough experimental evidence in its favor to silence most skeptics. However, as recent advances in microfabrication techniques have made it possible to examine flows in well-defined geometries at progressively smaller length scales, the no-slip condition has again become an active area of research (Lauga *et al.*, 2007). It is generally agreed now that it breaks down in sufficiently small channels.

The situation for Newtonian liquids may be summarized as follows: the no-slip condition is accurate sometimes for system length scales on the order of a few nm (at or near the lower limit for continuum models), and almost always for length scales exceeding about 1 μm. Although it may fail at very small dimensions and in certain other situations (e.g., for Newtonian liquids in the vicinity of moving contact lines and for certain non-Newtonian liquids), no slip is valid for most applications.

[4] Exceptions include the accumulation of surfactants that may occur at fluid–fluid interfaces and the adsorption of gaseous species onto solid surfaces. Such problems, which involve mass transfer in mixtures, are beyond the scope of this book.

Stresses at phase boundaries

The normal and tangential components of the total stress tensor are denoted as σ_{nn} and σ_{nt}, respectively. In terms of the unit vectors in Fig. 6.5, these are calculated as $\sigma_{nn} = \mathbf{n} \cdot \boldsymbol{\sigma} \cdot \mathbf{n}$ and $\sigma_{nt} = \mathbf{n} \cdot \boldsymbol{\sigma} \cdot \mathbf{t}$, each evaluated at the interface. Likewise, the normal and shear components of the viscous stress tensor at the interface are τ_{nn} and τ_{nt}, respectively.

In the absence of surface tension, the stresses at an interface must balance, just as for any test surface within a fluid. Applying Eq. (6.4-12) to a thin control volume centered on the interface shows that $\mathbf{s}(\mathbf{n})^{(1)} = \mathbf{s}(\mathbf{n})^{(2)}$. Accordingly, the stress tensors must satisfy

$$(\mathbf{n} \cdot \boldsymbol{\sigma})^{(1)} = (\mathbf{n} \cdot \boldsymbol{\sigma})^{(2)}. \tag{6.6-16}$$

This will be used to derive the normal-stress and shear-stress relationships in the absence of surface tension, and then the modifications to each that are required by surface tension will be discussed.

Normal stresses. Keeping in mind that σ_{nn} has both pressure and viscous contributions [Eq. (6.3-12)], the dot product of $\mathbf{n}$ with both sides of Eq. (6.6-16) gives

$$-P^{(1)} + \tau_{nn}^{(1)} = -P^{(2)} + \tau_{nn}^{(2)}. \tag{6.6-17}$$

The difference in signs stems from the fact that positive pressures are compressive, whereas positive values of τ_{nn} are tensile. Normal viscous stresses are often negligible. Wherever an incompressible Newtonian fluid contacts a solid, the normal viscous stress is zero. A proof of that for objects of any shape, moving or stationary, is outlined in Deen (2012, p. 266); see also Problem 6.5. Thus, a common normal-stress condition is simply $P^{(1)} = P^{(2)}$.

For nonplanar interfaces and length scales small enough for surface tension to be important,

$$-P^{(1)} + \tau_{nn}^{(1)} = -P^{(2)} + \tau_{nn}^{(2)} + 2\mathcal{H}\gamma. \tag{6.6-18}$$

If the fluids are static or if the normal viscous stresses are otherwise negligible, this reduces to Eq. (4.4-6); for planar interfaces, $\mathcal{H} = 0$ (Section 4.4) and Eq. (6.6-17) is recovered.

Shear stresses. Because pressure does not contribute to shear stresses, $\sigma_{nt} = \tau_{nt}$. Accordingly, the dot product of $\mathbf{t}$ with both sides of Eq. (6.6-16) gives

$$\tau_{nt}^{(1)} = \tau_{nt}^{(2)}. \tag{6.6-19}$$

That is, the shear stresses on the two sides of an interface normally balance. This holds whenever the surface tension is constant, even if the interface is curved. An example of how to modify this result for a planar interface with variable surface tension follows. Relationships for nonplanar interfaces with variable γ are given in Deen (2012, pp. 245–246).

Example 6.6-2 Shear-stress boundary condition with variable surface tension. The effect of a nonuniform surface tension will be illustrated using the control volume in Fig. 6.6. It is centered on a planar interface at $y = 0$ and has lengths Δx and Δz, respectively, in the x and z directions. The y dimension of the control volume is made small enough to neglect gravity, inertia, and the pressure and viscous forces on its edges. In this system $\tau_{nt} = \tau_{yx}$. What we seek is the relationship between τ_{yx} in the two phases when γ varies with x.

For a control volume that is sufficiently thin, all that acts in the x direction are the shear forces on the top and bottom and the surface tension force at the contour where the

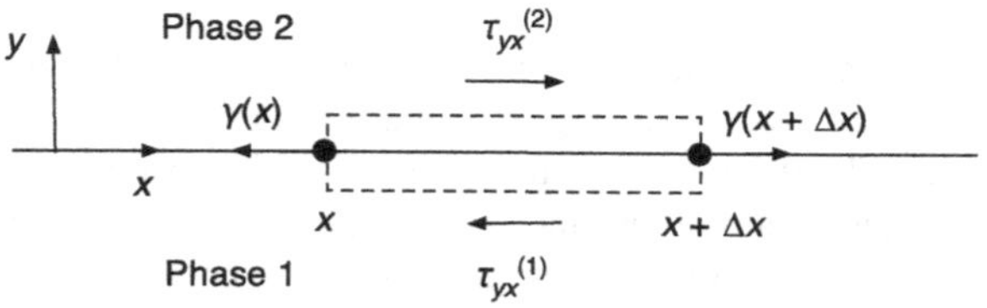

Figure 6.6 Control volume for examining the effects of surface tension variations at a planar interface.

control surface intersects the interface. With inertia negligible, these forces must balance. Thus,

$$\tau_{yx}^{(2)}\Delta x\,\Delta z - \tau_{yx}^{(1)}\Delta x\,\Delta z + \gamma(x+\Delta x)\Delta z - \gamma(x)\Delta z = 0. \qquad (6.6\text{-}20)$$

At the top surface, where $\mathbf{n} = \mathbf{e}_y$, a positive $\tau_{yx}^{(2)}$ gives a force in the $+x$ direction; at the bottom, where $\mathbf{n} = -\mathbf{e}_y$, a positive $\tau_{yx}^{(1)}$ acts in the $-x$ direction. That is the basis for the force arrows in Fig. 6.6 and the algebraic signs in Eq. (6.6-20). Dividing by $\Delta x\,\Delta z$ and letting $\Delta x \to 0$ gives

$$\tau_{yx}^{(2)} - \tau_{yx}^{(1)} + \frac{\partial \gamma}{\partial x} = 0. \qquad (6.6\text{-}21)$$

Thus, spatial variations in surface tension cause an imbalance in the shear stresses. If $\partial\gamma/\partial x \neq 0$, this relationship cannot be satisfied in a static system, where $\tau_{yx} = 0$ in both phases. Accordingly, variations in surface tension always cause fluid motion; such flows are called *Marangoni flows*. If γ is constant, Eq. (6.6-21) is equivalent to Eq. (6.6-19).

The velocity and stress boundary conditions presented above must be supplemented occasionally by ones based on the type of symmetry present in a particular problem. Because familiarity with examples of microscopic analysis is needed to fully appreciate the use of symmetry, such conditions are not discussed until Section 7.6.

Force calculations

The main result of this chapter is the Navier–Stokes equation. Together with the continuity equation and the boundary conditions just discussed, it provides a way to predict velocity and pressure variations within incompressible Newtonian fluids. Once the velocity and pressure have been found, it is desired often to calculate the force that the fluid exerts on an object. That force ($\mathbf{F}$) is the integral of the stress vector over the surface in question. Thus,

$$\mathbf{F} = \int_S \mathbf{n}\cdot\boldsymbol{\sigma}\,dS = -\int_S \mathbf{n}P\,dS + \int_S \mathbf{n}\cdot\boldsymbol{\tau}\,dS = \mathbf{F_P} + \mathbf{F}_\tau \qquad (6.6\text{-}22)$$

where $\mathbf{n}$ points into the *fluid* (i.e., the phase exerting the force). The pressure and viscous forces are denoted as $\mathbf{F_P}$ and $\mathbf{F}_\tau$, respectively. The expression for $\mathbf{F_P}$ was given in Section 4.3. In what follows, it is assumed that the object is completely surrounded by the fluid, so that S is a closed surface.

Contributing to $\mathbf{F_P}$ are static pressure variations, which create the buoyancy force $\mathbf{F_0}$ given by Eq. (4.3-19), and pressure variations caused by the fluid motion. Denoting the excess pressure force due to flow as $\mathbf{F}_\mathcal{P}$,

$$\mathbf{F_P} = \mathbf{F_0} + \mathbf{F}_\mathcal{P}. \qquad (6.6\text{-}23)$$

What is sought usually is just $\mathbf{F}_{\mathcal{P}}$. That is, Archimedes' law need not be derived repeatedly. How to calculate $\mathbf{F}_{\mathcal{P}}$ is revealed by recalling that $P = -\rho g h + \mathcal{P}$, where h is the height above a reference plane. The pressure force is then

$$\mathbf{F_P} = \rho g \int_S \mathbf{n} h \, dS - \int_S \mathbf{n} \mathcal{P} \, dS. \tag{6.6-24}$$

From Eq. (4.3-17), the second integral will vanish in a static fluid, where $\mathcal{P}$ is constant. Accordingly, the first term must equal $\mathbf{F_0}$. A comparison of Eqs. (6.6-23) and (6.6-24) indicates then that

$$\mathbf{F}_{\mathcal{P}} = -\int_S \mathbf{n} \mathcal{P} \, dS. \tag{6.6-25}$$

Thus, $\mathbf{F}_{\mathcal{P}}$ is calculated in the same manner as $\mathbf{F_P}$, with $\mathcal{P}$ simply replacing P.

As introduced in Chapter 3, the *drag* on an object is the component of the fluid-dynamic force that resists its translational motion. For a stationary object with an approaching flow in the z direction, it is the z component of the force that the fluid exerts on the object, buoyancy aside. Thus, the drag F_D is given by

$$F_D = \mathbf{e_z} \cdot (\mathbf{F}_{\mathcal{P}} + \mathbf{F}_{\boldsymbol{\tau}}) = -\int_S \mathbf{e_z} \cdot \mathbf{n} \mathcal{P} \, dS + \int_S \mathbf{n} \cdot \boldsymbol{\tau} \cdot \mathbf{e_z} \, dS. \tag{6.6-26}$$

If the flow direction is horizontal and it is desired to calculate the upward force (the *lift*), then $\mathbf{e_z}$ is replaced by a unit vector directed upward.

As with drag and lift, the fluid-dynamic *torque* is determined by $\mathcal{P}$ and $\boldsymbol{\tau}$. The torque vector $\mathbf{G}$ is given by

$$\mathbf{G} = -\int_S \mathbf{r} \times \mathbf{n} \mathcal{P} \, dS + \int_S \mathbf{r} \times (\mathbf{n} \cdot \boldsymbol{\tau}) \, dS \tag{6.6-27}$$

where the origin for the position vector $\mathbf{r}$ is typically the center of mass of the object. Usually, what is of interest is just the magnitude of the torque (G).

Example 6.6-3 General expression for the drag on a sphere. It is desired to calculate the drag for flow at velocity U relative to a sphere of radius R. The sphere might be a solid, a bubble, or a droplet. The coordinates to be used are shown in Fig. 6.7. It is assumed that the flow is axisymmetric (independent of ϕ) and that $\mathcal{P}(R, \theta)$ and $\boldsymbol{\tau}(R, \theta)$ are known from theory, numerical simulation, or experiment.

Equation (6.6-26) indicates that the dynamic pressure and certain parts of the viscous stress must be integrated over the sphere surface. Setting $r = R$ in Eq. (A.5-46), the differential area is $dS = R^2 \sin\theta \, d\theta \, d\phi$, where $0 \leq \theta \leq \pi$ and $0 \leq \phi \leq 2\pi$. The unit normal pointing into the surrounding fluid (the phase that is exerting the force) is $\mathbf{n} = \mathbf{e_r}$. As shown in Example 4.3-3, the dot product in the pressure term is $\mathbf{e_z} \cdot \mathbf{e_r} = \cos\theta$. The viscous part of the stress vector is

$$\mathbf{e_r} \cdot \boldsymbol{\tau} = \tau_{rr} \mathbf{e_r} + \tau_{r\theta} \mathbf{e_\theta} + \tau_{r\phi} \mathbf{e_\phi}. \tag{6.6-28}$$

What is needed is the z component of this vector, or

$$\mathbf{e_r} \cdot \boldsymbol{\tau} \cdot \mathbf{e_z} = \tau_{rr} \mathbf{e_r} \cdot \mathbf{e_z} + \tau_{r\theta} \mathbf{e_\theta} \cdot \mathbf{e_z} + \tau_{r\phi} \mathbf{e_\phi} \cdot \mathbf{e_z}. \tag{6.6-29}$$

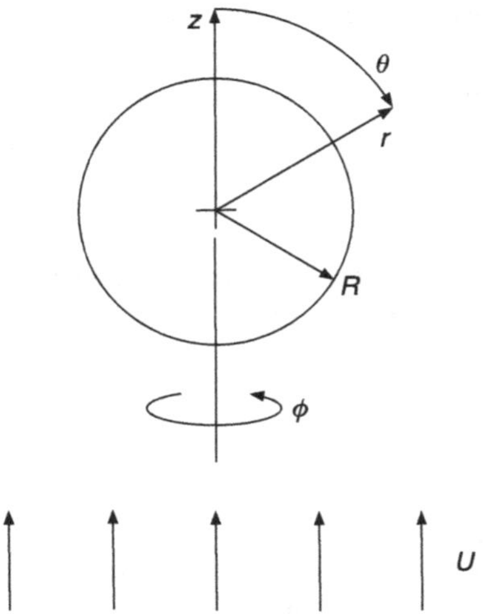

Figure 6.7 Flow at velocity U past a sphere of radius R.

Using Eq. (A.5-42), the additional unit-vector dot products are $\mathbf{e}_\theta \cdot \mathbf{e}_z = -\sin\theta$ and $\mathbf{e}_\phi \cdot \mathbf{e}_z = 0$. Substituting everything into Eq. (6.6-26) and performing the integration over ϕ gives

$$F_D = 2\pi R^2 \int_0^\pi [-\mathcal{P}(R,\theta)\cos\theta + \tau_{rr}(R,\theta)\cos\theta - \tau_{r\theta}(R,\theta)\sin\theta]\sin\theta\, d\theta. \quad (6.6\text{-}30)$$

It is seen that the only components of the viscous stress tensor that matter here are τ_{rr} and $\tau_{r\theta}$. For a *solid* sphere, the expression may be simplified further because there is no normal viscous stress at a solid surface. In that case, $\tau_{rr}(R,\theta) = 0$.

6.7 CONCLUSION

The stress vector $\mathbf{s}$ is a force per unit area that varies from point to point on a surface. Irrespective of whether the surface is an actual interface or an imaginary plane within a fluid, $\mathbf{s}$ depends on its orientation. The local orientation of a surface is described by a unit normal $\mathbf{n}$ that points into the material that is exerting the force. The need to keep track of two directions at once, those of $\mathbf{n}$ and of the resultant force, creates a need for tensor notation. The stress tensor $\boldsymbol{\sigma}$ is defined such that $\mathbf{s}(\mathbf{n}) = \mathbf{n} \cdot \boldsymbol{\sigma}$. The surface force per unit volume at any point in a fluid is $\nabla \cdot \boldsymbol{\sigma}$, and, including gravity, the total force per unit volume is $\rho\mathbf{g} + \nabla \cdot \boldsymbol{\sigma}$. Equating that force with inertial terms that involve temporal and spatial variations in $\mathbf{v}$ leads to the Cauchy momentum equation, the most general expression for conservation of linear momentum in a fluid. Decomposing $\boldsymbol{\sigma}$ into pressure and viscous contributions as $\boldsymbol{\sigma} = -P\boldsymbol{\delta} + \boldsymbol{\tau}$, where $\boldsymbol{\delta}$ is the identity tensor and $\boldsymbol{\tau}$ is the viscous stress tensor, leads to the "P–$\boldsymbol{\tau}$ form" of the Cauchy equation (Tables 6.1 through 6.3).

In a Newtonian fluid, each component of $\boldsymbol{\tau}$ is proportional to a rate of strain (Tables 6.4 through 6.6). By convention, the proportionality factor for an incompressible fluid is twice the shear viscosity or 2μ. If μ is independent of position, conservation of momentum leads to the Navier–Stokes equation (Tables 6.8 through 6.10). It and the continuity equation provide the usual starting point for analyzing incompressible flows.

Boundary conditions involving the normal component of **v** are derived by equating mass flow rates relative to interfaces. Boundary conditions involving tangential components of **v** come usually from the assumption of no slip, or no relative motion, where two phases come into contact. Additional boundary conditions arise from the need to balance normal and tangential components of stresses at interfaces, with or without the effects of surface tension.

The equations of motion in this chapter are comprehensive enough that the derivations need not be repeated when analyzing specific flows. Rather than begin each problem by deriving momentum and mass conservation statements anew, we can focus on identifying which terms (or even which equations) are not needed. By removing what equals zero or is otherwise negligible, the equations in the tables can be reduced to ones that are tractable. The simplifications possible for various geometric and dynamic conditions of engineering interest, and the physical insights that result, are the focus of Part III.

REFERENCES

Dahler, J. S. and L. E. Scriven. Theory of structured continua. I. General consideration of angular momentum and polarization. *Proc. Roy. Soc. Lond. A* 275: 504–527, 1963.

Deen, W. M. *Analysis of Transport Phenomena*, 2nd ed. Oxford University Press, New York, 2012.

Lauga, E., M. P. Brenner, and H. A. Stone. Microfluidics: the no-slip boundary condition. In *Handbook of Experimental Fluid Dynamics*, C. Tropea, A. Yarin, and J. F. Foss (Eds.). Springer, New York, 2007, pp. 1219–1240.

Rosensweig, R. E. *Ferrohydrodynamics*. Cambridge University Press, Cambridge, 1985.

Schlichting, H. *Boundary-Layer Theory*, 6th ed. McGraw-Hill, New York, 1968.

Whitaker, S. *Introduction to Fluid Mechanics*. Prentice-Hall, Englewood Cliffs, NJ, 1968.

PROBLEMS

6.1. Stress vector and tensor Suppose that the stress at point P in a fluid is determined by placing a small transducer there and measuring the force per unit area for different transducer orientations. The following time-independent results are obtained:

Surface orientation	Stress vector
$\mathbf{e}_x$	$\mathbf{e}_x + 2\mathbf{e}_y$
$\mathbf{e}_y$	$2\mathbf{e}_x + 3\mathbf{e}_y - \mathbf{e}_z$
$\mathbf{e}_z$	$-\mathbf{e}_y + 4\mathbf{e}_z$

(a) Determine the Cartesian components of the stress tensor, σ_{ij}.

(b) Evaluate the stress vector **s** at point P for a test surface with the orientation

$$\mathbf{n} = \frac{1}{3}(\mathbf{e}_x + 2\mathbf{e}_y + 2\mathbf{e}_z). \tag{P6.1.1}$$

(c) Calculate the normal component of the stress vector for the surface of part (b).

(d) Find the unit vector **t** that is tangent to the surface of part (b), lies in an x–y plane, and points in the $+x$ direction (i.e., $t_z = 0$ and $t_x > 0$). Compute the corresponding tangential component of the stress vector.

6.2. Effect of surface orientation on the stress vector Suppose that at point P in a fluid the Cartesian stress tensor is

$$\boldsymbol{\sigma} = \begin{pmatrix} 4 & 6 & 3 \\ 6 & 0 & 9 \\ 3 & 9 & 0 \end{pmatrix}. \tag{P6.2-1}$$

(a) Evaluate the stress vector **s** at point P for a test surface that is perpendicular to the x axis. Find the stress exerted by the fluid on the $+x$ side.
(b) Repeat part (a) for a test surface perpendicular to the y axis, finding the stress exerted by the fluid on the $+y$ side.
(c) With this stress tensor there is a particular surface orientation for which $\mathbf{s} = \mathbf{0}$. Find the unit normal **n** for the stress-free surface.

6.3. Force balance for plane Couette flow Figure P6.3 depicts flow in a horizontal parallel-plate channel in which $v_x = U$ at the upper wall, the lower wall is fixed, and P is independent of x. This is *plane Couette flow*, as introduced in Section 1.2. The wall spacing is H, the length under consideration is L, and the width (not shown) is W. Assuming that the flow is steady, laminar, and fully developed, the local velocity is found to be (Example 7.3-1)

$$v_x(y) = \frac{Uy}{H}, \quad v_y = 0 = v_z. \tag{P6.3-1}$$

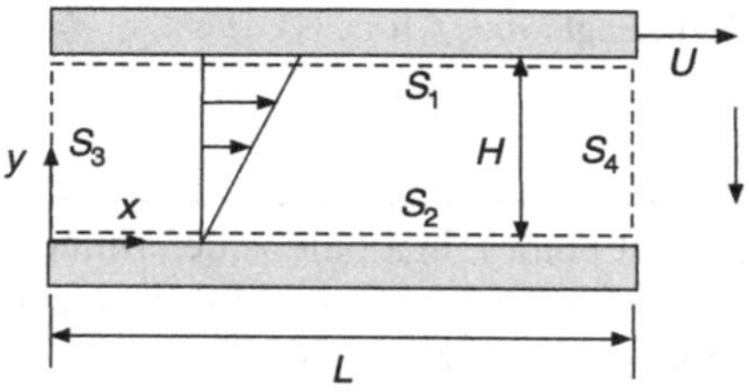

Figure P6.3 Plane Couette flow in a horizontal channel. The control volume for an overall force balance is shown by the dashed lines.

(a) Calculate each component of $\boldsymbol{\tau}$ for this flow.
(b) Use a force balance on the control volume in Fig. P6.3 to confirm that the average pressures at surfaces S_3 and S_4 ($x = 0$ and $x = L$, respectively) are equal. In using the results from part (a) to calculate the x component of the viscous force, be careful with the directions of the four unit normals and with the algebraic signs.

6.4. Force balance for plane Poiseuille flow Figure P6.4 depicts pressure-driven flow in a horizontal parallel-plate channel with wall spacing H, length L, and width W (not shown). This is called *plane Poiseuille flow*. Assuming that the flow is steady, laminar, and fully developed with a mean velocity u, the local velocity is found to be (Example 7.2-1)

$$v_x(y) = 6u\left[\left(\frac{y}{H}\right) - \left(\frac{y}{H}\right)^2\right], \quad v_y = 0 = v_z. \tag{P6.4-1}$$

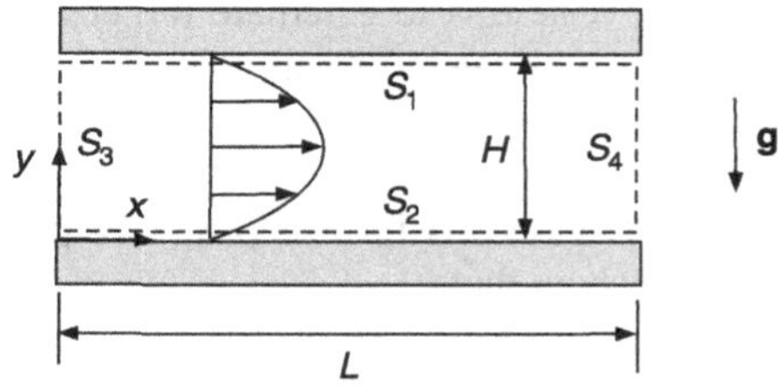

Figure P6.4 Plane Poiseuille flow in a horizontal channel. The control volume for an overall force balance is shown by the dashed lines.

(a) Calculate each component of $\boldsymbol{\tau}$ for this flow.
(b) Use a force balance on the control volume in Fig. P6.4 to evaluate the difference between the average pressures at surfaces S_3 and S_4 ($x = 0$ and $x = L$, respectively). In using the results from part (a), be careful with the directions of the four unit normals and with the algebraic signs. (You can check your pressure drop using information in Section 2.4.)

6.5. Normal viscous stress at a solid surface Show that for an incompressible Newtonian or generalized Newtonian fluid, the normal viscous stress vanishes at a solid surface. You may assume that the surface is stationary and that the no-slip condition holds. It is sufficient to consider planar, cylindrical, and spherical surfaces. (*Hint*: Use the continuity equation.)

6.6. Drag on a cylinder at high Reynolds number Consider flow at velocity U relative to a solid cylinder of radius R and length L whose axis is perpendicular to the approaching fluid, as in Fig. P6.6.

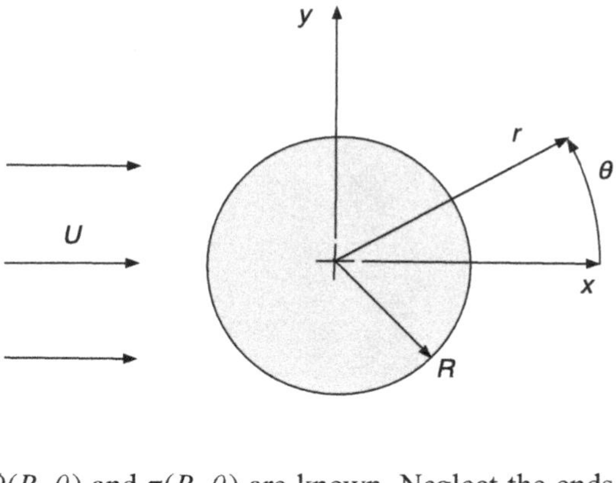

Figure P6.6 Flow at velocity U past a long cylinder of radius R.

(a) Show how to calculate the drag if $\mathcal{P}(R, \theta)$ and $\boldsymbol{\tau}(R, \theta)$ are known. Neglect the ends of the cylinder (i.e., assume that $L \gg R$) and exclude from your expression any components of $\boldsymbol{\tau}(R, \theta)$ that must be zero.
(b) A set of pressure measurements obtained for $\mathrm{Re} = 1.86 \times 10^5$ (Schlichting, 1968, p. 21) can be represented approximately as

$$\frac{\mathcal{P}(R,\theta)}{\rho U^2/2} = \begin{cases} -1 & \text{for } 0 \le \theta \le \dfrac{2\pi}{3} \\ 1 - \dfrac{8}{3}\sin^2\theta & \text{for } \dfrac{2\pi}{3} \le \theta \le \pi. \end{cases} \tag{P6.6-1}$$

Use this to calculate the contribution of *form drag* to C_D and compare your result with the information in Chapter 3. What does this comparison show about the relative importance of form drag and friction drag on a cylinder at high Re?

6.7. Pressure for creeping flow past a solid sphere Use the velocity field in Problem 5.4 to show that

$$\mathcal{P}(r, \theta) = -\frac{3}{2}\frac{\mu U}{R}\left(\frac{R}{r}\right)^2 \cos\theta \qquad \text{(P6.7-1)}$$

for flow at small Re (creeping flow) past a solid sphere. You may assume that all inertial terms in the Navier–Stokes equation are negligible for this flow.

6.8. Pressure between porous and solid disks Use the velocity field in Problem 5.6(b) to calculate $\mathcal{P}(r, z)$ for flow at small Re between porous and solid disks. You may assume that all inertial terms in the Navier–Stokes equation are negligible for this flow.

Part III

Microscopic analysis

7

Unidirectional flow

7.1 INTRODUCTION

A *unidirectional flow* is one that has a single nonzero velocity component. The resulting simplification of the continuity and momentum equations makes such flows a logical starting point for discussing how to predict velocity and pressure fields. Unidirectional flows are also very important in engineering. In particular, in a fluid-filled pipe or other conduit with a uniform cross-section, laminar flow becomes unidirectional if there is sufficient length. This chapter consists mainly of a set of examples of how to apply the governing equations of Chapter 6 in such situations. The topics include fully developed flow (including the derivation of Poiseuille's law), flows caused by moving surfaces, problems involving free surfaces (fluid–fluid interfaces), and flows of non-Newtonian fluids. Here and in Chapters 8 and 9 we are concerned only with laminar flow. Discussion of turbulent flow is deferred until Chapter 10.

7.2 FULLY DEVELOPED FLOW

Example 7.2-1 Velocity and pressure for plane Poiseuille flow. The objective is to predict the velocity and pressure fields for steady, fully developed flow in the parallel-plate channel in Fig. 7.1. We are given the channel height ($2H$), the length (L), and the inlet and outlet pressures ($\mathcal{P}_0$ and $\mathcal{P}_L$, respectively, where $\mathcal{P}_0 > \mathcal{P}_L$). The mean velocity U is an unknown.

What is meant in general by "fully developed" is that the velocity does not vary in the direction of flow; in this case $\mathbf{v}$ is independent of x. As in Chapter 2, we will assume that the entrance region is short enough that the flow is fully developed over nearly the entire channel length. As detailed in Section 12.2, this is a good approximation if $L/H \gg 0.5 + 0.1\,\text{Re}$, where $\text{Re} = 2UH/\nu$. The channel is assumed to be wide enough that its side walls can be ignored, making this a two-dimensional problem involving just x and y. In other words, $v_z = 0$ and the velocity and pressure do not depend on z. The accuracy of this approximation for rectangular channels of varying width was examined in Section 2.4. Pressure-driven flow in this geometry is called *plane Poiseuille flow*.

For fully developed, two-dimensional flow in such a channel, $v_y = 0$. This can be seen as follows. From Table 5.1, the Cartesian continuity equation for constant ρ is

$$\frac{\partial v_x}{\partial x} + \frac{\partial v_y}{\partial y} + \frac{\partial v_z}{\partial z} = 0. \tag{7.2-1}$$

Because the first and third terms are each zero from the problem statement, this indicates that $\partial v_y/\partial y = 0$. In other words, v_y is independent of y. Because the solid walls are

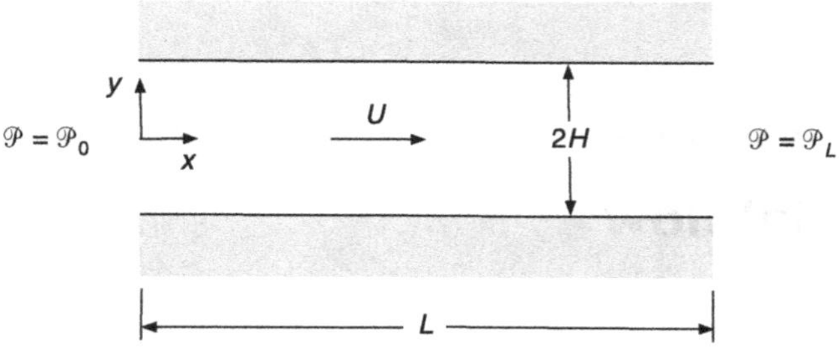

Figure 7.1 Pressure-driven flow in a parallel-plate channel.

stationary and impermeable, $v_y = 0$ at $y = \pm H$. With v_y vanishing at the walls and required by conservation of mass to be independent of y, we conclude that $v_y = 0$ everywhere. This leaves v_x as the only nonzero velocity component.

Each velocity term in the y and z components of the Navier–Stokes equation (Table 6.8) contains either v_y or v_z. Accordingly, $v_y = 0 = v_z$ implies that

$$\frac{\partial \mathcal{P}}{\partial y} = 0 = \frac{\partial \mathcal{P}}{\partial z}. \tag{7.2-2}$$

Thus, momentum conservation requires that $\mathcal{P}$ be independent of y and z, or that $\mathcal{P} = \mathcal{P}(x)$ only. Because it is independent of x and z, the velocity also depends on just one coordinate, such that $v_x = v_x(y)$ only.

Almost all terms in the x component of the Navier–Stokes equation are identically zero. What remains may be written as

$$\frac{d^2 v_x}{dy^2} = \frac{1}{\mu}\frac{d\mathcal{P}}{dx}. \tag{7.2-3}$$

Ordinary derivatives are appropriate now because v_x and $\mathcal{P}$ are each functions of just one variable (y and x, respectively). This second-order differential equation requires two boundary conditions, which are provided by no-slip conditions at the walls,

$$v_x(H) = 0 = v_x(-H). \tag{7.2-4}$$

Solving Eq. (7.2-3) subject to Eq. (7.2-4) will complete the evaluation of $v_x(y)$. Together with the known pressures at $x = 0$ and $x = L$, this will also reveal $\mathcal{P}(x)$.

The left-hand side of Eq. (7.2-3) depends at most on y, whereas the right-hand side depends at most on x. As there is no common variable, each side must equal a constant. Using the constancy of $d\mathcal{P}/dx$ and the given pressures at the ends, the local pressure is evaluated as

$$\mathcal{P}(x) = \mathcal{P}_0 + (\mathcal{P}_L - \mathcal{P}_0)\frac{x}{L}. \tag{7.2-5}$$

Thus, the pressure decline along the channel is linear. Equation (7.2-3) is rewritten now in terms of the inlet and outlet pressures as

$$\frac{d^2 v_x}{dy^2} = -\frac{|\Delta \mathcal{P}|}{\mu L}. \tag{7.2-6}$$

Two integrations in y give

$$v_x(y) = -\frac{|\Delta \mathcal{P}|}{2\mu L} y^2 + ay + b \tag{7.2-7}$$

where a and b are constants. The no-slip conditions [Eq. (7.2-4)] provide two algebraic equations for the two unknown constants. Satisfying them requires that

$$a = 0, \quad b = \frac{|\Delta \mathcal{P}|}{2\mu L} H^2. \tag{7.2-8}$$

The local velocity is then

$$v_x(y) = \frac{H^2 \, |\Delta \mathcal{P}|}{2\mu L} \left[1 - \left(\frac{y}{H} \right)^2 \right]. \tag{7.2-9}$$

This profile has the parabolic shape sketched in Fig. 2.1. It is symmetric about the channel midplane ($y = 0$) and has a maximum there, which is

$$v_{\max} = \frac{H^2 \, |\Delta \mathcal{P}|}{2\mu L}. \tag{7.2-10}$$

The velocity profile can be expressed also in terms of the mean velocity U, which is calculated by averaging v_x over the channel height. Because v_x is symmetric about $y = 0$, its average can be found by integrating over just the top half. Thus,

$$U = \frac{1}{H} \int_0^H v_x \, dy = \frac{H^2 \, |\Delta \mathcal{P}|}{3\mu L} = \frac{2}{3} v_{\max}. \tag{7.2-11}$$

An alternate form of the velocity profile is then

$$v_x(y) = \frac{3}{2} U \left[1 - \left(\frac{y}{H} \right)^2 \right]. \tag{7.2-12}$$

Although the problem was posed as if $|\Delta \mathcal{P}|$ were known, U could have been given instead. After essentially the same analysis, $|\Delta \mathcal{P}|$ then could have been calculated from U by rearranging Eq. (7.2-11). Thus, the pressure drop may be viewed either as the cause or as the result of the flow. In that only the dynamic pressure was involved, the results apply equally well to horizontal, inclined, or vertical channels (Section 2.3).

Example 7.2-2 Velocity and pressure for Poiseuille flow. The objective is to determine the velocity profile for steady, fully developed flow in a cylindrical tube of radius R. The system is like that in Fig. 7.1, except that R replaces H and the cylindrical coordinates z and r replace x and y, respectively. It is assumed that the flow is *axisymmetric*, which means that the velocity and pressure do not depend on the cylindrical angle θ and there is no swirling motion ($v_\theta = 0$). Pressure-driven flow in this geometry is called *Poiseuille flow*.

Arguments like those in Example 7.2-1 indicate that $v_r = 0$, $v_z = v_z(r)$ only, and $\mathcal{P} = \mathcal{P}(z)$ only. It is seen again that a fully developed flow is unidirectional. That is true for conduits of any cross-sectional shape, including but not limited to those in Table 2.1. The z component of the cylindrical Navier–Stokes equation (Table 6.9) reduces to

$$\frac{1}{r} \frac{d}{dr} \left(r \frac{dv_z}{dr} \right) = \frac{1}{\mu} \frac{d\mathcal{P}}{dz}. \tag{7.2-13}$$

The left-hand side depends at most on r and the right-hand side depends at most on z, so each must be constant. Thus, there is a linear decline in pressure along the tube, analogous to Eq. (7.2-5). Evaluating $d\mathcal{P}/dz$ and multiplying by r gives

$$\frac{d}{dr}\left(r\frac{dv_z}{dr}\right) = -\frac{r\,|\Delta\mathcal{P}|}{\mu L}. \tag{7.2-14}$$

Integrating once, dividing by r, and integrating again leads to

$$v_z(r) = -\frac{r^2\,|\Delta\mathcal{P}|}{4\mu L} + a\ln r + b \tag{7.2-15}$$

where a and b are constants. In the radial direction there is only one true boundary (the wall), but because the domain is $0 \leq r \leq R$ the tube axis is a mathematical boundary. To avoid the physical impossibility of an infinite velocity at $r = 0$, we set $a = 0$. Evaluating b using the no-slip condition $[v_z(R) = 0]$, the velocity is found to be

$$v_z(r) = \frac{R^2\,|\Delta\mathcal{P}|}{4\mu L}\left[1 - \left(\frac{r}{R}\right)^2\right]. \tag{7.2-16}$$

Again, the profile is parabolic. The maximum (centerline) value is

$$v_{\text{maz}} = \frac{R^2\,|\Delta\mathcal{P}|}{4\mu L}. \tag{7.2-17}$$

The mean velocity is[1]

$$U = \frac{2}{R^2}\int_0^R v_z r\,dr = \frac{R^2\,|\Delta\mathcal{P}|}{8\mu L} = \frac{1}{2}v_{\text{max}}. \tag{7.2-18}$$

Whereas for a parallel-plate channel it was found that $U/v_{\text{max}} = 2/3$, for a circular tube $U/v_{\text{max}} = 1/2$. Expressed in terms of U, the velocity profile is

$$v_z(r) = 2U\left[1 - \left(\frac{r}{R}\right)^2\right]. \tag{7.2-19}$$

The volume flow rate is

$$Q = \pi R^2 U = \frac{\pi R^4\,|\Delta\mathcal{P}|}{8\mu L}. \tag{7.2-20}$$

This is *Poiseuille's law*, expressed previously as Eq. (2.3-7). It is seen now that Poiseuille's empirical relationship is derivable from first principles. The theoretical analysis for flow through a tube seems to have been reported first by E. Hagenbach of Basel in 1860, some 20 years after Poiseuille's experiments (Sutera and Skalak, 1993).

Example 7.2-3 Friction factor for laminar tube flow. The friction factor for laminar flow in cylindrical tubes was presented in Chapter 2 as an empirical finding. However, it can be derived from first principles in either of two ways. One approach is to start with the relationship between f and the pressure drop [Eq. (2.3-6)], and then use Eq. (7.2-18)

[1] In a plane of constant z the differential area in cylindrical coordinates is $r\,dr\,d\theta$. Accordingly, the mean value of a function $f(r,\theta)$ over a circular area of radius R is

$$\frac{\int_0^{2\pi}\int_0^R f r\,dr\,d\theta}{\int_0^{2\pi}\int_0^R r\,dr\,d\theta}.$$

For $f = f(r)$ only, the θ integrations each give a factor of 2π, which cancels. The denominator then equals $R^2/2$.

to relate the pressure drop to the mean velocity. The other is to begin with the definition of f as a dimensionless wall shear stress [Eq. (2.2-4)], and then use the velocity profile in Eq. (7.2-19) to calculate that stress. To illustrate the calculation of viscous stresses, the latter approach is taken here.

As used in Part I, τ_w is the shear stress exerted on a fluid by a channel wall or other solid surface. Using the more systematic notation introduced in Chapter 6,

$$\tau_w = -\tau_{rz}(R) = -\mu \frac{dv_z}{dr}(R) \tag{7.2-21}$$

for a cylindrical tube of radius R. The minus sign is needed because $\tau_w > 0$ and τ_{rz} at the tube wall is negative. That is, the force exerted by the wall on the fluid acts in the $-z$ direction, resisting the fluid motion. The second equality comes from the stress relationships in Table 6.5. There is no $\partial v_r/\partial z$ term because $v_r = 0$ everywhere, making that contribution to τ_{rz} (or τ_{zr}) zero. Even if the flow were not fully developed and v_r vanished only at the wall, we would still have $\partial v_r/\partial z = 0$ at $r = R$ and Eq. (7.2-21) would still apply.

Differentiating Eq. (7.2-19) to find the wall shear stress gives

$$\tau_w = -\mu(2U)\left(-\frac{2r}{R^2}\right)\bigg|_{r=R} = \frac{4\mu U}{R}. \tag{7.2-22}$$

The friction factor is then evaluated from Eq. (2.2-4) as

$$f = \frac{2\tau_w}{\rho U^2} = \frac{2}{\rho U^2}\left(\frac{4\mu U}{R}\right) = \frac{8\mu}{\rho U R} = \frac{16}{\text{Re}} \tag{7.2-23}$$

in agreement with Eq. (2.2-6). The final factor of 2 comes from using the diameter (rather than R) as the length scale in the Reynolds number. Of course, the same result is obtained from Eqs. (2.3-6) and (7.2-18).

In the foregoing examples only one transverse coordinate (y or r) was needed to specify positions within a cross-section. With other conduit shapes two may be needed. For example, a rectangular channel of moderate width would require both y and z. Also, fully developed flows can be time-dependent, as when there is a pulsatile pump. In such cases the fluid at all axial positions speeds or slows in unison, such that the flow is still fully developed and unidirectional. The assertion that v_y and v_z are both zero remains consistent with the continuity equation and no-penetration conditions, even for complex cross-sectional shapes and time-dependent pressures.

The most general Cartesian form of the Navier–Stokes equation for fully developed flow in the x direction is

$$\frac{1}{\nu}\frac{\partial v_x}{\partial t} = -\frac{1}{\mu}\frac{\partial \mathcal{P}}{\partial x} + \frac{\partial^2 v_x}{\partial y^2} + \frac{\partial^2 v_x}{\partial z^2} \tag{7.2-24}$$

where $v_x = v_x(y, z, t)$ and $\mathcal{P} = \mathcal{P}(x, t)$. When z and t are absent, this partial differential equation becomes an ordinary differential equation, Eq. (7.2-3). Analytical solutions can still be obtained when there is more than one independent variable, although the methods needed are largely outside the scope of this book. For example, Eq. (7.2-24) can be solved for a rectangular channel using an *eigenfunction expansion*, in which v_x is represented as a Fourier series (see, for example, Chapter 5 of Deen, 2012). This is how Eq. (2.4-4) was derived. That method is applicable whenever each part of the channel wall is a coordinate surface. For more irregular shapes, Eq. (7.2-24) can be solved numerically using commercial finite-element packages.

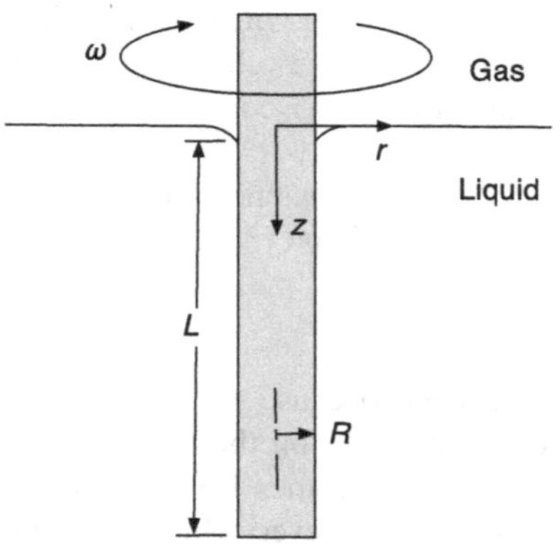

Figure 7.2 Vertical rod rotating in a liquid.

7.3 MOVING SURFACES

Example 7.3-1 Plane Couette flow. We return now to the shear flow used in Section 1.2 to introduce viscosity. A key aspect of that flow is the absence of a pressure gradient; the fluid is propelled only by the shear stress exerted on it by a moving surface. As shown in Fig. 1.1, a flat surface at $y = 0$ is stationary and one at $y = H$ moves at a constant speed U in the $+x$ direction. It is assumed that the flow is steady, two-dimensional, and fully developed. That is, the plates are considered to be long and wide enough to neglect the effects of ends or side walls. Surface-driven flow in this geometry is called *plane Couette flow*. The objective is to determine $v_x(y)$ and to verify Eq. (1.2-1).

In that this is a steady, unidirectional flow in Cartesian coordinates, the differential equation that governs $v_x(y)$ can be obtained by eliminating z and t from Eq. (7.2-24) and setting $\partial\mathcal{P}/\partial x = 0$. Thus, the governing equations are

$$\frac{d^2 v_x}{dy^2} = 0 \tag{7.3-1}$$

$$v_x(0) = 0, \quad v_x(H) = U. \tag{7.3-2}$$

Equation (7.3-1) indicates that $v_x(y)$ is linear. The linear profile that satisfies the no-slip conditions in Eq. (7.3-2) is

$$v_x(y) = \frac{Uy}{H} \tag{7.3-3}$$

as sketched in Fig. 1.1. The shear stress exerted on the fluid by the top wall is

$$\tau_{yx}(H) = \mu \frac{dv_x}{dy}(H) = \frac{\mu U}{H} \tag{7.3-4}$$

which is consistent with Eq. (1.2-1). Because dv_x/dy in this simple shear flow is independent of y, this is also the shear stress that the fluid exerts on the bottom wall.

Example 7.3-2 Rotating rod. Suppose that a vertical rod of radius R is rotated at a constant angular velocity ω in a large container of liquid, as shown in Fig. 7.2. The immersed length is L. We will assume that the flow is axisymmetric and unidirectional, with $v_\theta = v_\theta(r)$ and $v_r = 0 = v_z$. This is consistent with the cylindrical continuity equation (Table 5.1), and our ability to satisfy all three components of the Navier–Stokes equation will confirm that such a flow is feasible. Implicit in the assumption that v_θ is independent of z is that L is large enough to ignore the more complex fluid motion near the bottom of the rod. It is desired to determine the velocity, pressure, and applied torque.

The symmetry implies that $\partial\mathcal{P}/\partial\theta = 0$. With v_r and v_z also zero, the θ component of the cylindrical Navier–Stokes equation (Table 6.9) reduces to

$$\frac{d}{dr}\left(\frac{1}{r}\frac{d}{dr}(rv_\theta)\right) = 0. \tag{7.3-5}$$

Applying a no-slip condition at the rod and assuming that the fluid far away is at rest,

$$v_\theta(R) = \omega R, \quad v_\theta(\infty) = 0. \tag{7.3-6}$$

Because v_θ is a linear velocity (units of m/s), its value at the rod surface is the angular velocity times the radius. In the second boundary condition it is assumed that the container radius is practically infinite (i.e., large enough so as not to influence the flow near the rod).

Two integrations of Eq. (7.3-5) give

$$v_\theta(r) = ar + \frac{b}{r} \tag{7.3-7}$$

where a and b are constants. A static liquid far from the rod requires that $a = 0$, and the no-slip condition then gives $b = \omega R^2$. The final result for the velocity is

$$v_\theta(r) = \frac{\omega R^2}{r}. \tag{7.3-8}$$

The ability to satisfy both boundary conditions confirms that it is realistic to assume that the distant liquid is undisturbed by the rotation of the rod.

The torque that the liquid exerts on the rod, which must be balanced by that supplied by the motor, can be calculated using Eq. (6.6-27). The position vector that gives the torque about $r = 0$ is $\mathbf{r} = R\mathbf{e_r}$, and the unit normal into the fluid is $\mathbf{n} = \mathbf{e_r}$. The cross-product rules give $\mathbf{r} \times \mathbf{n} = \mathbf{0}$ and $\mathbf{r} \times (\mathbf{n} \cdot \boldsymbol{\tau}) = R(\tau_{r\theta}\mathbf{e_z} - \tau_{rz}\mathbf{e_\theta})$. Inspection of Table 6.5 indicates that $\tau_{rz} = 0$ in this system and that the nonzero component of the shear stress at the surface is

$$\tau_{r\theta}(R) = \mu r\frac{d}{dr}\left(\frac{v_\theta}{r}\right)\bigg|_{r=R} = -\frac{2\mu\omega R^2}{r^2}\bigg|_{r=R} = -2\mu\omega. \tag{7.3-9}$$

The surface area element is $dS = R\,d\theta\,dz$. Thus, the torque vector is

$$\mathbf{G} = \int_0^L \int_0^{2\pi} (-2\mu\omega R\mathbf{e_z})R\,d\theta\,dz = -4\pi\mu\omega R^2 L\mathbf{e_z}. \tag{7.3-10}$$

Alternatively, the torque magnitude G can be calculated simply as the lever arm times the force in the θ direction, which in turn is the shear stress times the lateral area. Thus,

$$G = R|\tau_{r\theta}(R)|(2\pi RL) = 4\pi\mu\omega R^2 L. \tag{7.3-11}$$

Although the flow near a rotating rod is similar to plane Couette flow in that it is caused by a shear stress imposed at a solid surface, it differs in that $\mathcal{P}$ is not constant in a rotating liquid. For an axisymmetric flow with $v_z = 0$, the z component of the cylindrical Navier–Stokes equation reduces to $\partial\mathcal{P}/\partial z = 0$. However, with $v_r = 0$ the r component becomes

$$\frac{d\mathcal{P}}{dr} = \frac{\rho v_\theta^2}{r} = \frac{\rho\omega^2 R^4}{r^3} \tag{7.3-12}$$

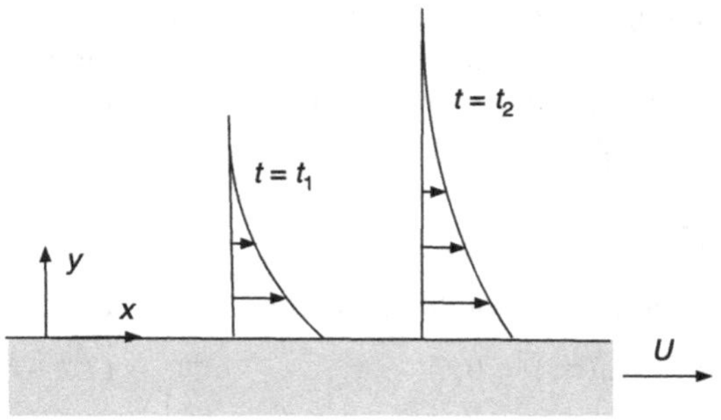

Figure 7.3 Flow near a suddenly accelerated plate. Velocity profiles are sketched for two times, with $t_2 > t_1 > 0$.

which indicates that $\mathcal{P}$ increases with r. This is a consequence of inertia. The radial pressure gradient is the force that keeps each fluid element on a circular path. Moving at a constant speed but with a continuously changing direction is, of course, a kind of acceleration. Integrating Eq. (7.3-12) and choosing $\mathcal{P}(\infty) = 0$ gives

$$\mathcal{P}(r) = -\frac{\rho\omega^2 R^4}{2r^2}. \tag{7.3-13}$$

The radial variation in P is the same as that in $\mathcal{P}$. Thus, at any given depth in the liquid, P near the rod will be lower than it is in the static fluid far away. If P is depressed below the vapor pressure of the liquid, bubbles will tend to form. The rapid formation and collapse of such bubbles, termed *cavitation*, can seriously damage rotating machinery. The pressure variations in this flow are examined further in Example 7.4-2.

Example 7.3-3 Plate suddenly set in motion. The time-dependent flow that is considered now is called the *Rayleigh problem* (Rayleigh, 1911). Suppose that a flat plate at $y = 0$ is suddenly accelerated from rest to a constant speed U in the $+x$ direction, as depicted in Fig. 7.3. The adjacent fluid, which occupies the space $y \geq 0$, is initially static. Over time, more and more of the fluid is set in motion, as indicated by the two velocity profiles that are sketched. It is assumed that $\mathcal{P}$ is constant and that v_x is independent of x and z, as in plane Couette flow. The objective is to determine $v_x(y, t)$. A special method for solving partial differential equations will be introduced and the time needed for viscous stresses to propagate within a fluid (the viscous time scale) will be found. Familiarity with the viscous time scale is needed in Chapter 8 and the solution method is used again in Chapter 9.

This is another fully developed flow in Cartesian coordinates. With v_x independent of z and no pressure gradient, Eq. (7.2-24) reduces to

$$\frac{\partial v_x}{\partial t} = \nu \frac{\partial^2 v_x}{\partial y^2}. \tag{7.3-14}$$

The initial condition and boundary conditions are

$$v_x(y, 0) = 0 \tag{7.3-15}$$

$$v_x(0, t) = U \tag{7.3-16}$$

$$v_x(\infty, t) = 0. \tag{7.3-17}$$

Equation (7.3-15) indicates that for all y the fluid is initially at rest, and Eq. (7.3-16) states that for all $t > 0$ the fluid contacting the plate moves at the speed U, consistent with no slip. Equation (7.3-17) embodies something more subtle, which is the assumption that any fluid sufficiently distant from the plate will remain static for finite values of t. The solution will confirm that the motion spreads gradually over time to larger y, rather than instantaneously.

The differential equation and auxiliary conditions are linear and contain just one nonhomogeneous term, U in Eq. (7.3-16). Consequently, $v_x(y, t)$ must always be proportional to U. Working with a normalized velocity, namely $F = v_x/U$, simplifies matters by eliminating U as a parameter. Dividing each governing equation by U replaces v_x by F and U by 1.

Dimensional analysis provides a key insight. The modified differential equation,

$$\frac{\partial F}{\partial t} = \nu \frac{\partial^2 F}{\partial y^2} \tag{7.3-18}$$

contains a dimensionless unknown (F) and three dimensional quantities on which it depends (y, t, and ν). The normalized auxiliary conditions contain only F, y, and t, so no additional variables or parameters are involved. The three dimensional quantities contain only two independent dimensions (L and T). Thus, from the Pi theorem (Section 1.4), they can form just one independent dimensionless group. That is, F must depend on a single group. Choosing y as the numerator of the group, its denominator must consist of a combination of t and ν that also has the dimension L. Setting

$$\mathrm{L} = \left[t^a \nu^b\right] = \mathrm{T}^a \left(\mathrm{L}^2 \mathrm{T}^{-1}\right)^b = \mathrm{T}^{a-b} \mathrm{L}^{2b} \tag{7.3-19}$$

leads to $a = b = 1/2$. We conclude that $F = F(\eta)$, where

$$\eta = \frac{y}{2(\nu t)^{1/2}}. \tag{7.3-20}$$

The factor 2 merely helps put the solution in the simplest form, as will be seen. What is important is that F does not depend on y and t separately. Accordingly, it should be possible to rewrite Eq. (7.3-18) as an ordinary differential equation that governs $F(\eta)$.

The partial derivatives are transformed as

$$\frac{\partial F}{\partial t} = \frac{dF}{d\eta} \frac{\partial \eta}{\partial t} = \frac{dF}{d\eta} \left(-\frac{\eta}{2t} \right) \tag{7.3-21}$$

$$\frac{\partial F}{\partial y} = \frac{dF}{d\eta} \frac{\partial \eta}{\partial y} = \frac{dF}{d\eta} \frac{1}{2(\nu t)^{1/2}} \tag{7.3-22}$$

$$\frac{\partial^2 F}{\partial y^2} = \frac{d^2 F}{d\eta^2} \left(\frac{\partial \eta}{\partial y} \right)^2 = \frac{d^2 F}{d\eta^2} \frac{1}{4\nu t}. \tag{7.3-23}$$

Substituting these expressions into Eq. (7.3-18) gives

$$\frac{d^2 F}{d\eta^2} + 2\eta \frac{dF}{d\eta} = 0. \tag{7.3-24}$$

The initial and boundary conditions [Eqs. (7.3-15)–(7.3-17)] transform to

$$F(0) = 1 \tag{7.3-25}$$

$$F(\infty) = 0. \tag{7.3-26}$$

Notice that because $t = 0$ and $y = \infty$ each correspond to $\eta = \infty$, the initial condition and distant boundary condition become one, Eq. (7.3-26). Thus, although Eq. (7.3-24) can satisfy only two boundary conditions, and we started with three auxiliary conditions, no information has been discarded. In that η is the only independent variable in Eqs. (7.3-24)–(7.3-26), the conclusion from dimensional analysis is confirmed.

Equation (7.3-24) is equivalent to a linear, first-order equation for $dF/d\eta$, the solution of which is

$$\frac{dF}{d\eta} = Ce^{-\eta^2} \tag{7.3-27}$$

where C is a constant.[2] Another integration gives

$$F(\eta) - F(0) = C\int_0^\eta e^{-s^2}\,ds \tag{7.3-28}$$

where s is a dummy variable. Employing a definite integral makes $F(0)$ the second integration constant. Evaluating Eq. (7.3-28) at $\eta = \infty$ and using Eqs. (7.3-25) and (7.3-26) gives

$$C = -\left(\int_0^\infty e^{-s^2}\,ds\right)^{-1} = -\frac{2}{\sqrt{\pi}}. \tag{7.3-29}$$

The final result for the velocity is

$$F(\eta) = \frac{v_x}{U} = 1 - \mathrm{erf}(\eta) = \mathrm{erfc}(\eta) \tag{7.3-30}$$

$$\mathrm{erf}(x) = \frac{2}{\sqrt{\pi}}\int_0^x e^{-s^2}\,ds. \tag{7.3-31}$$

The function erf(x) defined by Eq. (7.3-31) is the *error function*. It is encountered so often in probability theory and other fields that it is widely available in spreadsheets, math software, and published tables. Closely related is the *complementary error function*, $\mathrm{erfc}(x) = 1 - \mathrm{erf}(x)$.

The fluid velocity is plotted in Fig. 7.4. Combining y and t into η superimposes velocity profiles like those in Fig. 7.3 onto the single curve shown. The results confirm that, at any instant, there are measurable velocities only within a region of finite thickness next to the plate. That thickness grows over time. Among the many ways that growth might be described, let $y = \delta_{50}(t)$ and $\delta_5(t)$ be the positions at which the velocity is 50% and 5%, respectively, of that at the plate. In that $\mathrm{erfc}(0.477) = 0.500$ and $\mathrm{erfc}(1.386) = 0.0500$, Eq. (7.3-20) indicates that $\delta_{50}(t) = 0.95(\nu t)^{1/2}$ and $\delta_5(t) = 2.8(\nu t)^{1/2}$. Such results may be summarized as $\delta(t) = c(\nu t)^{1/2}$, where c is a somewhat arbitrary constant that has order of magnitude unity. Setting $c = 1$, the time required for velocity changes to propagate over a distance δ is estimated as

$$t_v = \frac{\delta^2}{\nu}. \tag{7.3-32}$$

In that the flow is due entirely to viscous stresses, this is called the *viscous time scale*. As mentioned in Section 1.3, such a time scale is inferred also from dimensional arguments.

[2] This is where the 2 in Eq. (7.3-20) is helpful, as it makes the coefficient of $dF/d\eta$ in Eq. (7.3-24) exactly the derivative of η^2. That simplifies the argument of the exponential in Eq. (7.3-27).

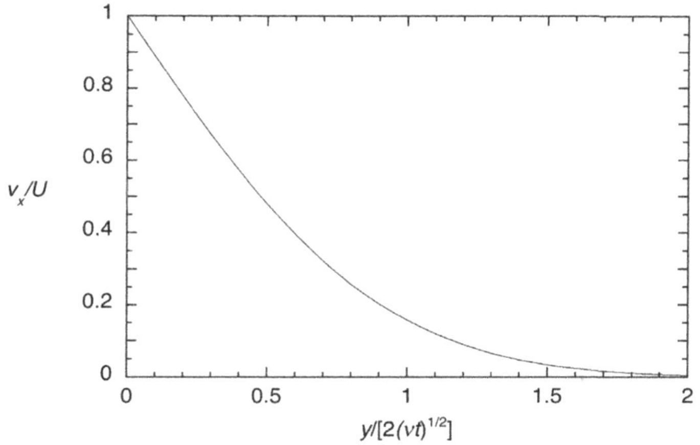

Figure 7.4 Fluid velocity in response to the sudden acceleration of a flat plate at $y = 0$ from rest to a speed U.

In this flow there is a gradual transfer of momentum in the y direction, due to viscous stresses. This is a dispersion process that is formally identical to diffusion of a solute or conduction of heat. Accordingly, ν is viewed as the diffusion coefficient for momentum (Deen, 2012, p. 7). It is the diffusion coefficient also for vorticity (Section 5.5), in that gradually increasing amounts of fluid acquire nonzero values of $w_z(= -\partial v_x/\partial y)$. The transport of vorticity is examined further in Chapter 9.

The Rayleigh problem resembles the startup of plane Couette flow, except that the second plate bounding the fluid is absent. Suppose that the plate spacing is H, as in Example 7.3-1. The present results indicate that at very early times, such that $(\nu t)^{1/2} \ll H$, the influence of the stationary plate will not yet be felt. In other words, if we wished to model the earliest transient stage of plane Couette flow, viewing the fluid region as semi-infinite would be a good approximation. For either a semi-infinite fluid or Couette flow at small t, the shear stress that the moving plate exerts on the fluid is

$$-\tau_{yx}|_{y=0} = -\mu \left.\frac{\partial v_x}{\partial y}\right|_{y=0} = -\frac{\mu U}{2(\nu t)^{1/2}} \frac{dF}{d\eta}(0) = \frac{\mu U}{(\pi \nu t)^{1/2}}. \tag{7.3-33}$$

The many analytical methods for solving partial differential equations all reduce the partial differential equation to one or more ordinary differential equations. The method used here does this in a special way, by combining two original independent variables into one. The reason this worked is that the velocity profile at each instant has exactly the same shape. That is, graphs of v_x vs. y at different values of t can be superimposed by using the factor $(\nu t)^{1/2}$ to account for the time-dependent length scale. Such profiles are referred to as *self-similar*, and the method is called a *similarity transformation* or *combination of variables*. Similarity solutions can be very elegant, as in this problem, but the method is applicable only to semi-infinite or infinite domains and to certain kinds of differential equations.

Dimensional analysis indicated that the similarity method would work and showed how to combine the independent variables. If the problem had involved plates spaced a finite distance H apart, it would have suggested (correctly) that the similarity method would fail. That is, having H in a boundary condition would create an additional dimensionless group, in which case y and t would no longer have to appear only in combination.

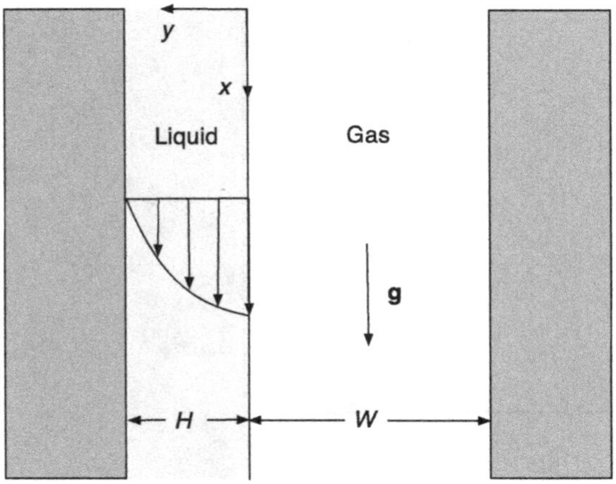

Figure 7.5 Liquid film of thickness H flowing down a flat, vertical surface.

Indeed, a natural choice of dimensionless independent variables would then be y/H and $t\nu/H^2$. Finally, note that, whereas the examples in Section 1.4 involved a certain amount of guesswork in assembling lists of relevant quantities, in the present problem we could be quite sure of what was involved. In general, the more one knows about the governing equations, the more confident one can be in drawing conclusions from dimensional analysis.

7.4 FREE SURFACES

In the examples in Sections 7.2 and 7.3 the governing equations could be formulated using only the dynamic pressure $\mathcal{P}$. This allowed the velocity fields to be found without explicit consideration of gravity. An aspect of problems involving free surfaces (fluid–fluid interfaces) is that knowing $\mathcal{P}$ is insufficient. In particular, information about the absolute pressure P is needed in order to apply the normal-stress boundary condition.

Example 7.4-1 Falling film on a vertical wall. A liquid film of constant thickness running down a flat, vertical surface, as in Fig. 7.5, is a prototype for many free-surface flows. The film thickness is H, the gas–liquid interface is at $y = 0$, and the width of the gas space is W. The objective is to determine $v_x(y)$ in the *liquid* for steady, fully developed flow. It will be assumed that pressure variations and viscous stresses in the *gas* are each negligible. Why such assumptions for the gas are usually valid will be discussed.

This unidirectional flow is steady and two-dimensional. Accordingly, in the liquid $v_x = v_x(y)$, $\mathcal{P} = \mathcal{P}(x)$, $d\mathcal{P}/dx$ is constant, and the Navier–Stokes equation reduces again to

$$\frac{d^2 v_x}{dy^2} = \frac{1}{\mu}\frac{d\mathcal{P}}{dx}. \tag{7.4-1}$$

The boundary conditions for v_x in the liquid are

$$\frac{dv_x}{dy}(0) = 0 \tag{7.4-2}$$

$$v_x(H) = 0. \tag{7.4-3}$$

Equation (7.4-2) is a consequence of the equality of shear stresses at the gas–liquid interface [Eq. (6.6-19)] and the assumption that viscous stresses in the gas are negligible. It comes from setting $\tau_{yx}(0) = 0$ and recognizing that $\partial v_y / \partial x = 0$ at $y = 0$.

The normal-stress boundary condition at the free surface [Eq. (6.6-18)] reveals the value of $d\mathcal{P}/dx$. In the liquid, $\tau_{yy}(0) = 0$ because $v_y = 0$ everywhere, and in the gas, τ_{yy} is assumed to be negligible. Because the interface is flat, the normal-stress balance is unaffected by surface tension. Thus, it reduces to equal P in the two phases. Referring to Eq. (6.6-4) and noting that $g_x = g$ and $g_y = 0$, the derivatives of P and $\mathcal{P}$ within the liquid are related as

$$\frac{\partial \mathcal{P}}{\partial x} = \frac{\partial P}{\partial x} - \rho_L g \tag{7.4-4}$$

$$\frac{\partial \mathcal{P}}{\partial y} = \frac{\partial P}{\partial y} \tag{7.4-5}$$

where ρ_L is the liquid density. Because the unidirectional flow implies that $\partial \mathcal{P}/\partial y = 0$ throughout the liquid, Eq. (7.4-5) indicates that both $\mathcal{P}$ and P are independent of y. It follows that, if P in the gas is constant along the surface $y = 0$, P will have that same constant value throughout the liquid. With $\partial P/\partial x = 0$ in the liquid, Equation (7.4-4) indicates that

$$\frac{d\mathcal{P}}{dx} = -\rho_L g \tag{7.4-6}$$

which is the information needed for the momentum equation.

Substitution of Eq. (7.4-6) into Eq. (7.4-1) gives

$$\frac{d^2 v_x}{dy^2} = \frac{1}{\mu_L}\frac{d\mathcal{P}}{dx} = -\frac{g}{\nu_L} \tag{7.4-7}$$

where μ_L and ν_L are the liquid shear viscosity and kinematic viscosity, respectively. Integrating twice and applying Eqs. (7.4-2) and (7.4-3) yields the velocity profile,

$$v_x(y) = \frac{gH^2}{2\nu_L}\left[1 - \left(\frac{y}{H}\right)^2\right]. \tag{7.4-8}$$

It is seen that the profile is parabolic, as sketched in Fig. 7.5, with a maximum value of $gH^2/(2\nu_L)$ at the gas–liquid interface. The mean velocity is

$$U = \frac{1}{H}\int_0^H v_x \, dy = \frac{gH^2}{3\nu_L}. \tag{7.4-9}$$

As in plane Poiseuille flow, the mean velocity is two-thirds of the maximum.

When analyzing liquid flows, it is common to neglect viscous stresses and pressure variations in adjacent gases, as was done here. What makes this reasonable is that the densities and viscosities of gases are relatively small. If the location of the gas–liquid interface is known, as in this example, that approximation completely uncouples the liquid model from events in the gas. Thus, the velocity, pressure, and material properties of the gas are all absent from the governing equations, as if the liquid were in contact with a vacuum. The great advantage of this is that the Navier–Stokes and continuity equations need not be solved for the gas. Instead of having two sets of differential equations (liquid and gas), coupled to one another via velocity and stress conditions at the free surface, there is only one. However, this simplification works only in one direction. That is, in

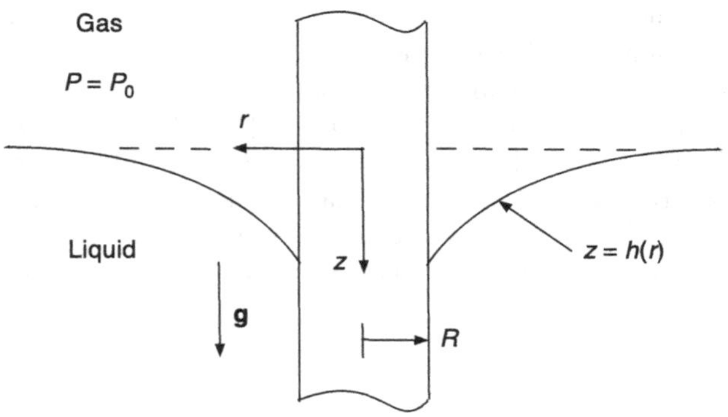

Figure 7.6 Gas–liquid interface near a rotating rod.

finding the velocity and pressure fields in a gas we cannot usually neglect the shear stress exerted on it by a flowing liquid. Although from a liquid perspective the gas viscosity is nearly zero, from a gas perspective the liquid viscosity is nearly infinite. The same is true for the densities.

If the gas in Fig. 7.5 is propelled only by the shear exerted on it by the liquid, the viscous and inertial stresses in the gas will be on the order of $\mu_G U/W$ and $\rho_G U^2$, respectively (Section 1.3). The shear stresses in the liquid are roughly $\mu_L U/H$. Thus, the stresses in the gas will be negligible relative to those in the liquid if $\mu_G/\mu_L \ll W/H$ and $\rho_G/\rho_L \ll 1/\mathrm{Re}_L$, respectively, where the liquid Reynolds number is $\mathrm{Re}_L = UH/\nu_L$. In that μ_G/μ_L is usually no larger than about 10^{-2} and ρ_G/ρ_L is typically about 10^{-3} (Section 1.2), ignoring the gas when analyzing the liquid flow will be acceptable unless W/H is very small or Re_L very large. If the gas flow were fully developed, inertia would be entirely absent and the restriction involving the densities and Re_L would be unnecessary. Then, ignoring the gas would require only that $\mu_G/\mu_L \ll W/H$.

Example 7.4-2 Surface of a stirred liquid. In general, the location of a free surface is determined by the normal-stress balance, which involves the velocities and pressures in both fluids. Thus, that location is often an additional unknown. The flow in a large body of liquid surrounding a rod of radius R that is rotating at an angular velocity ω was examined in Example 7.3-2. The objective now is to use those results to find the shape of the gas–liquid interface. As shown in Fig. 7.6, which focuses on the interface near the rod, coordinates are chosen in which the interface is at $z = h(r)$ and $h(\infty) = 0$. Viscous stresses and pressure variations within the gas are assumed to be negligible and the constant gas pressure is P_0. It is assumed also that surface tension is unimportant. Surface tension competes here with pressure variations caused either by gravity or inertia, and will be negligible if the Bond and Weber numbers are such that $\mathrm{Bo} = \rho g R^2/\gamma \gg 1$ and $\mathrm{We} = \rho U^2 R/\gamma \gg 1$ (Section 1.3).

In the absence of surface tension, the normal-stress balance at the free surface [Eq. (6.6-18)] involves only the viscous stresses and the pressures. The normal viscous stress in the gas is assumed to be negligible, as already indicated. To calculate the normal viscous stress in the liquid we would need the interface position $h(r)$ and unit normal $\mathbf{n}(r)$, both of which are unknown. However, the cylindrical symmetry indicates that $\mathbf{n}$ has only r and z components. It follows that $\tau_{nn} = \mathbf{n} \cdot \mathbf{t} \cdot \mathbf{n}$ will involve at most τ_{rr}, τ_{zz}, τ_{rz}, and τ_{zr}. In this unidirectional flow each of those stress components are zero (Table 6.5), so that $\tau_{nn} = 0$. Thus, the normal-stress balance reduces to an equality of pressures, as in Example 7.4-1.

To apply the normal-stress balance we need to evaluate $P(r, z)$ in the liquid. In this flow $\partial\mathcal{P}/\partial z = 0$, and $\partial\mathcal{P}/\partial r$ is given by Eq. (7.3-12). Accordingly,

$$\frac{\partial P}{\partial r} = \frac{\partial \mathcal{P}}{\partial r} + \rho g_r = \frac{\rho\omega^2 R^4}{r^3} \tag{7.4-10}$$

$$\frac{\partial P}{\partial z} = \frac{\partial \mathcal{P}}{\partial z} + \rho g_z = \rho g. \tag{7.4-11}$$

Integration of each expression gives

$$P(r, z) = -\frac{\rho\omega^2 R^4}{2r^2} + f(z) + C \tag{7.4-12}$$

$$P(r, z) = \rho g z + g(r) + C \tag{7.4-13}$$

where $f(z)$ and $g(r)$ are unknown functions and C is an unknown constant. Evidently, $f(z)$ must equal the z-dependent part of Eq. (7.4-13) and $g(r)$ must equal the r-dependent part of Eq. (7.4-12). The undisturbed interface is at $z = 0$ and the pressure there is $P(\infty, 0) = P_0$, so that $C = P_0$. The liquid pressure is then

$$P(r, z) = -\frac{\rho\omega^2 R^4}{2r^2} + \rho g z + P_0. \tag{7.4-14}$$

The location of the interface is dictated by the need for the liquid pressure to equal the constant pressure in the gas. Setting $P(r, h) = P_0$ and solving for h gives

$$h(r) = \frac{\omega^2 R^4}{2gr^2}. \tag{7.4-15}$$

This indicates that $h > 0$, which confirms that the interface is depressed by the rotation of the rod. The interface shape resembles that in Fig. 7.6.

7.5 NON-NEWTONIAN FLUIDS

In problems involving non-Newtonian fluids the Navier–Stokes equation is inapplicable and the Cauchy momentum equation is needed. That is true also for Newtonian fluids in which the viscosity varies with position, as when there are significant temperature variations.

Example 7.5-1 Poiseuille flow of a power-law fluid. It is desired to determine the velocity profile and volume flow rate for steady, fully developed flow of a power-law fluid in a long tube of radius R. What is given is the pressure drop per unit length, $|\Delta\mathcal{P}|/L$.

This is the same as Example 7.2-2, except that the viscosity now varies with the rate of strain according to Eq. (6.5-17). The usual continuity equation applies, but what is needed now is the cylindrical form of the Cauchy momentum equation (Table 6.2). Given that $v_z = v_z(r)$ and $v_r = 0 = v_\theta$, the inertial terms are all zero, as they were for the Newtonian problem. Moreover, from Table 6.5, in this unidirectional flow all viscous stress components are zero except τ_{rz} and τ_{zr}, which can depend at most on r. Thus, the

r and θ components of the Cauchy equation indicate that $\mathcal{P} = \mathcal{P}(z)$ only (as before), and the z component reduces to

$$\frac{1}{r}\frac{d}{dr}(r\tau_{rz}) = \frac{d\mathcal{P}}{dz} = -\frac{|\Delta\mathcal{P}|}{L}. \tag{7.5-1}$$

Multiplying by r and integrating gives

$$r\tau_{rz}(r) = -\frac{|\Delta\mathcal{P}|\,r^2}{2L} + c \tag{7.5-2}$$

where c is a constant. The cylindrical symmetry requires that $\tau_{rz}(0) = 0$, as shown in Section 7.6. Accordingly, $c = 0$ and

$$\tau_{rz}(r) = -\frac{|\Delta\mathcal{P}|\,r}{2L}. \tag{7.5-3}$$

Thus, the shear stress varies linearly from zero at the centerline to a maximally negative value at the wall. This holds for any generalized Newtonian fluid.

For this unidirectional flow, Table 6.7 indicates that the rate of strain is

$$\Gamma = \frac{1}{2}\left|\frac{dv_z}{dr}\right|. \tag{7.5-4}$$

The absolute value is needed because $dv_z/dr < 0$. The apparent viscosity of the power-law fluid is then

$$\mu = m(2\Gamma)^{n-1} = m\left|\frac{dv_z}{dr}\right|^{n-1} \tag{7.5-5}$$

and the local shear stress is

$$\tau_{rz}(r) = \mu\frac{dv_z}{dr} = m\left|\frac{dv_z}{dr}\right|^{n-1}\frac{dv_z}{dr} = -m\left|\frac{dv_z}{dr}\right|^{n}. \tag{7.5-6}$$

Using this to evaluate the stress in Eq. (7.5-3) and solving for the velocity derivative gives

$$\frac{dv_z}{dr} = -\left(\frac{|\Delta\mathcal{P}|}{2mL}\right)^{1/n} r^{1/n}. \tag{7.5-7}$$

Integrating and applying the no-slip condition at the wall ($v_z(R) = 0$) leads to

$$v_z(r) = \left(\frac{R^{n+1}\,|\Delta\mathcal{P}|}{2mL}\right)^{1/n}\left[1 - \left(\frac{r}{R}\right)^{(n+1)/n}\right] \tag{7.5-8}$$

which is the key result that was sought.

The mean velocity U is calculated as

$$U = \frac{2}{R^2}\int_0^R v_z r\,dr = \left(\frac{n}{3n+1}\right)\left(\frac{R^{n+1}\,|\Delta\mathcal{P}|}{2mL}\right)^{1/n}. \tag{7.5-9}$$

Accordingly, the volume flow rate is

$$Q = \pi R^2 U = \left(\frac{n\pi}{3n+1}\right)\left(\frac{R^{3n+1}\,|\Delta\mathcal{P}|}{2mL}\right)^{1/n} \tag{7.5-10}$$

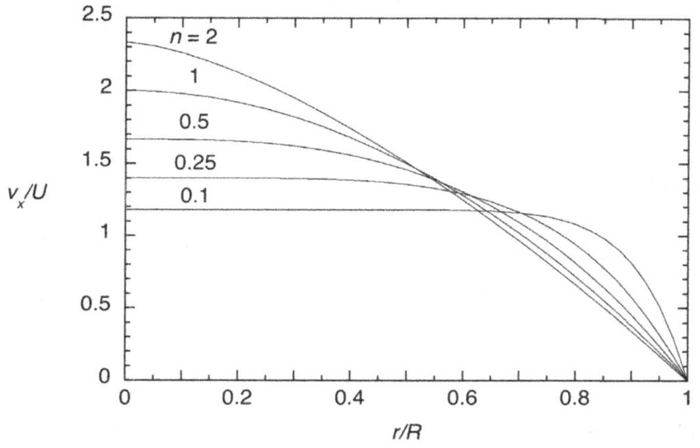

Figure 7.7 Velocity profiles for pressure-driven flow of power-law fluids in circular tubes, including ones that are shear-thickening ($n = 2$), Newtonian ($n = 1$), or shear-thinning ($n = 0.1$, 0.25, or 0.5).

and an alternative form for the velocity profile is

$$v_z(r) = \left(\frac{3n+1}{n+1}\right) U \left[1 - \left(\frac{r}{R}\right)^{(n+1)/n}\right]. \tag{7.5-11}$$

Equation (7.5-11) is plotted in Fig. 7.7 for several values of n. For the shear-thickening fluid ($n = 2$), the velocity profile is more sharply peaked than for the Newtonian case ($n = 1$), and for the shear-thinning ones ($n < 1$), it is more blunted. As the profiles flatten the local velocity approaches the mean velocity; that is, $v_z(r) \to U$ as $n \to 0$.

Example 7.5-2 Plane Couette flow of generalized Newtonian fluids. A remarkable aspect of plane Couette flow is that the linear velocity profile derived in Example 7.3-1 for a Newtonian fluid applies also to any generalized Newtonian fluid. The objective is to show this.

Given that $v_x = v_x(y)$ and $v_y = 0 = v_z$, Table 6.4 indicates that the only viscous stress components that are not zero in this flow are τ_{yx} and τ_{xy}, which depend at most on y. This holds for any generalized Newtonian fluid. With constancy of $\mathcal{P}$ being another feature of plane Couette flow, the x component of the Cauchy momentum equation (Table 6.1) reduces to

$$\frac{d\tau_{yx}}{dy} = 0. \tag{7.5-12}$$

Thus, the shear stress within the fluid is uniform. From Table 6.7, the local rate of strain in this unidirectional flow is

$$\Gamma = \frac{1}{2}\frac{dv_x}{dy}. \tag{7.5-13}$$

For an apparent viscosity that depends on the rate of strain, the shear stress is

$$\tau_{yx} = \mu(\Gamma)\frac{dv_x}{dy} = 2\mu(\Gamma)\Gamma = \text{constant}. \tag{7.5-14}$$

Because μ never varies as rapidly as $1/\Gamma$ (which would imply that $n = 0$ for a power-law fluid), constancy of $\mu\Gamma$ implies that Γ itself must be constant. Together with the no-slip boundary conditions at the plates, such constancy of dv_x/dy leads again to Eq. (7.3-4). The uniformity of both τ_{yx} and Γ enhances the value of Couette viscometers for rheological studies.

7.6 SYMMETRY CONDITIONS

The velocity and stress boundary conditions discussed in Section 6.6 must be supplemented occasionally. For instance, in Example 7.5-1 it was necessary to know τ_{rz} at $r = 0$. Although not a physical boundary, in that problem $r = 0$ was a mathematical one. Additional or alternative conditions can be inferred from whatever symmetry is present.

Cylindrical symmetry

Axisymmetric flow in tubes has *cylindrical symmetry*, meaning that the velocity, pressure, and viscous stresses do not depend on the cylindrical angle θ. That holds not just for fully developed flow, but also for flow in circular tubes of varying radius. For any flow with such symmetry, the velocity and stress conditions at $r = 0$ are derived by considering a control volume of radius r and length ℓ that is centered on the z axis. The aspect ratio r/ℓ is made so small that any flow through the ends of this control volume and any forces on them are each negligible. The volume flow rate out of the control volume, which must be zero for an incompressible fluid, is then its lateral area, $2\pi r\ell$, times the radial velocity averaged over its length, $\langle v_r \rangle$. Accordingly, for any fixed ℓ,

$$\lim_{r\to 0} (r \langle v_r \rangle) = 0. \tag{7.6-1}$$

If ℓ is now made small enough that $\langle v_r \rangle$ approaches v_r at a point, this indicates that $rv_r \to 0$ as $r \to 0$. Although any finite v_r at the symmetry axis would suffice, the usual consequence of this is that $v_r = 0$ at $r = 0$. For the special case of fully developed flow, where $v_r = 0$ everywhere, this does not provide new information.

The shear stress at $r = 0$ is inferred from the stress equilibrium condition, Eq. (6.4-12). We will focus on the z component of the surface force, again neglecting the ends of the cylindrical control volume. Replacing L^2 by $S = 2\pi r\ell$ and setting $dS = r\, dz\, d\theta$ gives

$$\lim_{r\to 0} \frac{1}{2\pi r\ell} \int_0^{2\pi} \int_0^{\ell} \tau_{rz} r\, dr\, d\theta = \lim_{r\to 0} \langle \tau_{rz} \rangle = 0 \tag{7.6-2}$$

where the angle brackets again denote an average over the length ℓ. Making ℓ small simplifies this to $\tau_{rz} = 0$ at $r = 0$. That holds for any cylindrically symmetric flow, and is the condition that was used in Example 7.5-1.

Reflective symmetry

Analogous conditions apply at a symmetry plane. An example of such a mathematical surface is the midplane ($y = 0$) of the parallel plate channel in Fig. 7.1. If the flow for $y > 0$ mirrors that for $y < 0$, there can be no flow across the imaginary dividing surface. That is, $v_y = 0$ at $y = 0$. Moreover, reflective symmetry about $y = 0$ requires that $\partial v_x/\partial y$ change sign there. Because velocities and velocity gradients within a fluid are continuous functions of position, $\partial v_x/\partial y$ must pass through zero to change sign. Thus, $\partial v_x/\partial y = 0$ at $y = 0$. That this holds for the parallel-plate channel may be confirmed by differentiating

Eq. (7.2-9). In that problem the same results would have been obtained if $dv_x/dy = 0$ at $y = 0$ had been used in place of one of the no-slip conditions. Because $\partial v_x/\partial y$ and $\partial v_y/\partial x$ both vanish at the symmetry plane, $\tau_{yx} = 0$ at $y = 0$. In a three-dimensional flow that is symmetric about $y = 0$, analogous arguments lead to $\partial v_z/\partial y = 0$ and $\tau_{yz} = 0$ at $y = 0$. The same conclusions hold for a symmetry plane at *any* constant value of y. That is, although $r = 0$ in cylindrical coordinates is a special location, $y = 0$ in Cartesian ones is not. In summary, at any plane with reflective symmetry, the velocity normal to the plane and the shear stresses are each zero.

7.7 CONCLUSION

Laminar unidirectional flow, in which there is just one nonzero component of $\mathbf{v}$, yields the greatest simplification of the continuity and momentum equations and often permits an analytical solution to be obtained. Pressure-driven flow in long conduits of uniform cross-section has this characteristic (Section 7.2). In such flows, where the velocity profile is fully developed (unchanging along the conduit), the pressure decline is linear and the rate of change in the pressure is proportional to the mean velocity. Unidirectional flow can result also from the tangential motion of a surface without an applied pressure (Section 7.3). For a planar surface moving parallel to itself or a cylindrical one rotating about its axis, the fluid may be propelled entirely by the imposed shear stress. When analyzing the flow in a space completely filled by a single fluid, the dynamic pressure $\mathcal{P}$ can be employed and the results so obtained are independent of whether the flow is horizontal, upward, or downward. However, when free surfaces (fluid–fluid interfaces) are present, information on the absolute pressure P is needed (Section 7.4). The location of such surfaces is sometimes an additional unknown that must be determined by using a normal stress balance. Whereas the Navier–Stokes equation is the starting point for predicting $\mathbf{v}$ and P for Newtonian fluids with constant properties, the Cauchy momentum equation is needed when the viscosity of a fluid depends on the rate of strain or otherwise varies with position (Section 7.5). The velocity and stress boundary conditions presented in Chapter 6 must sometimes be supplemented with symmetry conditions (Section 7.6).

REFERENCES

Bird, R. B., W. E. Stewart, and E. N. Lightfoot. *Transport Phenomena*, 2nd ed. Wiley, New York, 2002.

Deen, W. M. *Analysis of Transport Phenomena*, 2nd ed. Oxford University Press, New York, 2012.

Landau, L. D. and E. M. Lifshitz. *Fluid Mechanics*, 2nd ed. Pergamon Press, Oxford, 1987.

Merrill, E. W. Rheology of blood. *Physiol. Rev.* 49: 863–888, 1969.

Rayleigh, Lord. On the motion of solid bodies through viscous liquid. *Philos. Mag.* 21: 697–711, 1911.

Sutera, S. P. and R. Skalak. The history of Poiseuille's law. *Annu. Rev. Fluid Mech.* 25: 1–19, 1993.

PROBLEMS

7.1. Couette viscometer In a Couette viscometer the liquid to be studied fills the annular space between two cylinders, as shown in Fig. P7.1. The inner and outer radii are κR and R, respectively. The viscosity is determined by measuring the torque required to keep the inner cylinder stationary when the outer one is rotated at a constant angular velocity ω. The cylinders are long enough that the end effects associated with the top and bottom can be neglected and it can be assumed that $\mathbf{v}$ and $\mathcal{P}$ are independent of z as well as of θ.

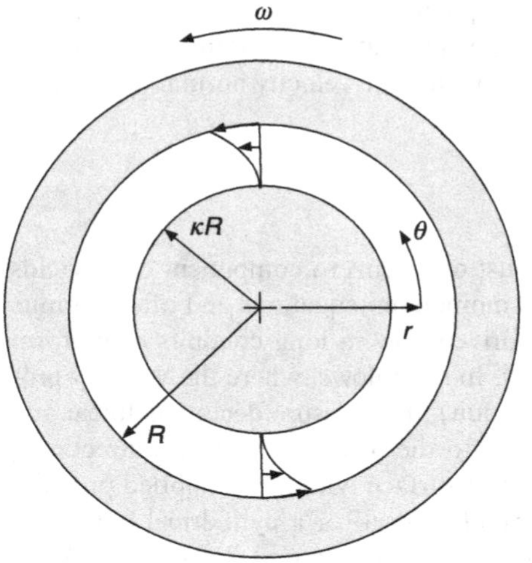

Figure P7.1 Couette viscometer. As the outer cylinder is rotated, the torque needed to hold the inner one in place is measured.

(a) Show that a unidirectional velocity field with $v_\theta = v_\theta(r)$ is consistent with continuity and determine $v_\theta(r)$.
(b) Relate the torque on the inner cylinder to ω, the dimensions, and μ.
(c) As the gap between cylinders is made smaller, the shear stress at the inner one approaches that for plane Couette flow. For the torque obtained from the planar result to be within 1% of the exact value, how large must κ be?

For more on the construction and operation of Couette viscometers and a summary of findings concerning the stability of laminar flow between rotating coaxial cylinders, see Bird *et al.* (2002, pp. 89–93).

7.2. Annular conduit Consider pressure-driven flow in an annular channel. The inner and outer radii are as in Fig. P7.1, but now both cylinders are stationary and there is a mean fluid velocity U in the z direction. The conduit length is L.

(a) Solve for $v_z(r)$ in terms of the pressure drop. It is convenient to lump the viscosity and pressure drop together as $B = |\Delta\mathcal{P}|/(\mu L)$ and write the final result in terms of a dimensionless radial coordinate $\eta = r/R$.
(b) Relate U to B and the two radii.
(c) Derive the result for the friction factor that is given by Eq. (2.4-5).

7.3. Triangular conduit An exact solution for the velocity in a conduit is usually obtainable only when each wall is a coordinate surface. An interesting exception is a duct with an equilateral triangular cross-section, as in Fig. P7.3. Adopting the coordinates in Bird *et al.* (2002, p. 106), the triangle side length is $2H/\sqrt{3}$ and the duct length is L. Although the approach described below for finding $v_z(x, y)$ seems as if it might be generalized to other polygons, unfortunately it works only for equilateral triangles and parallel plates.

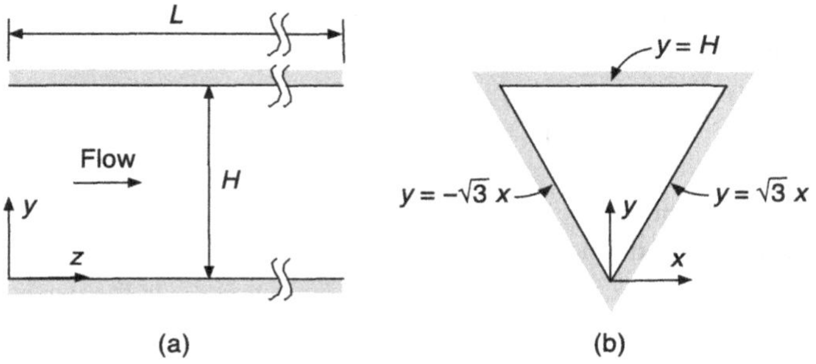

Figure P7.3 Duct with a triangular cross-section: (a) side view and (b) cross-sectional view.

(a) Show that, as suggested in Landau and Lifshitz (1987, p. 54), a solution for the velocity may be constructed as

$$v_z(x, y) = Ah_1(x, y)h_2(x, y)h_3(x, y) \tag{P7.3-1}$$

where A is a constant and h_i is the shortest distance of an interior point (x, y) to side i of the triangle. Find the functions h_i and then evaluate A in terms of the pressure drop.

(b) Relate the mean velocity U to the pressure drop.

(c) Show that the friction factor is given by $f\mathrm{Re}_H = 40/3$, as stated in Section 2.4.

7.4. Elliptical conduit Consider fully developed flow at mean velocity U in a conduit with the elliptical cross-section in Fig. P7.4. The semi-axes are a and b and the length is L. The location of the wall is governed by

$$\left(\frac{x}{a}\right)^2 + \left(\frac{y}{b}\right)^2 = 1. \tag{P7.4-1}$$

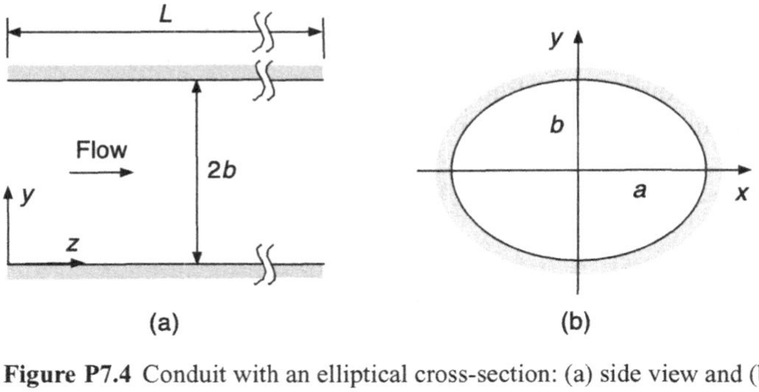

Figure P7.4 Conduit with an elliptical cross-section: (a) side view and (b) cross-sectional view.

(a) The function

$$v_z(x, y) = Ax^2 + By^2 + C \tag{P7.4-2}$$

where A, B, and C are any constants, is a solution of the Navier–Stokes equation for fully developed flow in the z direction. Show that, as pointed out in Landau and Lifshitz (1987, p. 53), if $A = -C/a^2$ and $B = -C/b^2$ it will also satisfy the no-slip condition everywhere on the wall of the elliptical conduit.

(b) Complete the derivation of $v_z(x, y)$ by using the Navier–Stokes equation to evaluate C in terms of the pressure drop.

(c) Show that $U = C/2$.

(d) The area of the ellipse is πab, but there is no exact, closed-form expression for its perimeter. A good approximation for the perimeter is

$$2\pi \left(\frac{a^2 + b^2}{2} \right)^{1/2} \tag{P7.4-3}$$

which reduces to the circumference of a circle when $a = b$. Show that the friction factor based on the hydraulic diameter is

$$f\, \mathrm{Re}_H = 16 \tag{P7.4-4}$$

as for a circular tube.

7.5. Slip in tube flow An alternative to the no-slip boundary condition at a stationary solid, proposed by Navier in the 1820s and revived in recent years for modeling liquid flows in extremely small channels, is to suppose that the tangential fluid velocity is proportional to the shear rate. For fully developed flow in a tube of radius R this may be written as

$$v_z(R) = L_s \left| \frac{dv_z}{dr}(R) \right| \tag{P7.5-1}$$

where the proportionality constant L_s is called the *slip length*.

Derive an expression for the volume flow rate Q in a tube with radius R, slip length L_s, and a given pressure drop per unit length. How is the dependence of Q on R affected by the ratio L_s/R?

7.6. Darcy permeability of a fibrous material It is desired to predict the Darcy permeability k of a porous material that consists of many cylindrical fibers arranged in parallel. For simplicity, assume that long fibers of radius R are equally spaced and that the flow is parallel to their axes. A particularly symmetric configuration is shown in Fig. P7.6(a), in which the cross-section is divided into identical hexagonal cells. It is sufficient then to consider a single fiber. A further simplification is to approximate the hexagons of side length H as circles of radius λR, as enlarged in Fig. P7.6(b). For equal areas, $\lambda R = 0.91H$.

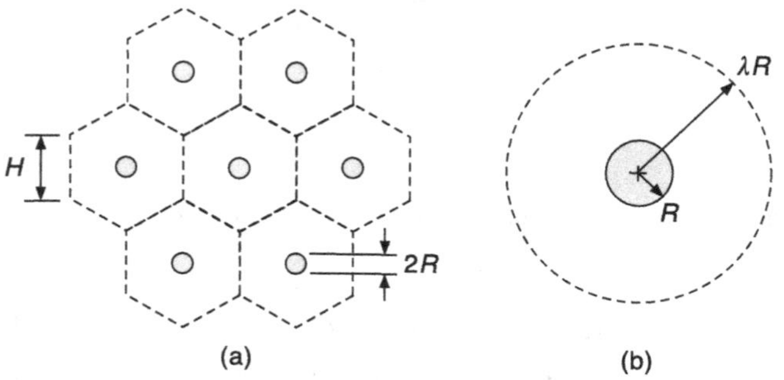

Figure P7.6 Material consisting of parallel fibers: (a) hexagonal arrangement of fiber axes, and (b) a hexagonal unit cell approximated as a circle.

(a) Determine the fully developed velocity profile, $v_z(r)$, for a given pressure drop per unit length, $|\Delta\mathcal{P}|/L$. Based on the symmetry, you may assume that $\tau_{rz} = 0$ at $r = \lambda R$.
(b) Calculate the volume flow rate per fiber and use that to find the superficial velocity.

(c) If the fibers are widely spaced ($\lambda \gg 1$), show that

$$\frac{k}{d^2} = \frac{1}{16\phi}\left(-\ln\phi - \frac{3}{2}\right) \tag{P7.6-1}$$

where $d(=2R)$ is the fiber diameter and $\phi(=1/\lambda^2)$ is the volume fraction of fibers. Note that this result has the same form as Eq. (3.4-12), which is for randomly oriented fibers. The parallel arrangement and flow direction assumed here merely affected the constants (1/16 vs. 3/80 and 3/2 vs. 0.931).

7.7. Surface of a liquid in rigid-body rotation Suppose that an open container of radius R is filled with liquid to an initial height h_0. It is then rotated at a constant angular velocity ω, as shown in Fig. P7.7. After an initial transient, the liquid rotates as if it were a rigid body.

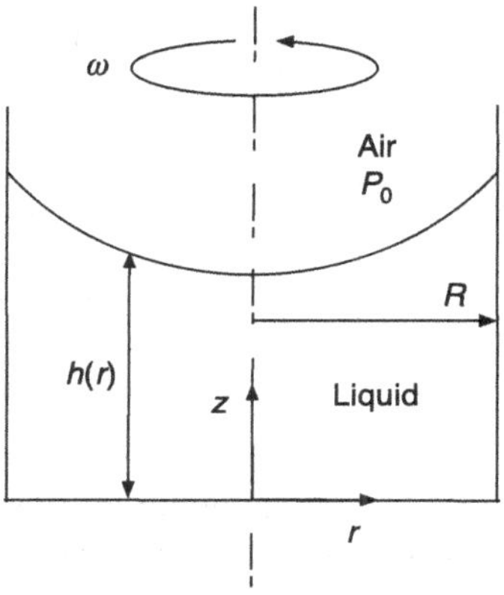

Figure P7.7 Rigid-body rotation of a liquid in an open container.

Assuming that the air pressure is constant at P_0 and that surface tension and viscous stresses in the air are negligible, determine the liquid pressure $P(r, z)$ and interface height $h(r)$.

7.8. Layered liquids on an inclined surface Simultaneous flow of two or more thin layers of liquid is needed to form various composite coatings. Suppose that immiscible liquids 1 and 2 flow down a surface that is inclined at an angle β relative to vertical, as shown in Fig. P7.8. The respective layer thicknesses are H_1 and H_2, the densities are ρ_1 and ρ_2, the viscosities are μ_1 and μ_2, and the fully developed velocities are $v_x^{(1)}(y)$ and $v_x^{(2)}(y)$. As usual, it may be assumed that the air pressure is constant at P_0 and that viscous stresses in the air are negligible.

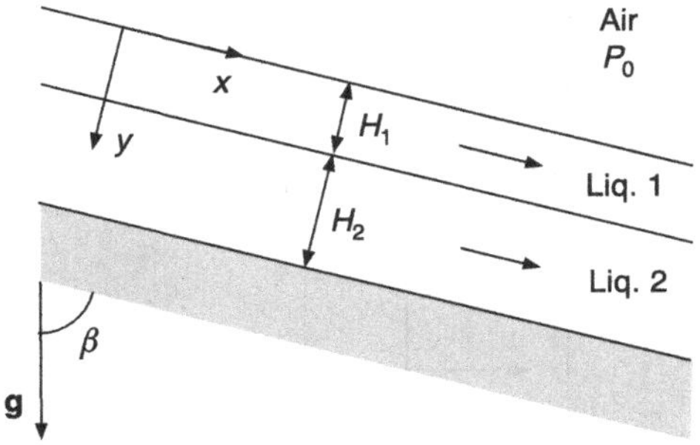

Figure P7.8 Flow of immiscible liquids down an inclined surface.

(a) What is $d\mathcal{P}/dx$ in liquid 1?
(b) What is $d\mathcal{P}/dx$ in liquid 2?
(c) State the differential equations and boundary conditions and find $v_x^{(1)}(y)$ and $v_x^{(2)}(y)$.

7.9. Liquid film outside a vertical tube Suppose that a liquid is pumped upward at constant volume flow rate Q through a tube of outer radius R, as shown in Fig. P7.9. The liquid overflows the top and runs down the outside. At a certain distance from the top the air–liquid interface reaches a constant radius λR and the flow becomes fully developed. The liquid film is not necessarily thin relative to R.

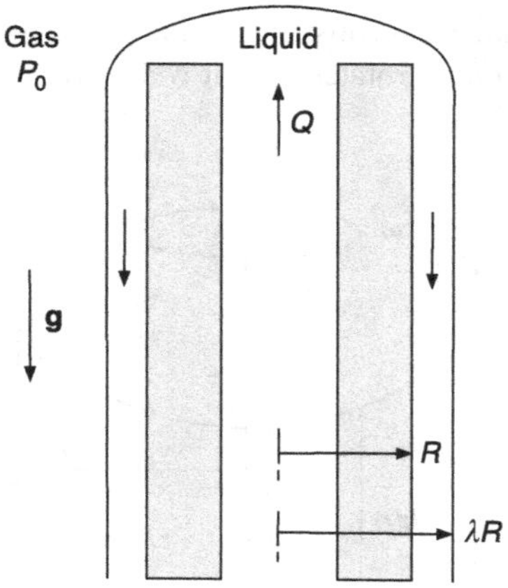

Figure P7.9 Liquid overflowing a vertical tube and flowing down the outside.

(a) Assuming for the moment that λ is known, determine the velocity profile in the film.
(b) Derive an expression that would allow you to calculate λ, assuming that Q, R, and the liquid properties are given.

7.10. Film on an upward-moving surface Dip coating is a process in which an immersed substrate is withdrawn through a gas–liquid interface. Problem 1.6 focused on what occurs near the surface of a pool of liquid, as a solid sheet is pulled upward at a constant velocity V. The concern now is what happens farther up. Above a certain height, the liquid film on the sheet will achieve a constant thickness H, as shown in Fig. P7.10. At

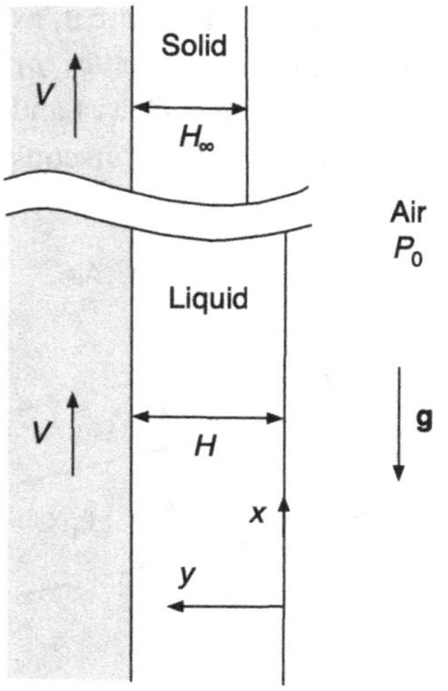

Figure P7.10 Liquid and solid films on an upward-moving surface.

even larger x, suppose that evaporation or reaction causes the film to solidify. It is desired to predict the thickness H_∞ of the solid film.

(a) Assuming that the air pressure is constant at P_0 and that viscous stresses in the gas are negligible, determine the local liquid velocity $v_x(y)$ and the mean liquid velocity U.
(b) If the liquid and solid densities are the same, how is H_∞/H related to H, V, and the liquid properties?
(c) Use the result of part (b) and Eq. (P1.6-2) to evaluate H_∞/H for the two extremes of the capillary number, $\text{Ca} \to 0$ and $\text{Ca} \to \infty$.

7.11. Slot coating In the coating process in Fig. P7.11 a liquid is extruded through a slot in a die onto a solid sheet moving horizontally at velocity V. The pressure at the slot exit is P_1 and there is a constant gap thickness H between the die and sheet. Within the gap there is forward flow of liquid in a region of length L_1 and no net flow in a region of length L_2. Both regions are long enough that the flow in them may be approximated as fully developed. After exiting the die the liquid thickness reaches another constant value, H_∞. The upstream and downstream air pressures are uniform at P_0 and surface tension, viscous stresses in the air, and static pressure variations are all negligible. Assume that P_1, V, H, and L_1 are given, whereas L_2 and H_∞ are unknown.

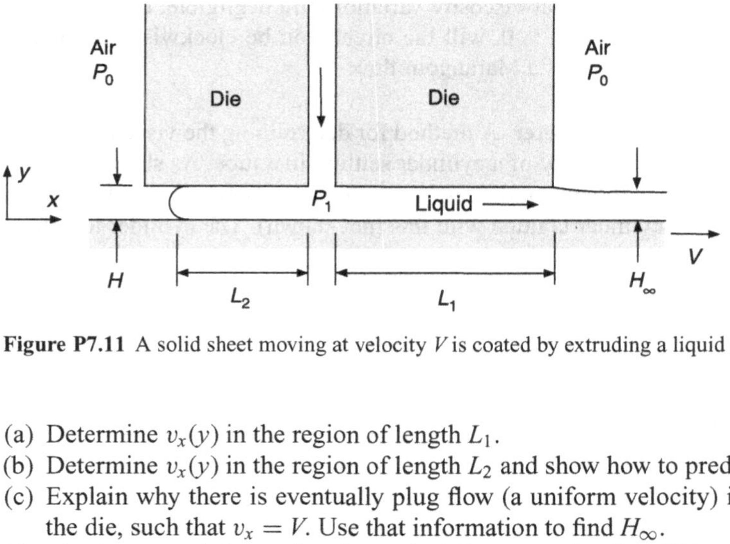

Figure P7.11 A solid sheet moving at velocity V is coated by extruding a liquid through a slot.

(a) Determine $v_x(y)$ in the region of length L_1.
(b) Determine $v_x(y)$ in the region of length L_2 and show how to predict L_2.
(c) Explain why there is eventually plug flow (a uniform velocity) in the liquid beyond the die, such that $v_x = V$. Use that information to find H_∞.
(d) The force needed to pull a sheet of width W past the die will be affected by the shear force F that the liquid exerts on the sheet. Calculate F/W. Does the force from the liquid assist or retard the movement of the sheet?

7.12. Flow in a cavity Suppose that a long and shallow cavity of length L and depth H is filled with liquid, as in Fig. P7.12. If something at the top surface causes the liquid there to move from left to right (i.e., if $v_x > 0$ at $y = H$), a circulating flow is created, as indicated by the arrows. However, if L/H is large enough, the flow in the central region will be fully developed. This problem focuses on that region, centered on $x = 0$, where $v_x = v_x(y)$.

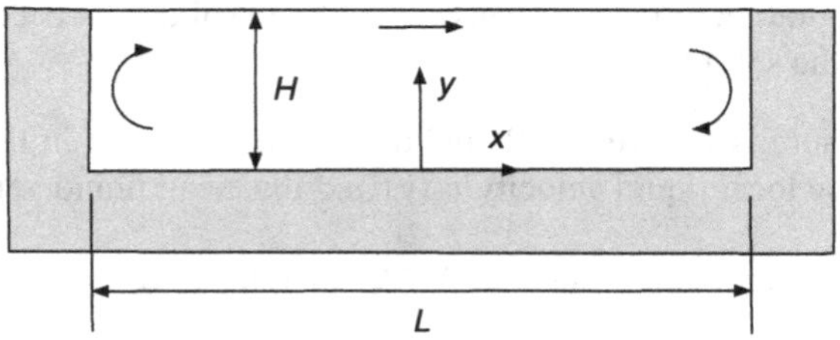

Figure P7.12 Flow in a cavity.

(a) By choosing an appropriate control volume, show that v_x averaged over the depth of the cavity must be zero for all x.
(b) Determine $v_x(y)$ for the case in which the top surface is a solid plate sliding past the cavity at velocity U in the $+x$ direction.
(c) Suppose now that $y = H$ is a gas–liquid interface at which the temperature varies as

$$T(x) = T_0 + (T_L - T_0)\frac{x}{L} \qquad \text{(P7.12-1)}$$

and that the temperature-dependence of the surface tension is described by

$$\gamma(T) = \gamma_0\,[1 - \beta(T - T_0)] \qquad \text{(P7.12-2)}$$

where $\beta > 0$. Assuming that viscosity variations are negligible, determine $v_x(y)$ for this case. If $\Delta T = T_L - T_0 > 0$, will the circulation be clockwise or counterclockwise? This is an example of a Marangoni flow.

7.13. Falling-cylinder viscometer A method for determining the viscosity of a liquid is to measure the terminal velocity of a cylinder settling in a tube. As shown in Fig. P7.13, the cylinder radius is a, the tube radius is b, and the velocity relative to the tube is U. To keep it centered, the cylinder is fitted with fins (not shown). The cylinder length is L and

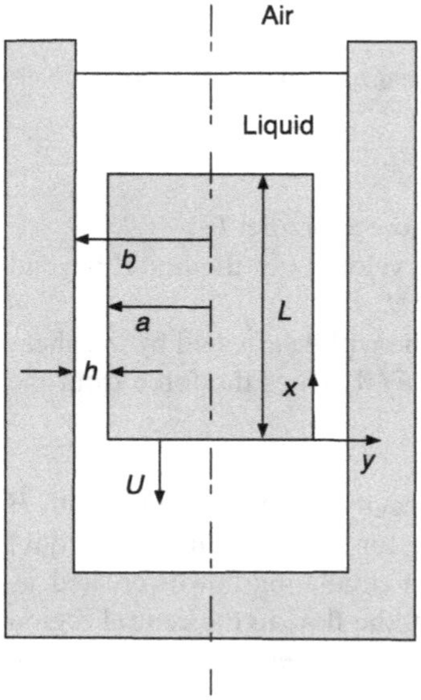

Figure P7.13 Falling cylinder viscometer. The relative lengths have been distorted for clarity in labeling; actually $h/a \ll 1$ and $h/L \ll 1$.

the gap between the cylinder and the tube wall is $h = b - a$. Assume that h/L is small enough that the flow in the gap can be modeled as fully developed. Moreover, assume that the pressure variation above and below the cylinder is very nearly that in a static fluid. For simplicity, assume also that $h/a \ll 1$, allowing the annular gap to be approximated as the space between parallel plates.

(a) Solve for $v_x(y)$ in the gap in terms of $d\mathcal{P}/dx$. A convenient reference frame is one where the cylinder is stationary and the tube wall moves upward at velocity U.
(b) By considering the rate at which the falling cylinder displaces liquid, evaluate the mean velocity in the gap and find $d\mathcal{P}/dx$.
(c) Use an overall force balance on the cylinder (which has density ρ_0) to show how to calculate μ from U and the dimensions.

7.14. Bubble rising in a tube Suppose that an elongated bubble is rising at its terminal velocity U in an open, liquid-filled tube, as in Fig. P7.14. The bubble is shaped as a cylinder of radius a and length L with hemispherical caps, such that $L \gg a$. The tube radius is b. Between the bubble and tube wall is a liquid film of thickness $h = b - a$. Assume that h/L is small enough that the flow is fully developed in almost the entire film, and that the liquid pressure elsewhere is nearly static. For simplicity, assume also that $h/a \ll 1$, in which case the boundaries of the film can be approximated as planar. The objective is to predict U for a given set of linear dimensions.

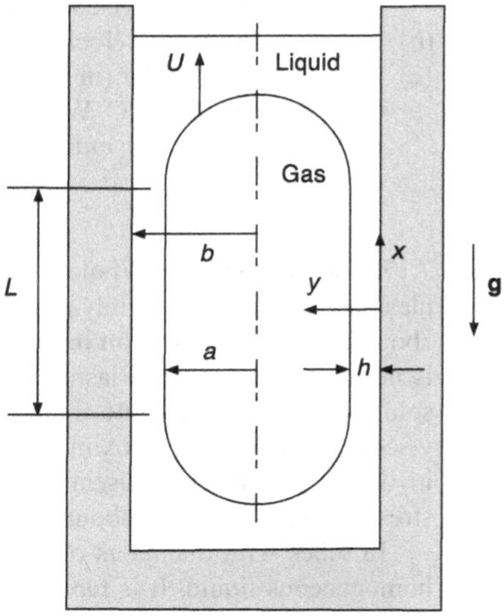

Figure P7.14 Long bubble rising in a liquid-filled tube. The relative lengths have been distorted for clarity in labeling; actually, $h \ll a \ll L$.

(a) It is helpful to use Cartesian coordinates fixed on the tube wall, as shown. Solve for $v_x(y)$ in the liquid film in terms of $d\mathcal{P}/dx$. Explain why the liquid velocity at $y = h$ will *not* equal the mean gas velocity U. (Sketch the circulating flow pattern within the bubble.)
(b) By considering the rate at which the bubble displaces liquid, evaluate the mean velocity in the film and find $d\mathcal{P}/dx$.
(c) Use an overall force balance on the bubble to relate U to the fluid properties and the dimensions.

7.15. Paint film House paint is formulated so that, if applied properly, it will not run down a vertical surface before it dries. It can be modeled as a Bingham fluid.

(a) Suppose that a latex paint with a density like that of water is brushed on a wall at a thickness $H = 0.5$ mm. What is the minimum yield stress τ_0 that is needed to avoid running?
(b) Assume now that the paint is applied carelessly, such that H exceeds the film thickness that can be immobilized by the yield stress. Determine $v_x(y)$ for steady, fully developed flow, using the coordinates in Fig. 7.5. What is the maximum velocity?

7.16. Temperature-dependent viscosity Suppose that temperature variations within the parallel-plate channel of Fig. 7.1 noticeably alter the viscosity. In particular, assume that

$$T(y) = T_0 + (T_1 - T_0)\frac{y}{H}, \quad \mu = \frac{\mu_0}{1 + \alpha(T - T_0)} \qquad \text{(P7.16-1)}$$

where $\alpha > 0$. That is, there is a linear temperature variation across the channel, with $T = T_0$ at the center and $T = T_1$ at the upper wall. The viscosity at $T = T_0$ is μ_0 and, as is typical for liquids, μ decreases with increasing T.

(a) Determine $v_x(y)$. It is convenient to work with a dimensionless coordinate $\eta = y/H$ and a dimensionless viscosity-temperature parameter $A = \alpha(T_1 - T_0)$.
(b) How does the mean velocity U compare with its value for $\mu = \mu_0$?
(c) The viscosity of water (in Pa · s) is 1.139×10^{-3} at 15 °C, 1.002×10^{-3} at 20 °C, and 0.8904×10^{-3} at 25 °C. What value of α best fits these data? If the wall temperatures span this range, how much would U for water differ from that for an isothermal channel at 20 °C?

7.17. Blood rheology Whole blood consists of a concentrated suspension of cells in plasma. The cells are mainly red blood cells (RBCs), which are flexible, rounded disks about 2 μm thick and 8 μm in diameter. Their volume fraction ϕ, called the *hematocrit*, is normally about 0.45. Plasma is an aqueous solution containing electrolytes, organic solutes of low to moderate molecular weight, and proteins. It is Newtonian and has a viscosity μ_p that is approximately 1.6 times that of water. When fibrinogen (a protein involved in clotting) is present, the RBCs tend to aggregate, causing blood to have a yield stress τ_0 that is typically about 0.004 Pa.

In tubes with diameters of at least a few hundred μm, blood can be modeled as a homogeneous liquid. It is represented well as a *Casson fluid*, a generalized Newtonian fluid for which

$$\mu = \left\{ \begin{array}{ll} \infty, & \tau < \tau_0 \\ \mu_0 \left[1 + \left(\dfrac{2\tau_0}{\mu_0 \Gamma} \right)^{1/2} + \dfrac{\tau_0}{2\mu_0 \Gamma} \right], & \tau > \tau_0 \end{array} \right\}. \qquad \text{(P7.17-1)}$$

The parameter μ_0 is the apparent viscosity at high shear rates (Merrill, 1969), given by

$$\mu_0 = \mu_p(1 + 2.5\phi + 7.35\phi^2). \qquad \text{(P7.17-2)}$$

(a) For fully developed flow in a tube with $\tau > \tau_0$, show that the Casson equation reduces to

$$|\tau_{rz}|^{1/2} = \tau_0^{1/2} + \left(\mu_0 \left|\frac{dv_z}{dr}\right|\right)^{1/2}. \tag{P7.17-3}$$

(b) Determine $v_z(r)$ for blood in a tube of radius R. In the momentum equation it is convenient to use the wall shear stress τ_w instead of $d\mathcal{P}/dz$. The velocity can be expressed most easily in term of a dimensionless radial position $\eta = r/R$ and the parameters $v_w = \tau_w R/\mu_0$ and $v_0 = \tau_0 R/\mu_0$, both of which have the dimension of velocity.

(c) Time-averaged shear rates at the walls of human blood vessels range from about 50 s^{-1} in large veins to 700 s^{-1} in large arteries. In which type of large vessel will the yield stress be more noticeable? Plot $v_z(\eta)/v_z(0)$ for both in comparison with the result for a Newtonian fluid. You may assume that τ_w equals μ_0 times the given shear rate.

8

Approximations for viscous flows

8.1 INTRODUCTION

A special aspect of unidirectional flow is that the continuity and Navier–Stokes equations often can be solved exactly, as in Chapter 7. After making the initial assumptions that define the problem (e.g., that the flow is fully developed), tractable differential equations are obtained by simply eliminating all terms that are identically zero. Although unidirectional flow has important applications, flows of engineering interest are often more complex. Fortunately, there are numerous other practical situations in which the Navier–Stokes equation can be simplified enough to permit either analytical or elementary numerical solutions to be obtained. Here and in Chapter 9 we explore flows in which some of the terms that were exactly zero in Chapter 7 are merely small. The present chapter focuses on several types of flow in which inertial effects, while not entirely absent, are much less important than viscous stresses. What these flows have in common mathematically is that the $\mathbf{v} \cdot \nabla \mathbf{v}$ term in the Navier–Stokes equation is negligible, which leads to differential equations that are linear. Chapter 9 focuses on laminar flows where inertia is prominent and the momentum equations are nonlinear.

The chapter begins with flows at small or moderate Reynolds number that are *nearly unidirectional*. That is, one velocity component is much larger than the others. The resulting simplification is called the *lubrication approximation*. Although named for the kinds of applications where it was first used, this approximation has broad utility for thin films or narrow channels. The second major topic is *creeping flow*, in which inertia is made negligible by the fact that Re is very small. Flows with Re $\to 0$ have special features, including the rather counterintuitive property of reversibility. The third major topic is *pseudosteady flow*, in which the velocity and pressure are time-dependent, but the time derivative in the momentum equation is negligible. Creeping flow is typically pseudosteady, as will be shown. The chapter closes with a reexamination of the various approximations using order-of-magnitude (OM) analysis. Building on the concept of scales that was introduced in Section 1.3, OM analysis provides a way to anticipate when certain simplifications will be valid.

8.2 LUBRICATION APPROXIMATION

Nearly unidirectional flows at small-to-moderate Re are characterized by the almost complete absence of inertia and of certain viscous stresses. They are governed by the *lubrication approximation*, in which the x-momentum equation for a two-dimensional steady problem reduces to

$$\frac{\partial^2 v_x}{\partial y^2} = \frac{1}{\mu}\frac{d\mathcal{P}}{dx}, \quad \mathcal{P} = \mathcal{P}(x) \text{ only.} \tag{8.2-1}$$

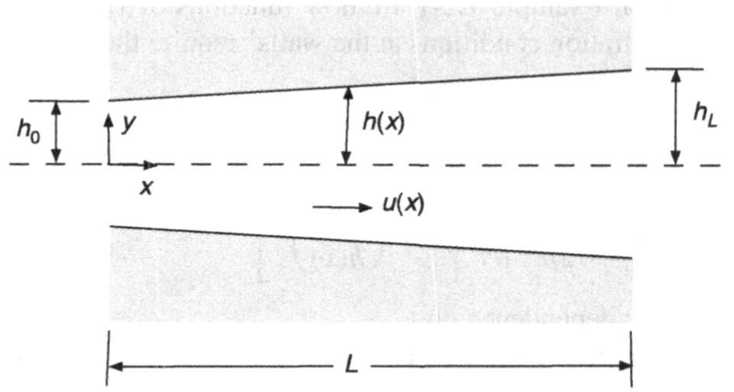

Figure 8.1 Flow in a tapered channel with planar walls.

This differs from Eq. (7.2-3), which applies to exactly unidirectional flow, only in that a partial derivative is needed now on the left because $v_x = v_x(x, y)$. The dependence of v_x on x implies that both sides of the equation are functions of x. Although $d\mathcal{P}/dx$ is no longer constant, we can proceed much as in Chapter 7. The approach is to first solve this equation for v_x and $\mathcal{P}$, and then (if desired) use the continuity equation to calculate the small velocity component, $v_y(x, y)$. This sequential procedure works because v_y does not appear in Eq. (8.2-1).

Example 8.2-1 Tapered channel. It is desired to determine the velocity and pressure in a channel with walls that are flat but not quite parallel. The symmetric surfaces might taper outward, as in Fig. 8.1, or inward; the same solution applies to both. The volume flow rate is Q and the channel width (in the z direction) is W. In two-dimensional problems such as this it is convenient to work with the volume flow rate per unit width, $q = Q/W$. The half-height of the channel is given by

$$h(x) = h_0 + (h_L - h_0)\frac{x}{L} \tag{8.2-2}$$

and the local mean velocity (the average of v_x over y) is $u(x)$.

It is reasonable to suppose that the velocity and pressure fields in a gradually tapered channel closely resemble those for plane Poiseuille flow. In particular, we might guess that the velocity profile will be parabolic, as in Eq. (7.2-12). Allowing for the varying half-height and mean velocity, this suggests that

$$v_x(x, y) = \frac{3}{2}u(x)\left[1 - \left(\frac{y}{h(x)}\right)^2\right] = \frac{3q}{4h(x)}\left[1 - \left(\frac{y}{h(x)}\right)^2\right]. \tag{8.2-3}$$

The second equality uses overall conservation of mass, $q = 2u(x)h(x)$, to eliminate $u(x)$ as an unknown.

The guess represented by Eq. (8.2-3) is consistent with the lubrication approximation. That is, with $\mathcal{P}$ a function of x only, two integrations of Eq. (8.2-1) give

$$v_x(x, y) = \frac{y^2}{2\mu}\frac{d\mathcal{P}}{dx} + a(x)y + b(x). \tag{8.2-4}$$

What were integration constants in Example 7.2-1 are now functions of x, namely $a(x)$ and $b(x)$. The no-slip and no-penetration conditions at the walls[1] require that

$$v_x(x, \pm h) = 0. \tag{8.2-5}$$

Using this to evaluate $a(x)$ and $b(x)$ gives

$$v_x(x, y) = -\frac{h^2(x)}{2\mu}\frac{d\mathcal{P}}{dx}\left[1 - \left(\frac{y}{h(x)}\right)^2\right] \tag{8.2-6}$$

which has the postulated parabolic dependence on y.

If the parabolic profile is correct, a comparison of Eqs. (8.2-3) and (8.2-6) implies that

$$\frac{d\mathcal{P}}{dx} = -\frac{3}{2}\frac{\mu q}{h^3(x)}. \tag{8.2-7}$$

This indicates that the magnitude of the pressure gradient decreases as the channel widens and increases as it narrows. Integration gives

$$\mathcal{P}(0) - \mathcal{P}(x) = \frac{3}{2}\mu q \int_0^x \frac{dX}{h^3(X)} \tag{8.2-8}$$

where X is a dummy variable. Evaluating the integral using Eq. (8.2-2) for $h(x)$ leads to

$$\mathcal{P}(0) - \mathcal{P}(x) = \frac{3}{2}\frac{\mu q L}{h_0^3}\left[\frac{\left(\frac{x}{L}\right) + \frac{\Delta h}{2h_0}\left(\frac{x}{L}\right)^2}{\left(1 + \frac{\Delta h}{h_0}\left(\frac{x}{L}\right)\right)^2}\right] \tag{8.2-9}$$

where $\Delta h = h_L - h_0$. The overall pressure drop is

$$\mathcal{P}(0) - \mathcal{P}(L) = \frac{3}{2}\mu q L\left(\frac{h_0 + h_L}{2h_0^2 h_L^2}\right) \tag{8.2-10}$$

which reduces to Eq. (7.2-11) for the special case of parallel walls ($h_0 = h_L = H$).

With v_x known, the two-dimensional continuity equation,

$$\frac{\partial v_x}{\partial x} + \frac{\partial v_y}{\partial y} = 0 \tag{8.2-11}$$

can be used to find v_y. Recognizing that $v_y(x, 0) = 0$ from symmetry, integration gives

$$v_y(x, y) = -\int_0^y \frac{\partial v_x}{\partial x}\, dY \tag{8.2-12}$$

where Y is a dummy variable. Differentiating Eq. (8.2-3) and writing $h(x)$ now simply as h,

$$\frac{\partial v_x}{\partial x} = -\frac{3q}{4h^2}\frac{dh}{dx}\left[1 - 3\left(\frac{y}{h}\right)^2\right]. \tag{8.2-13}$$

[1] These are not simply no-slip conditions because $\mathbf{e_x}$ does not quite parallel either wall. Thus, the tangential velocity has both x and y components, as does the normal velocity. However, zero tangential velocity and zero normal velocity together imply that v_x and v_y each vanish at $y = \pm hy = \pm h$.

Evaluating the integral in Eq. (8.2-12) gives

$$v_y(x,y) = \frac{3q}{4h}\frac{dh}{dx}\left[\left(\frac{y}{h}\right) - \left(\frac{y}{h}\right)^3\right]. \tag{8.2-14}$$

As we might have suspected, v_y is proportional to dh/dx, and thus will be small if the change in wall spacing is gradual. Except at the exact center and the walls (where $v_y = 0$), there is an outward velocity if the channel widens and an inward one if it narrows.

Although the above solution is plausible, we should examine more carefully when it will be accurate. That requires consideration of all terms omitted from the Navier–Stokes equation that are not identically zero. For a two-dimensional steady problem, the complete x and y components are

$$\rho\left(\underset{\text{T4}}{v_x\frac{\partial v_x}{\partial x}} + \underset{\text{T3}}{v_y\frac{\partial v_x}{\partial y}}\right) = -\frac{\partial \mathcal{P}}{\partial x} + \mu\left(\underset{\text{T2}}{\frac{\partial^2 v_x}{\partial x^2}} + \underset{\text{T1}}{\frac{\partial^2 v_x}{\partial y^2}}\right) \tag{8.2-15}$$

$$\rho\left(\underset{\text{T8}}{v_x\frac{\partial v_y}{\partial x}} + \underset{\text{T7}}{v_y\frac{\partial v_y}{\partial y}}\right) = -\frac{\partial \mathcal{P}}{\partial y} + \mu\left(\underset{\text{T6}}{\frac{\partial^2 v_y}{\partial x^2}} + \underset{\text{T5}}{\frac{\partial^2 v_y}{\partial y^2}}\right). \tag{8.2-16}$$

The various viscous and inertial terms have been numbered. Using Eqs. (8.2-1) and (8.2-7), the dominant viscous term (T1) is evaluated as

$$\text{T1} = \mu\frac{\partial^2 v_x}{\partial y^2} = \frac{\partial \mathcal{P}}{\partial x} = -\frac{3}{2}\frac{\mu q}{h^3}. \tag{8.2-17}$$

Because T1 is the only term containing v_x that was retained in Eq. (8.2-1), it is what the neglected terms should be compared with. Differentiating Eq. (8.2-13) by x, the viscous term that was neglected in the x component is

$$\text{T2} = \mu\frac{\partial^2 v_x}{\partial x^2} = \frac{3}{2}\frac{\mu q}{h^3}\left(\frac{dh}{dx}\right)^2\left[1 - 6\left(\frac{y}{h}\right)^2\right]. \tag{8.2-18}$$

To facilitate comparisons, we will average each term over the channel height and consider only absolute values. Using angle brackets to denote such averages, we find that

$$\frac{\langle \text{T2}\rangle}{\langle \text{T1}\rangle} = \left|\frac{dh}{dx}\right|^2. \tag{8.2-19}$$

Thus, T2 will be negligible, as assumed, if $|dh/dx| \ll 1$. Averaging the two inertial terms in the x component gives

$$\langle \text{T3}\rangle = \frac{3}{20}\frac{\rho q^2}{h^3}\left|\frac{dh}{dx}\right| = \langle \text{T4}\rangle\,. \tag{8.2-20}$$

Although unequal locally, T3 and T4 have the same sign and turn out to be identical on average, as shown. Comparing their sum with the dominant viscous term, we obtain

$$\frac{\langle \text{T3} + \text{T4}\rangle}{\langle \text{T1}\rangle} = \frac{1}{5}\frac{\rho q}{\mu}\left|\frac{dh}{dx}\right| \equiv \frac{\text{Re}}{5}\left|\frac{dh}{dx}\right|. \tag{8.2-21}$$

This indicates that $\text{Re}\,|dh/dx| \ll 1$ is sufficient to make inertia negligible.

The other part of the lubrication approximation is the assumption that $\mathcal{P}$ depends on x only. Stated another way, $|\partial\mathcal{P}/\partial y| \ll |\partial\mathcal{P}/\partial x|$. Just as T1 is the dominant velocity term in Eq. (8.2-15), T5 is the dominant one in Eq. (8.2-16). As it is the largest other term in the equation, T5 places an upper bound on $|\partial\mathcal{P}/\partial y|$. It follows that comparing T5 with T1 is a way to compare the pressure derivatives. Using Eq. (8.2-14),

$$\mathrm{T5} = \mu \frac{\partial^2 v_y}{\partial y^2} = -\frac{9}{2}\frac{\mu q}{h^3}\frac{dh}{dx}\frac{y}{h}. \tag{8.2-22}$$

Averaging this over y, we find that

$$\frac{|\partial\mathcal{P}/\partial y|}{|\partial\mathcal{P}/\partial x|} \cong \frac{\langle \mathrm{T5}\rangle}{\langle \mathrm{T1}\rangle} = \frac{3}{2}\left|\frac{dh}{dx}\right|. \tag{8.2-23}$$

This indicates that what is needed to neglect the y-dependence of $\mathcal{P}$ is essentially the same as what is needed to neglect T2, the smaller viscous term.

We conclude that the lubrication approximation will be accurate if $|dh/dx| \ll 1$ and $\mathrm{Re}|dh/dx| \ll 1$, where $\mathrm{Re} = \rho q/\mu$. If dh/dx were not constant, the *maximum* values of $|dh/dx|$ and $\mathrm{Re}|dh/dx|$ would need to be small. For curved or even wavy walls the velocity expressions would remain the same, provided the wall spacing were still symmetric and gradually varying. However, nonplanar walls would alter the pressure calculated using Eq. (8.2-8).

A comment on the use of double inequality signs is in order. A restriction written as $x \ll 1$ and read as "x is much smaller than one" means that what is being referred to becomes exact as $x \to 0$. Likewise, $x \gg 1$ implies approach to an exact solution as $x \to \infty$. In practice, $x < 0.1$ and $x > 10$, respectively, are often sufficient to give acceptable results.

Example 8.2-2 Permeable tube. Nearly unidirectional flows can occur also in channels of constant cross-section that have permeable walls. For example, in filtration devices that consist of bundles of hollow fibers whose walls are porous membranes, a form of the lubrication approximation is usually applicable. Suppose that the pressure outside a porous, cylindrical tube of radius R and length L is adjusted to give a radial velocity at the wall, $v_r(R, z) = v_w$, that is constant. The objective is to determine the internal velocity and pressure fields, given that the mean velocity at the inlet is u_0.

If v_w is small, velocity variations along the tube should be gradual enough for a lubrication approximation. The cylindrical analog of Eq. (8.2-1) is

$$\frac{1}{r}\frac{\partial}{\partial r}\left(r\frac{\partial v_z}{\partial r}\right) = \frac{1}{\mu}\frac{d\mathcal{P}}{dz}, \qquad \mathcal{P} = \mathcal{P}(z)\,\text{only}. \tag{8.2-24}$$

As before, a single viscous term is retained and transverse pressure variations are neglected. The no-slip boundary condition usually remains applicable at a porous surface. Accordingly, two integrations give

$$v_z(r, z) = -\frac{R^2}{4\mu}\frac{d\mathcal{P}}{dz}\left[1 - \left(\frac{r}{R}\right)^2\right] = 2u(z)\left[1 - \left(\frac{r}{R}\right)^2\right]. \tag{8.2-25}$$

This is the same parabolic velocity profile as for Poiseuille flow (Example 7.2-2), except that the mean velocity is now a function of position along the tube, $u(z)$.

As shown in Example 5.2-3, the mean velocity for constant v_w is

$$u(z) = u_0 - \frac{2v_w z}{R}. \tag{8.2-26}$$

Substituting this into Eq. (8.2-25) completes the determination of $v_z(r, z)$. Comparing the two expressions for the centerline velocity indicates that the axial pressure gradient is

$$\frac{d\mathcal{P}}{dz} = -\frac{8\mu}{R^2}\left(u_0 - \frac{2v_w z}{R}\right). \tag{8.2-27}$$

Integrating to find the local pressure gives

$$\mathcal{P}(0) - \mathcal{P}(z) = \frac{8\mu u_0 z}{R^2}\left(1 - \frac{v_w z}{u_0 R}\right). \tag{8.2-28}$$

It is seen that outward flow ($v_w > 0$) reduces the pressure drop and inward flow ($v_w < 0$) increases it, relative to that for an impermeable tube. The solution is completed by using the continuity equation to obtain

$$v_r(r) = v_w\left[2\left(\frac{r}{R}\right) - \left(\frac{r}{R}\right)^3\right] \tag{8.2-29}$$

as detailed in Example 5.2-3. It is seen that v_r is independent of z.

Again, the solution allows us to examine when the lubrication approximation will be accurate. Using a numbering scheme analogous to that in Eqs. (8.2-15) and (8.2-16), the dominant viscous term in the z component of the Navier–Stokes equation is

$$\mathrm{T1} = \frac{\mu}{r}\frac{\partial}{\partial r}\left(r\frac{\partial v_z}{\partial r}\right) = -\frac{8\mu u}{R^2}. \tag{8.2-30}$$

The viscous term T2 that contains $\partial^2 v_z/\partial z^2$ happens to be zero because v_z is linear in z. When averaged over the cross-section, the two inertial terms again are equal,

$$\langle \mathrm{T3}\rangle = \frac{8}{3}\frac{\rho v_w u}{R} = \langle \mathrm{T4}\rangle . \tag{8.2-31}$$

Accordingly, the ratio of inertial to viscous terms is

$$\frac{\langle \mathrm{T3} + \mathrm{T4}\rangle}{\langle \mathrm{T1}\rangle} = \frac{2}{3}\frac{\rho|v_w|R}{\mu} \equiv \frac{2}{3}\mathrm{Re}_w. \tag{8.2-32}$$

Thus, neglecting inertia requires only that the wall Reynolds number, $\mathrm{Re}_w = \rho|v_w|R/\mu$, be small. To assess how accurate it is to ignore radial pressure variations, we compare the dominant viscous terms in the r and z momentum equations, which dictate the size of the corresponding pressure derivatives. The largest term in the r component is

$$\mathrm{T5} = \mu\frac{\partial}{\partial r}\left(\frac{1}{r}\frac{\partial}{\partial r}(r v_r)\right) = -\frac{8\mu v_w r}{R^2}. \tag{8.2-33}$$

Averaging this over the cross-section and comparing the result with Eq. (8.2-30) gives

$$\frac{|\partial\mathcal{P}/\partial r|}{|\partial\mathcal{P}/\partial z|} \cong \frac{\langle \mathrm{T5}\rangle}{\langle \mathrm{T1}\rangle} = \frac{2}{3}\frac{|v_w|}{u}. \tag{8.2-34}$$

In summary, the two criteria for applying the lubrication approximation are $\mathrm{Re}_w \ll 1$ and $|v_w|/u \ll 1$, where $\mathrm{Re}_w = \rho|v_w|R/\mu$. These conditions are often satisfied in hollow-fiber devices.

Example 8.2-3 Slider bearing. The application of Eq. (8.2-1) to lubrication was pioneered by Reynolds (1886). The key principles are illustrated by the thrust bearing in Fig. 8.2, in which the objects might be machine parts bathed in oil. A horizontal surface at the bottom moves to the right at a constant speed U, and a flat but inclined one at the top

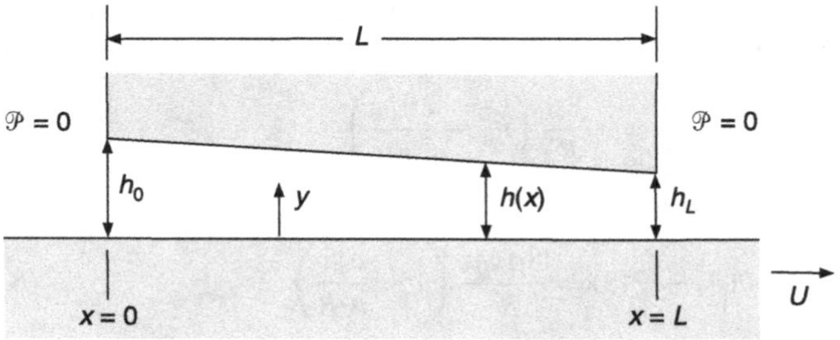

Figure 8.2 Slider bearing.

is stationary. The position of the top surface is described by Eq. (8.2-2) and the upstream and downstream pressures are equal. The requirements for the lubrication approximation are the same as in Example 8.2-1. However, whereas pressure-driven flow in a gradually widening channel is much the same as in a gradually narrowing one, a crucial feature of the slider bearing is that $h_0 > h_L$.

Starting with Eq. (8.2-4) and using the boundary conditions $v_x(x, 0) = U$ and $v_x(x, h) = 0$ to evaluate a and b, we find that

$$v_x(x, y) = U\left[1 - \left(\frac{y}{h}\right)\right] - \frac{h^2}{2\mu}\frac{d\mathcal{P}}{dx}\left[\left(\frac{y}{h}\right) - \left(\frac{y}{h}\right)^2\right]. \tag{8.2-35}$$

Integrating over the gap height, the pressure gradient is related to the volume flow rate per unit width (q) by

$$q = \int_0^h v_x\, dy = \frac{Uh}{2} - \frac{h^3}{12\mu}\frac{d\mathcal{P}}{dx}. \tag{8.2-36}$$

Using this to eliminate $d\mathcal{P}/dx$ from Eq. (8.2-35) gives

$$v_x(x, y) = U\left[1 - 4\left(\frac{y}{h}\right) + 3\left(\frac{y}{h}\right)^2\right] + \frac{6q}{h}\left[\left(\frac{y}{h}\right) - \left(\frac{y}{h}\right)^2\right]. \tag{8.2-37}$$

Having an unknown constant (q) in v_x is better than having an unknown function ($d\mathcal{P}/dx$).

The overall pressure difference provides the remaining information needed to find q. Rearrangement of Eq. (8.2-36) indicates that

$$\frac{d\mathcal{P}}{dx} = \frac{6\mu U}{h^2} - \frac{12\mu q}{h^3}. \tag{8.2-38}$$

Integrating over the length of the fluid-filled gap gives

$$\Delta\mathcal{P} = \mathcal{P}(L) - \mathcal{P}(0) = 6\mu U \int_0^L h^{-2}\, dx - 12\mu q \int_0^L h^{-3}\, dx. \tag{8.2-39}$$

What this shows in general is that, in a two-dimensional lubrication analysis, only two of the quantities $\Delta\mathcal{P}$, U, and q are independent. Noting that $\Delta\mathcal{P} = 0$ in the present problem

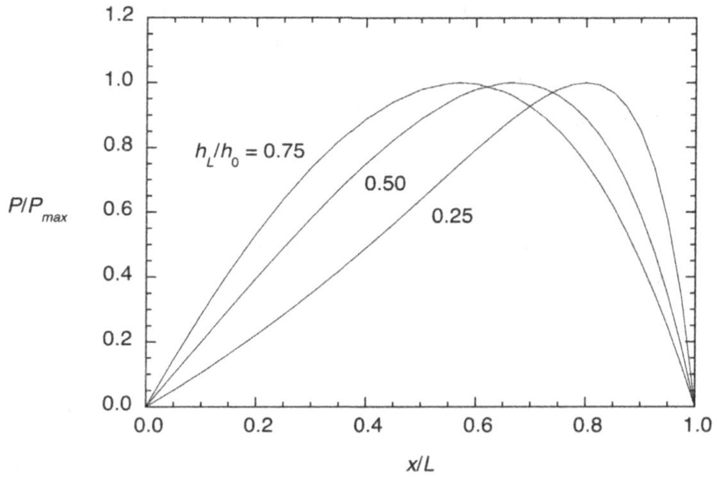

Figure 8.3 Pressure in a slider bearing as a function of position.

and evaluating the two integrals using Eq. (8.2-2), we find that

$$q = U\left(\frac{h_0 h_L}{h_0 + h_L}\right) \equiv U\bar{h}. \tag{8.2-40}$$

Substituting this into Eq. (8.2-37) yields the final expression for the major velocity component,

$$v_x(x, y) = U\left[1 - 4\left(\frac{y}{h}\right)\left(1 - \frac{3\bar{h}}{2h}\right) + 3\left(\frac{y}{h}\right)^2\left(1 - \frac{2\bar{h}}{h}\right)\right]. \tag{8.2-41}$$

The minor velocity component, obtained as usual from the continuity equation, is

$$v_y(x, y) = 2U\frac{dh}{dx}\left(\frac{3\bar{h}}{h} - 1\right)\left[\left(\frac{y}{h}\right)^2 - \left(\frac{y}{h}\right)^3\right]. \tag{8.2-42}$$

The pressure is found by substituting Eq. (8.2-40) into Eq. (8.2-38) and integrating from $x = 0$ to an arbitrary position, which gives

$$\mathcal{P}(x) = 6\mu UL\left[\frac{(h_0 - h)(h - h_L)}{\left(h_0^2 - h_L^2\right)h^2}\right]. \tag{8.2-43}$$

Given that $h_0 \geq h \geq h_L$, it is evident that $\mathcal{P} > 0$ within the gap. That is the result of fluid being dragged into an ever-narrowing space. The maximum pressure, which can be very large, is

$$\mathcal{P}_{\max} = \frac{3}{2}\frac{\mu UL}{h_0 h_L}\frac{(h_0 - h_L)}{(h_0 + h_L)}. \tag{8.2-44}$$

This is the pressure at $x/L = h_0/(h_0 + h_L)$. Figure 8.3 shows pressure profiles for three values of h_L/h_0. As h_L/h_0 decreases, the location at which the maximum pressure occurs shifts downstream.

The force that the fluid exerts on the upper surface is calculated as described in Example 8.5-4. The force components F_x and F_y for a bearing of width W are

$$\frac{F_x}{W} = \frac{6\mu UL}{(h_0 - h_L)}\left[\frac{2}{3}\ln\left(\frac{h_0}{h_L}\right) - \left(\frac{h_0 - h_L}{h_0 + h_L}\right)\right] \tag{8.2-45}$$

$$\frac{F_y}{W} = \frac{6\mu UL^2}{(h_0 - h_L)^2}\left[\ln\frac{h_0}{h_L} - 2\left(\frac{h_0 - h_L}{h_0 + h_L}\right)\right]. \tag{8.2-46}$$

The ability of a bearing to support a large load (y component of force) with relatively little drag (x component of force) is a hallmark of effective lubrication. For $h_0/h_L = 2$, the force ratio is

$$\frac{F_y}{F_x} = 0.41\frac{L}{h_0}. \tag{8.2-47}$$

This will be quite large when the gap between objects is relatively thin.

Many other kinds of bearings have been analyzed (Cameron, 1981; Pinkus and Sternlicht, 1961). Other applications for the lubrication approximation include coating processes and polymer processing operations such as calendering (Middleman, 1977; Tanner, 2000). Analyses of flow in permeable tubes or other channels have applications to physiological processes and microfluidic devices. Although we have discussed only two-dimensional problems in which the flow is nearly unidirectional, there is a three-dimensional form of the lubrication approximation (Cameron, 1981, pp. 46–47; Tanner, 2000, pp. 312–313). It applies to thin films in which there are two major velocity components and a third one that is small.

8.3 CREEPING FLOW

Stokes' equation

The defining characteristic of *creeping flow*, also called *Stokes flow*, is that the Reynolds number is practically zero. Because Re is the ratio of inertial to viscous stresses, the inertial terms on the left-hand side of the Navier–Stokes equation are ordinarily negligible for Re $\to$ 0 and

$$0 = -\nabla\mathcal{P} + \mu\nabla^2\mathbf{v}. \tag{8.3-1}$$

This simplified form of the momentum equation for incompressible Newtonian fluids is called *Stokes' equation* (after G. G. Stokes). It is formally the same as setting $\rho = 0$ in Eq. (6.6-5). Because $\rho\mathbf{v}$ is the momentum per unit volume and ρ no longer appears in the governing equations, in creeping-flow problems momentum is essentially absent. With momentum changes being negligible, the forces due to pressure, gravity, and viscosity are in near-perfect balance at each point in the fluid, as indicated by the right-hand side of Eq. (8.3-1) summing to zero.

Because our expectations about fluid behavior are based largely on everyday experiences with air and water at large Re, creeping flow has strikingly nonintuitive features. These stem largely from its *kinematic reversibility*, which means that, if the driving force for fluid motion is suddenly reversed, each fluid element promptly reverses direction and starts to retrace its previous path.[2] For instance, suppose that the rod in Example 7.3-2 is

[2] Kinematic reversibility, or reversibility of motion, should not be confused with thermodynamic reversibility. Mechanical energy is dissipated into heat during both forward and reverse flow at low Re, consistent with the second law of thermodynamics.

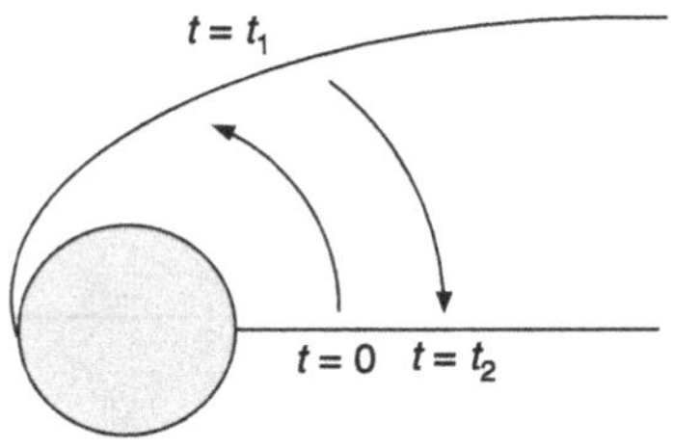

Figure 8.4 Dye streak for creeping flow near a rotating rod: initially (at $t = 0$); after one half turn counterclockwise (at $t = t_1$); and after a subsequent half turn clockwise (at $t = t_2$).

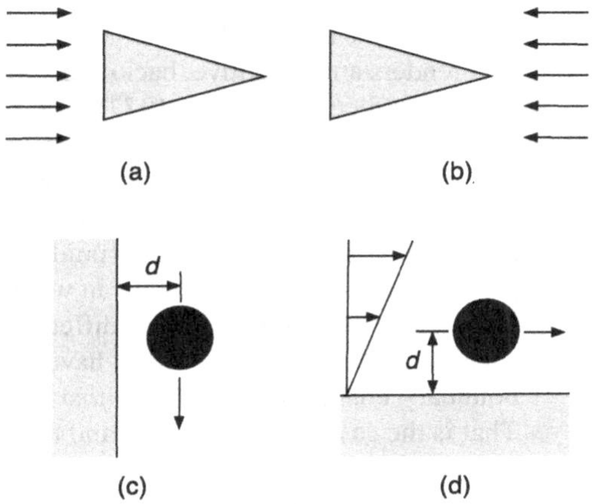

Figure 8.5 Some consequences of the reversibility of creeping flow: the drag coefficient for a cone or wedge oriented as in (a) is the same as that in (b), and a sphere moves parallel to a wall, whether settling as in (c) or neutrally buoyant and propelled by the fluid as in (d).

rotated slowly in a very viscous liquid, such that $\mathrm{Re} = \omega R^2/\nu \ll 1$. Imagine that at $t = 0$ the surface of the liquid is marked with a line of dye, as in Fig. 8.4. If diffusion of the dye is negligible, after one-half turn the straight line will become a spiral. If the rotation is then reversed one-half turn, the dye streak will return exactly to its original shape and position. This will occur even if the forward and reverse speeds differ. Mathematically, the linearity of Eq. (8.3-1) is what makes the motion reversible, and the absence of a time derivative is what makes the fluid response immediate.

Kinematic reversibility permits a variety of inferences to be made without the need for mathematical analysis. One consequence of it is that the streamlines for flow past any solid object remain the same if the flow direction is reversed. Thus, photographs of streaklines at low Re do not reveal the direction of flow. That is true even if the object lacks fore-and-aft symmetry (e.g., is blunt on the leading side and "streamlined" on the trailing side). Because the velocity components all simply change sign, so do the shear stresses and pressure gradient. Consequently, the drag coefficient remains the same if the flow direction is reversed. For example, the drag on the cone or wedge in Fig. 8.5(a) is the same as that in Fig. 8.5(b). Unchanging streamlines imply also that a rigid particle will retrace its path. For example, in creeping flow a solid sphere settles parallel to a vertical wall, as in Fig. 8.5(c), because only if the distance d is constant can it retrace its path if gravity is reversed by turning the container upside down. Similarly, a neutrally buoyant sphere in a shear flow will also move parallel to a wall, as in Fig. 8.5(d). In contrast, at high Re there are readily identifiable wakes and the drag coefficients of asymmetric objects are sensitive to the direction of flow.

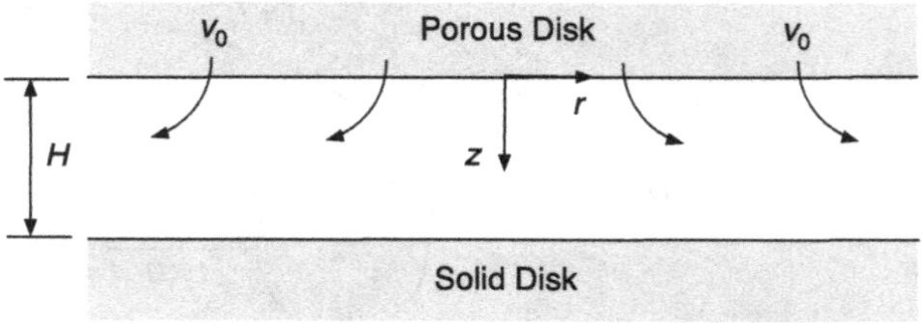

Figure 8.6 Flow due to fluid injection into the space between porous and solid disks.

Kinematic reversibility at small Re limits how microorganisms can propel themselves, as it renders any repetitive, back-and-forth motion of appendages (such as rowing underwater) ineffective (Purcell, 1977). Such inertia-free conditions are experienced also by aerosol particles, blood cells in the microcirculation, and particles, bubbles, or droplets in various microfluidic reactors or separation devices.

Unlike the lubrication formulations in Section 8.2, the use of Stokes' equation does not require the flow to be nearly unidirectional, although it can be. The following examples involve axisymmetric flows in which there are two nonzero velocity components of comparable magnitude. Partial differential equations that are linear and homogeneous, such as Stokes' equation, often have *separable* solutions. That is, for certain kinds of boundary conditions, a variable that depends on both x and y takes the form $f(x)g(y)$. That is the case for the pressure and the velocity components in these examples.

Example 8.3-1 Flow between porous and solid disks. Suppose that a fluid is injected steadily into a space of thickness H between two disks of indefinitely large radius, as shown in Fig. 8.6. The inward velocity at the upper (porous) surface is constant at v_0. It is desired to find $v_r(r, z)$ and $v_z(r, z)$ when $\text{Re} = v_0 H/\nu \ll 1$.

We begin with the hypothesis that v_r is separable, such that $v_r(r, z) = g(r)f(z)$. The same might be assumed for v_z, but we will go further and suppose that $v_z(r, z) = h(z)$. In this regard, notice that neither of its boundary conditions, $v_z(r, 0) = v_0$ and $v_z(r, H) = 0$, requires that v_z depend on r. The cylindrical continuity equation indicates that

$$\frac{\partial v_z}{\partial z} = -\frac{1}{r}\frac{\partial}{\partial r}(r v_r). \tag{8.3-2}$$

If v_z is indeed a function of z only, then both sides must be independent of r. That will be true only if $r v_r \propto r^2$ or $g(r) = r$. Including the boundary conditions, what we have then is that

$$v_r(r, z) = r f(z), \quad f(0) = 0 = f(H) \tag{8.3-3}$$

$$v_z(r, z) = h(z), \quad h(0) = v_0, \quad h(H) = 0. \tag{8.3-4}$$

Substituting this form of the velocity field into Eq. (8.3-2) shows that $h(z)$ and $f(z)$ are related as

$$\frac{dh}{dz} = -\frac{1}{r}\frac{\partial}{\partial r}(r^2 f) = -2f. \tag{8.3-5}$$

Putting the pressure terms on the left, the r and z components of Stokes' equation are

$$\frac{1}{\mu}\frac{\partial \mathcal{P}}{\partial r} = \frac{\partial}{\partial r}\left(\frac{1}{r}\frac{\partial}{\partial r}(r v_r)\right) + \frac{\partial^2 v_r}{\partial z^2} \tag{8.3-6}$$

$$\frac{1}{\mu}\frac{\partial \mathcal{P}}{\partial z} = \frac{1}{r}\frac{\partial}{\partial r}\left(r\frac{\partial v_z}{\partial r}\right) + \frac{\partial^2 v_z}{\partial z^2}. \tag{8.3-7}$$

According to the assumed forms of v_r and v_z, the r derivatives on the right-hand sides of both equations vanish. Differentiating Eq. (8.3-7) with respect to r indicates then that $\partial^2\mathcal{P}/\partial r\,\partial z = 0$. Differentiating Eq. (8.3-6) with respect to z yields another expression for $\partial^2\mathcal{P}/\partial r\,\partial z$. Setting that to zero gives

$$\frac{d^3 f}{dz^3} = 0. \tag{8.3-8}$$

This and Eq. (8.3-5) provide two linear differential equations that govern $f(z)$ and $h(z)$. Their solution is simplified by the fact that h is absent from Eq. (8.3-8). In that the pressure has been eliminated from the governing equations, it is unnecessary to assume a functional form for $\mathcal{P}(r, z)$. Three integrations of Eq. (8.3-8) lead to

$$f(z) = az^2 + bz + c. \tag{8.3-9}$$

Evaluating two of the three constants using the boundary conditions in Eq. (8.3-3) gives

$$f(z) = -aH^2\left[\left(\frac{z}{H}\right) - \left(\frac{z}{H}\right)^2\right] \tag{8.3-10}$$

where a remains to be determined. Substituting this into Eq. (8.3-5) and integrating once more indicates that

$$h(z) = aH^3\left[\left(\frac{z}{H}\right)^2 - \frac{2}{3}\left(\frac{z}{H}\right)^3\right] + d. \tag{8.3-11}$$

The boundary conditions in Eq. (8.3-4) require that $a = -3v_0/H^3$ and $d = v_0$. The final results are

$$v_r(r, z) = \frac{3v_0 r}{H}\left[\left(\frac{z}{H}\right) - \left(\frac{z}{H}\right)^2\right] \tag{8.3-12}$$

$$v_z(z) = v_0\left[1 - 3\left(\frac{z}{H}\right)^2 + 2\left(\frac{z}{H}\right)^3\right]. \tag{8.3-13}$$

The ability to satisfy the continuity equation, Stokes' equation, and the boundary conditions confirms that the assumed forms of the velocity components were correct. If desired, $\mathcal{P}(r, z)$ could be found by substituting the velocities into Eqs. (8.3-6) and (8.3-7) and integrating (Problem 6.8).

Example 8.3-2 Flow past a sphere. The objective is to find the velocity and pressure fields for creeping flow past a sphere. The analysis of this problem by Stokes (1851) was the first published solution for a viscous flow, predating even that for Poiseuille flow. As shown in Fig. 8.7, it is convenient to align the z axis with the approaching flow. Accordingly, $v_z = U$ far from the sphere, in any direction. There is rotational symmetry about the z axis, such that the velocity and pressure are independent of the spherical angle ϕ and $v_\phi = 0$. We wish to determine $v_r(r, \theta)$, $v_\theta(r, \theta)$, and $\mathcal{P}(r, \theta)$ for $\mathrm{Re} = 2UR/\nu \ll 1$.

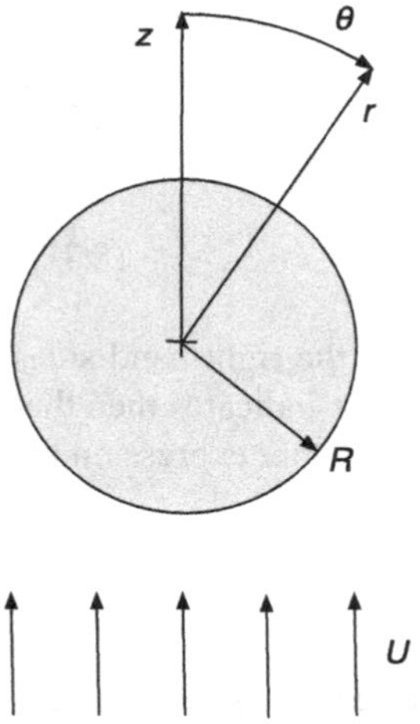

Figure 8.7 Streaming flow at velocity U past a sphere of radius R.

Using Eq. (A.5-43) to relate $\mathbf{e_z}$ to the spherical unit vectors, the unperturbed velocity far from the sphere is

$$\mathbf{v}(\infty, \theta) = U\mathbf{e_z} = U\cos\theta\,\mathbf{e_r} - U\sin\theta\,\mathbf{e_\theta} \tag{8.3-14}$$

which shows that $v_r = U\cos\theta$ and $v_\theta = -U\sin\theta$ at large r. If the velocity components are separable functions, the boundary conditions indicate that

$$v_r(r, \theta) = f(r)\cos\theta, \quad f(R) = 0, \quad f(\infty) = U \tag{8.3-15}$$

$$v_\theta(r, \theta) = g(r)\sin\theta, \quad g(R) = 0, \quad g(\infty) = -U. \tag{8.3-16}$$

The no-penetration and no-slip conditions at the sphere surface require only that $f(R) = 0$ and $g(R) = 0$, respectively. However, what is needed to obtain $v_z = U$ at $r = \infty$ reveals the θ-dependence of each velocity component. As will be seen shortly, the θ-dependence of $\mathcal{P}$ must mimic that of v_r. Thus,

$$\mathcal{P}(r, \theta) = h(r)\cos\theta, \quad h(\infty) = 0. \tag{8.3-17}$$

The spatially uniform velocity far from the sphere indicates that $\mathcal{P}$ is constant there, and that constant has been set to zero. The correctness of the three assumed functional forms will be confirmed by our ability to find functions $f(r)$, $g(r)$, and $h(r)$ that satisfy the continuity equation, Stokes' equation, and the boundary conditions.

The spherical continuity equation indicates that

$$\frac{1}{r^2}\frac{\partial}{\partial r}(r^2 v_r) = -\frac{1}{r\sin\theta}\frac{\partial}{\partial\theta}(v_\theta\sin\theta). \tag{8.3-18}$$

Substituting in the assumed forms for v_r and v_θ gives

$$\frac{1}{r^2}\frac{\partial}{\partial r}(r^2 f\cos\theta) = -\frac{1}{r\sin\theta}\frac{\partial}{\partial\theta}(g\sin^2\theta) = -\frac{2g}{r}\cos\theta. \tag{8.3-19}$$

Eliminating the common factor $\cos\theta$ and rearranging what remains, we find that

$$g = -\frac{r}{2}\frac{df}{dr} - f. \tag{8.3-20}$$

Putting the pressure terms on the left, the r and θ components of the spherical, axisymmetric form of Stokes' equation are

$$\frac{1}{\mu}\frac{\partial \mathcal{P}}{\partial r} = \nabla^2 v_r - \frac{2}{r^2}\left(v_r + \frac{\partial v_\theta}{\partial \theta} + v_\theta \cot\theta\right) \tag{8.3-21}$$

$$\frac{1}{\mu r}\frac{\partial \mathcal{P}}{\partial \theta} = \nabla^2 v_\theta + \frac{2}{r^2}\left(\frac{\partial v_r}{\partial \theta} - \frac{v_\theta}{2\sin^2\theta}\right) \tag{8.3-22}$$

where

$$\nabla^2 = \frac{1}{r^2}\frac{\partial}{\partial r}\left(r^2\frac{\partial}{\partial r}\right) + \frac{1}{r^2 \sin\theta}\frac{\partial}{\partial \theta}\left(\sin\theta\frac{\partial}{\partial \theta}\right). \tag{8.3-23}$$

The $\nabla^2 v_r$ and other velocity terms in Eq. (8.3-21) indicate that $\mathcal{P}$ must vary as $\cos\theta$, as asserted in Eq. (8.3-17). Substituting in the three assumed functional forms and evaluating the derivatives, the components of Stokes' equation become

$$\frac{1}{\mu}\frac{dh}{dr} = \frac{1}{r^2}\left[\frac{d}{dr}\left(r^2\frac{df}{dr}\right) - 4(f+g)\right] \tag{8.3-24}$$

$$\frac{h}{\mu} = -\frac{1}{r}\left[\frac{d}{dr}\left(r^2\frac{dg}{dr}\right) - 2(f+g)\right]. \tag{8.3-25}$$

Together with Eq. (8.3-20), these provide three linear differential equations that govern the three unknown functions of r.

The terms in the momentum equations that involve h are eliminated by differentiating Eq. (8.3-25) with respect to r and equating the result with Eq. (8.3-24). Then, g is eliminated using Eq. (8.3-20). The resulting equation for f is

$$\frac{d^4 f}{dr^4} + \frac{8}{r}\frac{d^3 f}{dr^3} + \frac{8}{r^2}\frac{d^2 f}{dr^2} - \frac{8}{r^3}\frac{df}{dr} = 0. \tag{8.3-26}$$

This linear, fourth-order differential equation with variable coefficients is an *equidimensional* (or *Euler*) equation. Such equations have fundamental solutions of the form $f(r) = r^n$, where n is an integer. Substituting this trial solution into Eq. (8.3-26) gives the characteristic equation,

$$n(n-1)(n-2)(n-3) + 8n(n-1)(n-2) + 8n(n-2) = 0. \tag{8.3-27}$$

By inspection, two of the roots are $n = 2$ and $n = 0$. Eliminating the common factor $n(n-2)$ to obtain a quadratic, the remaining roots are found to be $n = -1$ and $n = -3$. Thus, the general solution of Eq. (8.3-26) is

$$f(r) = Ar^2 + B + \frac{C}{r} + \frac{D}{r^3}. \tag{8.3-28}$$

From Eqs. (8.3-20) and (8.3-25), the corresponding solutions for g and h are

$$g(r) = -B - \frac{1}{2}\frac{C}{r} + \frac{1}{2}\frac{D}{r^3} \tag{8.3-29}$$

$$\frac{h(r)}{\mu} = 2Ar + \frac{C}{r^2}. \tag{8.3-30}$$

Thus, the solution will be complete once the four constants (A, B, C, D) are determined.

The boundary conditions in Eqs. (8.3-15) and (8.3-16) require that

$$A = 0, \quad B = U, \quad C = -\frac{3}{2}UR, \quad D = \frac{UR^3}{2}. \tag{8.3-31}$$

The final results for the velocity components and pressure are then

$$\frac{v_r(r,\theta)}{U} = \left[1 - \frac{3}{2}\left(\frac{R}{r}\right) + \frac{1}{2}\left(\frac{R}{r}\right)^3\right]\cos\theta \tag{8.3-32}$$

$$\frac{v_\theta(r,\theta)}{U} = -\left[1 - \frac{3}{4}\left(\frac{R}{r}\right) - \frac{1}{4}\left(\frac{R}{r}\right)^3\right]\sin\theta \tag{8.3-33}$$

$$\frac{\mathcal{P}(r,\theta)}{\mu U/R} = -\frac{3}{2}\left(\frac{R}{r}\right)^2\cos\theta. \tag{8.3-34}$$

The absence of inertia in creeping flow ensures that the variations in $\mathcal{P}$ are comparable to the viscous pressure scale, which is $\mu U/R$. It is seen from Eq. (8.3-34) that $\mathcal{P}$ increases on the upstream side of the sphere (where $\pi/2 < \theta \leq \pi$) and decreases on the downstream side (where $0 \leq \theta < \pi/2$), relative to its value of zero in the unperturbed fluid. These pressure variations contribute to the drag, as does the shear stress at the surface of the sphere. The velocity and pressure results are used in Example 8.3-3 to derive Stokes' law for the drag.

The flow pattern can be visualized by plotting constant values of the stream function, $\psi(r,\theta)$. From Table 5.4 and the results for v_r and v_θ,

$$\frac{\partial\psi}{\partial\theta} = v_r r^2\sin\theta = Ur^2\sin\theta\cos\theta\left[1 - \frac{3}{2}\left(\frac{R}{r}\right) + \frac{1}{2}\left(\frac{R}{r}\right)^3\right] \tag{8.3-35}$$

$$\frac{\partial\psi}{\partial r} = -v_\theta r\sin\theta = Ur\sin^2\theta\left[1 - \frac{3}{4}\left(\frac{R}{r}\right) - \frac{1}{4}\left(\frac{R}{r}\right)^3\right]. \tag{8.3-36}$$

Integration results in two identical expressions for $\psi(r, \theta)$. Setting the shared integration constant to zero gives

$$\psi(r,\theta) = \frac{UR^2}{2}\sin^2\theta\left[\left(\frac{r}{R}\right)^2 - \frac{3}{2}\left(\frac{r}{R}\right) + \frac{1}{2}\left(\frac{R}{r}\right)\right]. \tag{8.3-37}$$

Streamlines corresponding to various values of ψ are shown in Fig. 8.8. As discussed in Section 5.6, the velocity is greatest where the streamlines are closest together.

The solutions of the two preceding problems involved the same steps: (1) using the boundary conditions as a guide, separable forms for the velocity components were postulated; (2) the continuity equation was used to relate one velocity component to the other; (3) by combining continuity with Stokes' equation, the pressure and one velocity component were eliminated; (4) the resulting third- or fourth-order differential equation was solved for the velocity component that was retained; (5) with one velocity component known, the other was found by returning to continuity; and (6) with both velocity components known, the pressure (if of interest) was calculated by returning to Stokes' equation.

Although beyond the scope of this book, it is worth noting that reformulating the continuity and Navier–Stokes equations in terms of a stream function often provides a more efficient way to solve fully bidirectional problems. The governing equations are

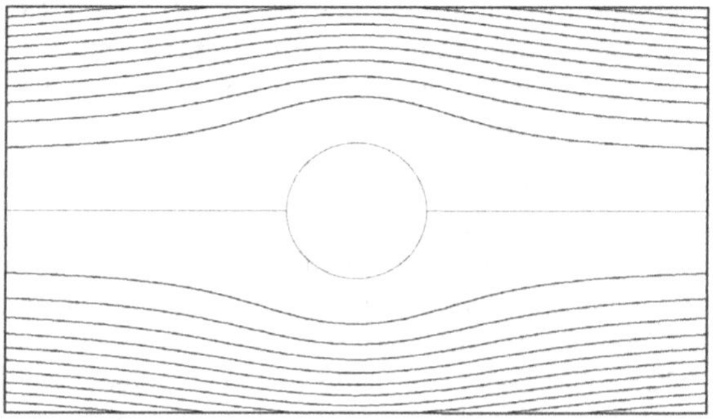

Figure 8.8 Streamlines for creeping flow past a sphere.

combined into a single fourth-order differential equation in which the stream function is the only dependent variable. Stream-function forms of the Navier–Stokes equation are found in many sources, including Deen (2012) and Leal (2007). Working with the stream-function formulation avoids the sometimes-tedious manipulations needed to eliminate the pressure and one velocity component. The stream-function solution for creeping flow past a sphere is detailed, for example, in Deen (2012, pp. 326–328).

Example 8.3-3 Stokes' law. The objective is to derive Stokes' law for the drag on a sphere at low Re, presented previously as Eq. (3.2-3).

From Example 6.6-3, for any axisymmetric flow past a solid sphere (with or without inertia) the drag is

$$F_D = -2\pi R^2 \int_0^{\pi} [\mathcal{P}(R,\theta)\cos\theta + \tau_{r\theta}(R,\theta)\sin\theta]\sin\theta\, d\theta. \tag{8.3-38}$$

Using Eqs. (8.3-32)–(8.3-34), the pressure and shear stress at the surface are evaluated as

$$\mathcal{P}(R,\theta) = -\frac{3}{2}\frac{\mu U}{R}\cos\theta \tag{8.3-39}$$

$$\tau_{r\theta}(R,\theta) = -\frac{3}{2}\frac{\mu U}{R}\sin\theta. \tag{8.3-40}$$

Substitution into Eq. (8.3-38) gives

$$F_D = 3\pi\mu U R \int_0^{\pi} (\cos^2\theta + \sin^2\theta)\sin\theta\, d\theta. \tag{8.3-41}$$

Keeping the pressure and viscous contributions separate, the two integrals are

$$\int_0^{\pi} \cos^2\theta \sin\theta\, d\theta = \frac{2}{3}, \quad \int_0^{\pi} \sin^3\theta\, d\theta = \frac{4}{3}. \tag{8.3-42}$$

Accordingly, the drag is

$$F_D = 6\pi\mu U R \tag{8.3-43}$$

which is Stokes' law written in terms of the radius. As may be seen from Eq. (8.3-42), one-third of the drag is from the pressure (form drag, the first integral) and two-thirds is from the shear stress (friction drag, the second integral). This is in contrast to flow past spheres or other bluff objects at high Re, where form drag predominates.

Any valid approximation to the Navier–Stokes equation should yield internally consistent results. That is, when evaluated using the solution, the neglected terms should prove to be small for the geometric and dynamic conditions being considered, as in Section 8.2. Surprisingly, given the innumerable experiments that have verified the accuracy of Eq. (8.3-43) for small Re, Stokes' solution for flow past a sphere fails this test at large r. In particular, using Eqs. (8.3-32) and (8.3-33), the sum of the viscous terms in the r component is found to be $3\mu U R \cos\theta / r^3$. The sum of the inertial terms approaches $(3/4)(\rho U^2 R/r^2)(3\cos^2\theta - 1)$ as $r/R \to \infty$. The ratio of inertial to viscous terms at large r is then

$$\frac{Ur}{\nu}\left|\frac{3\cos^2\theta - 1}{4\cos\theta}\right| = \mathrm{Re}\left(\frac{r}{R}\right)\left|\frac{3\cos^2\theta - 1}{8\cos\theta}\right|. \tag{8.3-44}$$

The absence of R in the first expression implies that the length scale for the region far from the sphere is r, not R. Because the function of θ in the second expression is not always small, if inertia is to be negligible everywhere it is necessary that $\mathrm{Re}(r/R) \ll 1$ or that $r/R \ll 1/\mathrm{Re}$. For any nonzero value of Re, however small, this requirement will be violated sufficiently far from the sphere. Fortunately, this failure of Stokes' equation at large r has little effect on the velocity and pressure near the sphere, and advanced analytical methods have been used to confirm that Eq. (8.3-43) is exact for $\mathrm{Re} \to 0$ (Deen, 2012, pp. 343–349). This good fortune does not extend to two-dimensional flows, such as that past a long cylinder, where inertia again becomes important at large r. Unlike the situation for a sphere, there is no solution of Stokes' equation for a cylinder that satisfies the boundary conditions for $r \to \infty$. Consequently, the drag on a cylinder depends on Re even for $\mathrm{Re} \to 0$ (Problem 8.12).[3]

Porous media

Creeping flow in porous media is often modeled by replacing Stokes' equation with Darcy's law, the vector form of which is

$$\mathbf{v} = -\frac{k}{\mu}\nabla\mathcal{P} \tag{8.3-45}$$

where k is the Darcy permeability. As discussed in Section 3.4, this is applicable only on length scales that are large compared to the pore dimensions, and $\mathbf{v}$ is a kind of superficial velocity. For a porous material that is homogeneous enough for k to be constant, combining Darcy's law with the continuity equation ($\nabla \cdot \mathbf{v} = 0$) gives

$$\nabla^2\mathcal{P} = 0. \tag{8.3-46}$$

If the pressure or the normal component of the pressure gradient is known at all boundaries, this can be solved for $\mathcal{P}$ (analytically or numerically), and then $\mathbf{v}$ can be found from Eq. (8.3-45).

[3] In his pioneering paper in 1851, Stokes noted the absence of a solution for flow past a cylinder. This failure of a seemingly correct set of governing equations to have a solution came to be known as *Stokes' paradox*. The subtle effects of inertia at low Re were largely unappreciated until the work of C. W. Oseen in 1910, and it was not until the development of singular perturbation theory in the 1950s that the associated mathematical difficulties were fully resolved.

The classic treatise on low-Reynolds-number flow is Happel and Brenner (1965). Leal (2007) also has an extensive discussion of creeping flow.

8.4 PSEUDOSTEADY FLOW

If a flow is time-dependent, but the time derivative in the momentum equation is negligible, it is *pseudosteady*. This occurs when a system responds immediately to the forces exerted on it. Even though a pressure or velocity imposed by the surroundings varies over time, the fast response makes a steady form of the momentum equation applicable at each instant. The time scale for the perturbations, or *process* time scale, is denoted as t_p, and when inertia is negligible the system response is governed by the *viscous* time scale, $t_v = L^2/\nu$ (Example 7.3-3). Thus, a viscous flow will tend to be pseudosteady when $t_v/t_p \ll 1$. This condition is satisfied at small length scales, provided that the perturbations are not extremely rapid. For example, with $L = 1$ mm and $\nu = 1 \times 10^{-6}$ m^2 s^{-1}, $t_v = 1$ s. Although we are focusing at present on flows that are dominated by viscous stresses, ones with inertia also can be pseudosteady.[4]

A consequence of the time derivative being negligible is that t acts only as a parameter. In effect, this reduces by one the number of independent variables in the momentum equation. Reducing the number of independent variables in any partial differential equation greatly facilitates solving it, so any such opportunity should be exploited.

Example 8.4-1 Parallel-plate channel with a decaying pressure drop. Assume that in a parallel-plate channel with wall spacing $2H$ the flow is fully developed, as in Example 7.2-1, but that the applied pressure decreases slowly over time. As a result, the mean velocity varies as

$$U(t) = U_0 e^{-t/t_0} \tag{8.4-1}$$

where U_0 and t_0 are constants. Of particular importance is the imposed time constant, t_0. It is desired to determine $v_x(y, t)$, with coordinates as in Fig. 7.1.

The time-dependence does not affect the conclusion from the two-dimensional continuity equation, which is that $v_y = 0$ everywhere if the flow is fully developed. Accordingly, $\partial\mathcal{P}/\partial y = 0$ and $\mathcal{P} = \mathcal{P}(x, t)$ only. If the changes in U are slow enough to make $\partial v_x/\partial t$ negligible, the x component of the Navier–Stokes equation simplifies to

$$\frac{\partial^2 v_x}{\partial y^2} = \frac{1}{\mu}\frac{\partial \mathcal{P}}{\partial x}. \tag{8.4-2}$$

With $\mathcal{P}$ independent of y, we can integrate, apply the no-slip conditions, and relate the pressure gradient to the mean velocity as in Example 7.2-1. The result is

$$v_x(y, t) = \frac{3}{2}U(t)\left[1 - \left(\frac{y}{H}\right)^2\right] \tag{8.4-3}$$

where $U(t)$ is given by Eq. (8.4-1). This differs from the steady-state solution [Eq. (7.2-12)] only in that a constant mean velocity has been replaced by a time-dependent one.

Because t_0 characterizes the rate at which the pressure drop and mean velocity vary, in this problem $t_p = t_0$. From Example 7.3-3, we expect the adjustment of the velocity

[4] Pseudosteady behavior occurs also in other dynamic systems. Time derivatives in the governing partial differential equations are negligible sometimes in systems with transient heat conduction, diffusion, chemical reactions, or current flow.

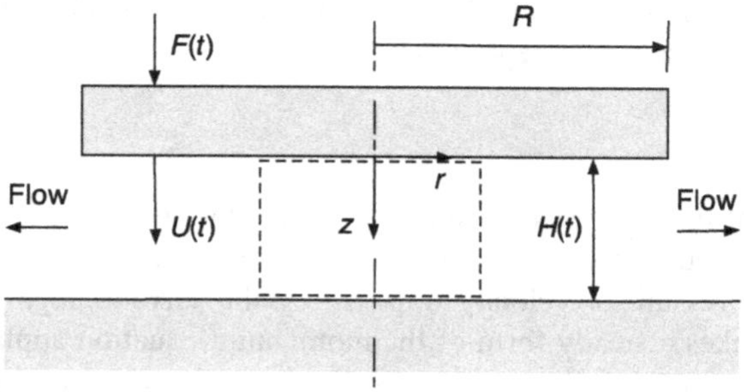

Figure 8.9 Squeeze flow. A force $F(t)$ on a disk of radius R pushes it at speed $U(t)$ toward a flat surface at a distance $H(t)$, squeezing the intervening fluid out radially.

profile to be limited by the rate at which changes in shear stress move across the channel. Accordingly, $t_v = H^2/\nu$ and we expect the flow to be pseudosteady if $t_v/t_0 \ll 1$. This can be checked by using Eq. (8.4-3) to estimate both the time derivative that was neglected and the viscous term that was retained. Using angle brackets to denote the absolute value of a term that has been averaged over y, we find that $\langle \partial v_x/\partial t \rangle = U/t_0$, $\langle \partial^2 v_x/\partial y^2 \rangle = 3U/H^2$, and

$$\frac{\langle \partial v_x/\partial t \rangle}{\nu \langle \partial^2 v_x/\partial y^2 \rangle} = \frac{1}{3}\frac{H^2}{\nu t_0} = \frac{1}{3}\frac{t_v}{t_0}. \tag{8.4-4}$$

This is consistent with our expectation that a small value of t_v/t_0 is what is needed to make the flow pseudosteady.

Example 8.4-2 Squeeze flow. Suppose that an applied force moves a disk of radius R toward a flat surface, as in Fig. 8.9. The intervening fluid is squeezed out as the gap between objects narrows. The external force is $F(t)$, the separation distance is $H(t)$, and the disk speed is $U(t)$. The motion is slow and the gap thin, such that $\text{Re} = U_0 H_0/\nu \ll 1$ and $H_0/R \ll 1$, where U_0 and H_0 are the maximum speed and maximum separation, respectively. The objective is to find the relationship among $F(t)$, $H(t)$, and $U(t)$.

This example is a kind of capstone, in that it involves each of the major approximations we have discussed. For reasons detailed in Example 8.5-3, the flow is nearly unidirectional, inertia is absent, and the motion is pseudosteady. That is, this is a lubrication problem made pseudosteady by creeping flow. The applicable form of the lubrication approximation is

$$\frac{\partial^2 v_r}{\partial z^2} = \frac{1}{\mu}\frac{\partial \mathcal{P}}{\partial r}, \quad \mathcal{P} = \mathcal{P}(r, t) \text{ only}. \tag{8.4-5}$$

Integrating twice and applying the no-slip conditions at $z = 0$ and $z = H(t)$ gives

$$v_r(r, z, t) = -\frac{H^2}{2\mu}\frac{\partial \mathcal{P}}{\partial r}\left[\left(\frac{z}{H}\right) - \left(\frac{z}{H}\right)^2\right]. \tag{8.4-6}$$

The dependence of v_r on r arises from the radial variation in $\partial \mathcal{P}/\partial r$, and the dependence of v_r on t is embedded in both $\partial \mathcal{P}/\partial r$ and H.

To evaluate the pressure gradient, consider the control volume of height $H(t)$ and arbitrary radius r that is depicted by the dashed rectangle in Fig. 8.9. The moving surface at the top acts as a piston, displacing volume at a rate $\pi r^2 U$. That fluid must exit radially.

Denoting the volume flow rate per unit of circumference as q, the balance is $\pi r^2 U = 2\pi r q$ or

$$q(r,t) = \frac{rU}{2}. \tag{8.4-7}$$

From Eq. (8.4-6),

$$q(r,t) = \int_0^H v_r\, dz = -\frac{H^3}{12\mu}\frac{\partial \mathcal{P}}{\partial r}. \tag{8.4-8}$$

Equating the two expressions for q gives

$$\frac{\partial \mathcal{P}}{\partial r} = -\frac{6\mu r U}{H^3}, \qquad \mathcal{P}(R) = 0 \tag{8.4-9}$$

where the pressure has been chosen as zero at the outer edge of the disk. Integration yields

$$\mathcal{P}(r,t) = \frac{3\mu R^2 U}{H^3}\left[1 - \left(\frac{r}{R}\right)^2\right]. \tag{8.4-10}$$

It is seen that the pressure increases sharply as the gap narrows, $\mathcal{P}$ varying as H^{-3}.

Neglecting the inertia of the disk itself, the net force on it is zero. Neglecting also gravity, buoyancy, and the small fluid forces outside the gap, the applied force must be balanced by the pressure at $z = 0$. Because the normal component of the viscous stress is zero at a solid surface (Problem 6.5), τ_{zz} does not affect the force balance. That leaves only the dynamic part of the pressure force. Accordingly,

$$F(t) = 2\pi \int_0^R \mathcal{P}(r,t) r\, dr = \frac{3\pi}{2}\frac{\mu R^4 U(t)}{H^3(t)}. \tag{8.4-11}$$

This relationship, obtained originally by J. Stefan in 1877, is called the *Stefan equation*. It shows that, if F is constant, U will decrease over time (varying as H^3), and that, to keep U constant, F must increase over time (varying as H^{-3}). Eliminating the speed as a variable using $U = -dH/dt$ yields a differential equation that could be solved to determine $H(t)$ for a specified $F(t)$.

8.5 ANTICIPATING APPROXIMATIONS

Quantitative justification for the approximations discussed earlier in the chapter was provided only for specific examples, and then only after the fact. That is, after a simplified form of the Navier–Stokes equation was employed in a particular problem, the solution was used to estimate the terms that had been neglected. Conclusions were then drawn concerning when that solution would be accurate. Such tests of consistency are informative. However, it would be better to have a way to assess potential approximations *before* investing the effort to obtain complete analytical or numerical solutions. Being able to generalize conclusions to families of problems is also highly desirable. The methodology described in this section is directed at those goals. The crucial element is the use of scales to estimate the orders of magnitude of derivatives and integrals. After describing the estimation methods, we will review more generally when each of the major approximations is valid.

Order-of-magnitude estimation

What is needed to simplify the Navier–Stokes equation at the outset of an analysis is to estimate the velocity components and their various first and second derivatives. In that the velocity field is not yet known, that may seem impossible. However, either the maximum or the average value of at least one velocity component is ordinarily given, and problem statements also contain clues about length and time scales. Moreover, in that the first task in modeling is only to identify what is negligible and what is not, order-of-magnitude (OM) precision is sufficient. It will be shown that adequate estimates actually can be obtained simply.

Consider a function $f(x, y, t)$, which will typically be a velocity component. In estimating the OM of f and its various partial derivatives, we are concerned only with absolute values and not algebraic signs. The rules for estimating the x derivatives are

$$\frac{\partial f}{\partial x} \sim \frac{(\Delta f)_x}{\Delta x} \tag{8.5-1}$$

$$\frac{\partial^2 f}{\partial x^2} \sim \frac{(\Delta f)_x}{(\Delta x)^2}. \tag{8.5-2}$$

In these and subsequent equations "$\sim$" denotes an OM estimate. It is assumed here that f varies smoothly and monotonically by an amount $(\Delta f)_x$ in the x direction over an interval Δx. In that $(\Delta f)_x/\Delta x$ equals the average of $\partial f/\partial x$ over the interval, Eq. (8.5-1) will always give the correct OM. Implicit in Eq. (8.5-2) is that $\partial f/\partial x$ varies from nearly zero to a maximum value over the interval Δx. That is, it supposes that $\partial^2 f/\partial x^2 \sim (|\partial f/\partial x|_{\max} - 0)/\Delta x$. Although often quite accurate, Eq. (8.5-2) will overestimate the second derivative of a function that has little or no curvature. The uncertainty in second derivatives notwithstanding, these expressions rarely cause misleading conclusions. For y or t derivatives, Δy or Δt replaces Δx and the numerators become $(\Delta f)_y$ and $(\Delta f)_t$, respectively. No subscripts are needed in the denominators because the velocity components and pressure are closely connected by the continuity and momentum equations. Accordingly, all tend to vary over similar intervals in position and time.

There is a need also for OM estimates of integrals. If $f(x)$ varies monotonically over the interval $[0, \Delta x]$, then

$$\int_0^{\Delta x} f\,dx \sim f_s\,\Delta x \tag{8.5-3}$$

where f_s is the *scale* for $f(x)$ within that interval. As introduced in Section 1.3, the scale of a variable is its maximum value. Within the precision of such estimates, the scale of something, its mean value, and its OM are all the same. Using the mean for f_s in Eq. (8.5-3) would make the equality exact.

In Eqs. (8.5-1) and (8.5-2), $(\Delta f)_x$ and Δx are scales for *variations* in f and x, respectively. If f varies from zero to some maximum, as is typical for velocity components, the scale for Δf is the same as that for f. Identifying the scales for independent variables (position or time) can be more subtle. For example, in the derivative formulas Δx is usually a distance over which there is a significant change in f. This dynamically relevant interval in x might differ from the total range of the x coordinate. Likewise, Δt may or may not be the total duration of a process. The estimation of scales is illustrated in the next example.

Table 8.1 Velocity components and their derivatives in a tapered channel

Quantity	OM estimate	OM/exact	Quantity	OM estimate	OM/exact
$\langle v_x \rangle$	u	1	$\langle v_y \rangle$	$u\left\lvert\frac{dh}{dx}\right\rvert$	$\frac{8}{3}$
$\langle \partial v_x/\partial y \rangle$	$\frac{u}{h}$	$\frac{2}{3}$	$\langle \partial v_y/\partial y \rangle$	$\frac{u}{h}\left\lvert\frac{dh}{dx}\right\rvert$	–
$\langle \partial^2 v_x/\partial y^2 \rangle$	$\frac{u}{h^2}$	$\frac{1}{3}$	$\langle \partial^2 v_y/\partial y^2 \rangle$	$\frac{u}{h^2}\left\lvert\frac{dh}{dx}\right\rvert$	$\frac{2}{9}$
$\langle \partial v_x/\partial x \rangle$	$\frac{u}{h}\left\lvert\frac{dh}{dx}\right\rvert$	–	$\langle \partial v_y/\partial x \rangle$	$\frac{u}{h}\left\lvert\frac{dh}{dx}\right\rvert^2$	–
$\langle \partial^2 v_x/\partial x^2 \rangle$	$\frac{2u}{h^2}\left\lvert\frac{dh}{dx}\right\rvert^2$	$\frac{2}{3}$	$\langle \partial^2 v_y/\partial x^2 \rangle$	$\frac{2u}{h^2}\left\lvert\frac{dh}{dx}\right\rvert^3$	$\frac{2}{21}$

Example 8.5-1 Order-of-magnitude analysis for a tapered channel. The objective is to predict the magnitudes of the velocity components and their various spatial derivatives for the tapered channel in Fig. 8.1. We will proceed as if the solutions for $v_x(x, y)$ and $v_y(x, y)$ in Example 8.2-1 were unavailable. Those solutions will be used only to check the reliability of the OM estimates. Comparisons will be made between height-averaged quantities at a given position x along the channel. As in Example 8.2-1, the absolute value of something that has been averaged over y will be denoted using angle brackets.

A suitable scale for v_x at a given position along the channel is the mean velocity; that is, $v_x \sim u(x)$. Because v_x varies from zero at the wall to roughly u in the center, $(\Delta v_x)_y = u$ and $\Delta y = h$. It follows from Eqs. (8.5-1) and (8.5-2) that $\partial v_x/\partial y \sim u/h$ and $\partial^2 v_x/\partial y^2 \sim u/h^2$. Less obvious are the values of $(\Delta v_x)_x$ and Δx (see below). However, the x derivatives can be inferred from the constancy of $q = 2u(x)h(x)$. With $d(uh)/dx = 0$,

$$\frac{du}{dx} = -\frac{u}{h}\frac{dh}{dx} \tag{8.5-4}$$

$$\frac{d^2u}{dx^2} = \frac{2u}{h^2}\left(\frac{dh}{dx}\right)^2. \tag{8.5-5}$$

The second expression is restricted to constant dh/dx (planar walls). Accordingly, $\partial v_x/\partial x \sim (u/h)|dh/dx|$ and $\partial^2 v_x/\partial x^2 \sim (2u/h^2)|dh/dx|^2$. The OM results for v_x are summarized in the second column of Table 8.1.

Just as known scales provide estimates of derivatives, known derivatives can be used to infer unknown scales. The result for $\partial v_x/\partial x$ can be rewritten as

$$\frac{\partial v_x}{\partial x} \sim \frac{(\Delta v_x)_x}{\Delta x} \sim \frac{u}{h/|dh/dx|}. \tag{8.5-6}$$

If Δx is defined as the channel length needed for a sizable change in v_x, such as a doubling or halving, then $(\Delta v_x)_x \sim u$ and $\Delta x \sim h/|dh/dx| = (h/|\Delta h|)L$, where L is the channel length and $|\Delta h| = |h_L - h_0|$ is the change in its half-height (Fig. 8.1). Thus, if $|\Delta h|$ is only a small fraction of h_0 or h_L, then $\Delta x \gg L$. In other words, the dynamically significant length scale can greatly exceed the actual length of the channel. There are other situations, such as boundary layers, where the dynamic length scale is much smaller than the geometric dimension.

Using the continuity equation and the rule for estimating integrals [Eq. (8.5-3)], the OM of v_y is evaluated as

$$v_y \sim \int_0^h \frac{\partial v_x}{\partial x}\,dy \sim \frac{\partial v_x}{\partial x} h \sim u\left|\frac{dh}{dx}\right|. \tag{8.5-7}$$

This is also the scale for $(\Delta v_y)_y$. Thus, dividing by h or h^2 gives estimates of $\partial v_y/\partial y$ and $\partial^2 v_y/\partial y^2$, respectively. The x derivatives of v_y are obtained by differentiating Eq. (8.5-7) by x, either once or twice. The OM estimates of v_y and its derivatives are summarized in the fifth column of Table 8.1. Notice that each v_y entry is the corresponding v_x entry times $|dh/dx|$.

The columns in Table 8.1 labeled "OM/exact" are the ratios of the OM estimates to the height-averaged values derived from Eqs. (8.2-3) and (8.2-14). It is seen that the ratio varies from about 0.1 to 3. Where values are absent, the "exact" derivative changed sign in such a way that its height-average was zero. In each of those three cases, the OM estimate was 1/3 of the "exact" maximum. Overall, the comparisons confirm that only a rudimentary knowledge of the velocity profile is needed to estimate the various derivatives to within OM precision. From the OM estimates alone, we would have predicted correctly that Eq. (8.2-1) is applicable to the tapered channel when $|dh/dx| \ll 1$ and $\mathrm{Re}|dh/dx| \ll 1$.

Example 8.5-2 Order-of-magnitude analysis for Stokes flow past a sphere. The objective is to show that the failure of Stokes' equation at large r (Section 8.3) could have been anticipated by realizing that the characteristic length far from the sphere is r, not R. We will examine the r component of the spherical Navier–Stokes equation, although the θ component would serve just as well. At the OM level, $\nabla^2 v_r$ amounts to $\partial^2 v_r/\partial r^2$ plus $(1/r^2)\partial^2 v_r/\partial\theta^2$. With $v_r \sim U$ and $\Delta r \sim r$, these terms are each $\sim U/r^2$. Each of the terms that accompany $\nabla^2 v_r$ on the right-hand side of the equation is also $\sim U/r^2$. Thus, barring an improbable cancellation, the sum of the viscous terms is $\sim \mu U/r^2$. Turning to the left-hand side, $v_r\,\partial v_r/\partial r \sim U^2/r$ and all terms that accompany it are of similar size. Again barring an unlikely cancellation, the sum of the inertial terms is $\sim \rho U^2/r$. The ratio of inertial to viscous terms is then

$$\frac{\rho U^2/r}{\mu U/r^2} = \frac{Ur}{\nu} \sim \mathrm{Re}\left(\frac{r}{R}\right). \tag{8.5-8}$$

The agreement of this with Eq. (8.3-44) is evidence that r is indeed the correct length scale in the region far from the sphere.

Lubrication approximation

As seen in Section 8.2, the lubrication approximation for two-dimensional, steady flows rests on two conditions: the flow is nearly unidirectional and inertia is negligible. Let U be the scale for v_x and changes in v_x, and let V be the scale for v_y and changes in v_y. The continuity equation indicates then that

$$\frac{\partial v_x}{\partial x} \sim \frac{\partial v_y}{\partial y} \quad \text{or} \quad \frac{U}{\Delta x} \sim \frac{V}{\Delta y} \quad \text{or} \quad \frac{V}{U} \sim \frac{\Delta y}{\Delta x}. \tag{8.5-9}$$

Accordingly, nearly unidirectional flows ($V/U \ll 1$) typically arise in systems that are relatively long and narrow ($\Delta y/\Delta x \ll 1$). (Example 8.2-2 is an exception, in which V/U is governed by permeation through the wall, rather than the geometry.) Such "thinness" has two consequences. The first is that one of the viscous stress terms in the x component of the Navier–Stokes equation is negligible. That is,

$$\frac{\partial^2 v_x}{\partial x^2} \sim \frac{U}{(\Delta x)^2} \ll \frac{U}{(\Delta y)^2} \sim \frac{\partial^2 v_x}{\partial y^2}. \tag{8.5-10}$$

The second consequence is that the pressure is nearly independent of y. As mentioned in Section 8.2, the largest velocity term in a momentum equation places an upper bound on the pressure derivative. In a nearly unidirectional flow, each term in the y component of the Navier–Stokes equation is smaller than the corresponding one in the x component by a factor V/U (e.g., Table 8.1). Accordingly, in such a flow the ratio of the pressure changes in the two directions is

$$\frac{(\Delta \mathcal{P})_y}{(\Delta \mathcal{P})_x} \sim \frac{(\partial \mathcal{P}/\partial y)\,\Delta y}{(\partial \mathcal{P}/\partial x)\,\Delta x} \sim \frac{\left(\dfrac{\mu V}{(\Delta y)^2}\right)\Delta y}{\left(\dfrac{\mu U}{(\Delta y)^2}\right)\Delta x} \sim \frac{V}{U}\frac{\Delta y}{\Delta x} \sim \left(\frac{\Delta y}{\Delta x}\right)^2. \tag{8.5-11}$$

The consideration of inertia is simplified by the fact that continuity causes the two terms in $\mathbf{v}\cdot\nabla\mathbf{v}$ to be indistinguishable at the OM level, whether or not a flow is nearly unidirectional. That is, using Eq. (8.5-9),

$$v_x\frac{\partial v_x}{\partial x} \sim \frac{U^2}{\Delta x} \sim \frac{VU}{\Delta y} \sim v_y\frac{\partial v_x}{\partial y}. \tag{8.5-12}$$

Thus, it is sufficient to compare either inertial term with the dominant viscous term. In a nearly unidirectional flow, the ratio of inertial to viscous effects is then

$$\left|\frac{\rho v_x\,\partial v_x/\partial x}{\mu\,\partial^2 v_x/\partial y^2}\right| \sim \frac{\rho U^2/\Delta x}{\mu U/(\Delta y)^2} = \mathrm{Re}\frac{\Delta y}{\Delta x} \tag{8.5-13}$$

where $\mathrm{Re} = U\Delta y/\nu$. In summary, the lubrication approximation typically requires that $\Delta y/\Delta x \ll 1$ and $\mathrm{Re}\,\Delta y/\Delta x \ll 1$. Boundary-layer flows (Chapter 9) are also nearly unidirectional, but are distinguished by the fact that a large Re makes inertia important.

Creeping-flow approximation

In a steady creeping flow, $\rho\mathbf{v}\cdot\nabla\mathbf{v}$ is negligible relative to $\mu\nabla^2\mathbf{v}$. Such flows are not necessarily unidirectional, so $\Delta x \sim \Delta y$ in general. Setting $\Delta x = \Delta y$ in Eq. (8.5-13) leads to the criterion in Section 8.3 for the use of Stokes' equation, namely $\mathrm{Re} = UL/\nu \ll 1$. For flow past a sphere, $L = R$ for fluid that is not too far away. Thus, $UR/\nu \ll 1$ suffices to make creeping flow a good approximation in that region. However, far from the sphere $L = r$ and neglecting inertia requires that $Ur/\nu \ll 1$ (Example 8.5-2).

Pseudosteady approximation

Pseudosteady viscous flows are characterized by the fact that $\rho\partial\mathbf{v}/\partial t$ is negligible relative to $\mu\nabla^2\mathbf{v}$. In that such flows are not necessarily unidirectional, $\Delta x \sim \Delta y \sim L$ again. If t_p is the characteristic time for the velocity variations, then $\Delta t = t_p$ and

$$\left|\frac{\rho\,\partial v_x/\partial t}{\mu\,\partial^2 v_x/\partial y^2}\right| \sim \frac{\rho U/t_p}{\mu U/L^2} = \frac{L^2/\nu}{t_p} = \frac{t_v}{t_p}. \tag{8.5-14}$$

Thus, as discussed in Section 8.4, what is needed to neglect the time derivative is $t_v/t_p \ll 1$. This requirement may be restated in terms of the Reynolds and Strouhal numbers. As mentioned in Section 1.3, $\mathrm{Sr} = t_p/t_c$, where $t_c = L/U$ is the convective time scale, and $\text{Re} = t_v/t_c$. What is needed then to make a viscous flow pseudosteady is $\text{Re}/\text{Sr} \ll 1$. In the absence of an imposed time scale that is created by a decaying, growing, or oscillating applied force, $t_p = t_c$ and $\text{Sr} = 1$. In such problems $\text{Re} \ll 1$ is sufficient to give pseudosteady behavior. Accordingly, creeping flows are ordinarily pseudosteady. Only when a system is subjected to very rapid forcing is it necessary to augment Stokes' equation by including $\rho\partial\mathbf{v}/\partial t$ on the left-hand side.

Example 8.5-3 Order-of-magnitude analysis for squeeze flow. The objective is to examine the approximations made in Example 8.4-2. For the system in Fig. 8.9 it is assumed that $H_0/R \ll 1$ and $\text{Re} = U_0H_0/\nu \ll 1$, where H_0 and U_0 are the maximum separation and maximum velocity, respectively. Those maxima would occur at $t = 0$ for either a constant applied force or a constant velocity. Thus, $H(t) \leq H(0) = H_0$ and, typically, $U(t) \leq U(0) = U_0$.

Length and velocity scales are needed to estimate the spatial derivatives. Moving outward in r from the symmetry axis, v_r varies from 0 to roughly $\langle v_r\rangle$, its average over z. Given the no-slip conditions at the top and bottom, this is also the change in v_r in the z direction at a fixed value of r. From the overall mass balance in Eq. (8.4-7), $\langle v_r\rangle = q/H = rU/2H$. As z varies from 0 to H, v_z varies from U to 0. Accordingly, the scales for the variations are $\Delta r = r$, $\Delta z = H$, $(\Delta v_r)_r = (\Delta v_r)_z = rU/(2H)$, and $(\Delta v_z)_z = U$. The scales for v_r and v_z themselves are the same as those for their variations. The value for $(\Delta v_z)_r$ is not obvious from such an overview, but it is not needed in what follows.

As evidenced by $v_z/v_r \sim 2H/r$, the flow will be nearly unidirectional in most of the region modeled, at all times, if $H_0/R \ll 1$. This will make $\mathcal{P}$ nearly independent of z and cause one of the viscous terms in the r-momentum equation to be negligible. Notice that the OM of the neglected viscous term, given by

$$\frac{\partial}{\partial r}\left(\frac{1}{r}\frac{\partial}{\partial r}(rv_r)\right) = \frac{\partial^2 v_r}{\partial r^2} + \frac{1}{r}\frac{\partial v_r}{\partial r} - \frac{v_r}{r^2} \sim \frac{(\Delta v_r)_r}{r^2} \tag{8.5-15}$$

is unaffected by the curvature corrections in cylindrical coordinates. That is, the sum of the three contributions, each of which is $\sim(\Delta v_r)_r/r^2$, is still $\sim(\Delta v_r)_r/r^2$. The retained second derivative is $\partial^2 v_r/\partial z^2 \sim (\Delta v_r)_z/H^2$. Thus, the ratio of the neglected term to the one retained is $\sim(H/r)^2$, which will be negligible at most radial positions.

The nonlinear inertial terms in the r-momentum equation are

$$\rho v_r\frac{\partial v_r}{\partial r} \sim \frac{\rho\langle v_r\rangle^2}{r} = \frac{\rho r U^2}{4H^2} \tag{8.5-16}$$

$$\rho v_z\frac{\partial v_r}{\partial z} \sim \frac{\rho U\langle v_r\rangle}{H} = \frac{\rho r U^2}{2H}. \tag{8.5-17}$$

These are indistinguishable at the OM level. Using the larger estimate to be conservative, the ratio of inertial to viscous effects is

$$\left|\frac{\rho v_z\partial v_r/\partial z}{\mu\partial^2 v_r/\partial z^2}\right| \sim \frac{\rho r U^2/(2H)}{\mu r U/(2H^2)} = \frac{UH}{\nu} = \text{Re}\left(\frac{U}{U_0}\right)\left(\frac{H}{H_0}\right). \tag{8.5-18}$$

Accordingly, inertia will be negligible at all t if $\text{Re} \ll 1$.

Finally, the process time scale in this problem is how long it will take for the surfaces to come together, which is roughly H_0/U_0. This is also the convective time scale. With $\Delta t = t_p = t_c$, $\partial v_r/\partial t \sim \langle v_r\rangle U_0/H_0$, and

$$\left|\frac{\rho\partial v_r/\partial t}{\mu\partial^2 v_r/\partial z^2}\right| \sim \frac{\rho\langle v_r\rangle U_0/H_0}{\mu\langle v_r\rangle/H^2} = \text{Re}\left(\frac{H}{H_0}\right)^2. \tag{8.5-19}$$

Thus, Sr $= 1$ in this system and Re $\ll 1$ is sufficient to neglect the time derivative at all t. In summary, knowing that $H_0/R \ll 1$ and Re $\ll 1$ is enough to anticipate the applicability of the pseudosteady lubrication approximation represented by Eq. (8.4-5).

Example 8.5-4 Force on a slider bearing. In addition to identifying approximations that help in finding the velocity and pressure fields, OM analysis can simplify subsequent force calculations. That will be illustrated by evaluating the fluid-dynamic force on the upper surface in Fig. 8.2, the stationary part of the slider bearing in Example 8.2-3.

As shown in Example A.4-5, the unit normal that points into the fluid is

$$\mathbf{n} = \frac{1}{g}\frac{dh}{dx}\mathbf{e_x} - \frac{1}{g}\mathbf{e_y}, \quad g(x) = \left[1 + \left(\frac{dh}{dx}\right)^2\right]^{1/2}. \tag{8.5-20}$$

To focus on the force due to the flow, the stress vector is written as $\mathbf{s} = -\mathbf{n}\mathcal{P} + \mathbf{n} \cdot \boldsymbol{\tau}$. The stress components are then

$$s_x = n_x(-\mathcal{P} + \tau_{xx}) + n_y\tau_{yx} = \frac{1}{g}\left[(-\mathcal{P} + \tau_{xx})\frac{dh}{dx} - \tau_{yx}\right] \tag{8.5-21}$$

$$s_y = n_x\tau_{xy} + n_y(-\mathcal{P} + \tau_{yy}) = \frac{1}{g}\left[\tau_{xy}\frac{dh}{dx} + \mathcal{P} - \tau_{yy}\right]. \tag{8.5-22}$$

For a bearing of width W, a differential surface element is $dS = Wg\,dx$ (Example A.4-5). The force components are then

$$\frac{F_x}{W} = \int_0^L \left[(-\mathcal{P} + \tau_{xx})\frac{dh}{dx} - \tau_{yx}\right]_{y=h} dx \tag{8.5-23}$$

$$\frac{F_y}{W} = \int_0^L \left[\tau_{xy}\frac{dh}{dx} + \mathcal{P} - \tau_{yy}\right]_{y=h} dx. \tag{8.5-24}$$

While complete, Eqs. (8.5-23) and (8.5-24) are unnecessarily complicated. That some terms are negligible may be seen from OM estimates of the viscous stresses and pressure. As with the tapered channel (Example 8.5-1), the scales for Δx and Δy in the slider bearing are $h/|dh/dx|$ and h, respectively, where $|dh/dx| \ll 1$. With $v_x \sim U$ (the velocity of the lower bearing surface), continuity indicates that $v_y \sim U|dh/dx|$. Accordingly, we can simply replace u by U in the expressions in Table 8.1. The viscous stress components are

$$\tau_{xx} = 2\mu\frac{\partial v_z}{\partial x} \sim \frac{\mu U}{h}\left|\frac{dh}{dx}\right| \tag{8.5-25}$$

$$\tau_{yx} = \tau_{xy} = \mu\left(\frac{\partial v_x}{\partial y} + \frac{\partial v_y}{\partial x}\right) \sim \frac{\mu U}{h}\left(1 + \left|\frac{dh}{dx}\right|^2\right) \sim \frac{\mu U}{h} \tag{8.5-26}$$

$$\tau_{yy} = 2\mu\frac{\partial v_y}{\partial y} \sim \frac{\mu U}{h}\left|\frac{dh}{dx}\right|. \tag{8.5-27}$$

From the lubrication form of the momentum equation, $d\mathcal{P}/dx = \mu\, \partial^2 v_x/\partial y^2$. Thus, $\mathcal{P}/\Delta x = (\mathcal{P}/h)|dh/dx| \sim \mu U/h^2$ or

$$\mathcal{P} \sim \frac{\mu U}{h\,|dh/dx|}. \tag{8.5-28}$$

It is apparent now that $\mathcal{P}dh/dx$ and τ_{yx} are comparable; that occurs because $\mathcal{P}$ is large enough to compensate for the small value of $|dh/dx|$. Moreover, it is seen that within τ_{yx}, only $\partial v_x/\partial y$ is important. It is also evident that both viscous contributions to F_y are negligible. Accordingly, the force expressions simplify to

$$\frac{F_x}{W} = -\int_0^L \left[\mathcal{P}\frac{dh}{dx} + \mu\frac{\partial v_x}{\partial y}(x,h)\right] dx \tag{8.5-29}$$

$$\frac{F_y}{W} = \int_0^L \mathcal{P}\, dx. \tag{8.5-30}$$

Using Eq. (8.2-41) for v_x, the one required velocity derivative is

$$\frac{\partial v_x}{\partial y}(x,h) = \frac{2U}{h}\left(1 - \frac{3\bar{h}}{h}\right). \tag{8.5-31}$$

Substituting Eqs. (8.2-43) and (8.5-31) into Eqs. (8.5-29) and (8.5-30), and evaluating the integrals, gives the force results in Eqs. (8.2-45) and (8.2-46).

8.6 CONCLUSION

The Navier–Stokes equation (NSE) is simplified considerably whenever inertia is much less important than viscous stresses. If the $\mathbf{v}\cdot\nabla\mathbf{v}$ term is negligible, the resulting linearity of the NSE provides numerous opportunities for analytical solutions. Three major types of approximations were discussed, all for predominantly viscous flows. First was the lubrication approximation [Eqs. (8.2-1) or (8.2-24)], which is applicable to two-dimensional or axisymmetric flows that are nearly unidirectional, as occurs in long, narrow channels with gradually changing cross-sections or moderately permeable walls. In a flow at small or moderate Re that is nearly unidirectional, inertia is negligible, as is one of the viscous terms, and the simplified NSE is like that for fully developed flow. However, with the lubrication approximation the pressure gradient varies in the flow direction, rather than being constant. A sequential solution procedure can be used, in which the major velocity component is evaluated first, followed by P and (if desired) the minor velocity component.

The second type of approximation discussed was creeping flow or Stokes flow, which applies when Re is very small. For Re $\to 0$, the NSE reduces to Stokes' equation [Eq. (8.3-1)]. There are no geometric restrictions in this case. A distinctive property of creeping flow is that streamlines and drag coefficients remain the same if the direction of flow is reversed. The best-known creeping-flow result is Stokes' law for the drag on a sphere [Eq. (8.3-43)].

Pseudosteady flow, the third type of approximation, applies to time-dependent problems in which the $\partial\mathbf{v}/\partial t$ term in the NSE is negligible. The mathematical consequence of this is that the number of independent variables in the NSE is effectively reduced, t acting only as a parameter. Pseudosteady flow occurs when a system responds rapidly to changes in the applied forces. When viscous stresses are dominant, the necessary condition is that the viscous time scale be small relative to that of the imposed changes. Creeping flows are ordinarily pseudosteady.

The use of approximate forms of the NSE can be justified in either of two ways. One is to choose an approximation that seems reasonable, obtain a solution, and use that solution to check the magnitudes of the terms that were neglected. The other approach is to identify the relevant velocity, length, and time scales in advance, and use those scales to obtain order-of-magnitude (OM) estimates of the competing terms in the NSE. The needed scales usually can be inferred without advance knowledge of the flow details, allowing reliable estimates to be obtained. The use of OM estimation to anticipate which approximations will be valid is a key tool in fluid-dynamics research.

REFERENCES

Batchelor, G. K. *An Introduction to Fluid Dynamics*. Cambridge University Press, Cambridge, 1967.

Baxter, L. T. and R. K. Jain. Transport of fluid and macromolecules in tumors, I: Role of interstitial pressure and convection. *Microvasc. Res.* 37: 77–104, 1989.

Cameron, A. *Basic Lubrication Theory*, 3rd ed. Halsted Press, New York, 1981.

Deen, W. M. *Analysis of Transport Phenomena*, 2nd ed. Oxford University Press, New York, 2012.

Happel, J. and H. Brenner. *Low Reynolds Number Hydrodynamics*. Prentice-Hall, Englewood Cliffs, NJ, 1965 [reprinted by Martinus Nijhoff, The Hague, 1983].

Leal, L. G. *Advanced Transport Phenomena*. Cambridge University Press, Cambridge, 2007.

Middleman, S. *Fundamentals of Polymer Processing*. McGraw-Hill, New York, 1977.

Pinkus, O. and B. Sternlicht. *Theory of Hydrodynamic Lubrication*. McGraw-Hill, New York, 1961.

Purcell, E. M. Life at low Reynolds number. *Am. J. Phys.* 45: 3–11, 1977.

Randall, G. C. and P. S. Doyle. Permeation-driven flow in poly(dimethylsiloxane) microfluidic devices. *Proc. Natl. Acad. Sci. U.S.A.* 102: 10813–10818, 2005.

Reynolds, O. On the theory of lubrication and its application to Mr. Beauchamp Tower's experiments, including an experimental determination of the viscosity of olive oil. *Philos. Trans. R. Soc. Lond. A* 177: 157–234, 1886.

Stokes, G. G. On the effect of the internal friction of fluids on the motion of pendulums. *Trans. Camb. Phil. Soc.* 9: 8–106, 1851.

Tanner, R. I. *Engineering Rheology*, 2nd ed. Oxford University Press, Oxford, 2000.

Walker, G. M. and D. J. Beebe. A passive pumping method for microfluidic devices. *Lab Chip* 2: 131–134, 2002.

PROBLEMS

8.1. Imperfect parallel-plate channel It is desired to predict the effects of imperfections that might occur in fabricating a parallel-plate channel. One possibility is a sinusoidal variation in the half-height,

$$h(x) = h_0 \left[1 + \varepsilon \sin \left(\frac{2\pi x}{L} \right) \right] \quad \text{(P8.1-1)}$$

as depicted in Fig. P5.2. That is, h might vary from its mean value h_0 by $\pm \varepsilon h_0$ over a length L. Suppose that a steady flow rate q (volume flow per unit width) is specified.

(a) For the lubrication approximation to apply, what constraints must be satisfied by q and the linear dimensions?

(b) Assuming that the conditions in part (a) hold, determine $v_x(x, y)$.

(c) Will the pressure drop over a length L be less than, equal to, or greater than that for a channel of constant half-height h_0? (*Hint*: The integral that determines $|\Delta \mathcal{P}|$ in a sinusoidal channel is very tedious to evaluate analytically. It is sufficient to calculate the ratio of the pressure drops numerically for $\varepsilon = 0.1$.)

8.2. Permeable closed-end tube Consider a small, permeable tube of radius R and length L that is closed at one end, as in Fig. P8.2. Suppose that a high external pressure creates a constant inward velocity v_w at the wall, but with no flow across the closed end.

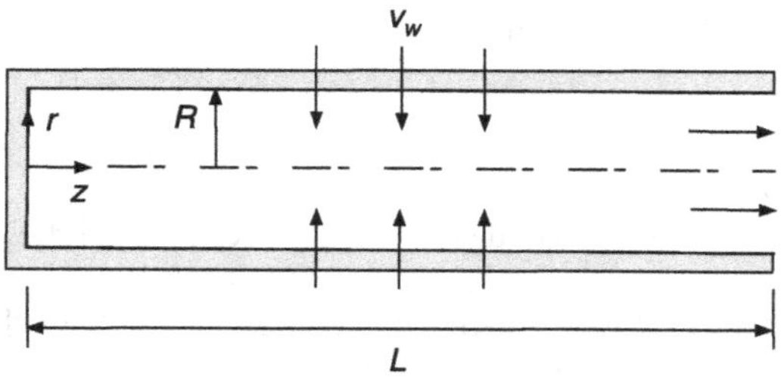

Figure P8.2 Tube with one end closed and a constant inward velocity at the wall.

(a) Relate $u(z)$, the local mean velocity, to v_w.
(b) For the lubrication approximation to apply, what constraints must be satisfied by v_w and the linear dimensions?
(c) Assuming that the conditions in part (b) hold, determine $v_z(r, z)$, $v_r(r, z)$, and $\mathcal{P}(z)$.

8.3. Permeation-driven flow in a microchannel Randall and Doyle (2005) discovered that diffusion of water through the poly(dimethylsiloxane) (PDMS) that formed the top of a microfluidic channel produced an observable flow. Their device is shown schematically in Fig. P8.3. A channel of thickness H and length $2L$ was connected at each end to reservoirs filled to equal heights. The channel width W greatly exceeded H, which justifies a two-dimensional model. The upper wall of the channel was PDMS and the lower wall was glass, which is impermeable to water. Diffusion of water through the PDMS and into the surrounding air created a velocity v_w normal to the upper wall. Although v_w was actually a function of x, you may assume here that it is constant.

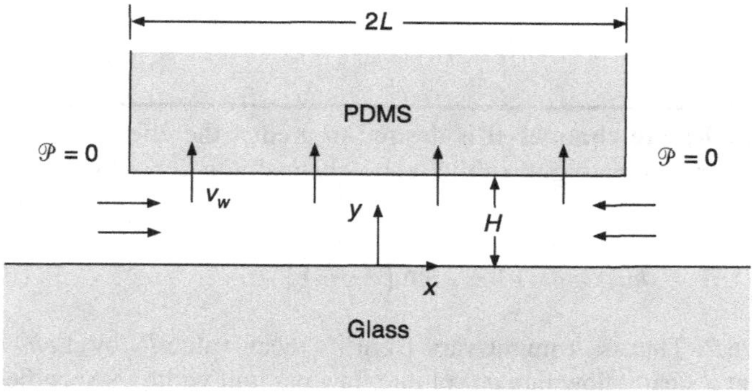

Figure P8.3 Permeation-driven flow of water in a microchannel.

(a) Relate the local mean velocity $u(x)$ to v_w.
(b) Typical channel dimensions were $H = 2\ \mu\text{m}$, $W = 50\ \mu\text{m}$, and $L = 12.5\ \text{mm}$, and it was found that $u = 35\ \mu\text{m/s}$ at $|x| = 10\ \text{mm}$. Estimate the average value of v_w and explain why a lubrication approximation is justified.
(c) Derive expressions for $v_x(x, y)$ and $v_y(x, y)$.

8.4. Candy manufacturing A manufacturer is planning to coat candy centers with thin films of chocolate. This is to be done by flowing molten chocolate onto a solid center by gravity, and then cooling quickly to solidify the film. A lubrication analysis of steady, gravity-driven flow of a thin film on a hemisphere is suggested as a way to explore the relationships among the process variables. Of particular interest are the thickness and uniformity of the coating. As shown in Fig. P8.4(a), a volume flow Q directed downward onto a hemisphere of radius a will create a liquid film of varying thickness, $h(\theta)$. It is expected that $h \ll a$ for most of the range of θ, so that within the film the solid surface will appear flat. This motivates using the local Cartesian coordinates in Fig. 8.4(b), which are related to the spherical ones as $x = a\theta$ and $y = r - a$. The thin-film assumption may be used throughout the analysis.

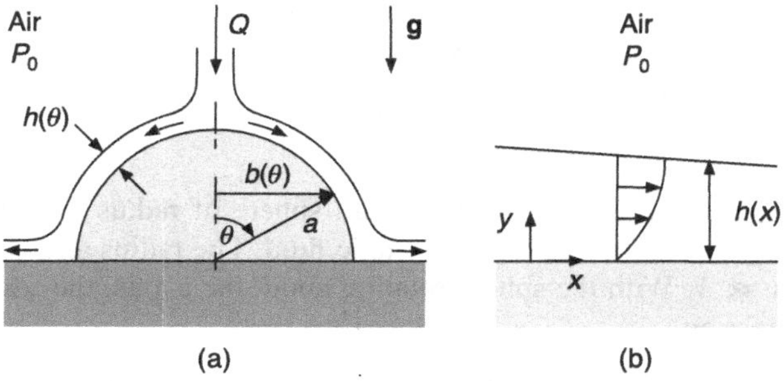

Figure P8.4 Candy coating: (a) molten chocolate flowing at rate Q over a solid hemisphere of radius a to create a liquid film of thickness $h(\theta)$, and (b) an enlargement showing local Cartesian coordinates x and y within the film.

(a) Neither $h(\theta)$ nor the mean velocity $u(\theta)$ is known in advance. Use a mass balance to relate these quantities to Q. (*Hint*: The radius $b(\theta)$ in Fig. 8.4(a) is helpful in this derivation.)
(b) Determine $v_x(\theta, y)$ and use it to find a second relationship between $h(\theta)$ and $u(\theta)$.
(c) Solve for $h(\theta)$ and $u(\theta)$. How is the film thickness predicted to vary with the operating parameters and with position?

8.5. Blade coating In *blade coating*, which is widely used for sheet materials such as paper, a substrate passes through an opening of fixed height as it is pulled from a pool of the coating liquid. The opening can have various shapes. For the one in Fig. P8.5, the gap between blade and substrate decreases linearly from h_0 to h_L over a distance L, as in Eq. (8.2-2). The substrate velocity is U. Of interest is the final coating thickness h_∞, which generally differs from h_L. The pressure P_1 in the upstream liquid may exceed the pressure P_0 in the air. Static pressure variations over the small heights involved can be ignored. You may assume that the lubrication approximation is applicable and that surface tension is negligible.

(a) Explain why there is plug flow far from the opening, such that $v_x = U$ for all y when $h = h_\infty$.
(b) Relate h_∞ to the dimensions, U, and the pressure drop ($|\Delta P| = P_1 - P_0$), and show that it can be less than, equal to, or greater than h_L. (*Hint*: The first part of the analysis in Example 8.2-3 is applicable again here.)

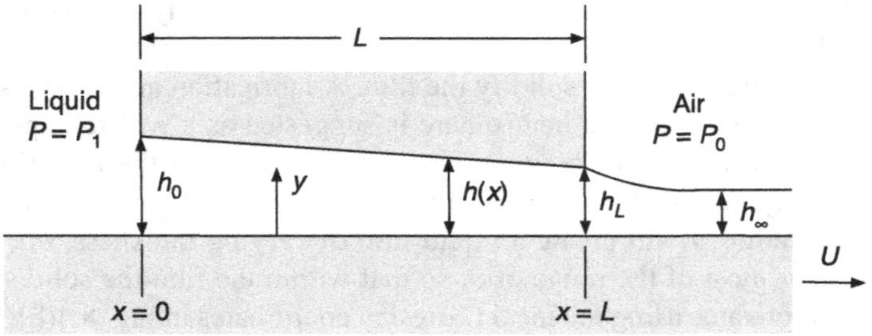

Figure P8.5 Blade coating.

(c) Under certain conditions, flow reversal in part of the gap will carry some of the liquid back to the upstream reservoir. This recirculation might greatly increase how long some of the liquid stays in the system, which might in turn degrade the coating material. How should the dimensions and operating conditions be chosen to ensure that $v_x \geq 0$ for all y?

8.6. Torque on a rotating sphere Suppose that a small sphere of radius R rotates at angular velocity ω in a large container of otherwise static fluid. The radius and velocity are such that $\text{Re} = \omega R^2/\nu \ll 1$. With the sphere rotating about the z axis, the resulting fluid motion is in the ϕ direction.

(a) Show that there is a solution to the ϕ component of Stokes' equation of the form $v_\phi(r, \theta) = f(r) \sin\theta$ and determine $f(r)$.
(b) Show that all the other equations are satisfied if $v_r = v_\theta = 0$ and $\mathcal{P}$ is constant. This confirms that this creeping flow is unidirectional (i.e., purely rotational).
(c) Show that the torque on the sphere is

$$G = 8\pi \mu \omega R^3. \tag{P8.6-1}$$

8.7. Velocity and pressure for flow past a bubble Consider creeping flow of a liquid at velocity U relative to a spherical bubble of radius R, with coordinates as in Fig. 8.7. Show that the velocity and pressure in the surrounding liquid are

$$v_r(r, \theta) = U \cos\theta \left(1 - \frac{R}{r}\right) \tag{P8.7-1}$$

$$v_\theta(r, \theta) = -U \sin\theta \left(1 - \frac{1}{2}\frac{R}{r}\right) \tag{P8.7-2}$$

$$\mathcal{P}(r, \theta) = -\frac{\mu U R}{r^2} \cos\theta \tag{P8.7-3}$$

where μ is the viscosity of the *liquid*. [*Hint*: The general solution given by Eqs. (8.3-28)–(8.3-30) is applicable again here.]

8.8. Terminal velocity of a small bubble

(a) Use the velocity and pressure in Problem 8.7 to calculate the drag coefficient for a bubble in creeping flow. How do the pressure and viscous contributions compare?
(b) Derive an expression for the terminal velocity of the bubble.

8.9. Rotating and stationary disks Suppose that an extremely viscous liquid fills a space of thickness H between two disks of radius R, as shown in Fig. P8.9. The upper disk rotates at a constant angular velocity ω and the lower one is fixed. Inertia is negligible because $\mathrm{Re} = \omega R^2/\nu \ll 1$.

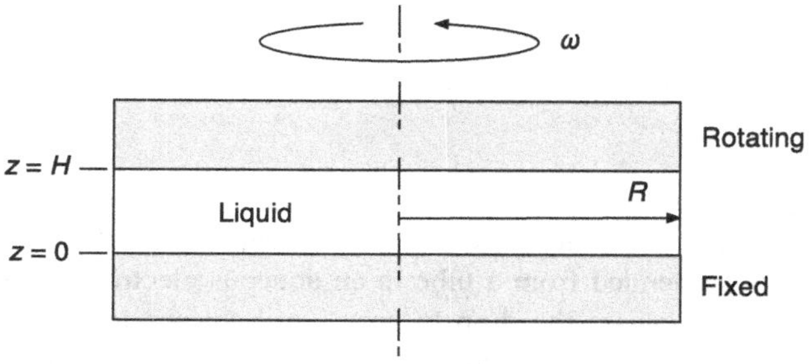

Figure P8.9 Flow between two disks, one fixed and the other rotating.

(a) Show that there is a solution to the θ component of Stokes' equation of the form $v_\theta(r, z) = rf(z)$ and determine $f(z)$.
(b) Show that all the other equations are satisfied if $v_r = v_z = 0$ and $\mathcal{P}$ is constant. This confirms that this creeping flow is unidirectional (i.e., purely rotational).
(c) Calculate the torque that must be applied to the upper disk to maintain its rotation.

8.10. Cone-and-plate viscometer The arrangement in Fig. P8.10 is well suited for determining the viscosity of small liquid samples. An inverted cone of radius R and small angle β is brought into contact with a pool of the liquid on a flat plate, and μ is determined by measuring the torque required to rotate the cone at a constant angular velocity ω. Stokes' equation is applicable and spherical coordinates are useful, as shown.

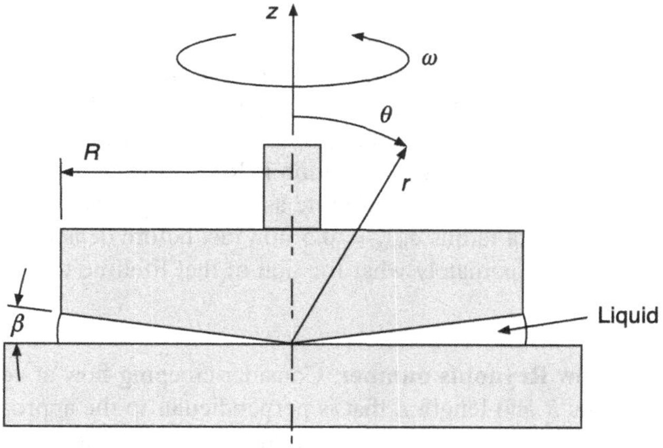

Figure P8.10 Cone-and-plate viscometer. The torque on the cone is measured as it is rotated on top of the stationary plate.

(a) Show that a velocity field with $v_\phi(r, \theta) = rf(\theta)$ and $v_r = v_\theta = 0$ is consistent with conservation of mass, conservation of momentum, and the boundary conditions at the solid surfaces. Derive the differential equation and boundary conditions for $f(\theta)$.

(b) Simplify the governing equations for $f(\theta)$ for $\beta \ll 1$ and evaluate $v_\phi(r, \theta)$.

(c) The torque applied to the cone must equal that on the plate. Show that the torque G on the plate is

$$G = \frac{2\pi}{3} \frac{\mu \omega R^3}{\beta}. \tag{P8.10-1}$$

8.11. Growing mercury drop *Dropping-mercury electrodes* have been used extensively in analytical chemistry and electrochemical research. As shown in Fig. P8.11, a small mercury drop of radius a is suspended from a tube in an aqueous electrolyte solution. A constant volume flow rate Q causes the drop to grow until surface tension can no longer support it and it detaches. The time-dependent current caused by an electrochemical reaction at the mercury–water interface is measured up to the moment of detachment. Knowledge of the velocity in the surrounding solution is needed to fully interpret such data.

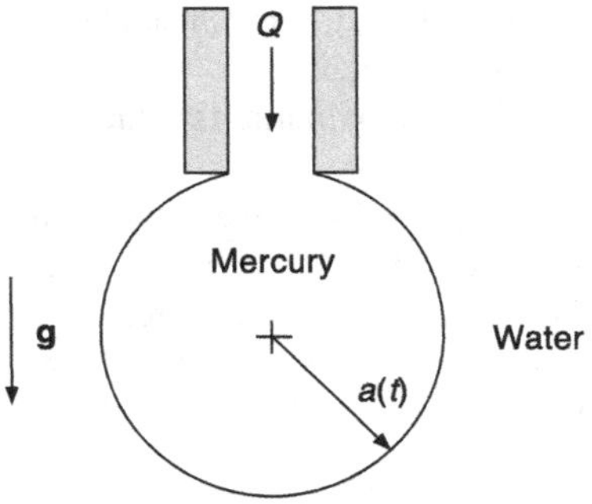

Figure P8.11 A growing mercury drop suspended from a tube in an aqueous solution.

(a) Assuming that Stokes' equation applies in the electrolyte solution and that v_θ and v_ϕ inside the mercury drop are each negligible, show that the external flow is time-independent and purely radial. Determine $v_r(r)$ and $\mathcal{P}(r)$ in the electrolyte solution.

(b) A reasonable criterion for the use of Stokes' equation is $\text{Re}(t) = a(t)v_r(a)/\nu < 0.1$. Will Stokes' equation become less or more accurate as the drop grows? In a typical application, a mercury drop has a radius $a_{\max} = 0.5$ mm just before detachment and a lifetime $t_{\max} = 3$ s. Over approximately what fraction of that lifetime will Stokes' equation apply?

8.12. Drag on a cylinder at low Reynolds number Consider creeping flow at velocity U past a long cylinder of radius R and length L that is perpendicular to the approaching fluid, as in Fig. P6.6. Unlike the case of flow past a sphere, there are no simple expressions for the velocity and pressure throughout the fluid. However, adequate approximations for the region near the cylinder when $\text{Re} \to 0$ are (Batchelor, 1967, pp. 244–246)

$$v_r(r, \theta) = -\frac{C}{2}\left[1 - \left(\frac{R}{r}\right)^2 - 2\ln\left(\frac{r}{R}\right)\right]\cos\theta \tag{P8.12-1}$$

$$v_\theta(r,\theta) = -\frac{C}{2}\left[1 - \left(\frac{R}{r}\right)^2 - 2\ln\left(\frac{r}{R}\right)\right]\sin\theta \tag{P8.12-2}$$

$$\mathcal{P}(r,\theta) = -\frac{2\mu C}{r}\cos\theta \tag{P8.12-3}$$

$$C = \frac{U}{\ln(7.4/\text{Re})}, \qquad \text{Re} = \frac{2RU}{\nu}. \tag{P8.12-4}$$

Use these results to show that the drag is

$$F_D = \frac{4\pi\mu UL}{\ln(7.4/\text{Re})} \tag{P8.12-5}$$

as implied by Eq. (3.2-10). What are the relative contributions of pressure and viscous stresses (form and friction drag)?

8.13. Darcy flow in a tumor The *interstitium* of a body tissue, which is the space outside blood vessels and cells, contains fibers of collagen, glycosaminoglycans, and other biopolymers that are surrounded by water. Fluid movement in that space affects the distribution of nutrients and drugs. As in other porous materials, flow in the interstitium can be described using Darcy's law. The model outlined here is a simplified version of one in Baxter and Jain (1989), which was used to explore the therapeutic consequences of interstitial pressure and velocity distributions in "solid" tumors.

Figure P8.13(a) depicts a spherical tumor of radius R that is large enough to contain many blood vessels. The enlarged volume element in Fig. P8.13(b) shows a few blood capillaries and the associated interstitium. In general, there is net flow out of the capillaries. Neglecting the drainage that might occur in lymphatic vessels, the fluid leaving the capillaries must move to the periphery of the tumor via interstitial flow.

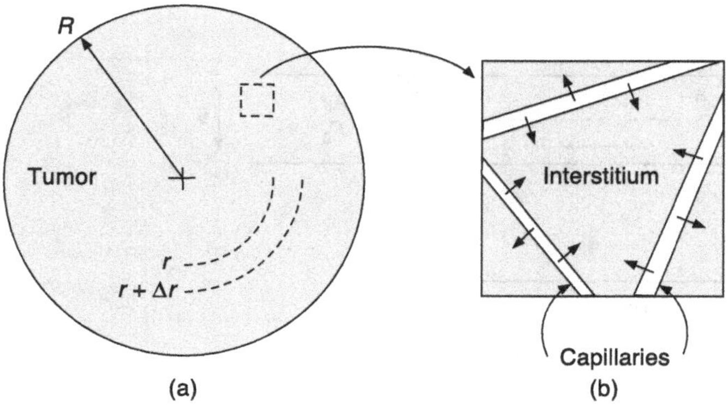

Figure P8.13 Spherical tumor: (a) overall view, and (b) enlargement of a small region.

(a) Let W be the volume of water moving from the capillaries to the interstitium per unit time, per unit volume of tumor (units of s^{-1}). Assuming that the flow is steady and purely radial, use a balance on a spherical shell of thickness Δr to derive the differential equation that relates W to $v_r(r)$. Combine that result (a kind of continuity equation) with Darcy's law to show that

$$\frac{1}{r^2}\frac{d}{dr}\left(r^2\frac{d\mathcal{P}}{dr}\right)+\frac{\mu W}{k}=0. \tag{P8.13-1}$$

This is the spherical form of Eq. (8.3-46), but with an added source term.

(b) If W is constant, determine $\mathcal{P}(r)$ and $v_r(r)$. Assume that the Darcy permeability of the interstitium (k) is known and that $\mathcal{P}(R)=0$.

(c) Because flow across the permeable wall of a capillary is caused by a pressure difference, a more realistic assumption is that W increases as $\mathcal{P}$ decreases. Assume that

$$W(r)=\alpha[\mathcal{P}_0-\mathcal{P}(r)] \tag{P8.13-2}$$

where α is the hydraulic permeability of the capillary walls times the capillary surface area per volume of tumor, and $\mathcal{P}_0$ is a constant determined by the intracapillary pressure and the osmotic pressure of dissolved proteins. Again determine $\mathcal{P}(r)$ and $v_r(r)$. [*Hint*: The differential equation

$$\frac{1}{x^2}\frac{d}{dx}\left(x^2\frac{dy}{dx}\right)-m^2y=0 \tag{P8.13-3}$$

where m is a real constant, has the general solution

$$y(mx)=A\frac{\sinh(mx)}{mx}+B\frac{\cosh(mx)}{mx}.] \tag{P8.13-4}$$

8.14. Washburn's law Liquid uptake by porous media is typically driven by surface tension. Suppose that a long, horizontal pore of radius a is initially filled with air, and that at $t=0$ one end is immersed in a liquid. If the contact angle α is $<\pi/2$, there will be a liquid column of growing length $L(t)$, as shown in Fig. P8.14. Assume that the far end of the pore is vented, keeping the air pressure constant at P_0, and that the liquid outside the pore is also at P_0.

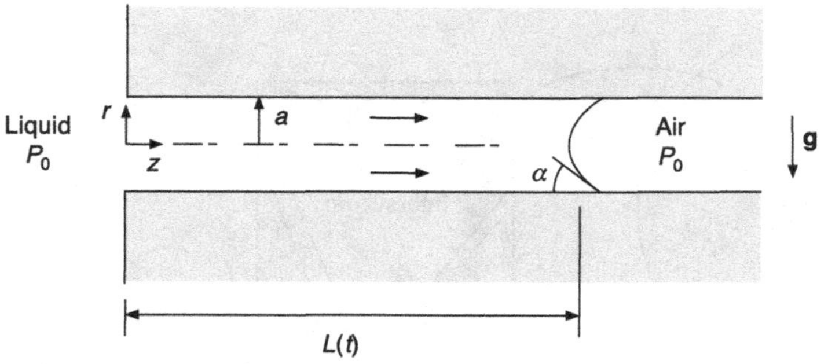

Figure P8.14 Capillary flow into a horizontal pore.

(a) Relate dL/dt to the volume flow rate $Q(t)$ of the liquid.

(b) Assuming that the flow in the pore is pseudosteady and fully developed, show that

$$L(t)=\left[\frac{a\gamma\cos\alpha}{2\mu}t\right]^{1/2} \tag{P8.14-1}$$

which is called *Washburn's law*.

(c) Estimate how large t must be for the pseudosteady approximation to become valid. Washburn's law will be accurate if the observation time greatly exceeds that.

8.15. Injection molding *Injection molding* is a manufacturing process in which a mold is filled with a liquid that solidifies after cooling. The mold is then opened and the product removed. Such molds have openings for injecting the liquid and venting the displaced air. Consider the simple prototype in Fig. P8.15, in which the space to be filled is a gap of thickness H between disks of radius R_2. Applying a constant pressure P_{in} at an injection port of radius R_1 in one of the disks results in a volume flow rate $Q(t)$. The radius of the filled region at a given instant is $R(t)$. Assume that the flow is pseudosteady, that inertia, surface tension, and gravity are all negligible, and that $R_2 \gg R_1$ or H. It is desired to relate the filling time t_f to the pressure drop $|\Delta P| = P_{in} - P_0$, the viscosity, and the linear dimensions.

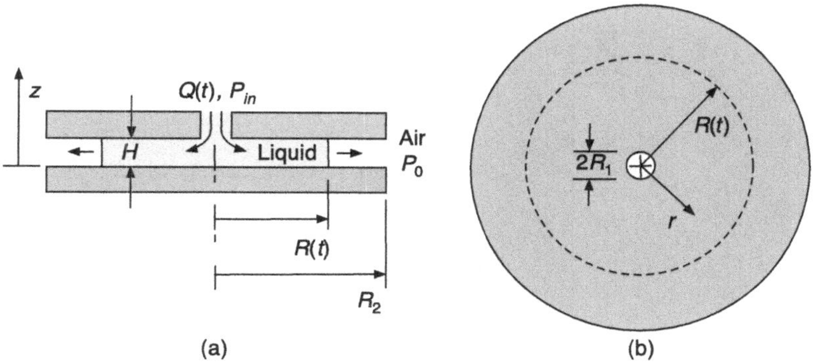

Figure P8.15 Injection of a liquid at volume flow rate Q between disks of radius R_2 that are separated by a distance H. The radius of the injection port is R_1 and the pressure there is P_{in}.

(a) Relate the height-averaged radial velocity $U(r, t)$ to $Q(t)$ and the dimensions.
(b) Solve for $v_r(r, z, t)$ in terms of $\partial P/\partial r$.
(c) Evaluate $\partial P/\partial r$ and relate $Q(t)$ to $|\Delta P|$ and $R(t)$.
(d) Obtain the differential equation that governs $R(t)$ and find the relationship between t_f and $|\Delta P|$.
(e) Approximately how large must t_f be to justify the pseudosteady approximation?

8.16. Capillary pump* Walker and Beebe (2002) demonstrated a way to pump liquid through small tubes without the need for moving parts, and their idea has been applied to point-of-care diagnostics and other microfluidic devices. The concept is illustrated in Fig. P8.16. Liquid drops of radius R_0 and R_L are in contact with the ends of a horizontal capillary tube of radius R and length L. The flow is created by the unequal drop radii.

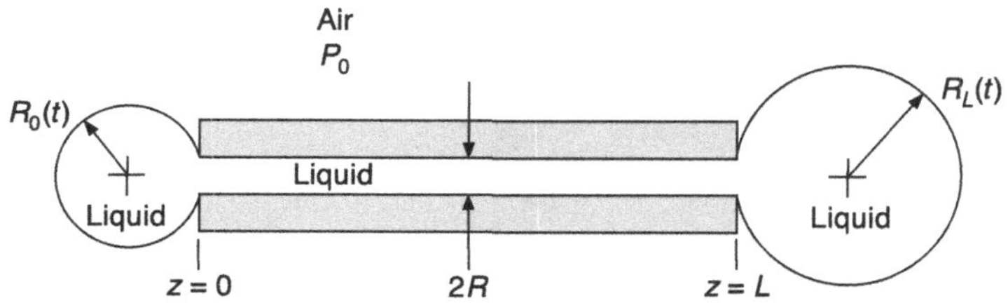

Figure P8.16 Liquid-filled capillary tube with spherical drops at each end.

(a) If $R_L > R_0$, as shown, in which direction will the liquid flow?
(b) Assuming that the flow in the tube is fully developed and pseudosteady, determine the instantaneous flow rate $Q(t)$.
(c) If $R_0 \gg R$ and $R_L \gg R$, such that both drops are nearly complete spheres, derive the differential equations that govern $R_0(t)$ and $R_L(t)$. (You are not asked to solve them.)
(d) Sketch the final appearance of the two drops. Would the equations in part (c) be sufficient to determine the time t_p at which the flow will stop?

* This problem was suggested by P. S. Doyle.

9

Laminar flow with inertia

9.1 INTRODUCTION

The discussion of approximations for laminar flow begun in Chapter 8 is extended now to situations where inertia is prominent. Inertia tends to be more important than viscosity when the Reynolds number is large, and flows with Re > 100 are common in nature and in engineering. A logical starting point for analyzing such flows is the extreme in which the fluid is *inviscid*, meaning that it is idealized as having zero viscosity. In that Re $\rightarrow \infty$ as $\mu \rightarrow 0$, inviscid flow is the converse of creeping flow. As mentioned in Chapter 5, an irrotational flow is one in which the vorticity is zero everywhere, and if a flow is irrotational a velocity potential exists. Thus, irrotational flow is also called potential flow. Although irrotational flow might occur in principle at any Re, it is favored if the viscosity is negligible, as will be discussed. The velocity potential is attractive analytically because it is governed by a partial differential equation (Laplace's equation) that is linear, unlike the Navier–Stokes equation. The analysis of such problems began in the 1700s, well before viscous stresses were fully understood. Potential-flow theory has provided numerous results that are useful in engineering, but also has important limitations.

The most troubling aspect of potential-flow theory is that it predicts that the drag on any solid object will be zero. That was shown by J. R. d'Alembert in 1752 and became known as *d'Alembert's paradox*. After it had damaged the credibility of theoretical fluid mechanics for generations, the paradox was resolved in 1904 by L. Prandtl (Prandtl, 1904). He realized that, even for Re $\rightarrow \infty$, viscous stresses will be significant in a thin boundary layer next to an object. The key insight was that the dynamics must be examined at two different length scales, one corresponding to the overall size of the object and the other determined by the thickness of the boundary layer. Outside the boundary layer, the flow at large Re may be nearly inviscid, as had always been assumed; inside, a different approximation to the Navier–Stokes equation is needed, as will be described. Boundary layers exist not just at solid surfaces, but also where fluids moving at different velocities are brought into contact. The concept of dynamically distinct regions with different length scales has paved the way for numerous advances in fluid dynamics, heat transfer, and mass transfer. The analogous concept of multiple time scales has applications in chemical kinetics.

9.2 INVISCID AND IRROTATIONAL FLOW

Inviscid flow

The momentum equation for an inviscid fluid is obtained either by setting $\mu = 0$ in the Navier–Stokes equation or $\boldsymbol{\tau} = \mathbf{0}$ in the Cauchy momentum equation. For steady flow it is

$$\mathbf{v} \cdot \nabla \mathbf{v} + \frac{1}{\rho} \nabla \mathcal{P} = 0. \tag{9.2-1}$$

Although seldom used as is, this leads to one of the best-known equations of fluid dynamics. A vector identity [obtained by setting $\mathbf{a} = \mathbf{b} = \mathbf{v}$ in (9) of Table A.1] is

$$\mathbf{v} \cdot \nabla \mathbf{v} = \nabla \left(\frac{v^2}{2} \right) - \mathbf{v} \times (\nabla \times \mathbf{v}) \tag{9.2-2}$$

where v is the magnitude of the velocity vector. Forming the dot product of $\mathbf{v}$ with both sides,

$$\mathbf{v} \cdot (\mathbf{v} \cdot \nabla \mathbf{v}) = \mathbf{v} \cdot \nabla \left(\frac{v^2}{2} \right) - \mathbf{v} \cdot [\mathbf{v} \times (\nabla \times \mathbf{v})] = \mathbf{v} \cdot \nabla \left(\frac{v^2}{2} \right). \tag{9.2-3}$$

The last equality follows from the fact that $\mathbf{v} \times \mathbf{a}$ is orthogonal to $\mathbf{v}$, and thus $\mathbf{v} \cdot (\mathbf{v} \times \mathbf{a}) = 0$ for any $\mathbf{a}$; in this case $\mathbf{a} = \nabla \times \mathbf{v}$. For constant ρ, the dot product of $\mathbf{v}$ with Eq. (9.2-1) gives

$$\mathbf{v} \cdot \nabla \left(\frac{v^2}{2} + \frac{\mathcal{P}}{\rho} \right) = \mathbf{v} \cdot \nabla \left(\frac{v^2}{2} + \frac{P}{\rho} + gh \right) = 0 \tag{9.2-4}$$

where h is height above a reference plane. Recalling that $\mathbf{v} \cdot \nabla f$ is the rate of change of the function f along a streamline (Section 5.3), this may be rewritten as

$$\Delta \left(\frac{v^2}{2} + \frac{\mathcal{P}}{\rho} \right) = \Delta \left(\frac{v^2}{2} + \frac{P}{\rho} + gh \right) = 0 \tag{9.2-5}$$

where Δ denotes a difference between any two positions on the same streamline. This is *Bernoulli's equation*, a cornerstone of classical fluid dynamics.[1] It indicates that in a steady, inviscid, incompressible flow the sums in parentheses remain constant along any streamline. Thus, $\mathcal{P}$ decreases where v rises and increases where v falls. Although the sums are constant along a given streamline, they may vary from one streamline to another.

Although derived from conservation of momentum, Eq. (9.2-5) is a kind of mechanical energy balance. It describes the reversible interconversion of kinetic energy ($v^2/2$ per unit mass) and gravitational potential energy (gh per unit mass), as influenced by pressure–volume work. In an inviscid fluid there is no internal friction and thus no dissipation of mechanical energy as heat. When ρ is constant, there is also no loss or gain of mechanical energy due to compression or expansion of the fluid. Bernoulli's equation is generalized in Section 11.4 to include these other sources or sinks for mechanical energy.

Vorticity transport

Examining how vorticity ($\mathbf{w} = \nabla \times \mathbf{v}$) moves through a fluid reveals the connection between inviscid flow and irrotational flow, which will be explained shortly. The full Navier–Stokes equation is

$$\frac{\partial \mathbf{v}}{\partial t} + \mathbf{v} \cdot \nabla \mathbf{v} = -\frac{1}{\rho} \nabla \mathcal{P} + \nu \nabla^2 \mathbf{v}. \tag{9.2-6}$$

[1] Daniel Bernoulli (1700–1782) was a Swiss mathematician and physicist who made pioneering contributions to fluid mechanics, statistics, probability theory, and other subjects. He was a contemporary and friend of the prolific mathematician Leonhard Euler (1707–1783), who (in 1755) formulated the continuity and momentum equations for an inviscid fluid.

An equivalent relationship involving **w** is obtained by taking the curl of each term. Using Eq. (9.2-2) and other identities from Table A.1, it is found that

$$\nabla \times \frac{\partial \mathbf{v}}{\partial t} = \frac{\partial \mathbf{w}}{\partial t} \tag{9.2-7}$$

$$\nabla \times (\mathbf{v} \cdot \nabla \mathbf{v}) = \mathbf{v} \cdot \nabla \mathbf{w} - \mathbf{w} \cdot \nabla \mathbf{v} \tag{9.2-8}$$

$$\nabla \times \nabla \mathcal{P} = \mathbf{0} \tag{9.2-9}$$

$$\nabla \times \nabla^2 \mathbf{v} = \nabla^2 \mathbf{w}. \tag{9.2-10}$$

Collecting these results gives

$$\frac{\partial \mathbf{w}}{\partial t} + \mathbf{v} \cdot \nabla \mathbf{w} - \mathbf{w} \cdot \nabla \mathbf{v} = \nu \nabla^2 \mathbf{w} \tag{9.2-11}$$

which is called the *vorticity transport equation.*

Irrotational flow

The fact that Eq. (9.2-11) is homogeneous in **w** indicates that $\mathbf{w} = \mathbf{0}$ is a solution. Because the vorticity-transport and Navier–Stokes equations are equivalent, any irrotational velocity field evidently satisfies the full Navier–Stokes equation. It is noteworthy that the viscous term in Eq. (9.2-6) vanishes when $\mathbf{w} = \mathbf{0}$, even if $\nu \neq 0$. That may be seen by setting $\mathbf{a} = \mathbf{v}$ in (8) of Table A.1, which gives $\nabla^2 \mathbf{v} = \nabla(\nabla \cdot \mathbf{v}) - \nabla \times \mathbf{w}$. Thus, $\nabla^2 \mathbf{v} = 0$ for any flow that is both incompressible ($\nabla \cdot \mathbf{v} = 0$) and irrotational ($\mathbf{w} = \mathbf{0}$).

In an inviscid fluid $\nu = 0$ and Eq. (9.2-11) simplifies to

$$\frac{D\mathbf{w}}{Dt} = \frac{\partial \mathbf{w}}{\partial t} + \mathbf{v} \cdot \nabla \mathbf{w} = \mathbf{w} \cdot \nabla \mathbf{v}. \tag{9.2-12}$$

Recall that D/Dt gives the rate of change of anything at a material point, a tiny element of fluid being carried along with the flow (Section 5.3). Thus, the rate of change of **w** at such a point is $\mathbf{w} \cdot \nabla \mathbf{v}$. Suppose now that the fluid element is initially irrotational. Because $D\mathbf{w}/Dt = \mathbf{0}$ when $\mathbf{w} = \mathbf{0}$, it will remain irrotational a moment later at its new position. But with the rate of change of **w** always zero, $\mathbf{w} = \mathbf{0}$ will hold indefinitely. Thus, *in an inviscid fluid, an irrotational flow remains irrotational.* A different proof of the persistence of irrotational flow in an inviscid fluid may be found in Lamb (1945, pp. 17–18). Physically, vorticity is associated with rotation (Section 5.5), a torque is needed to make a fluid element rotate, a shear stress is needed to create a torque, and without viscosity there are no shear stresses. Accordingly, an inviscid fluid lacks an internal mechanism for creating vorticity.

Steady streaming flow of an inviscid fluid past a stationary object is irrotational. The uniform velocity far from the object indicates that the vorticity is zero there. Because all streamlines extend to such positions, $\mathbf{w} = \mathbf{0}$ somewhere on each. The persistence of irrotationality ensures then that $\mathbf{w} = \mathbf{0}$ everywhere on all streamlines, even where they curve around the object. There are numerous other inviscid flows in which all fluid elements are irrotational at some position or time, and thus remain so. Accordingly, inviscid and irrotational flow often coincide.

In any flow that is irrotational, inviscid or not, a stronger form of Bernoulli's equation is obtained by returning to the full Navier–Stokes equation. In such flows $\mathbf{v} = \nabla \phi$, where

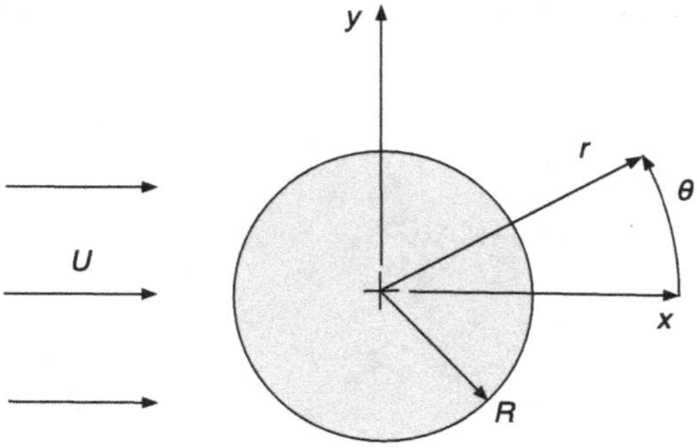

Figure 9.1 Steady flow past a long cylinder of radius *R*.

ϕ is the velocity potential (Section 5.5). Rewriting the time derivative in terms of ϕ, evaluating $\mathbf{v}\cdot\nabla\mathbf{v}$ using Eq. (9.2-2), and recalling that $\nabla^2\mathbf{v}=\mathbf{0}$, Eq. (9.2-6) reduces to

$$\nabla\left(\frac{\partial\phi}{\partial t}+\frac{v^2}{2}+\frac{\mathcal{P}}{\rho}\right)=\nabla\left(\frac{\partial\phi}{\partial t}+\frac{v^2}{2}+\frac{P}{\rho}+gh\right)=0. \tag{9.2-13}$$

Whereas in Eq. (9.2-5) the sums are constant only along a given streamline, here they are constant throughout the fluid. Given that uniformity, Eq. (9.2-13) may be written also as

$$\frac{\partial\phi}{\partial t}+\frac{v^2}{2}+\frac{\mathcal{P}}{\rho}=F(t) \tag{9.2-14}$$

which we will refer to as the *irrotational Bernoulli equation*. The function *F*(*t*), which arises when integrating over position, is arbitrary and is typically chosen to be constant. Equation (9.2-14) tends to be the more useful form of Bernoulli's equation for microscopic analysis, whereas Eq. (9.2-5) has numerous applications in macroscopic problems (Chapters 11 and 12).

Example 9.2-1 Velocity for potential flow past a cylinder. It is desired to find the velocity field for irrotational flow past a long cylinder of radius *R*. As shown in Fig. 9.1, far from the cylinder there is a uniform velocity in the *x* direction, such that $\mathbf{v}\to U\mathbf{e_x}$ as $r\to\infty$.

From Eq. (5.5-5), the velocity potential $\phi(r,\theta)$ is governed by

$$\nabla^2\phi=\frac{1}{r}\frac{\partial}{\partial r}\left(r\frac{\partial\phi}{\partial r}\right)+\frac{1}{r^2}\frac{\partial^2\phi}{\partial\theta^2}=0. \tag{9.2-15}$$

The no-penetration and no-slip conditions on the cylinder surface are

$$v_r(R,\theta)=\frac{\partial\phi}{\partial r}(R,\theta)=0 \tag{9.2-16}$$

$$v_\theta(R,\theta)=\frac{1}{R}\frac{\partial\phi}{\partial\theta}(R,\theta)=0. \tag{9.2-17}$$

Using $\mathbf{e_x}=\cos\theta\,\mathbf{e_r}-\sin\theta\,\mathbf{e_\theta}$ [Eq. (A.5-36a)], the unperturbed velocity far from the cylinder is converted to cylindrical coordinates as

$$v_r(\infty,\theta)=\lim_{r\to\infty}\frac{\partial\phi}{\partial r}=U\cos\theta \tag{9.2-18}$$

$$v_\theta(\infty,\theta)=\lim_{r\to\infty}\frac{1}{r}\frac{\partial\phi}{\partial\theta}=-U\sin\theta. \tag{9.2-19}$$

Assuming that the velocity potential is separable, such that $\phi(r,\theta) = f(r)g(\theta)$, Eqs. (9.2-18) and (9.2-19) each indicate that $g(\theta) = \cos\theta$. When $\phi = f(r)\cos\theta$ is substituted into Eq. (9.2-15) there is a common factor $\cos\theta$ that cancels. The resulting differential equation for $f(r)$ is

$$\frac{d^2 f}{dr^2} + \frac{1}{r}\frac{df}{dr} - \frac{f}{r^2} = 0 \tag{9.2-20}$$

which is an equidimensional (or Euler) equation. Substituting $f(r) = r^n$ yields $n^2 - 1 = 0$, or $n = \pm 1$. The general solution for $\phi(r, \theta)$ is then

$$\phi(r,\theta) = \left(Ar + \frac{B}{r}\right)\cos\theta \tag{9.2-21}$$

and the corresponding velocity components are

$$v_r(r,\theta) = \frac{\partial\phi}{\partial r}(r,\theta) = \left(A - \frac{B}{r^2}\right)\cos\theta \tag{9.2-22}$$

$$v_\theta(r,\theta) = \frac{1}{r}\frac{\partial\phi}{\partial r}(r,\theta) = -\left(A + \frac{B}{r^2}\right)\sin\theta. \tag{9.2-23}$$

What remains is to evaluate the constants A and B. A comparison of Eqs. (9.2-22) and (9.2-23) with Eqs. (9.2-18) and (9.2-19), respectively, indicates that $A = U$ will give the proper uniform velocity at large r. However, the boundary conditions at the cylinder surface present a difficulty. Equation (9.2-16) requires that $B = UR^2$, whereas Eq. (9.2-17) requires that $B = -UR^2$. The problem as posed is evidently overdetermined, and one boundary condition at $r = R$ must be disregarded. Equation (9.2-16) is more essential because, without the no-penetration condition, fluid would pass through what is supposed to be a solid cylinder. Choosing $B = UR^2$ to satisfy no penetration, the final expressions for the velocity components are

$$v_r(r,\theta) = U\left[1 - \left(\frac{R}{r}\right)^2\right]\cos\theta \tag{9.2-24}$$

$$v_\theta(r,\theta) = -U\left[1 + \left(\frac{R}{r}\right)^2\right]\sin\theta. \tag{9.2-25}$$

The flow pattern may be visualized using the stream function $\psi(r,\theta)$. From Table 5.4,

$$\frac{\partial\psi}{\partial\theta} = rv_r = Ur\cos\theta\left[1 - \left(\frac{R}{r}\right)^2\right] \tag{9.2-26}$$

$$\frac{\partial\psi}{\partial r} = -v_\theta = U\sin\theta\left[1 + \left(\frac{R}{r}\right)^2\right]. \tag{9.2-27}$$

Integrating these two equations yields two identical expressions for $\psi(r, \theta)$. Setting the shared integration constant to zero gives

$$\psi(r,\theta) = UR\sin\theta\left[\frac{r}{R} - \frac{R}{r}\right]. \tag{9.2-28}$$

Streamlines computed using equal intervals in ψ are shown in Fig. 9.2. The linearity of Laplace's equation makes potential flows reversible, like creeping flows (Section 8.3).

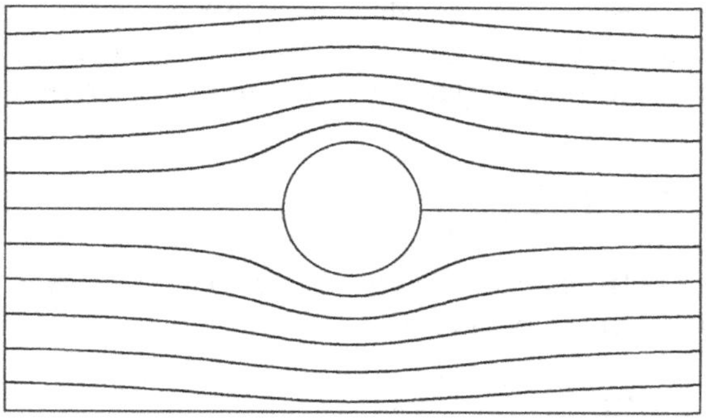

Figure 9.2 Streamlines for potential flow past a cylinder.

Accordingly, the streamlines for flow from the left are the same as those for flow from the right.

From Eq. (9.2-25), the tangential velocity at the cylinder surface is

$$v_\theta\,(R,\theta) = -2U\sin\theta. \tag{9.2-29}$$

It is seen that, except at the stagnation points ($\theta = 0$ and π), the no-slip condition is violated. This inability of a potential-flow solution to satisfy no slip is typical. This failing is a clue that the inviscid approximation breaks down near solid surfaces, even when $\text{Re} \to \infty$.

Example 9.2-2 Pressure and drag for inviscid and irrotational flow past a cylinder. The objective is to use the results of the preceding example to find the pressure and drag, assuming that the fluid is inviscid.

The pressure is evaluated from the irrotational Bernoulli equation. Setting $F(t) = U^2/2$ in Eq. (9.2-14), which is the same as setting $\mathcal{P} = 0$ far from the cylinder,

$$\frac{v^2(r,\theta)}{2} + \frac{\mathcal{P}(r,\theta)}{\rho} = \frac{U^2}{2} \tag{9.2-30}$$

where $v^2 = v_r^2 + v_\theta^2$. Substitution of the velocity components from Eqs. (9.2-24) and (9.2-25) gives

$$\mathcal{P}(r,\theta) = \frac{\rho U^2}{2}\left(\frac{R}{r}\right)^2\left[2 - 4\sin^2\theta - \left(\frac{R}{r}\right)^2\right]. \tag{9.2-31}$$

The pressure at the surface is then

$$\mathcal{P}(R,\theta) = \frac{\rho U^2}{2}(1 - 4\sin^2\theta). \tag{9.2-32}$$

This result is shown by the dashed curve in Fig. 9.3, where the pressure is expressed relative to $\rho U^2/2$ and angles are measured from the leading stagnation point (i.e., 0° at $\theta = \pi$ and 180° at $\theta = 0$). It is seen that $\mathcal{P}$ has a maximum of $\rho U^2/2$ at either stagnation point and a minimum of $-3\rho U^2/2$ halfway around the cylinder. Why the potential-flow prediction deviates so much from what is observed experimentally at high Re (the solid curve) is discussed in Section 9.4.

In an inviscid fluid a pressure imbalance is the only possible source of drag. However, Fig. 9.3 shows that $\mathcal{P}$ for potential flow around a cylinder is symmetric about the angle 90°, which corresponds to the plane $x = 0$ in Fig. 9.1. Thus, for each point on the upstream

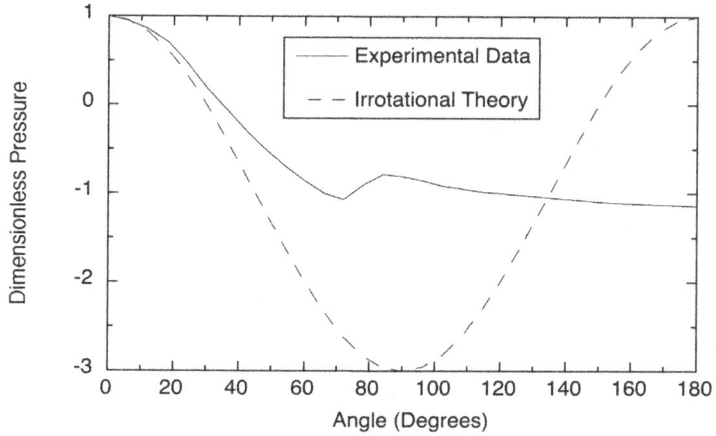

Figure 9.3 Pressure at the surface of a cylinder at high Reynolds number. The ordinate is $\mathcal{P}/(\rho U^2/2)$ and angles are measured from the leading stagnation point. The prediction for irrotational flow is from Eq. (9.2-32) and the data (at Re $= 1.86 \times 10^5$) are due to O. Flachsbart, as presented in Schlichting (1968, p. 21).

side of the cylinder, there is one on the downstream side where the pressure force is equal and opposite. Consequently, there will be zero drag. This is confirmed by using Eq. (6.6-26) to evaluate the drag force. That is,

$$F_D = -\int_S \mathbf{e_x} \cdot \mathbf{n}\mathcal{P}\, dS = -\frac{\rho U^2 RL}{2} \int_0^{2\pi} (1 - 4\sin^2\theta)\cos\theta\, d\theta = 0 \qquad (9.2\text{-}33)$$

where L is the cylinder length.

The finding that $F_D = 0$ for inviscid flow past a cylinder is an example of d'Alembert's paradox. This disturbing result holds not just when there is fore-and-aft symmetry, as with a cylinder or sphere, but for objects of any shape (Deen, 2012, pp. 374–375). Whereas a drag force is parallel to the approaching flow, a *lift* force is perpendicular to it. Although potential-flow theory fails to predict the existence of drag, it provides insight into lift and is valuable in airfoil design. Aircraft wings have a top-to-bottom asymmetry that causes the air velocity near the top to exceed that near the bottom. From Bernoulli's equation, that will make the pressure at the top less than at the bottom and create an upward force. Desirable wing shapes have high lift-to-drag ratios. Another kind of application in which potential-flow theory has been useful is wave mechanics, as illustrated in the next example.

Example 9.2-3 Water waves. The objective is to predict wave speeds in large bodies of water. The waves are assumed to have the regular, two-dimensional form shown in Fig. 9.4. The air–water interface is at $y = H + h(x, t)$, where H is the mean depth and

$$h(x, t) = h_0 \sin[k(x - ct)]. \qquad (9.2\text{-}34)$$

The wave amplitude and speed are h_0 and c, respectively, and the wavenumber k is related to the wavelength λ as $k = 2\pi/\lambda$. The water depth and wave dimensions are given, but the wave speed is unknown. After c has been found, $v_x(x, y, t)$ and $v_y(x, y, t)$ will be determined.

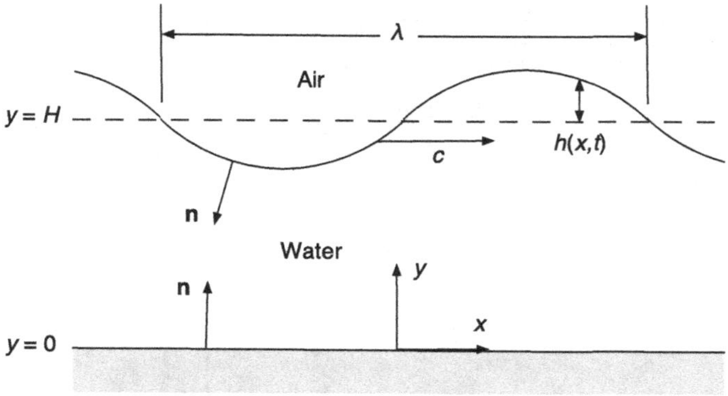

Figure 9.4 Two-dimensional waves of speed c and length λ in water of mean depth H. The deviation of the depth from that of still water is $h(x, t)$.

The Reynolds number will be large enough to neglect viscous stresses relative to inertia, and the Bond number large enough to neglect surface tension relative to gravity (Example 1.3-1). It is assumed that the velocity field was irrotational prior to the creation of the waves and that it remains so under these nearly inviscid conditions. For potential flow in Cartesian coordinates, Eq. (5.5-5) becomes

$$\frac{\partial^2 \phi}{\partial x^2} + \frac{\partial^2 \phi}{\partial y^2} = 0. \tag{9.2-35}$$

In place of boundary conditions in x, we will assume that the solution is spatially periodic,

$$\phi(x + \lambda, y, t) = \phi(x, y, t). \tag{9.2-36}$$

There is no flow normal to the bottom, where $\mathbf{n} = \mathbf{e_y}$. Thus, one boundary condition in y is

$$v_y(x, 0, t) = \frac{\partial \phi}{\partial y}(x, 0, t) = 0. \tag{9.2-37}$$

The boundary condition at the top is complicated by the fact that the free surface is neither planar nor stationary. Neglecting evaporation or condensation, there is no water flow across that interface. For there to be no flow relative to the moving boundary, the normal component of the water velocity must equal the normal component of the velocity of the interface itself [Eq. (6.6-12)]. Letting $\mathbf{u}$ be the interface velocity and $\mathbf{n}$ the unit normal pointing into the liquid,

$$\mathbf{n} \cdot \mathbf{v}|_{y=H+h} = \mathbf{n} \cdot \mathbf{u}. \tag{9.2-38}$$

For simplicity, we will assume now that the waves are long relative to their height, such that $h_0/\lambda \ll 1$ (or $kh_0 \ll 1$). Accordingly, the top surface is nearly horizontal and $\mathbf{n} \cong -\mathbf{e_y}$ there. Moreover, we will assume that the water is deep enough relative to the wave height to make $h_0/H \ll 1$, in which case $y \cong H$ at the free surface. Recognizing that $u_y = \partial h/\partial t$, Eq. (9.2-38) becomes

$$\frac{\partial \phi}{\partial y}(x, H, t) = \frac{\partial h}{\partial t} = -h_0 kc \cos[k(x - ct)]. \tag{9.2-39}$$

The last expression comes from differentiating Eq. (9.2-34).

Assuming that the velocity potential is separable, Eq. (9.2-39) suggests that

$$\phi(x, y, t) = h_0 kc \cos[k(x - ct)]G(y). \tag{9.2-40}$$

The part of ϕ involving x and t is consistent with the periodicity condition, Eq. (9.2-36). When Eq. (9.2-40) is substituted into Eqs. (9.2-35), (9.2-37), and (9.2-39), the cosine terms factor out and $G(y)$ is found to be governed by

$$\frac{d^2G}{dy^2} = k^2 G, \quad \frac{dG}{dy}(0) = 0, \quad \frac{dG}{dy}(H) = -1. \tag{9.2-41}$$

The finite interval in y makes hyperbolic functions more convenient than exponentials. Thus, the general solution is written as

$$G(y) = A \sinh(ky) + B \cosh(ky). \tag{9.2-42}$$

The boundary conditions indicate that $A = 0$ and $B = -1/[k \sinh(kh)]$. The final expression for the velocity potential is then

$$\phi(x, y, t) = -h_0 c \cos[k(x - ct)] \frac{\cosh(ky)}{\sinh(kH)}. \tag{9.2-43}$$

However, c is still unknown.

The additional information needed to determine the wave speed comes from applying the irrotational Bernoulli equation along the free surface. Noting that $P = P_0$ (the ambient air pressure), choosing $y = H$ as the reference height for $\mathcal{P}$, and setting $F(t) = 0$ in Eq. (9.2-14),

$$\frac{\partial \phi}{\partial t}(x, H, t) + \frac{v^2(x, H, t)}{2} + \frac{P_0}{\rho} + gh(x, t) = 0. \tag{9.2-44}$$

As a final simplification, we suppose that the kinetic energy (the $v^2/2$ term) is negligible. Solving Eq. (9.2-44) for h gives

$$h(x, t) = -\frac{1}{g} \frac{\partial \phi}{\partial t}(x, H, t) - \frac{P_0}{\rho g} \tag{9.2-45}$$

and differentiating this with respect to time shows that

$$\frac{\partial h}{\partial t} = -\frac{1}{g} \frac{\partial^2 \phi}{\partial t^2}(x, H, t). \tag{9.2-46}$$

Equating this expression for $\partial h/\partial t$ with that in Eq. (9.2-39) yields

$$\frac{\partial \phi}{\partial y}(x, H, t) = -\frac{1}{g} \frac{\partial^2 \phi}{\partial t^2}(x, H, t). \tag{9.2-47}$$

Evaluating the derivatives using Eq. (9.2-43) reveals that the wave speed is

$$c^2 = \frac{g \tanh(kH)}{k} \tag{9.2-48}$$

a result obtained by G. B. Airy in 1845.

It is informative to examine the limits of Eq. (9.2-48) when the water is either shallow or deep, relative to the wavelength. For $kH \to 0$, $\tanh(kH) \to kH$ and

$$c = \sqrt{gH}. \tag{9.2-49}$$

Thus, in shallow water the wave speed depends only on the mean depth. For $kH \to \infty$, $\tanh(kH) \to 1$ and

$$c = \sqrt{\frac{g}{k}} = \sqrt{\frac{g\lambda}{2\pi}}. \qquad (9.2\text{-}50)$$

This indicates that in deep water the wave speed depends only on the wavelength. All of the results are restricted to wave amplitudes that are small relative to both the wavelength and mean depth, as already noted. Thus, for shallow water the analysis requires that $h_0 \ll H \ll \lambda$ and for deep water it is necessary that $h_0 \ll \lambda \ll H$.

The velocity components obtained by differentiating Eq. (9.2-43) are

$$v_x(x, y, t) = \frac{\partial \phi}{\partial x} = h_0 kc \sin[k(x - ct)] \frac{\cosh(ky)}{\sinh(kH)} \qquad (9.2\text{-}51)$$

$$v_y(x, y, t) = \frac{\partial \phi}{\partial y} = -h_0 kc \cos[k(x - ct)] \frac{\sinh(ky)}{\sinh(kH)}. \qquad (9.2\text{-}52)$$

Equation (9.2-51) indicates that $v_x \neq 0$ at the bottom surface, $y = 0$. Thus, as usual for potential flow, the no-slip condition is not satisfied. In contrast to what occurs in drag calculations, this has little effect on the wave results, which are still used in ocean engineering.

Neglecting kinetic energy relative to potential energy at the free surface, as was done in simplifying Eq. (9.2-44), is justified whenever the geometric requirements for the analysis are met. This may be seen from OM estimates. Equations (9.2-51) and (9.2-52) indicate that $v_x \sim h_0 kc/\tanh(kH)$ and $v_y \sim h_0 kc$. Thus, $v_y \ll v_x$ in shallow water, $v_y \sim v_x$ in deep water, and $v \sim v_x$ always. Using Eq. (9.2-48) to evaluate c, what is needed to neglect kinetic energy at the free surface is $v^2 \sim gh_0^2 k/\tanh(kH) \ll gh_0$, or $h_0 k \ll \tanh(kH)$. This will hold for shallow water if $h_0 \ll H$ and for deep water if $h_0 \ll \lambda$. Those conditions were needed already to approximate the location of the free surface as $y = H$.

9.3 BOUNDARY LAYERS: DIFFERENTIAL ANALYSIS

Boundary-layer approximation

Boundary layers are thin regions in which viscosity is as important as inertia, even though Re is extremely large. Coexisting with them and comprising the bulk of the fluid are "outer" regions in which the flow is nearly inviscid. Thus, any boundary-layer problem involves at least two dynamically distinct regions, a boundary layer and an outer region. Potential flow (Section 9.2) is often a good approximation in outer regions. In this section the momentum equation for a two-dimensional boundary layer is presented and the relationship between a boundary layer and the adjacent outer region is discussed. The solution of these multi-region problems is illustrated by examples here and in Section 9.4.

The regions for steady flow at high Reynolds number past a fairly "streamlined" (as opposed to bluff) object are shown in Fig. 9.5. As indicated by the dashed curves, there are boundary layers at the top and bottom surfaces. They start at the leading stagnation point, grow in thickness along the surface, and join to exit as a wake. (The thicknesses of the boundary layers have been exaggerated.) Outside the boundary layers and wake is a region in which viscous stresses are negligible. In that outer region the relevant length scale is the overall size of the object. This is its cross-sectional dimension L, which is also roughly the contour length along the surface. The Reynolds number for such a flow is $\text{Re} = UL/\nu$,

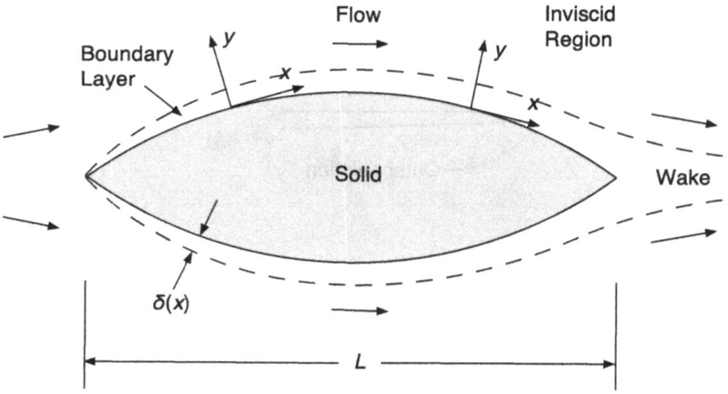

Figure 9.5 Regions and length scales for flow past an object at high Reynolds number.

where U is the velocity of the approaching fluid. Most relevant to the boundary layer is its thickness δ, which is usually small relative to the radius of curvature of the surface. Thus, from the boundary-layer perspective the surface of the object is nearly planar. Local Cartesian coordinates can be employed in the boundary layer, in which x is tangent to the surface and y is normal to it. In general, another coordinate system is needed for the outer region, such as the cylindrical one in Fig. 9.1. Of importance, $\delta(x)/L \ll 1$ in any boundary-layer problem, by definition.

Because a boundary layer is thin relative to its length, if it is two-dimensional the flow is nearly unidirectional. Using the same reasoning as for lubrication problems (Section 8.5), $v_y/v_x \sim \delta/L \ll 1$. If follows that in boundary layers $\mathcal{P}$ is approximately a function of x only and $\partial^2 v_x/\partial x^2 \ll \partial^2 v_x/\partial y^2$, as seen before for nearly unidirectional flow. What distinguishes boundary-layer from lubrication problems is that Re for boundary layers is large enough that inertia cannot be neglected. The boundary-layer momentum equation is thus

$$v_x \frac{\partial v_x}{\partial x} + v_y \frac{\partial v_x}{\partial y} = -\frac{1}{\rho}\frac{d\mathcal{P}}{dx} + \nu \frac{\partial^2 v_x}{\partial y^2}. \tag{9.3-1}$$

The scale for δ can be inferred by comparing the inertial and viscous terms in Eq. (9.3-1). With L the scale for variations in x, δ the scale for variations in y, and U the scale for v_x and for variations in v_x in either direction, the inertial and viscous terms will be comparable when $U^2/L \sim \nu U/\delta^2$, or

$$\frac{\delta}{L} \sim \left(\frac{\nu}{UL}\right)^{1/2} = \text{Re}^{-1/2}. \tag{9.3-2}$$

This confirms that δ/L is small if Re is large. Of great importance, it shows that viscous stresses do not disappear as Re is increased; instead, they act within progressively thinner regions. For a no-slip condition to be satisfied at a boundary, viscous stresses must persist there, even as Re $\to \infty$. Mathematically, the absence of a second derivative such as $\partial^2 v_x/\partial y^2$ in the inviscid momentum equation makes it impossible to satisfy the no-slip condition when using that idealization. Physically, the tendency of a fluid to adhere to a solid surface creates shear rates near the surface that are large enough to make viscous stresses competitive with inertia, even if the viscosity is small.

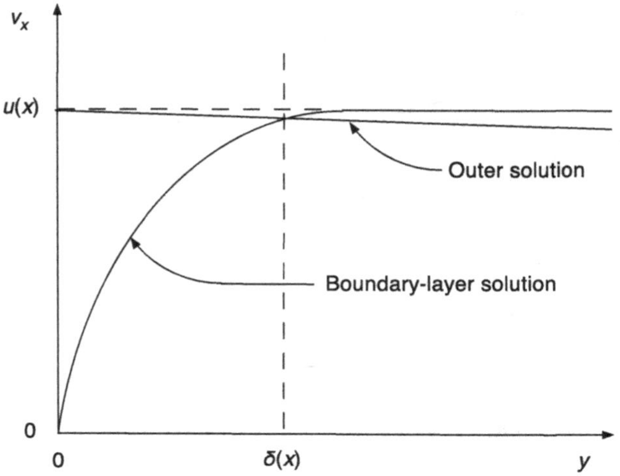

Figure 9.6 Relationship between the tangential velocities in a boundary layer and an adjacent outer region.

Joining the regions

To fully define the flow problem within a boundary layer, information from the adjacent outer region is required. Two things are needed: a boundary condition for v_x at the outer edge of the boundary layer, and a way to evaluate $d\mathcal{P}/dx$. This information may be obtained in principle by first solving the corresponding outer problem. If $\delta/L \to 0$, the inviscid region will be nearly the same size and shape as if the boundary layer were absent, allowing the outer problem to be formulated without knowledge of what occurs within the boundary layer. As discussed in Section 9.2, the velocity field for potential flow can be obtained by solving Laplace's equation, either analytically or numerically. What is important for the boundary layer is the tangential component of the outer velocity evaluated at the surface, which is denoted hereafter as $u(x)$. That velocity is not zero, because potential flow does not satisfy the no-slip condition. For flow past a cylinder (Example 9.2-1), $u(x)$ corresponds to $-v_\theta(R, \theta) = 2U \sin\theta$ and the boundary-layer coordinate is related to the cylindrical angle as $x = R(\pi - \theta)$. We will return to the cylinder later (Example 9.4-2).

The link between the velocities calculated in the two regions is illustrated in Fig. 9.6. Over a few boundary-layer thicknesses, v_x from the outer solution will vary only slightly as the surface is approached. For the hypothetical situation shown, the outer v_x increases, reaching $u(x)$ at $y = 0$. What is assumed for the boundary layer is that v_x increases from zero at $y = 0$ to a limiting value of $u(x)$ as y becomes large, with most of the change occurring over a distance δ. Where the boundary-layer solution intersects the outer one, labeled here as $y = \delta(x)$, is just one of several ways to define the boundary-layer thickness. Following the boundary-layer curve to that point, and then changing to the outer one, provides a good approximation to the actual variation in v_x. The larger the value of Re, the more the intersection is shifted to the left and the better the approximation becomes. In summary, $u(x)$ has a dual significance: it is both the outer v_x as $y/L \to 0$ and the boundary-layer v_x as $y/\delta \to \infty$. What permits y to be small in one sense and large in the other is that $\delta/L \to 0$ as $\text{Re} \to \infty$.[2]

[2] What is discussed here qualitatively can be made rigorous using *singular perturbation theory*. Developed mainly to analyze boundary-layer phenomena in fluid mechanics and other applications, it came into use in

The finding that $\mathcal{P}$ in a boundary layer is independent of y implies that the pressure is determined by the flow in the outer region. Because the pressure in the boundary layer is imposed on it by the outer flow, $d\mathcal{P}/dx$ can be evaluated by applying Bernoulli's equation to the outer solution at $y = 0$. Equation (9.2-5) is applicable because the outer flow is approximated as inviscid and an impermeable surface is a streamline. Setting $v = u(x)$ and differentiating with respect to x gives

$$-\frac{1}{\rho}\frac{d\mathcal{P}}{dx} = \frac{d}{dx}\left(\frac{u^2}{2}\right) = u\frac{du}{dx}. \tag{9.3-3}$$

Substituting this into Eq. (9.3-1) gives the working form of the boundary-layer momentum equation,

$$v_x\frac{\partial v_x}{\partial x} + v_y\frac{\partial v_x}{\partial y} = u\frac{du}{dx} + \nu\frac{\partial^2 v_x}{\partial y^2}. \tag{9.3-4}$$

In the boundary layer there are just two unknowns, $v_x(x, y)$ and $v_y(x, y)$, which are governed by Eq. (9.3-4) and the continuity equation. The requisite information on $\mathcal{P}$ is provided by $u(x)$, which is obtained by solving a separate problem for the outer region. Because Cartesian coordinates can be used for any boundary layer that is sufficiently thin, objects of varying shape are distinguished only by the differences in $u(x)$.

Example 9.3-1 Blasius solution for a flat plate. The objective is to find the velocity field and drag coefficient for flow parallel to a thin, flat plate. The approach velocity is U, the plate length is L, and the leading edge is at $x = 0$. It is sufficient to consider only one side of the plate ($y \geq 0$).

An infinitesimally thin plate does not disturb an inviscid fluid moving parallel to it, because the no-slip condition is not enforced. Accordingly, $v_x = U$ throughout the outer region, the pressure is constant everywhere, and Eq. (9.3-4) reduces to

$$v_x\frac{\partial v_x}{\partial x} + v_y\frac{\partial v_y}{\partial y} = \nu\frac{\partial^2 v_x}{\partial y^2}. \tag{9.3-5}$$

It will be shown that this, along with the continuity equation, can be solved using a similarity transformation that combines x and y into a single independent variable.

Unlike what happened with the suddenly accelerated plate in Example 7.3-3, dimensional analysis does not indicate how to combine the independent variables. The primary dimensional quantities are now x, y, v_x, ν, and U.[3] These five quantities have just two independent dimensions (i.e., L and LT^{-1}), so there are three independent dimensionless groups. An obvious one is v_x/U. Among the possibilities for the other two are Ux/ν and Uy/ν. Thus, the groups that v_x/U depends on do not necessarily contain x and y only in combination. However, something that the stationary and suddenly accelerated plates have in common is that viscous stresses and vorticity created at a surface spread gradually into an adjacent fluid of indefinitely large dimensions. In Example 7.3-3 the thickness of the affected layer of fluid varied as $(\nu t)^{1/2}$. By analogy, the boundary-layer thickness

the 1950s. An introduction to perturbation methods in the analysis of transport problems is provided, for example, in Chapter 4 of Deen (2012).

[3] Because v_y can be calculated from v_x using continuity, it is excluded. As will be seen, L influences neither v_x nor v_y, although it does affect the shear stress averaged over the plate.

might vary as $(\nu x/U)^{1/2}$, x/U being the transit time from the leading edge to position x. The boundary-layer thickness is the scale for y. It might be supposed then that

$$\eta = \frac{y}{(2\nu x/U)^{1/2}} \tag{9.3-6}$$

is a suitable combination of x and y. Our ability to transform the original problem into one involving η alone, with neither x nor y appearing separately, will confirm this speculation to be correct. The factor 2 in Eq. (9.3-6) merely puts the final equation in the simplest form.

A dimensionless function $f(\eta)$ is defined such that

$$v_x = U\frac{df}{d\eta} = Uf'. \tag{9.3-7}$$

Using primes to denote derivatives of f with respect to η, the various derivatives of v_x in the momentum equation are transformed as

$$\frac{\partial v_x}{\partial x} = U\frac{\partial \eta}{\partial x}f'' = -\frac{U\eta}{2x}f'' \tag{9.3-8}$$

$$\frac{\partial v_x}{\partial y} = U\frac{\partial \eta}{\partial y}f'' = \left(\frac{U^3}{2x\nu}\right)^{1/2} f'' \tag{9.3-9}$$

$$\frac{\partial^2 v_x}{\partial y^2} = U\left(\frac{\partial \eta}{\partial y}\right)^2 f''' = \frac{U^2}{2x\nu}f'''. \tag{9.3-10}$$

Using Eq. (9.3-8) in the continuity equation and integrating by parts gives

$$v_y = -\int \frac{\partial v_x}{\partial x}\,dy = \left(\frac{U\nu}{2x}\right)^{1/2} \int \eta f''\,d\eta = \left(\frac{U\nu}{2x}\right)^{1/2} (\eta f' - f). \tag{9.3-11}$$

The no-slip condition requires that $f'(0) = 0$ and the no-penetration condition indicates then that $f(0) = 0$.

Using Eqs. (9.3-7)–(9.3-11) in Eq. (9.3-5), the momentum equation for the flat-plate boundary layer becomes

$$f''' + ff'' = 0. \tag{9.3-12}$$

(If the 2 had been omitted from Eq. (9.3-6), the coefficient of ff'' would have been $1/2$ instead of 1.) The boundary conditions for this third-order differential equation are

$$f(0) = 0, \quad f'(0) = 0, \quad f'(\infty) = 1. \tag{9.3-13}$$

The first two conditions correspond to no penetration and no slip, respectively. The third represents the matching of the boundary-layer and outer velocities that is depicted in Fig. 9.6. Equations (9.3-12) and (9.3-13) are sufficient to determine $f(\eta)$, and x and y no longer appear separately. This confirms that the independent variables can be combined.

The boundary-value problem just formulated is nonlinear and has no closed-form analytical solution. An equivalent problem was solved by Blasius (1908) using a lengthy procedure that combined a power-series expansion for small η with asymptotic results for large η. His results were refined later by others using numerical methods. The curve in Fig. 9.7 is based on computational results tabulated in Jones and Watson (1963). The theory is in excellent agreement with a comprehensive set of air-flow data in Nikuradse

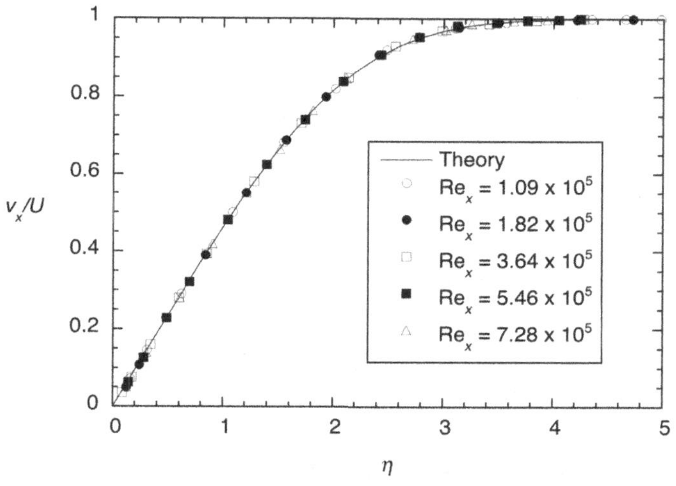

Figure 9.7 Velocity profile for flow past a flat plate. The theoretical curve is the solution of Eqs. (9.3-12) and (9.3-13) and the data points (for air flow) are from Nikuradse (1942).

(1942, p. 37), as shown. The data are for various values of the Reynolds number based on distance from the leading edge, $\mathrm{Re}_x = Ux/\nu$.

In that $v_x \to U$ only as $\eta \to \infty$, there is not a specific distance from the surface at which the transition from boundary layer to outer region is complete. However, Fig. 9.7 indicates that the transition is nearly complete when y exceeds a small multiple of $(\nu x/U)^{1/2}$. The boundary-layer thickness can be expressed as

$$\delta(x) = c\left(\frac{\nu x}{U}\right)^{1/2} \tag{9.3-14}$$

where the constant c depends on how close it is desired that v_x approach U. For $v_x/U = 0.50$, 0.90, and 0.99 at $y = \delta$, $c = 1.1$, 2.4, and 3.5, respectively. However c is chosen, it is evident that δ grows as $x^{1/2}$. Dividing both sides of Eq. (9.3-14) by x shows that $\delta/x = c\mathrm{Re}_x^{-1/2}$. Setting $x = L$ to obtain the maximum boundary-layer thickness gives $\delta/L = c\mathrm{Re}^{-1/2}$, where $\mathrm{Re} = UL/\nu$. This is consistent with the general prediction of Eq. (9.3-2).

Using Eq. (9.3-9) and the finding that $f''(0) = 0.46960$ (Jones and Watson, 1963), the local shear stress exerted on the plate is calculated as

$$\tau_0(x) \equiv \mu \left.\frac{\partial v_x}{\partial y}\right|_{y=0} = \mu\left(\frac{U^3}{x\nu}\right)^{1/2}\frac{f''(0)}{\sqrt{2}} = 0.3321\frac{\mu U}{x}\mathrm{Re}_x^{1/2}. \tag{9.3-15}$$

The absence of L shows that the distance x from the leading edge, not L, is the natural length scale for the viscous stress and Reynolds number. Because τ_0 is proportional to u/δ, u is constant at U, and δ varies as $x^{1/2}$, the shear stress varies as $x^{-1/2}$. To calculate the average shear stress over a plate of length L, we rearrange Eq. (9.3-15) as

$$\tau_0\,(x) = 0.3321\frac{\mu U}{L}\mathrm{Re}^{1/2}\left(\frac{x}{L}\right)^{-1/2}. \tag{9.3-16}$$

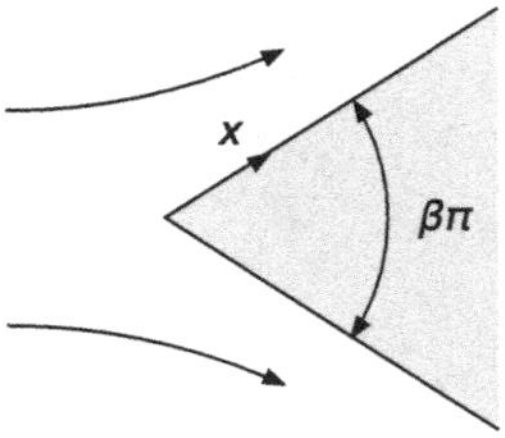

Figure 9.8 Flow past a wedge.

The average shear stress and drag coefficient are

$$\langle \tau_0 \rangle = \frac{1}{L}\int_0^L \tau_0 \, dx = 0.6642 \frac{\mu U}{L} \mathrm{Re}^{1/2} \tag{9.3-17}$$

$$C_f = \frac{2\langle \tau_0 \rangle}{\rho U^2} = 1.328\, \mathrm{Re}^{-1/2}. \tag{9.3-18}$$

The result for C_f was presented earlier [Eq. (3.2-15)].

Wedge flows

More generally, similarity transformations work whenever $u(x) = ax^m$, where a and m are constants. Such solutions correspond to flow past wedge-shaped objects, as depicted in Fig. 9.8. The exponent m is related to the wedge angle β as $m = \beta/(2-\beta)$. Thus, $\beta = 0$ and $m = 0$ for a flat plate, $\beta = 1/2$ and $m = 1/3$ for a right angle, and $\beta = 1$ and $m = 1$ for planar stagnation flow.

The dependence of δ on x in wedge flows can be deduced from the need to balance inertial and viscous stresses. That is, $v_x\, \partial v_x/\partial x \sim \nu\, \partial^2 v_x/\partial y^2$ at any given position along the surface. Using $\Delta x \sim x$, $\Delta y \sim \delta$, and $v_x \sim u(x) \sim \Delta v_x$, this gives $a^2 x^{2m-1} \sim \nu a x^m/\delta^2$ or

$$\delta(x) \sim \left(\frac{\nu}{a}\right)^{1/2} x^{(1-m)/2}. \tag{9.3-19}$$

Thus, δ varies as $x^{1/2}$ for a flat plate [as in Eq. (9.3-14)] and $x^{1/3}$ for a right-angle wedge, and is constant for planar stagnation flow. In general, because the similarity variable is proportional to $y/\delta(x)$, it is of the form $\eta = byx^{(m-1)/2}$, where b is a positive constant. Choosing

$$\eta = \left(\frac{(1+m)\,a}{2\nu}\right)^{1/2} y x^{(m-1)/2} \tag{9.3-20}$$

transforms the momentum equation for wedge flows to

$$f''' + ff'' + \beta[1 - (f')^2] = 0 \tag{9.3-21}$$

which is called the *Falkner–Skan equation*. The boundary conditions are given again by Eq. (9.3-13). Numerical solutions of the Falkner–Skan equation have been obtained for various values of β (Hartree, 1937; Jones and Watson, 1963). Derivations of Eq. (9.3-21) and summaries of the numerical results may be found in Schlichting (1968, pp. 137–141 and 150–151) or Deen (2012, pp. 381–383).

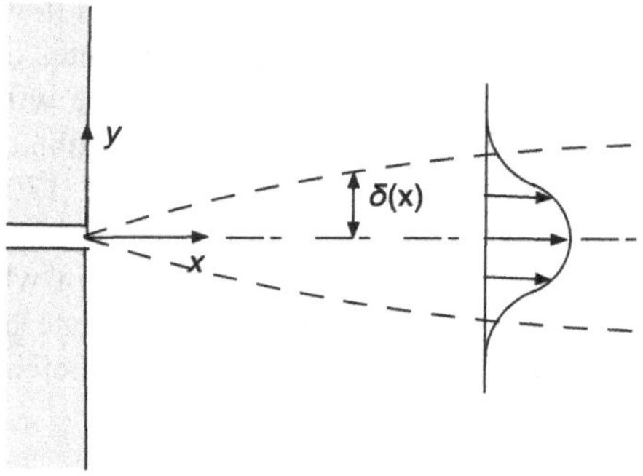

Figure 9.9 A two-dimensional jet passing from a slit into a large volume of the same fluid.

Internal boundary layers

Boundary layers exist at large Re not just at solid surfaces, but anywhere shear rates are large enough. Thin regions with shear rates high enough to make viscous stresses important are created when streams moving at different velocities are brought into contact. The same or different fluids might be involved, and their flow directions might be the same or opposite.

Example 9.3-2 Planar jet. Suppose that a two-dimensional jet emerges from a narrow slit into a large, stagnant volume of the same fluid, as shown in Fig. 9.9. The jet gradually slows and widens as its momentum is transferred to the outer fluid. The strength of the jet is characterized by its *kinematic momentum*,

$$K = \int_{-\infty}^{\infty} v_x^2 \, dy \tag{9.3-22}$$

which is assumed to be a known constant. As shown by the dashed curves, the half-thickness of the jet is $\delta(x)$. Its maximum velocity (at $y = 0$) is $v_m(x)$ and the flow rate per unit width in the z direction is $q(x)$. The objective is to estimate δ, v_m, and q to within OM precision. After that it will be shown that K is indeed constant.

As it is a thin region with high shear rates, the jet is a boundary layer. Accordingly, $v_x \partial v_x / \partial x \sim \nu \partial^2 v_x / \partial y^2$ within it. Using $\Delta x \sim x$, $\Delta y \sim \delta$, and $v_x \sim v_m \sim \Delta v_x$, this implies that $v_m^2 / x \sim \nu v_m / \delta^2$ or

$$v_m \sim \frac{\nu x}{\delta^2}. \tag{9.3-23}$$

The OM estimate from Eq. (9.3-22) is

$$v_m^2 \delta \sim K. \tag{9.3-24}$$

Substituting this into Eq. (9.3-23) and solving for δ gives

$$\delta(x) \sim \frac{(\nu x)^{2/3}}{K^{1/3}}. \tag{9.3-25}$$

The velocity and flow rate are then

$$v_m(x) \sim \frac{K^{2/3}}{(\nu x)^{1/3}} \tag{9.3-26}$$

$$q(x) = \int_{-\infty}^{\infty} v_x \, dy \sim v_m \delta \sim (K \nu x)^{1/3}. \tag{9.3-27}$$

The jet thickness, velocity, and flow rate are seen to vary as $x^{2/3}$, $x^{-1/3}$, and $x^{1/3}$, respectively. The increase in q indicates that there is entrainment of the surrounding fluid. The result for the flow rate may be written also as $q(x) = c(K\nu x)^{1/3}$, where c is a constant that is ~ 1. An exact solution, obtained using a similarity transformation, shows that $c = 3.3019$ (Schlichting, 1968, pp. 170–174). Experimental evidence in support of the predicted velocity field is cited in that source.

What remains is to explain why the kinematic momentum of this system is constant. An expression for the rate of change of K is obtained by integrating each term in Eq. (9.3-4) over y. It is sufficient to consider only $y \geq 0$. The integral of the viscous term is

$$\int_0^\infty \nu \frac{\partial^2 v_x}{\partial y^2}\, dy = \nu \frac{\partial v_x}{\partial y}\bigg|_0^\infty = 0 \qquad (9.3\text{-}28)$$

which vanishes because of the symmetry at $y = 0$ and the absence of shear at $y = \infty$. Because the outer fluid is stagnant ($u = 0$), the pressure term vanishes also and only the inertial terms remain. Using continuity and integration by parts, the second inertial term is evaluated as

$$\begin{aligned}\int_0^\infty v_y \frac{\partial v_x}{\partial y}\, dy &= \int_0^\infty \left(-\int_0^y \frac{\partial v_x}{\partial x}\, dY\right) \frac{\partial v_x}{\partial y}\, dy = -\, v_x \int_0^y \frac{\partial v_x}{\partial x}\, dY\bigg|_0^\infty + \int_0^\infty v_x \frac{\partial v_x}{\partial x}\, dy \\ &= \int_0^\infty (v_x - u)\frac{\partial v_x}{\partial x}\, dy.\end{aligned} \qquad (9.3\text{-}29)$$

Setting $u = 0$, the integrated momentum equation is

$$0 = \int_0^\infty \left(v_x \frac{\partial v_x}{\partial x} + v_y \frac{\partial v_x}{\partial y}\right) dy = \int_0^\infty 2v_x \frac{\partial v_x}{\partial x}\, dy = \frac{\partial}{\partial x}\int_0^\infty v_x^2\, dy = \frac{1}{2}\frac{dK}{dx}. \qquad (9.3\text{-}30)$$

This confirms that K is constant, as assumed.

9.4 BOUNDARY LAYERS: INTEGRAL ANALYSIS

Integral momentum equation

Boundary layers can be analyzed also from a more macroscopic viewpoint, in which the momentum equation is integrated over y and conservation of momentum is satisfied only in an overall sense. This greatly reduces the computational effort, because it replaces two partial differential equations (continuity and momentum) with one first-order, ordinary differential equation, as will be seen. At the expense of some precision, analytical or elementary numerical solutions can be obtained for what are otherwise very difficult problems. Although no longer widely used in research, the insights it provides make the integral method worthy of study.

The integration of Eq. (9.3-4) that follows is like that at the end of Example 9.3-2, except that $y = \infty$ is replaced by $y = \delta(x)$. That is, the outer edge of the boundary layer is viewed now as a definite location. Also, in what follows $y = 0$ is a solid surface, instead of the symmetry plane that was present in the planar jet. Noting that the shear stress is negligible at the outer edge of the boundary layer,

$$\int_0^\delta \nu \frac{\partial^2 v_x}{\partial y^2}\,dy = -\nu \frac{\partial v_x}{\partial y}\bigg|_{y=0} = -\frac{\tau_0}{\rho} \tag{9.4-1}$$

where $\tau_0(x)$ is the shear stress at $y = 0$. Using Eq. (9.3-29) to help evaluate the inertial terms, the integrated momentum equation is

$$\int_0^\delta (2v_x - u)\frac{\partial v_x}{\partial x}\,dy = \int_0^\delta u\frac{du}{dx}\,dy - \frac{\tau_0}{\rho}. \tag{9.4-2}$$

This may be rearranged as

$$\frac{\tau_0}{\rho} = \int_0^\delta \frac{\partial}{\partial x}\left[v_x(u - v_x)\right]dy + \frac{du}{dx}\int_0^\delta (u - v_x)\,dy. \tag{9.4-3}$$

To obtain this form, $v_x\,du/dx$ was subtracted from both integrands in Eq. (9.4-2). Because the term in square brackets vanishes at both limits, the derivative can be moved outside the integral to give

$$\frac{\tau_0}{\rho} = \frac{d}{dx}\int_0^\delta v_x(u - v_x)\,dy + \frac{du}{dx}\int_0^\delta (u - v_x)\,dy. \tag{9.4-4}$$

This is the *integral momentum equation* or *Kármán integral equation*.

The tangential velocity in the boundary layer may be written generally as

$$\frac{v_x}{u(x)} = F(x, \eta), \qquad \eta = \frac{y}{\delta(x)}. \tag{9.4-5}$$

The dimensionless coordinate η is analogous to the similarity variable in Eq. (9.3-20), but ranges now only from 0 to 1. If the velocity profiles are self-similar, as for wedge flows, then $F = F(\eta)$ only. A more compact version of Eq. (9.4-4) is

$$\frac{\tau_0}{\rho} = \frac{d}{dx}(u^2\delta_2) + u\frac{du}{dx}\delta_1. \tag{9.4-6}$$

The functions $\delta_1(x)$ and $\delta_2(x)$, which have the dimension of length, are called the *displacement thickness* and *momentum thickness*, respectively. They are related to δ and F as

$$\frac{\delta_1}{\delta} = \int_0^1 (1 - F)\,d\eta, \qquad \frac{\delta_2}{\delta} = \int_0^1 F(1 - F)\,d\eta. \tag{9.4-7}$$

For self-similar profiles, δ_1/δ and δ_2/δ are constant, but in general, both ratios depend on x.[4] The shear stress at the surface is given by

$$\frac{\tau_0}{\rho} = \frac{\nu u}{\delta}\frac{\partial F}{\partial \eta}(x, 0) = \frac{\nu u}{\delta}F'(x, 0). \tag{9.4-8}$$

[4] In more exact analyses, where $v_x \to u$ asymptotically as $y \to \infty$,

$$\delta_1 = \int_0^\infty (1 - F)\,dy, \qquad \delta_2 = \int_0^\infty F(1 - F)\,dy.$$

For a flat plate, the displacement and momentum thicknesses both obey Eq. (9.3-14). Evaluating the integrals using the similarity solution in Example 9.3-1, $c = 1.7208$ for δ_1 and $c = 0.6641$ for δ_2 (Jones and Watson, 1963).

Table 9.1 Properties of boundary-layer velocity profiles

(a) $v_x\|_{y=0} = 0$	$F(x, 0) = 0$
(b) $v_x\|_{y=\delta} = u(x)$	$F(x, 1) = 1$
(c) $\left.\frac{\partial v_x}{\partial y}\right\|_{y=\delta} = 0$	$F'(x, 1) = 0$
(d) $\left.\frac{\partial^2 v_x}{\partial y^2}\right\|_{y=0} = -\frac{u}{\nu}\frac{du}{dx}$	$F''(x, 0) = -\frac{\delta^2}{\nu}\frac{du}{dx} \equiv -\Lambda(x)$
(e) $\left.\frac{\partial^2 v_x}{\partial y^2}\right\|_{y=\delta} = 0$	$F''(x, 1) = 0$

Primes are used again to denote derivatives with respect to η, although they are now partial derivatives.

In problem solving we first choose a simple form for the η-dependence of F, such as a polynomial, which has certain attributes of an exact solution. The integrals in Eq. (9.4-7) are then evaluated, and Eq. (9.4-8) is used to relate the shear stress to $\delta(x)$ and $u(x)$. Substitution of those results into Eq. (9.4-6) gives a first-order differential equation involving the unknown boundary-layer thickness $\delta(x)$ and the known velocity $u(x)$. Solving that for $\delta(x)$ allows $\tau_0(x)$ to be calculated.

Needed to implement this method is a functional form for F that is simple to integrate and differentiate, yet realistic enough to yield good results. Guidance in choosing such a function is provided by the properties of an exact solution. For a stationary solid at $y = 0$, the characteristics of v_x and the corresponding requirements for F are given in Table 9.1. Conditions (a) and (b) are indispensable. Condition (c), which embodies the fact that viscous stresses are negligible at the outer edge of the boundary layer, is very desirable. Condition (d) comes from evaluating Eq. (9.3-4) at the surface. It is crucial for boundary layers in which the velocity profile changes shape with increasing x, because it is the only condition that leads to profiles that are not self-similar. The dimensionless function $\Lambda(x)$, which is related to the shape of the object through $u(x)$, is called the *shape factor*. Condition (e) comes from evaluating Eq. (9.3-4) at the outer edge of the boundary layer and recognizing that $v_x \partial v_x/\partial x = u\,du/dx$ and $\partial v_x/\partial y = 0$ there.

The more conditions in Table 9.1 that are satisfied by $F(x, \eta)$, the more accurate the results tend to be, although the improvements are not entirely predictable. In the *Kármán–Pohlhausen method*, a fourth-order polynomial is postulated and all five conditions are imposed (Schlichting, 1968, pp. 192–206). As seen in the next example, simpler functions may suffice.

Example 9.4-1 Integral solution for a flat plate. The objective is to use the integral approach to calculate the drag on a flat plate. The results for various assumed velocity profiles will be compared with the exact result given in Example 9.3-1.

In this flow $u(x) = U$ (a constant) and we have seen already that the velocity profiles are self-similar. Thus, the second term on the right-hand side of either Eq. (9.4-4) or (9.4-6) is zero and the integral momentum equation reduces to

$$\frac{\nu U}{\delta} F'(0) = U^2 \frac{d\delta}{dx} \int_0^1 F(1-F)\,d\eta \tag{9.4-9}$$

Table 9.2 Drag coefficient for a flat plate obtained from integral approximations

$F(\eta)$	Properties from Table 9.1	b	% Error
(1) η	(a) and (b)	1.155	−13.1
(2) $2\eta - \eta^2$	(a)–(c)	1.461	+10.0
(3) $\frac{3}{2}\eta - \frac{1}{2}\eta^3$	(a)–(d)	1.293	−2.6
(4) $\sin\left(\frac{\pi\eta}{2}\right)$	(a)–(d)	1.310	−1.3
(5) $2\eta - 2\eta^3 + \eta^4$	(a)–(e)	1.371	+3.2

where $F = F(\eta)$ only. This nonlinear but separable equation for δ is rewritten as

$$\frac{d(\delta^2)}{dx} = \frac{2\nu}{U} \frac{F'(0)}{\int_0^1 F(1-F)\,d\eta}. \tag{9.4-10}$$

The solution that satisfies $\delta(0) = 0$ is

$$\delta(x) = \left[\frac{2F'(0)}{\int_0^1 F(1-F)\,d\eta}\right]^{1/2} \left(\frac{\nu x}{U}\right)^{1/2} \equiv c\left(\frac{\nu x}{U}\right)^{1/2}. \tag{9.4-11}$$

Thus, the approximate integral analysis reveals that δ for a flat plate varies as $x^{1/2}$, which is consistent with the exact solution in Example 9.3-1. The constant c in Eq. (9.4-11) is analogous to that in Eq. (9.3-14), but c now has a unique value for any $F(\eta)$ that is chosen.

Evaluating $\tau_0(x)$ using Eq. (9.4-8) and averaging it over the plate length L as before,

$$C_f = \left[8F'(0)\int_0^1 F(1-F)\,d\eta\right]^{1/2} \mathrm{Re}^{-1/2} \equiv b\,\mathrm{Re}^{-1/2}. \tag{9.4-12}$$

Table 9.2 compares the results from several assumed velocity profiles. Included are the conditions satisfied by $F(\eta)$, the value of b from Eq. (9.4-12), and the percentage error in b relative to the exact value of 1.328 from Eq. (9.3-18). Given that it satisfies only conditions (a) and (b) in Table 9.1, it is remarkable that profile (1) gives an error of only 13%. Profile (2), satisfying conditions (a)–(c), provides a slight improvement. When conditions (a)–(d) are satisfied, as with profiles (3) and (4), the error is <3%. Imposing also condition (e), as with profile (5), does not improve the flat-plate result, although it is important for certain other flows. As shown by profile (4), $F(\eta)$ need not be a polynomial. That a sine would be slightly better than a third-order or fourth-order polynomial could not have been predicted. Overall, it is seen that the integral method can give quite accurate results with little computation.

Boundary-layer separation

On the downstream side of a bluff object there is a tendency for fluid that had been moving along the surface to abruptly depart from it. This phenomenon, called *boundary-layer separation*, creates complex vortices on the back side of the object. That is catastrophic analytically, in that there is now a significant volume in which the flow is not irrotational. Because the size and shape of this new region are not known a priori, it is unclear where the assumption of potential flow might still be valid. Moreover, the boundary-layer approximation is inaccurate beyond the separation point, and in the vortical region downstream there is no comparable simplification to take its place. Nonetheless, solving the boundary-layer equations up to the point at which they fail provides insight into why separation occurs. Understanding the causes and consequences of separation was one of the keys to resolving d'Alembert's paradox.

Example 9.4-2 Integral solution for a cylinder. The objective is to show that boundary-layer separation will occur during high-Re flow past a cylinder. The existence of separation is not assumed at the outset, and the outer velocity is obtained from Example 9.2-1. In addition to setting $\eta = y/\delta(x)$, it is convenient to use a dimensionless x coordinate given by $X = x/R$. Converting Eq. (9.2-29) from cylindrical coordinates using $u(X) = -v_\theta(R, \theta)$ and $X = \pi - \theta$ gives

$$u(X) = 2U \sin X, \quad \frac{du}{dx} = \frac{2U}{R} \cos X. \tag{9.4-13}$$

This shows that the flow accelerates along the upstream surface ($du/dx > 0$ for $0 \le X < \pi/2$) and decelerates along the downstream one ($du/dx < 0$ for $\pi/2 < X \le \pi$). These changes coincide with a pressure gradient that favors forward flow on the upstream side ($d\mathcal{P}/dx < 0$) and opposes it on the downstream side ($d\mathcal{P}/dx > 0$) [Eq. (9.3-3)].

A convenient dependent variable is a dimensionless boundary-layer thickness defined as

$$\Phi(X) = \left(\frac{\delta}{R}\right)^2 \left(\frac{UR}{\nu}\right) = \left(\frac{\delta}{R}\right)^2 \text{Re}_R. \tag{9.4-14}$$

where Re_R is the Reynolds number based on the cylinder radius. Equation (9.3-2) indicates that $\Phi \sim 1$, which is desirable for numerical solutions. From its definition in Table 9.1 and the expression for du/dx in Eq. (9.4-13), the shape factor is related to Φ and X as

$$\Lambda(X) = \frac{\delta^2}{\nu}\frac{du}{dx} = 2\Phi \cos X. \tag{9.4-15}$$

A polynomial that satisfies all five conditions in Table 9.1 is

$$F(X, \eta) = \left(2 + \frac{\Lambda}{6}\right)\eta - \frac{\Lambda}{2}\eta^2 - \left(2 - \frac{\Lambda}{2}\right)\eta^3 + \left(1 - \frac{\Lambda}{6}\right)\eta^4. \tag{9.4-16}$$

With this velocity profile, the left-hand side of Eq. (9.4-6) becomes

$$\frac{\tau_0}{\rho} = \frac{\nu u}{\delta} F'(X, 0) = \left(\frac{\nu U^3}{R}\right)^{1/2} \Phi^{-1/2} \sin X \left(4 + \frac{\Lambda}{3}\right). \tag{9.4-17}$$

Evaluating the integrals in Eq. (9.4-7), the displacement and momentum thicknesses are

$$\frac{\delta_1}{\delta} = \frac{3}{10} - \frac{1}{120}\Lambda \tag{9.4-18}$$

$$\frac{\delta_2}{\delta} = \frac{37}{315} - \frac{1}{945}\Lambda - \frac{1}{9072}\Lambda^2. \tag{9.4-19}$$

Table 9.3 Constants used to calculate the boundary-layer thickness for a cylinder

$a_1 = \dfrac{4}{945} = 0.004233$	$a_5 = \dfrac{2}{567} = 0.003527$
$a_2 = \dfrac{232}{315} = 0.7365$	$a_6 = \dfrac{37}{315} = 0.1175$
$a_3 = \dfrac{1}{567} = 0.001764$	$a_7 = \dfrac{2}{315} = 0.006349$
$a_4 = \dfrac{29}{630} = 0.04603$	$a_8 = \dfrac{5}{2268} = 0.002205$

Substituting these results into Eq. (9.4-6), canceling the common factors, and using Eq. (9.4-15) to relate Λ to Φ gives

$$\frac{d\Phi}{dX} = \frac{2 - a_1\Phi^2 - (a_2 + a_3\Phi^2)\Phi\cos X + a_4\Phi^2\cos^2 X + a_5\Phi^3\cos^3 X}{(a_6 - a_7\Phi\cos X - a_8\Phi^2\cos^2 X)\sin X}. \tag{9.4-20}$$

The values of the constants a_i are listed in Table 9.3.

An initial value for Φ is needed to complete the formulation. Because the cylinder is rounded and $\delta \ll R$, the velocity field for $X \to 0$ will resemble that for planar stagnation flow, where δ is constant [Eq. (9.3-19)]. This implies that $d\Phi/dX = 0$ at $X = 0$. Setting the numerator of Eq. (9.4-20) equal to zero at $X = 0$ yields a cubic,

$$(a_5 - a_3)\Phi_0^3 + (a_4 - a_1)\Phi_0^2 - a_2\Phi_0 + 2 = 0 \tag{9.4-21}$$

where $\Phi_0 = \Phi(0)$. This has one root within a physically realistic range,

$$\Phi_0 = \Phi(0) = 3.5262 \tag{9.4-22}$$

which provides the initial condition that was sought.

Equation (9.4-20) is neither linear nor separable, so a numerical solution is needed.[5] The dimensionless boundary-layer thickness and shear stress so obtained are shown as functions of position in Fig. 9.10. Along the upstream side of the cylinder ($X/\pi < 0.5$) there is a modest increase in boundary-layer thickness (expressed as $\Phi^{1/2}$). On that side the shear stress, expressed as $F'(X, 0)\sin X/\Phi^{1/2}$, first grows and then falls, the increasing u predominating at first and the increasing δ taking over later. On the downstream side the boundary-layer thickness increases much more rapidly. That and the decrease in u cause a continuing decline in the shear stress, which reaches zero at $X/\pi = 0.597$. Beyond that position, which corresponds to an angle of 107° from the leading stagnation point, the shear stress becomes negative and the calculated boundary-layer thickness soon becomes infinite.

The change from positive to negative τ_0 implies that there is flow reversal near the surface, which is a hallmark of boundary-layer separation. Streamlines in the vicinity of a separation point are depicted qualitatively in Fig. 9.11(a). The one marked by * branches away from the surface at the separation point (position 2). To the left of this dividing streamline, including position 1, all flow is forward; to its right, including position 3, there is reverse flow near the surface and forward flow farther away. Velocity profiles at

[5] The fourth-order Runge–Kutta method was used. To avoid the singularity at $X = 0$, the integration was started at $X = 0.01$.

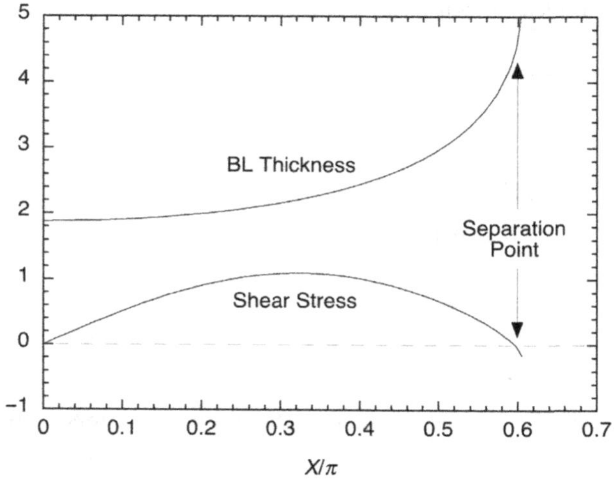

Figure 9.10 Dimensionless boundary-layer thickness and shear stress for a cylinder, based on the solution of Eq. (9.4-20). Plotted are $\Phi^{1/2}$ (BL thickness) and $F'(X,0)\sin X/\Phi^{1/2}$ (shear stress), where $F'(X,0) = 2 + (\Phi \cos X)/3$.

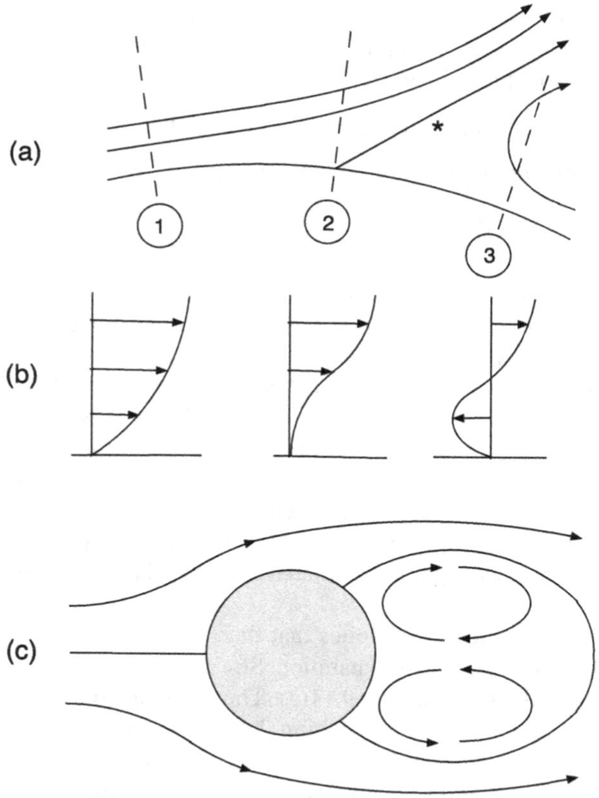

Figure 9.11 Boundary-layer separation on a cylinder: (a) streamlines in the vicinity of the separation point; (b) $v_x(y)$ at locations 1, 2, and 3; and (c) downstream vortices accompanying separation.

the three positions are sketched in Fig. 9.11(b). What characterizes a separation point is that $\partial v_x/\partial y = 0$ at $y = 0$. The overall view in Fig. 9.11(c) shows vortices caused by the separation. Sketched here is an approximation of just one of several vortex patterns that are observed, which depend on Re. As Re is increased, the flow eventually loses the top-to-bottom symmetry shown, and vortices periodically detach from the cylinder and move downstream in a time-dependent manner. Whatever the precise flow pattern in the downstream region, separation disrupts the fore-and-aft symmetry that characterizes potential flow (Fig. 9.2).

Physically, viscous friction tends to slow a fluid element moving along a surface, whereas inertia tends to sustain its motion. If there is a sufficiently rapid pressure decrease in the direction of flow, the balance of forces is such that the fluid element can continue on a path near the surface. However, once the pressure begins to increase in the flow direction, the inertia (or kinetic energy) of the fluid element is soon depleted and the fluid is pushed away from the surface by the increasing pressure.

At Reynolds numbers near the upper end of the laminar range ($1 \times 10^4 < \text{Re} < 1 \times 10^5$), where laminar boundary-layer theory should be most accurate, separation actually occurs at an angle of 81° (Churchill, 1988, p. 317), in contrast to the 107° predicted from Eq. (9.4-20). This large discrepancy is not due to the limitations of the integral method; a numerical solution of Eq. (9.3-4) that uses Eq. (9.4-13) for the outer velocity and pressure gives nearly the same result, 104.5° (Schlichting, 1968, p. 203). Rather, the problem is that the downstream vortices distort the entire streamline pattern, invalidating Eq. (9.4-13) even upstream from the separation point. Figure 9.3 compares pressures at the surface predicted from potential flow with those measured for $\text{Re} = 1.86 \times 10^5$. It is seen that the actual decline in $\mathcal{P}$ on the upstream side of the cylinder is more gradual than for potential flow, and that $\mathcal{P}$ remains nearly constant on the downstream side, rather than rising back to its initial value. This asymmetry of the pressure field is what is responsible for nearly all of the drag. Interestingly, if a measured pressure profile is used as an input to the boundary-layer equations, the separation point is predicted quite accurately (Schlichting, 1968, p. 202). Theoretical and experimental results for cylinders are reviewed in much more detail in Schlichting (1968) and Churchill (1988).

In general, the relative importance of form drag and friction drag at high Re may be understood as follows. Combining the pressure and viscous stress scales for laminar boundary layers with the projected areas defined in Section 3.2,

$$F_D = F_D(\text{form}) + F_D(\text{friction}) = C_1 A_\perp \rho U^2 + C_2 A_\parallel \frac{\mu U}{\delta} \tag{9.4-23}$$

where C_1 and C_2 are dimensionless constants that are ~ 1. For a cylinder oriented crosswise or for a sphere, $A_\perp = A_\parallel$. If $A_\perp \sim A_\parallel$, as is true also for many other bluff objects, then

$$\frac{F_D(\text{friction})}{F_D(\text{form})} \sim \frac{\mu U/\delta}{\rho U^2} = \left(\frac{\mu}{\rho U L}\right)\left(\frac{L}{\delta}\right) \sim \text{Re}^{-1/2} \tag{9.4-24}$$

where L is the characteristic dimension of the object. Thus, friction drag becomes negligible as $\text{Re} \to \infty$, as had always been supposed. At the root of d'Alembert's paradox was the failure to perceive the indirect way in which viscosity is responsible for form drag. That is, viscosity gives rise to boundary layers, and boundary-layer separation on bluff objects creates the pressure imbalance that underlies form drag.

9.5 CONCLUSION

Insight into flows where Re is large and inertia is prominent is provided by the inviscid idealization, in which $\mu = 0$ and Re $= \infty$. In inviscid flow there is a tradeoff between velocity and pressure along a given streamline, one increasing as the other decreases, as described by Bernoulli's equation [Eq. (9.2-5)]. Often associated with inviscid flow is irrotational flow (also called potential flow), in which vorticity is absent ($\mathbf{w} = \nabla \times \mathbf{v} = \mathbf{0}$). This association stems from the fact that, in an inviscid fluid, an irrotational velocity field remains irrotational. Potential flow is physically interesting, because any irrotational velocity field satisfies the Navier–Stokes equation, and mathematically attractive, because the existence of a velocity potential facilitates solving for $\mathbf{v}$. In such flows a stronger form of Bernoulli's equation applies [Eq. (9.2-14)]. Among its uses, the irrotational Bernoulli equation provides a shortcut for calculating $\mathcal{P}$ (or P) from $\mathbf{v}$.

Potential-flow theory has numerous uses, but also a major failing. Namely, it predicts an absence of drag during high-speed flow past any object. This longstanding problem, called d'Alembert's paradox, was resolved by realizing that the no-slip boundary condition holds at solid surfaces, even when velocities are large. The absence of slip creates a thin layer of fluid next to the surface, a boundary layer, in which viscous and inertial effects are comparable even for arbitrarily large Re. This balance between inertia and viscosity is reflected in the boundary-layer momentum equation [Eq. (9.3-4)]. As Re is increased the boundary-layer thickness decreases, varying as $\mathrm{Re}^{-1/2}$, but the effects of viscosity do not become negligible. Coexisting with any boundary layer is at least one other region, of relatively large size, in which the flow is nearly inviscid. This makes it necessary to consider at least two dynamically distinct regions with very different length scales. Friction drag on bluff objects becomes negligible at large Re, as had always been supposed. However, viscosity gives rise to boundary layers, and boundary-layer separation creates the pressure imbalance that underlies form drag. Thus, viscosity is indirectly responsible for all of the drag.

REFERENCES

Blasius, H. Grenzschichten in Flüssigkeiten mit kleiner Reibung. *Z. Math. Phys.* 56: 1–37, 1908. [Translated into English as NACA Technical Memorandum No. 1256, 1950.]

Churchill, S. W. *Viscous Flows*. Butterworth, Boston, MA, 1988.

Clift, R., J. R. Grace, and M. E. Weber. *Bubbles, Drops, and Particles*. Academic Press, New York, 1978.

Davies, R. M. and G. I. Taylor. The mechanics of large bubbles rising through extended liquids and through liquids in tubes. *Proc. Roy. Soc. Lond. A* 200: 375–390, 1950.

Deen, W. M. *Analysis of Transport Phenomena*, 2nd ed. Oxford University Press, New York, 2012.

Hartree, D. R. On an equation occurring in Falkner and Skan's approximate treatment of the equations of the boundary layer. *Proc. Cambr. Philos. Soc.* 33: 223–239, 1937.

Jones, C. W. and E. J. Watson. Two-dimensional boundary layers. In *Laminar Boundary Layers*, L. Rosenhead (Ed.), Clarendon Press, Oxford, 1963, pp. 198–257.

Lamb, H. *Hydrodynamics*, 6th ed. Dover Publications, New York, 1945.

Nikuradse, I. *Laminare Reibungsschichten und der längs angeströmten Platte*. Zentrale für wissenschaftliches Berichtswesen, Berlin, 1942.

Prandtl, L. Über Flüssigkeitsbewegung bei sehr kleiner Reibung. *Proceedings of the 3rd International Mathematical Congress*, Heidelberg, 1904, pp. 484–491. [Reprinted in *Vier Abhandlungen zur Hydrodynamik und Aerodynamik*, Göttingen, 1927, translated into English as NACA Technical Memorandum No. 452, 1928.]

Rogers, M. H. and G. N. Lance. The rotationally symmetric flow of a viscous fluid in the presence of an infinite rotating disk. *J. Fluid Mech.* 7: 617–631, 1960.
Schlichting, H. *Boundary-Layer Theory*, 6th ed. McGraw-Hill, New York, 1968.

PROBLEMS

9.1. Potential flow past a sphere Consider irrotational flow at velocity U past a stationary sphere of radius R, with coordinates as in Fig. 8.7.

(a) As in Example 9.2-1, the velocity potential is a separable function. Show that

$$\phi(r,\theta) = U\cos\theta\left(r + \frac{1}{2}\frac{R^3}{r^2}\right) \tag{P9.1-1}$$

and find $v_r(r,\theta)$ and $v_\theta(r,\theta)$.

(b) Show that the pressure at the sphere surface is

$$\mathcal{P}(R,\theta) = \frac{\rho U^2}{2}\left(1 - \frac{9}{4}\sin^2\theta\right). \tag{P9.1-2}$$

(c) Confirm that $F_D = 0$ if the fluid is inviscid.

9.2. Lift on a half-cylinder* Consider irrotational flow at velocity U past a half-cylinder of radius R and length L, as in Fig. P9.2. A numerical solution for the velocity potential would be needed to obtain precise results for the velocity and pressure fields. However, as a first approximation, suppose that the flow for $y > 0$ is like that past a complete cylinder and the flow for $y < 0$ is like that past a flat plate. Calculate the lift force, F_L.

* This problem was suggested by R. G. Larson.

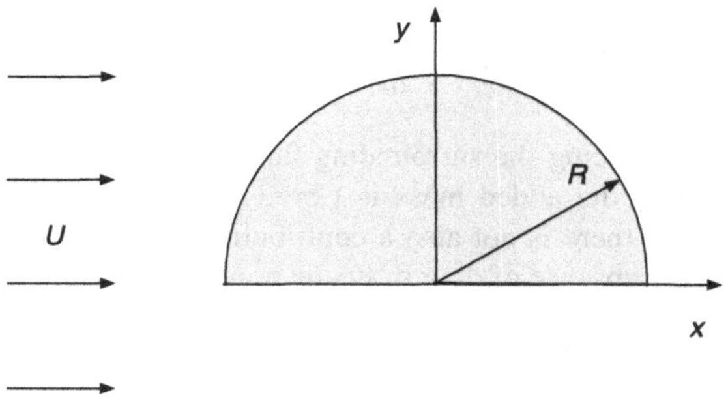

Figure P9.2 Potential flow past a half-cylinder.

9.3. Axisymmetric stagnation flow An *axisymmetric stagnation flow* results from directing a circular stream perpendicular to a planar surface, which deflects the fluid outward radially. Suppose that the surface is at $z = 0$, and the approaching flow is in the $-z$ direction.

(a) Show that the velocity field

$$v_r(r,z) = Cr, \quad v_z(r,z) = -2Cz \tag{P9.3-1}$$

where C is any positive constant and $v_\theta = 0$, is irrotational.

(b) Determine $\mathcal{P}(r, z)$.
(c) Confirm that this velocity and pressure satisfy the full Navier–Stokes equation.

9.4. Opposed circular jets Circular jets directed at one another have been used to study gas-phase combustion. Typically, one jet contains a fuel such as methane and the other contains oxygen. Upon ignition, a stationary flame front is created where the jets meet. This arrangement is advantageous because opposed jets are straightforward to set up and the velocity field, which is needed to analyze the mass and heat transfer, is relatively simple. Suppose that jets coming from the $+z$ and $-z$ directions have identical flow rates and properties and that the resulting symmetry plane is at $z = 0$.

Show that the irrotational velocity field in Eq. (P9.3-1) applies throughout each jet. In particular, explain why there is no boundary layer at $z = 0$, even though the velocities might be large.

9.5. Added mass for a sphere An object moving through a static fluid sets part of the fluid in motion, and if it accelerates, the neighboring fluid must accelerate as well. The acceleration of the fluid tends to slow the acceleration of the object. As discussed in Example 3.3-3, this may be accounted for by assigning to the object an effective mass that exceeds its actual mass. The increment, or *added mass*, is calculated most easily for potential flow, and the objective is to evaluate it for a solid sphere of radius R. Assume that the sphere moves at a velocity $U(t)$ in the z direction and that the inviscid fluid of density ρ is otherwise at rest.

(a) Supposing that $U(t)$ is known, find the velocity potential $\phi(r, \theta, t)$. (*Hint*: The general solution for ϕ is the same as in Problem 9.1, but now the boundary conditions in r must correspond to a sphere that is moving and distant fluid that is stationary.)
(b) Evaluate $\mathcal{P}(R, \theta, t)$ by applying Eq. (9.2-14) at the sphere surface.
(c) Show that the z component of the fluid-dynamic force on the sphere is

$$F_z(t) = -\frac{2}{3}\pi R^3 \rho \frac{dU}{dt}. \qquad \text{(P9.5-1)}$$

The negative sign confirms that displacing the surrounding fluid slows the sphere, and the coefficient of dU/dt shows that the added mass is $(2\pi/3)R^3\rho$, or half the mass of the displaced fluid. The fact that there is not also a contribution to F_z that involves U (as opposed to dU/dt) reflects the absence of drag in steady potential flow.

9.6. Spin coating A uniform coating of liquid on a flat surface can be created by rapidly spinning the solid substrate after some liquid is applied. It is found that such films level quickly and then gradually become thinner. The thinning phase of the process is shown in Fig. P9.6. The film thickness at a given instant is uniform at $h(t)$. Spinning the horizontal substrate at angular velocity ω causes the liquid to flow outward, slowly decreasing h. It is desired to predict $h(t)$.

(a) Assume that $v_\theta(r) = \omega r$ and that the r component of the Navier–Stokes equation can be simplified to

$$\frac{\partial^2 v_r}{\partial z^2} = -\frac{\omega^2 r}{\nu}. \qquad \text{(P9.6-1)}$$

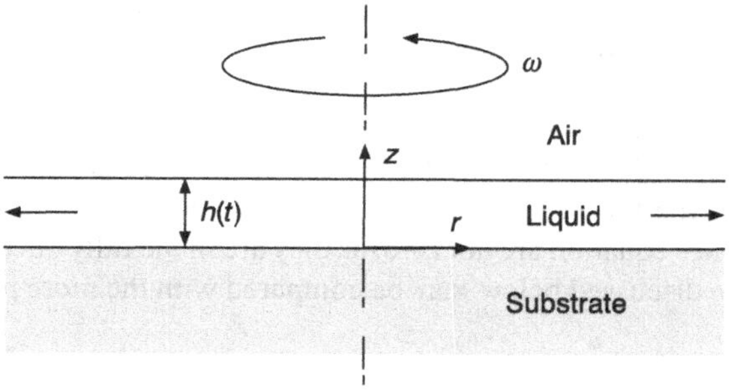

Figure P9.6 Spin coating. Rotation of the substrate gradually decreases the thickness of the liquid film, which remains spatially uniform.

That is, suppose that the thinning is slow enough to be pseudosteady and that h is small enough to neglect the other viscous and inertial terms, as in a lubrication problem with $\text{Re} = \omega h^2/\nu \ll 1$. However, inertia is crucial here, in that the centrifugal force $(\rho v_\theta{}^2/r)$ is what causes the radial flow. Find $v_r(r, z, t)$.

(b) Evaluate the radial flow rate per unit of circumference,

$$q(r,t) = \int_0^{h(t)} v_r \, dz. \tag{P9.6-2}$$

(c) Relate dh/dt to q. (*Hint*: Use a mass balance on a control volume that includes the liquid between $r = 0$ and an arbitrary radial position.)

(d) Solve for $h(t)$, assuming that $h(0) = h_0$.

9.7. Bubble growing in a liquid Suppose that a tube is immersed in a liquid and that air flow through the tube creates a bubble at its end, as shown in Fig. P9.7. Beginning at $t = 0$, when the bubble radius is R_0, the volume flow rate of air is held constant at Q. This is done in a gravity-free environment. Thus, the bubble remains spherical and static pressure variations are absent, even though the bubble is not necessarily small. Assume that the motion in the liquid is purely radial.

Figure P9.7 Air bubble growing at the end of a tube.

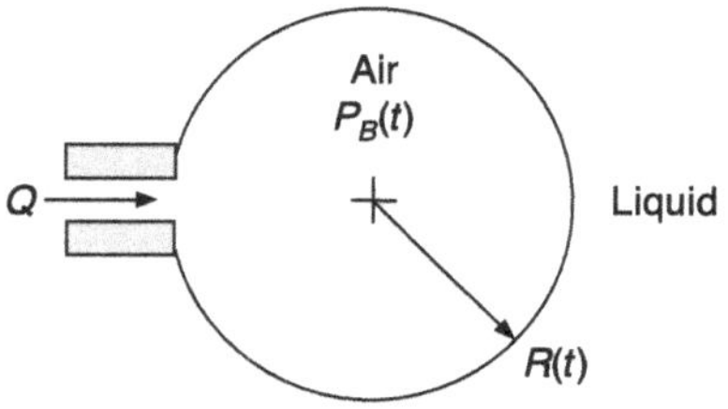

(a) Determine the radius $R(t)$ of the growing bubble, assuming that R_0 and Q are given.

(b) Find v_r in the liquid for all $r \geq R(t)$.

(c) Determine P in the liquid for all $r \geq R(t)$, assuming that $P = P_\infty$ far from the bubble. (*Hint*: Purely radial flow is irrotational.)

(d) Derive a general expression for the bubble pressure $P_B(t)$, including the effects of surface tension, viscosity, and inertia.

9.8. Entrance length In any long pipe or other conduit of constant cross-section the velocity profile eventually becomes fully developed, as in Chapter 7. The distance from the inlet at which this occurs is the *entrance length*, L_E. Within the entrance region part of the fluid accelerates and part decelerates, depending on the initial velocity profile. If there is plug flow at the inlet, as in Fig. P9.8, the fluid at the center speeds up and that near the wall slows down, until the final parabolic profile is achieved. Accordingly, the inertial terms in the Navier–Stokes equation are not zero, as they are in the fully developed region. The OM estimates of L_E discussed below may be compared with the more precise Eq. (12.2-1).

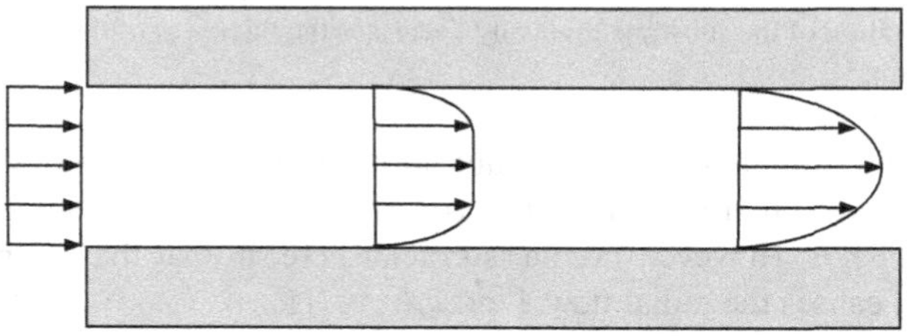

Figure P9.8 Development of the velocity profile in the entrance region of a tube.

(a) At large Re, what distinguishes the entrance region is that inertia is as important there as viscosity. Assuming that Re $\gg$ 1, obtain an OM estimate of L_E for a tube of radius R with a mean velocity U. The flow in such an entrance region is nearly unidirectional, as in boundary-layer or lubrication flows. Why?
(b) At small Re, inertia will be negligible everywhere. What then distinguishes the entrance region from the fully developed one? Obtain an OM estimate of L_E for creeping flow.

9.9. Axisymmetric jet Analogous to the planar jet in Example 9.3-2 is one that emerges from a circular opening. The boundary-layer momentum equation for an axisymmetric flow is

$$v_r \frac{\partial v_z}{\partial r} + v_z \frac{\partial v_z}{\partial z} = \frac{\nu}{r} \frac{\partial}{\partial r} \left(r \frac{\partial v_z}{\partial r} \right) \qquad \text{(P9.9-1)}$$

and the kinematic momentum is defined now as

$$K = 2\pi \int_0^\infty v_z^2 r \, dr. \qquad \text{(P9.9-2)}$$

(a) Show that K is constant. (*Hint*: Integrate the momentum equation from $r = 0$ to $r = \infty$ and use integration by parts to evaluate the contribution from $v_r \partial v_z / \partial r$.)
(b) Assuming that K is specified, obtain OM estimates for the jet radius $\delta(z)$, the maximum velocity $v_m(z)$, and the volume flow rate $Q(z)$. For comparison, the exact result for the flow rate is $Q = 8\pi\nu z$ (Schlichting, 1968, p. 221). Notice that, although K is the measure of jet strength, it does not affect Q.

9.10. Boundary layers in power-law fluids It is desired to determine the general features of boundary layers in fluids that obey Eq. (6.5-17). The velocity scale is U and the object dimension is L.

(a) Show that if $\delta/L \ll 1$, the momentum equation for a two-dimensional boundary layer is

$$v_x \frac{\partial v_x}{\partial x} + v_y \frac{\partial v_x}{\partial y} = u \frac{du}{dx} + \frac{m}{\rho} \frac{\partial}{\partial y} \left(\frac{\partial v_x}{\partial y} \right)^n. \quad \text{(P9.10-1)}$$

Explain why the other viscous terms are negligible.

(b) What dimensionless parameter (a generalization of the Reynolds number) must be large if δ/L is to be small? Use OM estimates to deduce the dependence of both δ/L and τ_0 on n and on the generalized Re.

9.11. Normal velocity component for a flat plate The fluid moving along a flat plate gradually slows, as shown in Example 9.3-1. That deceleration creates a positive v_y at the outer edge of the boundary layer. It is desired to compare the average value of that normal velocity (V) with the approach or free-stream velocity (U). The control volume that is suggested is shown by the dashed rectangle in Fig. P9.11, in which $h > \delta(x)$.

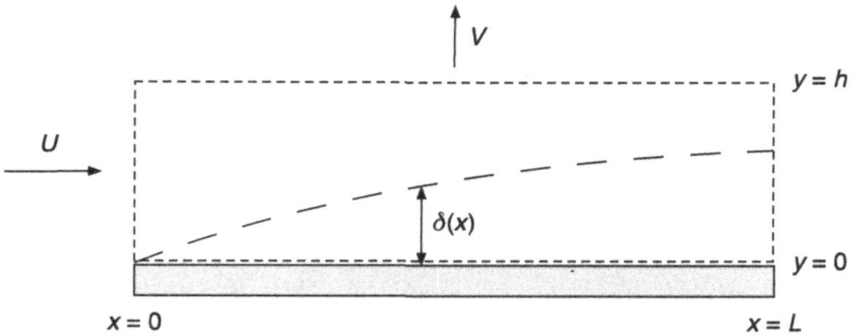

Figure P9.11 Control volume for calculating V, the average velocity normal to a flat plate of length L with free-stream velocity U. The height h exceeds the boundary layer thickness $\delta(x)$.

It is found from the similarity solution that (Jones and Watson, 1963, p. 224)

$$\int_0^\infty (1 - f')\, d\eta = 1.21678. \quad \text{(P9.11-1)}$$

Use this to evaluate V/U and plot the result as a function of Re for $10^2 \leq \text{Re} \leq 10^6$. Your results should confirm that V/U is quite small when Re is large enough for there to be a boundary layer.

9.12. Rotating disk *Rotating-disk electrodes* are widely used to study electrochemical kinetics. As depicted in Fig. P9.12, a metal (e.g., platinum) electrode is embedded in the end of an insulating rod of radius R, which is immersed in an electrolyte solution. Rotating the rod at an angular velocity ω causes the nearby liquid to rotate, while also drawing it toward the end of the rod and throwing it outward. Thus, all three velocity components (v_r, v_θ, v_z) are nonzero. Because the flow helps deliver the reacting ions to the surface, a detailed knowledge of $\mathbf{v}$ is needed to interpret measured currents.

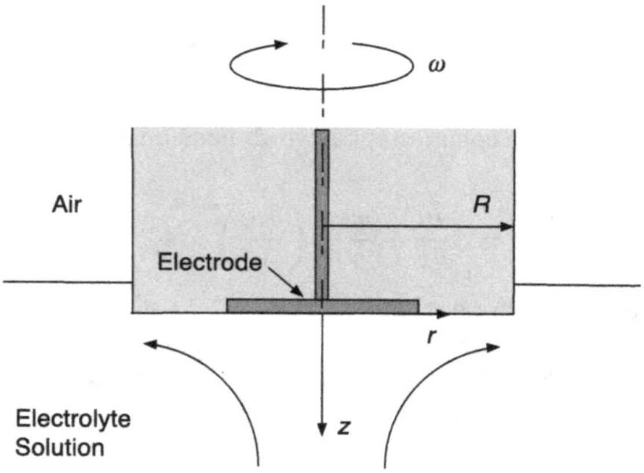

Figure P9.12 Rotating-disk electrode.

(a) In such systems the rod radius is large enough that the flow near the electrode is insensitive to R. If R is disregarded, show that the only way to make z dimensionless is

$$\zeta = z\sqrt{\omega/\nu}. \qquad \text{(P9.12-1)}$$

(b) Explain why the velocity might be assumed to be of the form

$$v_r = r\omega F(\zeta), \quad v_\theta = r\omega G(\zeta), \quad v_z = \sqrt{\nu\omega}H(\zeta). \qquad \text{(P9.12-2)}$$

In particular, show that the continuity equation becomes

$$H' + 2F = 0 \qquad \text{(P9.12-3)}$$

where the prime indicates differentiation with respect to ζ. Explain also why reasonable boundary conditions for the three unknown functions are

$$F(0) = 0, \quad G(0) = 1, \quad H(0) = 0, \quad F(\infty) = 0, \quad G(\infty) = 0. \qquad \text{(P9.12-4)}$$

(c) Show that, if it is assumed also that the pressure is of the form

$$\mathcal{P} = \mu\omega P(\zeta), \quad P(0) = 0 \qquad \text{(P9.12-5)}$$

the Navier–Stokes equation becomes

$$F'' - F^2 + G^2 - HF' = 0 \qquad \text{(P9.12-6)}$$

$$G'' - 2FG - HG' = 0 \qquad \text{(P9.12-7)}$$

$$H'' - HH' - P' = 0. \qquad \text{(P9.12-8)}$$

Thus, the continuity equation and three-dimensional Navier–Stokes equation are reduced to a set of four ordinary differential equations. These coupled, nonlinear equations must be solved numerically.

(d) Although it does not influence the velocity near the electrode, R affects the torque that must be applied. Given that $F'(0) = 0.5102$ and $G'(0) = -0.6159$ (Rogers and Lance, 1960), calculate the torque. You may assume that the immersion depth is small enough to enable you to neglect the shear stress on the side of the rod.

9.13. Flat plate with suction Applying suction on at least part of the surface of an object has been used to reduce drag. This can delay or prevent flow separation and also reduce the wall shear stress. This problem focuses on the latter effect. Suppose that suction on a porous flat plate creates a velocity normal to its surface that is given by $v_y(x, 0) = -v_0$, where v_0 is a positive constant. The no-slip condition continues to apply. It may be assumed that v_0/U is small enough that the boundary-layer approximation is still valid and that the outer velocity (U) remains nearly constant.

(a) The flow in the boundary layer eventually becomes fully developed. Determine $v_x(y)$ and the wall shear stress in that region. (You should find that τ_0 is independent of μ there!)
(b) If the flow over nearly the entire length L of a plate of width W were like that in part (a), how would the drag compare with that in the absence of suction?

9.14. Terminal velocity of a large bubble Bubbles at large Re are typically shaped as spherical caps, as shown in Fig. P9.14. The trailing surface is more irregular than what is depicted, but the leading surface is almost a perfect spherical section. It is desired to predict the terminal velocity U for a bubble of density ρ_o and cap radius R that is rising in a liquid of density ρ. There are boundary layers on both the liquid and gas sides of the spherical surface. Surface tension is negligible in such large bubbles.

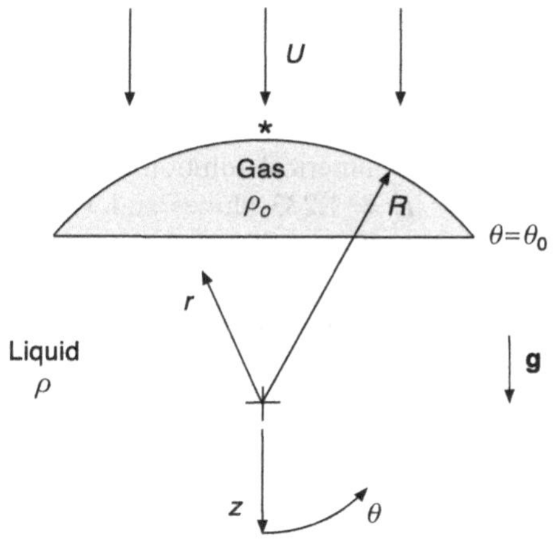

Figure P9.14 Spherical-cap bubble rising in a liquid. The coordinates are such that the bubble is stationary and the approach velocity is U. The "nose" of the bubble is marked by * and its base is at an angle θ_0 from the z axis.

(a) Assume that on the liquid side of the top surface $\mathcal{P}$ is that for potential flow past a complete sphere, as given in Eq. (P9.1-2). Convert $\mathcal{P}$ to P to show that

$$P(R, \theta) = \rho g R(1 + \cos\theta) - \frac{9}{8}\rho U^2 \sin^2\theta + P^* \tag{P9.14-1}$$

where P^* is the pressure at the "nose" ($\theta = \pi$).
(b) Even with a large tangential velocity at the top surface, most of the gas will circulate slowly. Thus, for the boundary layer on the gas side, the pressure will be nearly that in a static gas. Show that, in the gas,

$$P(R, \theta) = \rho_o g R(1 + \cos\theta) + P^*. \tag{P9.14-2}$$

(c) If the surface tension and normal viscous stresses are both negligible, in principle we could simply equate the pressures from parts (a) and (b) and solve for U. However, show that this gives

$$U^2 = \frac{8}{9}\frac{\Delta\rho}{\rho}gR\frac{(1+\cos\theta)}{\sin^2\theta} \tag{P9.14-3}$$

which cannot be satisfied for all θ on the top surface.

(d) Less ambitious is to expect the approximate pressures to match only near the nose. By expanding the trigonometric functions about $\theta = \pi$, show that this leads to

$$U = \frac{2}{3}\left(\frac{gR\,\Delta\rho}{\rho}\right)^{1/2} \tag{P9.14-4}$$

as in Davies and Taylor (1950).

(e) The Davies–Taylor equation is remarkably accurate for Re > 40 (Clift *et al.*, 1978, p. 206), where Re is based on the liquid properties and the diameter of a sphere of equivalent volume (D_E). For Re > 150, $\theta_0 \cong 130^\circ$ and $D_E/R = 0.77$. Calculate U for an air bubble in water with $R = 1.0$ cm.

9.15. Planar stagnation flow

(a) Using the integral method with profile (3) of Table 9.2, determine the boundary-layer thickness for planar stagnation flow, where $u(x) = ax$.

(b) With the wall shear stress written as

$$\frac{\tau_0(x)}{\rho u^2(x)} = K\,\mathrm{Re}_x^{-1/2} \tag{9.15-1}$$

a numerical solution of the Falkner–Skan equation for planar stagnation flow gives $K = 1.233$ (Jones and Watson, 1963, p. 232). How does your approximate value compare with this?

9.16. Flow past a right-angle wedge

(a) Using the integral method with profile (3) of Table 9.2, determine the boundary-layer thickness for flow past a 90° wedge, where $u(x) = ax^{1/3}$.

(b) With the wall shear stress written as

$$\frac{\tau_0(x)}{\rho u^2(x)} = K\,\mathrm{Re}_x^{-1/2} \tag{9.16-1}$$

a numerical solution of the Falkner–Skan equation for a right-angle wedge gives $K = 0.758$ (Jones and Watson, 1963, p. 237). How does your approximate value compare with this?

10

Turbulent flow

10.1 INTRODUCTION

As discussed in Chapters 2 and 3, laminar flows tend to be unstable at large Reynolds number. A transition to turbulence occurs when Re exceeds a critical value, which depends on the system but is often in the thousands to hundreds of thousands. Turbulence is ubiquitous in everyday life, the natural environment, and large-scale, continuous processes. One of its hallmarks is a marked increase in wall shear stresses. That is indicated by much higher friction factors for flow in pipes (Fig. 2.2) or past flat plates (Fig. 3.5), and is a reflection of the more rapid cross-stream transfer of momentum in turbulent than in laminar flow. In laminar flow, momentum transfer across streamlines is due only to the molecular-level friction that underlies viscosity. In turbulent flow, transient eddies or vortices of varying size are superimposed on the average motion. The eddies, which vary randomly from instant to instant, transport momentum at rates that greatly exceed those from the molecular mechanism.

The random velocity fluctuations associated with the eddies ensure that no two turbulent flows are ever exactly the same. Deterministic modeling of flow details must be abandoned and statistical descriptions used instead. In engineering, the momentary fluctuations are of secondary interest and it is desired mainly to predict time-smoothed or time-averaged quantities. Even so, all available tools must be brought to bear: dimensional analysis, experimentation, analytical theory, and computation. With the present state of knowledge, experimental findings are particularly crucial. Whereas the details of a laminar flow often can be predicted from first principles, turbulence calculations nearly always contain an element of empiricism.

This chapter is organized as follows. After several aspects of turbulence are described qualitatively, there is a discussion of insights into velocity and length scales that can be gained by combining key observations with dimensional analysis. How to reformulate the continuity and Navier–Stokes equations in terms of time-smoothed quantities, a procedure called *Reynolds averaging*, is then described. The concept of an *eddy diffusivity* is introduced, which is the classical approach for making turbulence calculations practical. Several examples then illustrate the application of this concept to shear flows in conduits and boundary layers.

10.2 CHARACTERISTICS AND SCALES

Basic features

The chaotic nature of turbulent flow promotes mixing. Evidence of a transition from one flow regime to another in tube flow is provided by dye visualization experiments, as in

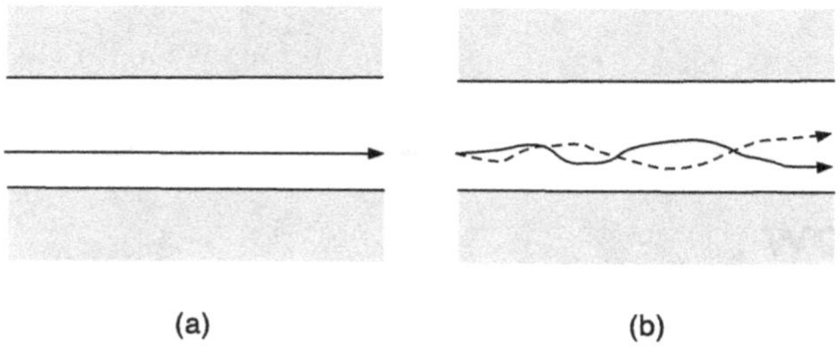

Figure 10.1 Paths of material points in a tube for (a) laminar flow and (b) turbulent flow. In both cases the flow is fully developed and steady on average. The two curves in (b) correspond to fluid elements starting at the same position at different times.

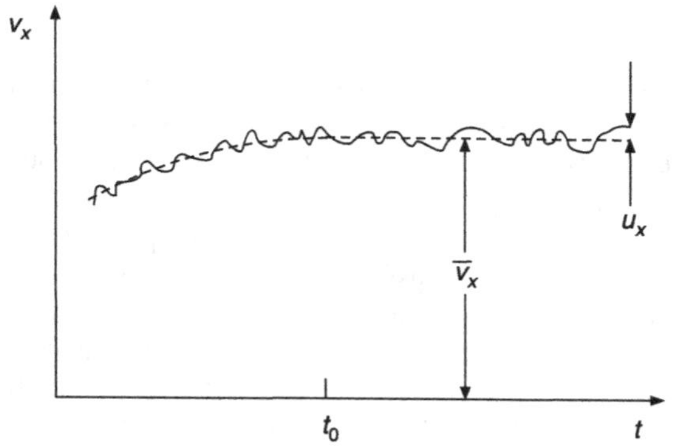

Figure 10.2 Velocity fluctuations at a fixed position in a turbulent flow. The instantaneous velocity is v_x (solid curve), its time-smoothed value is $\bar{v}_x$ (dashed curve), and the fluctuation is u_x.

the pioneering work of Reynolds (1883). In laminar flow, each material point follows a straight line, as in Fig. 10.1(a), and this can be seen by continuously injecting a slowly diffusing dye. In turbulent flow, the paths of fluid elements waver and vary over time. Figure 10.1(b) shows hypothetical streaklines for material points starting at the same position at different instants. Rapid dispersion of dye over the tube cross-section, due to the turbulent eddies, makes it difficult to actually see such streaklines.

All turbulent flows are three-dimensional, time-dependent, and irregular. Random fluctuations occur in all three velocity components and in the pressure. The fluctuations in the main velocity component are often several percent of its mean value. The instantaneous velocity v_x at a point within a hypothetical turbulent flow is shown by the solid curve in Fig. 10.2, and its time-smoothed value $\bar{v}_x$ is shown by the dashed curve. The difference between the two is the *velocity fluctuation* u_x. That is,

$$v_x = \bar{v}_x + u_x. \tag{10.2-1}$$

As detailed in Section 10.3, $\bar{v}_x$ is calculated by continually averaging v_x over a time interval that encompasses multiple fluctuations, but that is much smaller than the process time scale. Thus, $\bar{v}_x$ can vary with time, as shown in Fig. 10.2 for $t < t_0$. In this example the flow is steady overall for $t > t_0$, although still unsteady when examined on short time scales. The other velocity components also can be decomposed into a smoothed value plus a fluctuation. Even when $\bar{v}_y = 0 = \bar{v}_z$, as in a flow paralleling the x axis, $u_y \neq 0$ and $u_z \neq 0$.

The time-smoothed fluctuations are zero, by definition; that is, $\bar{u}_i = 0$ ($i = x, y$, or z). Accordingly, another measure of the size of the fluctuations is needed. What is usually

Table 10.1 Friction velocity and maximum wall coordinate for tube flow

Re	u_τ/U	R^+
1×10^4	0.062	310
1×10^5	0.047	2400
1×10^6	0.038	19,000
1×10^7	0.032	160,000

adopted is the root-mean-square (rms) value, $(\overline{u_i^2})^{1/2}$. The fluctuation (positive or negative) is squared, the result is smoothed, and then the square root is calculated. The rms fluctuations of all three velocity components tend to be similar, even when the values of $\bar{v}_i$ are not.

Wall variables

A distinctive aspect of turbulent flow is that the velocity profile near a solid surface has an almost universal form, as discussed in Section 10.5. It is affected little by conduit shape or even whether the flow is in a conduit or past an object such as a flat plate. The success of the hydraulic-diameter approach in extending results for cylindrical tubes to noncircular cross-sections (Section 2.4) is a consequence of this universality. This suggests that turbulent shear flows near walls are influenced mainly by local events. That is expressed by postulating that $\bar{v}_x$ near walls depends only on the distance from the surface (y), the fluid properties (μ and ρ), and the wall shear stress (τ_0).

The foregoing implies that there are local scales for $\bar{v}_x$ and y that each depend only on μ, ρ, and τ_0. The dimensional requirements are

$$[\bar{v}_x] = \text{LT}^{-1} = [\mu]^a[\rho]^b[\tau_0]^c = (\text{ML}^{-1}\text{T}^{-1})^a(\text{ML}^{-3})^b(\text{ML}^{-1}\text{T}^{-2})^c \quad (10.2\text{-}2)$$

$$[y] = \text{L} = [\mu]^A[\rho]^B[\tau_0]^C = (\text{ML}^{-1}\text{T}^{-1})^A(\text{ML}^{-3})^B(\text{ML}^{-1}\text{T}^{-2})^C. \quad (10.2\text{-}3)$$

Solving for the exponents gives $a = 0$, $b = -1/2$, $c = 1/2$, $A = 1$, $B = -1/2$, and $C = -1/2$. The velocity scale so obtained,

$$u_\tau = (\tau_0/\rho)^{1/2} \quad (10.2\text{-}4)$$

is called the *friction velocity*. The dimensionless velocity and length that result are

$$v_x^+ = \frac{\bar{v}_x}{u_\tau}, \quad y^+ = \frac{u_\tau y}{\nu} \quad (10.2\text{-}5)$$

which are the *wall variables*. Numerous studies have confirmed that, for turbulent flow near a solid boundary, v_x^+ depends only on y^+ (Section 10.5). Of special interest is that u_τ turns out to be the scale for the rms velocity fluctuations (e.g., McComb, 1990, p. 27).

For flow at mean velocity U in a tube of radius R, rearranging the definition of the friction factor [Eq. (2.2-4)] gives

$$\frac{u_\tau}{U} = \left(\frac{f}{2}\right)^{1/2}, \quad R^+ = \frac{u_\tau R}{\nu} = \text{Re}\left(\frac{f}{8}\right)^{1/2} \quad (10.2\text{-}6)$$

where R^+ is y^+ evaluated at the tube centerline. That is, $0 \leq y^+ \leq R^+$. Representative values of u_τ/U and R^+ are shown in Table 10.1. In that the rms fluctuations in the three velocity components are each comparable to u_τ, these results indicate that the fluctuations are several percent of U. Note also that, because R^+ is typically quite large, a value of y^+ in the hundreds or even thousands actually can correspond to a position very near the wall.

Kolmogorov scales

Turbulence is a high-energy state. The energy that creates the eddies is provided by the additional power input needed to sustain a turbulent flow. In a pipe, for example, the increment in $|\Delta\mathcal{P}|$ above that for laminar flow implies that more pressure–volume work is done on the fluid, which in turn requires more pumping power. The additional power is dissipated ultimately as heat, as is the pumping power in a fully developed laminar flow. The *additional* rate of energy dissipation in a turbulent flow, per unit mass, will be denoted as d. It has the dimension of energy per unit time, per unit mass, or $\mathrm{L^2T^{-3}}$.

The eddies in a turbulent flow have a wide range of sizes and lifetimes. Ones of varying size coexist at each location. They interact, in that kinetic energy is transferred from larger to successively smaller eddies (McComb, 1990; Tennekes and Lumley, 1972). Early in this cascade the eddies have Reynolds numbers large enough to make the effects of viscosity negligible. Only in the smallest eddies is viscous friction important, along with the consequent conversion of kinetic energy to heat.[1] The size of the largest eddies is limited only by the dimensions of the system. As suggested by an analysis done by A. N. Kolmogorov in Moscow in the 1940s, the size of the smallest eddies is determined by d.

The velocity, length, and time scales for the largest eddies will be denoted as u_1, L_1, and t_1, respectively, and those for the smallest eddies will be denoted as u_2, L_2, and t_2. The small-eddy scales can be inferred from dimensional analysis. The only assumption needed is that the characteristics of the smallest eddies are determined entirely by the additional dissipation d and the fluid properties. That is, the size and shape of the system are unimportant at this level. Because none of the other quantities contain the dimension M, μ and ρ must be combined as ν. The dimensional requirements are then

$$[u_2] = \mathrm{LT^{-1}} = [d]^a[\nu]^b = (\mathrm{L^2T^{-3}})^a(\mathrm{L^2T^{-1}})^b \qquad (10.2\text{-}7)$$

$$[L_2] = \mathrm{L} = [d]^A[\nu]^B = (\mathrm{L^2T^{-3}})^A(\mathrm{L^2T^{-1}})^B \qquad (10.2\text{-}8)$$

$$[t_2] = \mathrm{T} = [d]^\alpha[\nu]^\beta = (\mathrm{L^2T^{-3}})^\alpha(\mathrm{L^2T^{-1}})^\beta. \qquad (10.2\text{-}9)$$

Solving for the three pairs of exponents gives

$$u_2 = (\nu d)^{1/4}, \quad L_2 = \left(\frac{\nu^3}{d}\right)^{1/4}, \quad t_2 = \left(\frac{\nu}{d}\right)^{1/2}. \qquad (10.2\text{-}10)$$

These are called the *Kolmogorov scales* (or *microscales*). The corresponding Reynolds number is $\mathrm{Re}_2 = u_2L_2/\nu = 1$. This is consistent with the expectation that, for viscous dissipation of energy to become important, the eddy Reynolds number must be reduced to near unity.

Two additional assumptions allow the scales for the largest eddies to be related to those for the smallest ones. First, it is assumed that the lifetime of any eddy is its convective time scale. As may be confirmed from Eq. (10.2-10), $t_2 = L_2/u_2$; likewise, it is supposed that $t_1 = L_1/u_1$. Second, it is assumed that kinetic energy loss and viscous dissipation are always roughly in balance, even in an unsteady flow. The kinetic energy per

[1] This was summarized by L. F. Richardson as follows: "Big whirls have little whirls that feed on their velocity, and little whirls have lesser whirls and so on to viscosity – in the molecular sense" (Richardson, 1922, p. 66).

unit mass of the largest eddies is $u_1^2/2$ or $\sim u_1^2$. Equating the rate of kinetic energy loss from such eddies with the incremental dissipation rate leads to

$$d \sim \frac{u_1^2}{t_1} = \frac{u_1^3}{L_1}. \tag{10.2-11}$$

Using this to evaluate d in Eq. (10.2-10) gives

$$\frac{u_2}{u_1} \sim \mathrm{Re}_1^{-1/4}, \quad \frac{L_2}{L_1} \sim \mathrm{Re}_1^{-3/4}, \quad \frac{t_2}{t_1} \sim \mathrm{Re}_1^{-1/2} \tag{10.2-12}$$

where $\mathrm{Re}_1 = u_1 L_1/\nu$.

Equation (10.2-12) has profound implications. In particular, the result for L_2/L_1 indicates that the range of eddy sizes broadens as the Reynolds number increases; the turbulence becomes "finer-grained." Just as representing increasingly detailed images requires more and more pixels per unit length, computational investigations of turbulent flow must employ finer and finer grids or meshes as Re is increased. This greatly impacts attempts to predict the details of turbulent flow via direct numerical simulation (DNS). When practical, DNS provides a wealth of information on $\bar{\mathbf{v}}$ and $\mathbf{u}$ (Moin and Mahesh, 1998). However, its use has been limited to values of Re not much larger than those which produce turbulence ($<10^4$ in conduits). The problem is that the number of grid points needed to resolve the Kolmogorov scales increases with Re, causing the computational effort to vary approximately as Re^3 (Sandham, 2005).

Example 10.2-1 Turbulence scales for air flow in a pipe. The objective is to determine the scales of the largest and smallest eddies for air flow in a large pipe. The pipe diameter is $D = 0.50$ m, the overall Reynolds number (based on D and the mean velocity U) is $\mathrm{Re} = 1 \times 10^6$, and the air properties are $\rho = 1.2$ kg m^{-3} and $\nu = 1.6 \times 10^{-5}$ m^2 s^{-1}.

Eddies might be as large as the cross-sectional dimension, D. However, because velocity measurements suggest that $0.1D$ is more realistic (Problem 10.6), we set $L_1 = 0.050$ m. These eddies are associated with the largest velocity fluctuations, which are on the order of the friction velocity. Accordingly, $u_1 = u_\tau$. The mean velocity is

$$U = \frac{\nu\,\mathrm{Re}}{D} = \frac{(1.6 \times 10^{-5})(1 \times 10^6)}{0.50} = 32.0 \text{ m/s}.$$

Although large for a process stream, U is still only 9% of the speed of sound (340 m/s at sea level). From the Colebrook equation for smooth pipes [Eq. (2.2-8)],

$$f = \left[3.6 \log\left(\frac{1 \times 10^6}{6.9}\right)\right]^{-2} = 2.90 \times 10^{-3}.$$

From Eq. (10.2-6), the velocity scale for the largest eddies is then

$$u_1 = \left(\frac{2.90 \times 10^{-3}}{2}\right)^{1/2} 32.0 = 1.22 \text{ m/s}$$

which is about 4% of U. The time scale for these eddies is

$$t_1 = \frac{L_1}{u_1} = \frac{0.050}{1.22} = 0.041 \text{ s}$$

and their Reynolds number is

$$\mathrm{Re}_1 = \frac{(1.22)(0.050)}{1.6 \times 10^{-5}} = 3.81 \times 10^3.$$

Although Re_l is much smaller than Re, it is still large enough for viscosity to be negligible in the largest eddies.

The length and velocity scales for the smallest eddies are calculated from Re_l and Eq. (10.2-12) as

$$L_2 = (0.050)(3.81 \times 10^3)^{-3/4} = 1.0 \times 10^{-4}\ \text{m}$$

$$u_2 = (1.22)(3.81 \times 10^3)^{-1/4} = 0.16\ \text{m/s}.$$

Although much smaller than the pipe diameter, L_2 (= 100 μm) is still 1000 times the typical mean free path [Eq. (1.2-7)]. This indicates that the turbulence in such a pipe is a continuum phenomenon, not a molecular one; that is always the case. The small-eddy time scale is

$$t_2 = \frac{1.0 \times 10^{-4}}{0.16} = 6.7 \times 10^{-4}\ \text{s}$$

which suggests that instruments with a sub-millisecond response would be needed to fully resolve the velocity fluctuations in this flow.

If it had been assumed that $L_1 = D$, the results for the small eddies ($L_2 = 1.8 \times 10^{-4}$ m, $u_2 = 0.087$ m/s, $t_2 = 2.1 \times 10^{-3}$ s) would not have been different enough to alter the conclusions.

10.3 REYNOLDS AVERAGING

Although a fundamental aspect of turbulence, the velocity fluctuations are often of little interest in engineering. What are needed ordinarily in process design are just time-smoothed values. Smoothing of both the velocity and pressure was implicit in all calculations in Chapters 2 and 3 that involved turbulent flow. Reynolds (1895) showed how to rewrite the continuity and Navier–Stokes equations in terms of smoothed quantities. Ideally, knowledge of the fluctuations would not be needed to apply such equations. Although that is not the case, Reynolds averaging is still at the heart of common strategies for modeling turbulent flow.

Time-smoothed variables

The time-smoothed value of ξ, which could be a scalar, vector, or tensor, is defined as

$$\bar{\xi} = \frac{1}{t_a} \int_t^{t+t_a} \xi \, dt' \qquad (10.3\text{-}1)$$

where t_a is the averaging time. This average can be calculated repeatedly for various starting times t. Encompassing a representative number of fluctuations requires that $t_a \gg t_f$, where t_f is the time scale for one fluctuation, but modeling an unsteady process requires that $t_a \ll t_p$, where t_p is the process time scale. Ordinarily, such a value of t_a exists. The fact that the large-eddy time scale in Example 10.2-1 was only some 40 ms suggests that t_a for that flow would not need to be more than a few tenths of a second.

The manipulation of time-smoothed quantities is governed by two rules. The first is that a smoothed variable is unaffected by repeated smoothing. Taking the velocity vector as an example, $\bar{\bar{\mathbf{v}}} = \bar{\mathbf{v}}$. If the instantaneous velocity is decomposed as

$$\mathbf{v} = \bar{\mathbf{v}} + \mathbf{u} \qquad (10.3\text{-}2)$$

where $\mathbf{u}$ is the fluctuation, then smoothing it gives

$$\bar{\mathbf{v}} = \overline{(\bar{\mathbf{v}} + \mathbf{u})} = \bar{\bar{\mathbf{v}}} + \bar{\mathbf{u}} = \bar{\mathbf{v}} + \bar{\mathbf{u}}. \tag{10.3-3}$$

It follows that

$$\bar{\mathbf{u}} = \mathbf{0} \tag{10.3-4}$$

as noted in Section 10.2. Although the time-smoothed value of any fluctuation is zero, the smoothed values of products of fluctuations usually do not vanish, as will be seen.

The second rule is that the order of smoothing and differentiation of a variable can be interchanged. Again using the velocity,

$$\overline{\nabla \cdot \mathbf{v}} = \nabla \cdot \bar{\mathbf{v}} \tag{10.3-5}$$

$$\overline{\nabla^2 \mathbf{v}} = \nabla^2 \bar{\mathbf{v}} \tag{10.3-6}$$

$$\overline{\frac{\partial \mathbf{v}}{\partial t}} = \frac{\partial \bar{\mathbf{v}}}{\partial t}. \tag{10.3-7}$$

Equations (10.3-5) and (10.3-6) follow simply from the fact that any spatial derivative can be moved inside or outside the time integral in Eq. (10.3-1). For the time derivative, the left-hand side of Eq. (10.3-7) is evaluated as

$$\overline{\frac{\partial \mathbf{v}}{\partial t}} = \frac{1}{t_a} \int_t^{t+t_a} \frac{\partial \mathbf{v}}{\partial t} \, dt' = \frac{1}{t_a} [\mathbf{v}(t + t_a) - \mathbf{v}(t)]. \tag{10.3-8}$$

Using the Leibniz rule for differentiating an integral with variable limits, the right-hand side of Eq. (10.3-7) is

$$\begin{aligned} \frac{\partial \bar{\mathbf{v}}}{\partial t} = \frac{1}{t_a} \frac{\partial}{\partial t} \left[\int_t^{t+t_a} \mathbf{v} \, dt' \right] &= \frac{1}{t_a} \left[\mathbf{v}(t + t_a) \frac{d}{dt}(t + t_a) - \mathbf{v}(t) \frac{d}{dt}(t) \right] \\ &= \frac{1}{t_a} [\mathbf{v}(t + t_a) - \mathbf{v}(t)] \end{aligned} \tag{10.3-9}$$

which shows that smoothing and time-differentiation indeed can be interchanged.

Continuity equation

The instantaneous velocity in an incompressible fluid satisfies

$$\nabla \cdot \mathbf{v} = \mathbf{0}. \tag{10.3-10}$$

Smoothing both sides and using Eq. (10.3-5) gives

$$\nabla \cdot \bar{\mathbf{v}} = \mathbf{0}. \tag{10.3-11}$$

Substitution of Eq. (10.3-2) into Eq. (10.3-10) results in

$$\nabla \cdot \mathbf{v} = \nabla \cdot (\bar{\mathbf{v}} + \mathbf{u}) = \nabla \cdot \bar{\mathbf{v}} + \nabla \cdot \mathbf{u} = \mathbf{0}. \tag{10.3-12}$$

Together with Eq. (10.3-11), this indicates that

$$\nabla \cdot \mathbf{u} = \mathbf{0}. \tag{10.3-13}$$

Thus, $\bar{\mathbf{v}}$ and $\mathbf{u}$ each satisfy the usual continuity equation individually.

Navier–Stokes equation

In a Newtonian fluid with constant properties, the instantaneous velocity and pressure obey

$$\frac{\partial \mathbf{v}}{\partial t} + \mathbf{v} \cdot \nabla \mathbf{v} = -\frac{1}{\rho} \nabla \mathcal{P} + \nu \nabla^2 \mathbf{v}. \tag{10.3-14}$$

Smoothing the time derivative, pressure gradient, and viscous term simply changes $\mathbf{v}$ to $\bar{\mathbf{v}}$ and $\mathcal{P}$ to $\bar{\mathcal{P}}$. The nonlinear inertial term requires more attention. Using Eq. (10.3-2), it is expanded as

$$\mathbf{v} \cdot \nabla \mathbf{v} = \bar{\mathbf{v}} \cdot \nabla \bar{\mathbf{v}} + \mathbf{u} \cdot \nabla \bar{\mathbf{v}} + \bar{\mathbf{v}} \cdot \nabla \mathbf{u} + \mathbf{u} \cdot \nabla \mathbf{u}. \tag{10.3-15}$$

When this is smoothed, the first term on the right-hand side is unaffected because it already contains only smoothed variables. The second and third terms vanish because $\bar{\mathbf{u}} = \mathbf{0}$, but the fourth one does not. The result is

$$\overline{\mathbf{v} \cdot \nabla \mathbf{v}} = \bar{\mathbf{v}} \cdot \nabla \bar{\mathbf{v}} + \overline{\mathbf{u} \cdot \nabla \mathbf{u}}. \tag{10.3-16}$$

The smoothed form of the Navier–Stokes equation is then

$$\frac{\partial \bar{\mathbf{v}}}{\partial t} + \bar{\mathbf{v}} \cdot \nabla \bar{\mathbf{v}} = -\frac{1}{\rho} \nabla \bar{\mathcal{P}} + \nu \nabla^2 \bar{\mathbf{v}} - \overline{\mathbf{u} \cdot \nabla \mathbf{u}} \tag{10.3-17}$$

where the term containing $\mathbf{u}$ has been moved to the right-hand side. Except for that additional term, this differential equation has the same form as the instantaneous Navier–Stokes equation.

Closure problem

In laminar flow, Eqs. (10.3-10) and (10.3-14) provide a total of four differential equations governing four unknowns, the three velocity components and $\mathcal{P}$. Once an initial condition and boundary conditions are specified, such problems are fully defined. When using Reynolds averaging to model turbulent flow, the inability to fully eliminate velocity fluctuations from the smoothed Navier–Stokes equation adds the three components of $\mathbf{u}$ as unknowns. Governing the total of seven unknowns are the four equations provided by Eqs. (10.3-11) and (10.3-17), plus one more from Eq. (10.3-13). Thus, in general, conservation of mass and conservation of momentum provide only five equations for seven unknowns. The shortage of equations creates what is called the *closure problem*. The use of Reynolds averaging requires that the conservation equations be supplemented by other information, which inevitably is less rigorous.

Numerous strategies have been used to achieve closure (i.e., obtain a self-contained set of governing equations). While based on various plausible hypotheses, all contain constants that are fitted to experimental data for specific types of flow. In the turbulence literature, such partly theoretical and partly empirical constructs are called *models*. The simplest is the *eddy diffusivity* approach described in Section 10.4.

Reynolds stress

Equation (10.3-17) can be simplified by rearranging the viscous and velocity-fluctuation terms. From Eq. (6.6-1),

$$\nu \nabla^2 \mathbf{v} = \frac{1}{\rho} \nabla \cdot \boldsymbol{\tau} \tag{10.3-18}$$

from which it follows that

$$\nu\nabla^2\bar{\mathbf{v}} = \overline{\nu\nabla^2\mathbf{v}} = \overline{\frac{1}{\rho}\nabla\cdot\boldsymbol{\tau}} = \frac{1}{\rho}\nabla\cdot\bar{\boldsymbol{\tau}}. \tag{10.3-19}$$

The velocity-fluctuation term also can be expressed as the divergence of a tensor. From identity (10) in Table A.1 and Eq. (10.3-13),

$$\nabla\cdot(\mathbf{uu}) = (\nabla\cdot\mathbf{u})\mathbf{u} + \mathbf{u}\cdot\nabla\mathbf{u} = \mathbf{u}\cdot\nabla\mathbf{u}. \tag{10.3-20}$$

Accordingly,

$$\overline{\mathbf{u}\cdot\nabla\mathbf{u}} = \nabla\cdot(\overline{\mathbf{uu}}) = \frac{1}{\rho}\nabla\cdot(\rho\overline{\mathbf{uu}}). \tag{10.3-21}$$

The new tensor,

$$\boldsymbol{\tau}^* = -\rho\overline{\mathbf{uu}}. \tag{10.3-22}$$

is called the *Reynolds stress*. The symbol tau notwithstanding, it is unrelated to viscosity. In that it involves the density and products of velocity terms, it is evidently a kind of inertial stress. It represents the convective transport of momentum by the turbulent eddies.

Equation (10.3-17) may be written now as

$$\rho\left(\frac{\partial\bar{\mathbf{v}}}{\partial t} + \bar{\mathbf{v}}\cdot\nabla\bar{\mathbf{v}}\right) = -\nabla\bar{\mathcal{P}} + \nabla\cdot(\bar{\boldsymbol{\tau}} + \boldsymbol{\tau}^*) \tag{10.3-23}$$

which is analogous to the Cauchy momentum equation [Eq. (6.4-8)]. The enhanced cross-stream transport of momentum and the rapid mixing that characterize turbulent flows stem from the velocity fluctuations embedded in $\boldsymbol{\tau}^*$.

Although lumped together in Eq. (10.3-23), the viscous stress and Reynolds stress have very different properties. In a turbulent flow that is unidirectional on average, the viscous shear stress at a solid surface at $y = 0$ is

$$\bar{\tau}_{yx}|_{y=0} = \mu\left.\frac{\partial\bar{v}_x}{\partial y}\right|_{y=0} \tag{10.3-24}$$

which is analogous to that in laminar flow. However, the corresponding Reynolds stress is

$$\tau^*_{yx}|_{y=0} = -\rho\overline{u_y u_x}|_{y=0} = 0. \tag{10.3-25}$$

The reason why τ^*_{yx} vanishes is that the no-slip and no-penetration conditions require that v_x, v_y, $\bar{v}_x$, and $\bar{v}_y$ each be zero at $y = 0$. It follows that $u_x = 0 = u_y$ there at each instant. Similar reasoning leads to the conclusion that all components of $\boldsymbol{\tau}^*$ vanish at any solid surface.

10.4 CLOSURE SCHEMES

The essence of the closure problem is how to relate $\boldsymbol{\tau}^*$ to time-smoothed quantities. The difficulty is analogous to that encountered in evaluating viscous stresses prior to the early 1800s, before the constitutive equation for a Newtonian fluid was known. More than a century of turbulence research by engineers, physicists, and applied mathematicians has failed to yield an expression for $\boldsymbol{\tau}^*$ that is as general, simple, and accurate as Eq. (6.5-14). Nonetheless, progress has been made. In describing the most basic approaches we will focus on shear flows that are unidirectional or nearly unidirectional on average.

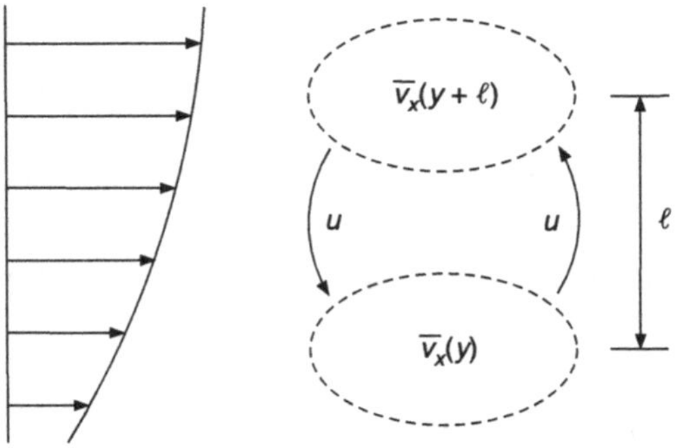

Figure 10.3 Conceptual model for momentum transfer by turbulent eddies, involving the interchange of fluid lumps positioned at y and $y + \ell$.

Eddy diffusivities

Eddy diffusivity models provide the simplest way to resolve the closure problem that has practical value. The shear stress in a unidirectional laminar flow can be written in terms of the kinematic viscosity as

$$\bar{\tau}_{yx} = \rho \nu \frac{\partial \bar{v}_x}{\partial y}. \tag{10.4-1}$$

By analogy, beginning with J. Boussinesq in 1877, many researchers have supposed that

$$\tau_{yx}^* = \rho \varepsilon \frac{\partial \bar{v}_x}{\partial y} \tag{10.4-2}$$

where ε is the *eddy diffusivity for momentum*. The eddy diffusivity has the same dimension as the kinematic viscosity, the diffusivity of a chemical species in a mixture, or the thermal diffusivity in heat conduction (L^2T^{-1}). Underlying Eq. (10.4-2), which bears no resemblance to Eq. (10.3-25), is an assumed analogy between the motion of turbulent eddies and the random movement of gas molecules. In a gas, the diffusion of molecules and the molecular-level transfer of energy and momentum are all governed by molecular velocities and by how far molecules travel between collisions. It is known from the statistics of random walks that, if a molecule makes randomly directed jumps of distance ℓ at speed u, in three dimensions, its diffusivity is $u\ell/6$ (Deen, 2012, p. 47). If u and ℓ are evaluated using the kinetic theory of gases, good estimates are obtained for ν and for the species and thermal diffusivities (Deen, 2012, pp. 15–16). One might hope, then, that estimates of eddy velocities and sizes would lead to equally good predictions for ε. It must be kept in mind that, unlike ν, which is a property of the *fluid*, ε is a property of the *flow*.

The model depicted in Fig. 10.3 supports Eq. (10.4-2) and suggests how ε might be evaluated. Imagine that eddy motion causes the fluid "lumps" depicted by the dashed ovals to exchange positions, each oval carrying with it an amount of x-momentum associated with its starting point. To do this, the lumps move a distance ℓ in the $\pm y$ direction at a speed u. Recalling from Section 6.4 that the flux of momentum due to bulk motion is a velocity times the concentration of momentum, the net rate of momentum transfer in the $-y$ direction is

$$\tau_{yx}^* = u\rho \bar{v}_x(y+\ell) - u\rho \bar{v}_x(y) = \rho u \, \Delta \bar{v}_x \tag{10.4-3}$$

where $\Delta \bar{v}_x$ is the velocity difference. Because τ_{yx}^* is a force per unit area exerted by the fluid at greater y, it is the rate at which eddies transfer x-momentum in the $-y$ direction. Thus, $\tau_{yx}^* > 0$ when $\Delta \bar{v}_x > 0$, as shown.

The fluid lumps are assumed to momentarily retain their initial values of $\bar{v}_x$, so the arrival of either at its new location causes a velocity fluctuation $u_x \sim \Delta\bar{v}_x$. As mentioned in Section 10.2, the rms fluctuations in all velocity components are similar. Accordingly, $u \sim u_y \sim u_x$ and Eq. (10.4-3) becomes

$$\tau^*_{yx} \sim \rho(\Delta\bar{v}_x)^2. \tag{10.4-4}$$

From the estimation rule for first derivatives,

$$\Delta\bar{v}_x \sim \ell\frac{\partial\bar{v}_x}{\partial y}. \tag{10.4-5}$$

The Reynolds stress is then

$$\tau^*_{yx} = \rho\ell^2\left(\frac{\partial\bar{v}_x}{\partial y}\right)^2. \tag{10.4-6}$$

An exact equality is used now, with the understanding that the unknown proportionality constant will be embedded in ℓ, which is called the *mixing length*. Equation (10.4-6) was obtained in this manner by Prandtl (1925). Using different reasoning, G. I. Taylor and T. von Kármán derived essentially equivalent expressions (Schlichting, 1968, pp. 545–553).

A comparison of Eqs. (10.4-6) and (10.4-2) indicates that

$$\varepsilon = \ell^2\left|\frac{\partial\bar{v}_x}{\partial y}\right|. \tag{10.4-7}$$

The absolute value ensures that $\varepsilon > 0$, which in turn ensures that τ^*_{yx} has the same sign as $\partial\bar{v}_x/\partial y$. In terms of the random-walk analogy, the scales for the jump length and jump velocity are seen to be ℓ and $\ell|\partial\bar{v}_x/\partial y|$, respectively.

Equation (10.4-6) almost provides closure for unidirectional or nearly unidirectional flows. All that is missing is how to evaluate the mixing length, which must be derived from experimental results. A significant complication is that ℓ is not constant. As shown by Eq. (10.3-25), τ^*_{yx} vanishes at a solid surface positioned at $y = 0$. Because the shear rate at such a surface is ordinarily not zero, it is necessary that $\ell = 0$ at $y = 0$. Thus, when modeling turbulent flows next to solid surfaces, the simplest realistic expression for the mixing length is

$$\ell = \kappa y \tag{10.4-8}$$

where κ is an empirical constant. This proportionality between ℓ and y was proposed by Prandtl (1933). In that a vortex centered a distance y from a surface cannot be much larger than y, it is reasonable to suppose that $\kappa \sim 1$. That is indeed the case, the best-fit value being $\kappa = 0.40$ (Example 10.5-1). Substituting Eq. (10.4-8) into Eq. (10.4-7) and changing to the wall variables gives

$$\frac{\varepsilon}{\nu} = (\kappa y^+)^2\left|\frac{\partial v_x^+}{\partial y^+}\right|. \tag{10.4-9}$$

This ratio of eddy diffusivity to kinematic viscosity is what is needed to predict turbulent velocity profiles, as will be seen.

Many other expressions for $\varepsilon(y)$ have been proposed. A continuous function that gives accurate velocity profiles is that of Van Driest (1956),

$$\frac{\varepsilon}{\nu} = (\kappa y^+)^2(1 - e^{-y^+/A})^2\left|\frac{\partial v_x^+}{\partial y^+}\right| \tag{10.4-10}$$

where $\kappa = 0.40$ as before and $A = 26$ for pipe flow. This mimics the behavior of Eq. (10.4-9) for $y^+ \to 0$ or $y^+ \gg A$. However, the Van Driest ε is significantly smaller than the Prandtl one for intermediate values of y^+.

Other approaches

In keeping with an introductory discussion, the examples in Sections 10.5 and 10.6 involve only the mixing-length concept. However, numerous other ways to obtain closure have been proposed, as reviewed in Speziale (1991). One that has been widely used is the *k–ε method*. It adds to the smoothed continuity and Navier–Stokes equations two partial differential equations, one involving the turbulent kinetic energy and the other the excess rate of energy dissipation. The need to specify a mixing length is eliminated by relating the eddy diffusivity to the kinetic energy and excess dissipation (Kays and Crawford, 1993, pp. 213–222). The resulting expression contains just one empirical constant and it is unnecessary to specify in advance how the eddy diffusivity varies with position.

10.5 UNIDIRECTIONAL FLOW

In a turbulent flow that is steady and unidirectional on average, the Cartesian form of Eq. (10.3-23) reduces to

$$0 = -\frac{d\bar{\mathcal{P}}}{dx} + \frac{d}{dy}(\bar{\tau}_{yx} + \tau^*_{yx}). \tag{10.5-1}$$

Although the inertial terms on the left-hand side are zero, inertial effects are embedded in τ^*_{yx}, which can greatly exceed $\bar{\tau}_{yx}$. It is assumed here that $\tau^*_{xx} = 0 = \tau^*_{zx}$, because the corresponding spatial derivatives in the time-smoothed velocity are zero.

Example 10.5-1 Velocity profile near a wall. The objective is to calculate the turbulent velocity profile near a tube wall. The predictions from Eqs. (10.4-9) and (10.4-10) will be compared with experimental results.

For fully developed flow in a cylindrical tube, the analog of Eq. (10.5-1) is

$$0 = -\frac{d\bar{\mathcal{P}}}{dz} + \frac{1}{r}\frac{d}{dr}[r(\bar{\tau}_{rz} + \tau^*_{rz})]. \tag{10.5-2}$$

Integrating over r and recognizing that $\bar{\tau}_{rz}(0) + \tau^*_{rz}(0) = 0$ from symmetry gives

$$\bar{\tau}_{rz}(r) + \tau^*_{rz}(r) = \frac{r}{2}\frac{d\bar{\mathcal{P}}}{dz}. \tag{10.5-3}$$

Relating the pressure gradient to the wall shear stress in the usual manner, this becomes

$$\bar{\tau}_{rz}(r) + \tau^*_{rz}(r) = -\frac{\tau_0 r}{R} \tag{10.5-4}$$

where τ_0 (>0) is the shear stress that the fluid exerts on the wall and R is the tube radius.

Letting $y = R - r$ to focus on the region near the wall, Eq. (10.5-4) is rewritten as

$$\bar{\tau}_{yz}(y) + \tau^*_{yz}(y) = \tau_0\left(1 - \frac{y}{R}\right). \tag{10.5-5}$$

Evaluating the viscous and Reynolds stresses as in Eqs. (10.4-1) and (10.4-2), respectively, and restricting the analysis to $y \ll R$, the momentum equation becomes

$$\rho(\nu + \varepsilon)\frac{d\bar{v}_z}{dy} = \tau_0. \tag{10.5-6}$$

Converting to wall variables simplifies this to

$$\left(1+\frac{\varepsilon}{\nu}\right)\frac{dv_z^+}{dy^+}=1, \quad v_z^+(0)=0. \tag{10.5-7}$$

This is the problem to be solved for certain choices of the function $\varepsilon(y^+)$.

Following Prandtl's suggestion concerning the mixing length, ε/ν is evaluated as in Eq. (10.4-9) and

$$\left[1+(\kappa y^+)^2\frac{dy_z^+}{dy^+}\right]\frac{dv_z^+}{dy^+}=1, \quad v_z^+(0)=0. \tag{10.5-8}$$

Although nonlinear, this first-order differential equation is separable and has an analytical solution. Before considering the full solution, which has a complicated form, it is helpful to examine the behavior of $v_z^+(y^+)$ for extremes of y^+. For $y^+ \to 0$, where the Reynolds stress becomes negligible, the differential equation reduces to

$$\frac{dv_z^+}{dy^+}=1 \tag{10.5-9}$$

and the solution is simply

$$v_z^+(y^+)=y^+ \quad (y^+ \to 0). \tag{10.5-10}$$

The same result is obtained by setting $\kappa=0$, as if there were no turbulent eddies. Accordingly, the region where Eq. (10.5-10) applies is called the *laminar sublayer*.

For $y^+ \to \infty$, the viscous stress is negligible and the differential equation becomes

$$\left(\kappa y^+\frac{dy_z^+}{dy^+}\right)^2=1. \tag{10.5-11}$$

Taking the square root of both sides and dividing by κy^+ gives

$$\frac{dv_z^+}{dy^+}=\frac{1}{\kappa y^+} \tag{10.5-12}$$

which has the solution

$$v_z^+(y^+)=\frac{1}{\kappa}\ln y^+ + c \quad (y^+ \to \infty). \tag{10.5-13}$$

The constant c is undetermined as yet, because the boundary condition at the wall is inapplicable when considering only large y^+. In summary, Prandtl's mixing length hypothesis predicts that $v_z^+(y^+)$ is linear for small y^+ and logarithmic for large y^+.[2] How small or large y^+ must be for these asymptotic results to apply will be discussed shortly.

The complete solution of Eq. (10.5-8) (which the author has not seen elsewhere) is

$$v_z^+(y^+)=\frac{1-\sqrt{1+(2\kappa y^+)^2}}{2\kappa^2 y^+}+\frac{1}{\kappa}\ln\left(2\kappa y^+ + \sqrt{1+(2\kappa y^+)^2}\right). \tag{10.5-14}$$

It may be confirmed that this has the limiting forms given by Eqs. (10.5-10) and (10.5-13). For $y^+ \to \infty$, a comparison of Eqs. (10.5-13) and (10.5-14) indicates that

$$c=\frac{1}{\kappa}\left[\ln(4\kappa)-1\right]. \tag{10.5-15}$$

[2] Equation (10.5-13) was derived also by T. von Kármán, as reviewed in Kármán (1934), and thus the logarithmic profile is associated with both him and Prandtl.

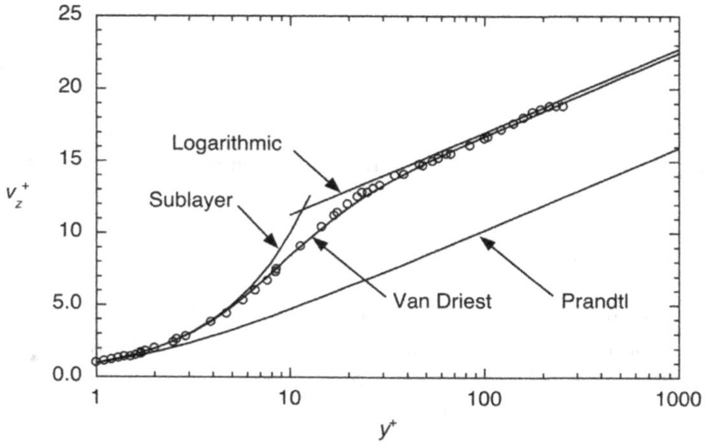

Figure 10.4 Velocity for turbulent flow near a wall. The curves labeled "sublayer," "logarithmic," and "Prandtl" are plots of Eqs. (10.5-10), (10.5-16), and (10.5-14), respectively. The "Van Driest" curve was obtained from a numerical solution of Eq. (10.5-17). The symbols represent data of Durst *et al.* (1993) obtained using laser-Doppler velocimetry.

Thus, in the Prandtl model, κ is the only independent constant that can be adjusted to fit velocity data.

The eddy-diffusivity predictions for the velocity profile near a wall are compared with experimental results from Durst *et al.* (1993) in Fig. 10.4. Note the semilogarithmic coordinates. It is seen that the full solution [Eq. (10.5-14)] with $\kappa = 0.40$ matches the data for small y^+ and gives the correct slope for large y^+. However, at most positions it greatly underestimates the velocity. The result for the laminar sublayer [Eq. (10.5-10)] is seen to be accurate for $0 \leq y^+ \leq 5$. For $y^+ > 30$, the data are represented quite well by

$$v_z^+(y^+) = 2.5 \ln y^+ + 5.5. \tag{10.5-16}$$

This has a slope that is consistent with Eq. (10.5-13) ($2.5 = 1/\kappa$ with $\kappa = 0.40$), but an intercept that is much larger than that calculated from Eq. (10.5-15) (5.5 vs. $c = -1.3$ with $\kappa = 0.40$).

A much better fit to the velocity data is obtained by evaluating the eddy viscosity using the Van Driest model. Using Eq. (10.4-10) in Eq. (10.5-7) gives

$$\left[1 + (\kappa y^+)^2 (1 - e^{-y^+/A})^2 \frac{dv_z^+}{dy^+}\right] \frac{dv_z^+}{dy^+} = 1, \quad v_z^+(0) = 0. \tag{10.5-17}$$

As before, this implies that $v_z^+(y^+)$ is linear for small y^+ and logarithmic for large y^+. However, as shown in Fig. 10.4, a numerical solution of Eq. (10.5-17) (with $\kappa = 0.40$ and $A = 26$) fits the data very well over the entire range of y^+. Because the total shear stress near the wall is approximately constant, smaller values of ε for intermediate y^+ lead to larger values of dv_z^+/dy^+. That is what elevates v_z^+ in the Van Driest model.

Overall, the prediction that the turbulent velocity profile is linear at first, and then logarithmic, is a triumph of the eddy-viscosity and mixing-length concepts. A remarkable aspect of the profile shown either by the data or by the Van Driest curve in Fig. 10.4 is its nearly universal character. That has led it to be called *the law of the wall*. Essentially the same velocity profile near a wall is found also in noncircular conduits and in boundary

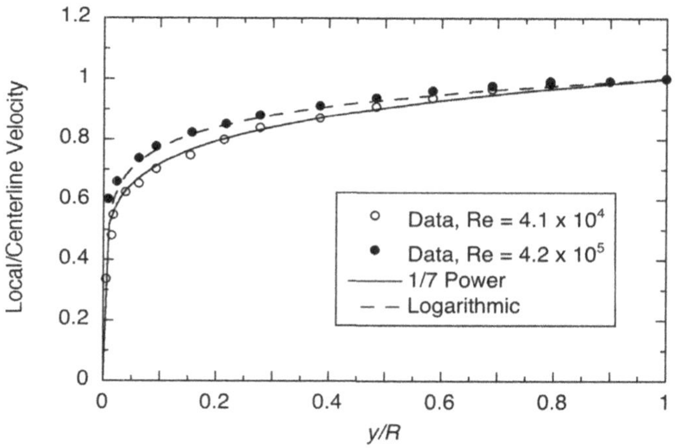

Figure 10.5 Velocity in a tube relative to its centerline value. The symbols represent the hot-wire anemometry data of Laufer (1954) at two Reynolds numbers. The "1/7 power" and "logarithmic" curves are based on Eqs. (10.5-18) and (10.5-19), respectively.

layers on submerged objects; for examples see Kim *et al.* (1987) and Kays and Crawford (1993, p. 211).

Complete velocity profile for tube flow

The velocities in Fig. 10.4 are for the region near a tube wall. More complete profiles are shown in Fig. 10.5, in which the velocity is normalized by its centerline value, $\bar{v}_{z,c}$. The symbols represent the data of Laufer (1954) at two values of Re. The experimental results (especially at the smaller Re) are well represented by a power-law relationship,

$$\frac{\bar{v}_z}{\bar{v}_{z,c}} = \left(\frac{y}{R}\right)^{1/7}. \tag{10.5-18}$$

Also close to the data is a curve derived from the logarithmic profile in Eq. (10.5-16). Noting that $y^+/R^+ = y/R$ and changing to $\bar{v}_z$ and y, the logarithmic expression becomes

$$\frac{\bar{v}_z}{\bar{v}_{z,c}} = \frac{2.5 \ln[R^+(y/R)] + 5.5}{2.5 \ln R^+ + 5.5}. \tag{10.5-19}$$

For clarity, what is shown in this case is only the average of the results obtained for the two values of Re. The accuracy of the logarithmic profile over most of the tube cross-section is surprising, in that it was derived assuming that $y/R \ll 1$ (Example 10.5-1). The fact that it does not satisfy the no-slip condition and otherwise loses accuracy for $y^+ < 30$ is not visually evident, because at these Reynolds numbers $y^+ < 30$ corresponds to $y/R < 0.03$.

Laminar and turbulent velocity profiles in tubes are compared in Fig. 10.6. In this plot the velocity is divided by the mean value, U. Using the 1/7-power relationship, the mean and centerline velocities are related as

$$\frac{U}{\bar{v}_{z,c}} = \frac{2}{R^2}\int_0^R \left(\frac{y}{R}\right)^{1/7}(R-y)\,dy = 2\int_0^1 \eta^{1/7}(1-\eta)\,d\eta = \frac{49}{60} \tag{10.5-20}$$

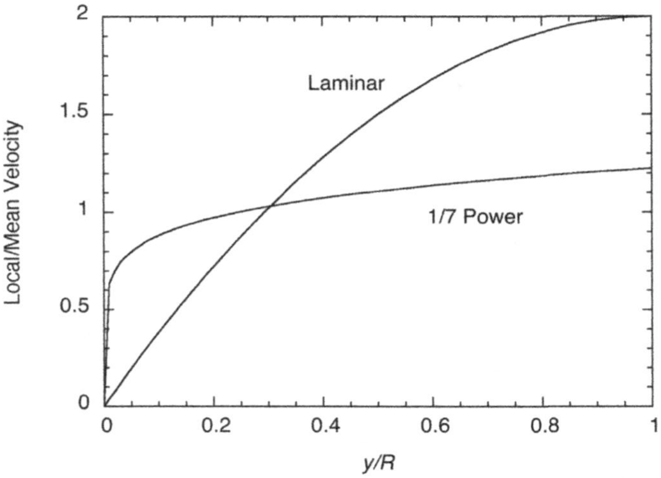

Figure 10.6 Velocity profiles for fully developed laminar or turbulent flow in a tube. The laminar and turbulent curves are based on Eqs. (7.2-19) and (10.5-21), respectively.

where $\eta = y/R$. Accordingly, Eq. (10.5-18) is rewritten as

$$\frac{\bar{v}_z}{U} = \frac{60}{49}\left(\frac{y}{R}\right)^{1/7}. \tag{10.5-21}$$

It is seen that the turbulent velocity profile is much blunter (closer to plug flow) than the parabolic profile for laminar flow.

Example 10.5-2 Prandtl–Kármán equation. It is desired to derive the friction-factor–Reynolds-number relationship for tube flow that is implied by the logarithmic velocity profile. Based on the comparisons in Figs. 10.4 and 10.5 and the large values of R^+ in Table 10.1, it will be assumed that Eq. (10.5-16) is satisfactory for nearly all radial positions. However, attempting to calculate the shear stress by differentiating the logarithmic profile at $y = 0$ would give $\tau_0 = \infty$. That reflects the fact that Eq. (10.5-16) is not valid at the wall and requires a more indirect approach. What will be obtained is essentially the Prandtl–Kármán equation, Eq. (2.2-7).

A relationship between f and Re can be found by calculating the mean velocity. Letting $\eta = y/R$, Eq. (10.5-16) is rewritten as

$$v_z^+ = \frac{\bar{v}_z}{u_\tau} = 2.5 \ln \eta + 2.5 \ln R^+ + 5.5. \tag{10.5-22}$$

Averaging as in Eq. (10.5-20), the mean velocity relative to the friction velocity is

$$\frac{U}{u_\tau} = 5.0 \int_0^1 \ln \eta (1-\eta)\, d\eta + 2.5 \ln R^+ + 5.5. \tag{10.5-23}$$

Using the fact that $x \ln x \to 0$ as $x \to 0$, the integral is evaluated as

$$\int_0^1 \ln \eta (1-\eta)\, d\eta = \left(\eta \ln \eta - \eta - \frac{\eta^2}{2} \ln \eta + \frac{\eta^2}{4} \right)\Bigg|_0^1 = -\frac{3}{4}. \tag{10.5-24}$$

Combining the numerical constants, the velocity ratio is

$$\frac{U}{u_\tau} = 2.5 \ln R^+ + 1.75. \tag{10.5-25}$$

Using Eq. (10.2-6) to change the variables to f and Re, we find that

$$\frac{1}{\sqrt{f}} = \frac{2.5(2.303)}{\sqrt{2}} \log\left(\mathrm{Re}\sqrt{f}\right) + \frac{1.75 - 2.5\sqrt{8}}{\sqrt{2}} = 4.1 \log\left(\mathrm{Re}\sqrt{f}\right) - 0.60. \tag{10.5-26}$$

This is the same as Eq. (2.2-7), except that the constants are somewhat different (4.1 vs. 4.0 and 0.60 vs. 0.40). The need to adjust the constants to optimize the fit to the $f(\mathrm{Re})$ data is not surprising, in that Eq. (10.5-16) is not an exact representation of the velocity profile. The basis for the peculiar form of the Prandtl–Kármán equation can be understood now.

10.6 BOUNDARY LAYERS

By analogy with the momentum equation for laminar boundary layers, that for steady turbulent flow is

$$\bar{v}_x \frac{\partial \bar{v}_x}{\partial x} + \bar{v}_y \frac{\partial \bar{v}_x}{\partial y} = u\frac{du}{dx} + \frac{1}{\rho}\frac{\partial}{\partial y}(\bar{\tau}_{yx} + \tau^*_{yx}). \tag{10.6-1}$$

As in Chapter 9, $u(x)$ is obtained by evaluating v_x from the outer solution at $y = 0$.[3] It is assumed that any effects of fluctuations in u are accounted for in τ^*_{yx}. Using Eqs. (10.4-1) and (10.4-2) for the shear stresses, the momentum equation becomes

$$\bar{v}_x \frac{\partial \bar{v}_x}{\partial x} + \bar{v}_y \frac{\partial \bar{v}_x}{\partial y} = u\frac{du}{dx} + \frac{\partial}{\partial y}\left((\nu + \varepsilon)\frac{\partial \bar{v}_x}{\partial y}\right). \tag{10.6-2}$$

Integrating either Eq. (10.6-1) or (10.6-2) over the boundary layer ($0 \le y \le \delta$) gives

$$\frac{\tau_0}{\rho} = \frac{d}{dx}\int_0^\delta \bar{v}_x(u - \bar{v}_x)\,dy + \frac{du}{dx}\int_0^\delta (u - \bar{v}_x)\,dy. \tag{10.6-3}$$

Again, this is analogous to the laminar equation. In the integration it was assumed that both parts of the shear stress vanish at $y = \delta$. As discussed in connection with Eq. (10.3-25), the Reynolds stress makes no contribution to τ_0, the shear stress that the fluid exerts on the solid surface at $y = 0$.

Example 10.6-1 Flat plate. A 1/7-power relationship like that discussed for tube flow has been used also to represent the velocity profile in a turbulent boundary layer next to a flat plate. Using wall coordinates, the empirical result is (Kays and Crawford, 1993, p. 207)

$$v_x^+ = 8.75(y^+)^{1/7}. \tag{10.6-4}$$

The objective is to see what this implies for the boundary-layer thickness and drag coefficient. As in Example 10.5-2, differentiating the given velocity profile at $y = 0$ gives

[3] The $u(x)$ in this section should not be confused with the velocity fluctuations discussed earlier.

$\tau_0 = \infty$ and a more indirect approach is needed. The integral form of the momentum equation will be used.

For a flat plate, $u = U$ (a constant) and Eq. (10.6-3) is rewritten as

$$\frac{\tau_0}{\rho} = \frac{d}{dx}\int_0^\delta \bar{v}_x(U - \bar{v}_x)\,dy = U^2\frac{d\delta}{dx}\int_0^1 \frac{\bar{v}_x}{U}\left(1 - \frac{\bar{v}_x}{U}\right)d\eta \qquad (10.6\text{-}5)$$

where $\eta = y/\delta(x)$. The given velocity profile implies that

$$\frac{\bar{v}_x}{U} = \eta^{1/7}. \qquad (10.6\text{-}6)$$

Using this to evaluate the integral in Eq. (10.6-5) gives

$$\frac{\tau_0}{\rho U^2} = \frac{d\delta}{dx}\left(\frac{7}{8}\eta^{8/7} - \frac{7}{9}\eta^{9/7}\right)\Bigg|_0^1 = \frac{7}{72}\frac{d\delta}{dx}. \qquad (10.6\text{-}7)$$

A second expression for τ_0 is obtained more directly from the velocity profile involving the wall coordinates. Noting that $\bar{v}_x = U$ at $y = \delta$,

$$\frac{U}{u_\tau} = 8.75\left(\frac{\delta u_\tau}{\nu}\right)^{1/7}. \qquad (10.6\text{-}8)$$

Solving for the friction velocity gives

$$u_\tau = \left(\frac{\tau_0}{\rho}\right)^{1/2} = \left(\frac{1}{8.75}\right)^{7/8} U^{7/8}\left(\frac{\nu}{\delta}\right)^{1/8}. \qquad (10.6\text{-}9)$$

Squaring both sides, the shear stress is evaluated as

$$\frac{\tau_0}{\rho U^2} = \left(\frac{1}{8.75}\right)^{7/4}\left(\frac{\nu}{U\delta}\right)^{1/4}. \qquad (10.6\text{-}10)$$

Notice that τ_0 is not simply proportional to $1/\delta$, as it would be if the boundary layer were laminar.

Equating the expressions for τ_0 in Eqs. (10.6-7) and (10.6-10) shows that the boundary-layer thickness is governed by

$$\delta^{1/4}\frac{d\delta}{dx} = \frac{72}{7}\left(\frac{1}{8.75}\right)^{7/4}\left(\frac{\nu}{U}\right)^{1/4}, \quad \delta(x_t) = \delta_t \qquad (10.6\text{-}11)$$

where x_t is the distance from the leading edge at which the laminar–turbulent transition occurs and δ_t is the boundary-layer thickness there. The solution is

$$\delta(x) = \left[\frac{90}{7}\left(\frac{1}{8.75}\right)^{7/4}\left(\frac{\nu}{U}\right)^{1/4}(x - x_t) + \delta_t^{5/4}\right]^{4/5}. \qquad (10.6\text{-}12)$$

For $x \gg x_t$ and $\delta \gg \delta_t$ this simplifies to

$$\delta(x) = 0.370\left(\frac{\nu x^4}{U}\right)^{1/5}. \qquad (10.6\text{-}13)$$

Thus, the 1/7-power velocity profile implies that the boundary-layer thickness grows as $x^{4/5}$. This is a more rapid increase than in the laminar region, where δ varies as $x^{1/2}$ [Eq. (9.3-14)]. This difference was depicted qualitatively in Fig. 3.4.

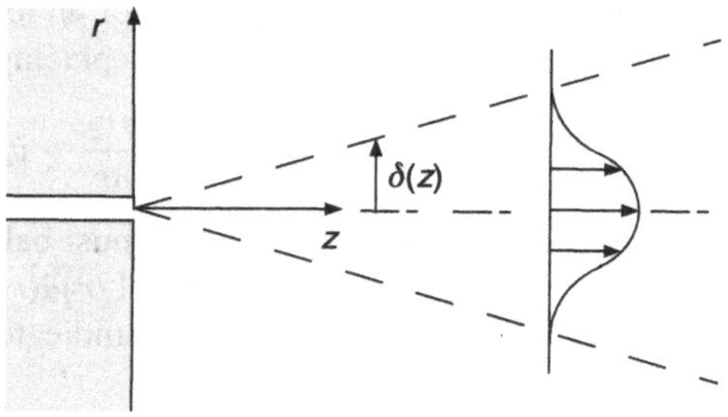

Figure 10.7 Circular turbulent jet within a large volume of the same fluid.

The local shear stress may be evaluated by substituting either Eq. (10.6-12) or (10.6-13) into Eq. (10.6-10). For $x \gg x_t$ this gives

$$\frac{\tau_0}{\rho U^2} = 0.0288\,\mathrm{Re}^{-1/5}\left(\frac{x}{L}\right)^{-1/5}. \tag{10.6-14}$$

If Re is large enough to make the laminar region negligible, the shear stress averaged over the plate length is

$$\frac{\langle \tau_0 \rangle}{\rho U^2} = 0.0288\,\mathrm{Re}^{-1/5}\frac{1}{L}\int_0^L \left(\frac{x}{L}\right)^{-1/5} dx = 0.0360\,\mathrm{Re}^{-1/5} \tag{10.6-15}$$

and the drag coefficient is

$$C_f = \frac{2\langle \tau_0 \rangle}{\rho U^2} = 0.0720\,\mathrm{Re}^{-1/5}. \tag{10.6-16}$$

Comparisons with Eq. (3.2-16) (with the B/Re term assumed to be negligible) indicate that Eq. (10.6-16) underestimates C_f by 4% at $\mathrm{Re} = 1 \times 10^7$, 15% at $\mathrm{Re} = 1 \times 10^8$, and 27% at $\mathrm{Re} = 1 \times 10^9$. The 1/7-power velocity profile evidently loses accuracy as Re increases. That is true also for tube flow, where the best-fit exponent decreases at higher Re, becoming as small as 1/10 at $\mathrm{Re} = 3 \times 10^6$ (Schlichting, 1968, p. 563).

Example 10.6-2 Axisymmetric jet. Suppose that a turbulent jet passes from a circular opening into a large volume of the same fluid, as in Fig. 10.7. The objective is to obtain OM estimates of the jet radius (δ), maximum velocity (v_m), and volume flow rate (Q), each of which depends on the axial position (z).

In that the jet does not flow along a solid surface, Eqs. (10.4-9) and (10.4-10) for the eddy diffusivity are inapplicable. Instead, it will be assumed that the mixing length is proportional to $\delta(z)$ and independent of r. Likewise, eddy velocities are assumed to be proportional to (although presumably much smaller than) the time-smoothed velocity at $r = 0$, which is $v_m(z)$. Based on the idea that $\varepsilon \sim \ell u$, where ℓ and u are the eddy length and velocity scales, respectively, it is postulated that

$$\varepsilon(z) = a\delta(z)v_m(z) \gg \nu \tag{10.6-17}$$

where a is a dimensionless constant. If $\varepsilon \gg \nu$ everywhere, as assumed, the viscous stresses can be neglected entirely. This makes turbulent jets and wakes simpler to model than flows with solid surfaces. That is, in a jet there is no laminar sublayer and no need to describe the intricacies of how ε varies with distance from a wall.

The momentum equation for an axisymmetric, turbulent boundary layer with $u = 0$ is analogous to Eq. (P9.9-1). Replacing ν by ε,

$$\bar{v}_r \frac{\partial \bar{v}_z}{\partial r} + \bar{v}_z \frac{\partial \bar{v}_z}{\partial z} = \frac{\varepsilon}{r} \frac{\partial}{\partial r}\left(r \frac{\partial \bar{v}_z}{\partial r} \right). \tag{10.6-18}$$

The inertial terms on the left must balance the Reynolds-stress terms on the right. With $\bar{v}_z \sim v_m$, $\partial \bar{v}_z/\partial z \sim v_m/z$, and $(1/r)\partial(r\,\partial \bar{v}_z/\partial r)/\partial r \sim \partial^2 \bar{v}_z/\partial r^2 \sim v_m/\delta^2$, and with ε evaluated using Eq. (10.6-17), this indicates that $v_m{}^2/z \sim a v_m^2/\delta$, or

$$\delta(z) \sim az. \tag{10.6-19}$$

Thus, the radius is predicted to be proportional to z, giving the jet a conical shape. Moreover, if a is a universal constant, as assumed, then all such jets will have the same set of cone angles. (Imaginary surfaces where the smoothed velocity is a specified fraction of its maximum correspond to specific cone angles; the smaller the fractional velocity, the larger the angle.)

The other needed information comes from the constancy of K, the kinematic momentum. By analogy with the laminar case (Problem 9.9),

$$K = 2\pi \int_0^\infty \bar{v}_z^2 r\, dr. \tag{10.6-20}$$

This implies that $K \sim 2\pi v_m^2 \delta^2$. When combined with Eq. (10.6-19), this gives

$$v_m(z) \sim \left(\frac{K}{2\pi} \right)^{1/2} \frac{1}{az}. \tag{10.6-21}$$

Thus, the eddy diffusivity is inferred to be

$$\varepsilon(z) = a\delta(z) v_m(z) \sim a \left(\frac{K}{2\pi} \right)^{1/2} \tag{10.6-22}$$

which is independent of position. Finally, the flow rate is predicted to be

$$Q(z) = 2\pi \int_0^\infty \bar{v}_z r\, dr \sim 2\pi v_m \delta^2 \sim (2\pi K)^{1/2} az. \tag{10.6-23}$$

The increase in Q with increasing z indicates that the external fluid is continually entrained.

The prediction of Eq. (10.6-19) that circular turbulent jets are conical and have universal cone angles is remarkably consistent with experimental observations (Abramovich, 1963, p. 15). The inverse relationship between v_m and z in Eq. (10.6-21) is also supported by many studies. For example, Panchapakesan and Lumley (1993) found that, far from a nozzle of diameter D placed at $z = 0$, the maximum velocity of air jets was fitted very well by

$$\frac{U}{v_m} = 0.165 \frac{z}{D}, \quad 30 \le \frac{z}{D} \le 160 \tag{10.6-24}$$

where U is the mean velocity at the nozzle. Moreover, if the value of a (which depends on the precise definition of δ) is chosen properly, a similarity solution of Eq. (10.6-18) yields a velocity profile that closely matches what has been measured (Schlichting, 1968, pp. 699–700).

Limitations of mixing-length concept

The mixing-length concept has had notable successes, as discussed above and in Section 10.5. However, the limitations of Eq. (10.4-6) were recognized even when it was first proposed. A fundamental problem is that calculations of velocity fields are not fully predictive, in that parameters such as κ must be fitted to experimental data. That is different than analyzing laminar flow in a viscometer and using data to calculate μ. Once measured, μ is applicable to any flow, whereas even the form of $\varepsilon(y)$ is system-specific and open to conjecture. Another problem is the incorrect prediction that the Reynolds stresses vanish whenever $\bar{\mathbf{v}}$ is spatially uniform. The k–ε method (Section 10.4) is one widely used approach for remedying some of these shortcomings. However, it too is based on the Reynolds-averaged Navier–Stokes (RANS) equation and introduces eddy diffusivities to achieve closure.

In principle, a way to avoid the empiricisms attached to the RANS equation is to solve the continuity and Navier–Stokes equations directly using numerical methods. However, the rather severe limitations on how large Re can be (Section 10.2) are likely to continue to limit direct numerical simulation (DNS) to use as a research tool for many more years. More promising for engineering design is a hybrid approach called *large-eddy simulation* (LES) (Frölich and Rodi, 2002; Lesieur, 2008; Sandham, 2005). In LES the larger scales of motion are simulated by solving a form of the Navier–Stokes equation, whereas models are used to describe the smaller scales. To the extent that the computational grid can be refined, LES becomes more like DNS. All of these approaches (RANS methods, DNS, and LES) are in use at the time of writing.

10.7 CONCLUSION

Random fluctuations in velocity and pressure are hallmarks of turbulent flow. The fluctuations in the main velocity component are typically several percent of its average value. Turbulent eddies, which have a range of sizes, velocities, and lifetimes, carry momentum across streamlines much more rapidly than it is transferred by viscous stresses. The ratio of the smallest to the largest eddy size decreases as Re is increased [Eq. (10.2-12)], giving the turbulence a progressively finer-grained structure. The chaotic nature of turbulent flow precludes the use of deterministic models. Statistical descriptions of turbulence focus on time-smoothed variables. Reformulating the continuity and Navier–Stokes equations in terms of such variables is called Reynolds averaging. An inherent limitation of the Reynolds-averaged equations of motion [Eqs. (10.3-11), (10.3-13), and (10.3-17)] is that there are more unknowns than equations, creating what is termed the closure problem. The momentum equation contains Reynolds stresses [Eq. (10.3-22)], which involve time-smoothed products of velocity fluctuations. Calculation of the Reynolds stresses usually requires that the fundamental equations be supplemented by models that are at least partly empirical.

The simplest approach for evaluating Reynolds stresses is the eddy-diffusivity model [Eq. (10.4-6)], which contains a mixing length. When the mixing length is assumed to be proportional to the distance y from a wall, the time-smoothed velocity is predicted to vary linearly with y at first, and then logarithmically [Eqs. (10.5-10) and (10.5-13)]. That form of velocity profile near solid surfaces is indeed widely observed, although two constants in the model must be fitted to experimental data. Complete velocity profiles measured for turbulent flow in tubes are much more blunt than the parabolic profile for laminar flow, and may be fitted using either power-law or logarithmic functions. A logarithmic profile leads to the Prandtl–Kármán equation for the friction factor. The eddy-diffusivity concept has been applied also to turbulent boundary layers, where the smoothed momentum equation is analogous to that for laminar flow [Eq. (10.6-1)]. Velocity profiles for boundary layers on solid surfaces are much like those near the walls of conduits. Good predictions for jets are obtained by assuming that the mixing length is proportional to the jet width or radius. A variety of computational approaches are being explored to better predict results for turbulent flows of engineering interest.

REFERENCES

Abramovich, G. N. *The Theory of Turbulent Jets*. MIT Press, Cambridge, MA, 1963.

Deen, W. M. *Analysis of Transport Phenomena*, 2nd ed. Oxford University Press, New York, 2012.

Dimotakis, P. E., R. C. Miake-Lye, and D. A. Papantoniou. Structure and dynamics of round turbulent jets. *Phys. Fluids* 26: 3185–3192, 1983.

Durst, F., J. Jovanovic, and J. Sender. Detailed measurements of the near wall region of turbulent pipe flows. In *Ninth Symposium on Turbulent Shear Flows*, Kyoto, 1993.

Fröhlich, J. and W. Rodi. Introduction to large eddy simulations of turbulent flows. In *Closure Strategies for Turbulent and Transitional Flows*, B. E. Launder and N. D. Sandham (Eds.). Cambridge University Press, Cambridge, 2002, pp. 267–298.

Kármán, T. von. Turbulence and skin friction. *J. Aero. Sci.* 1: 1–20, 1934.

Kays, W. M. and M. E. Crawford. *Convective Heat and Mass Transfer*, 3rd ed. McGraw-Hill, New York, 1993.

Kim, J., P. Moin, and R. Moser. Turbulence statistics in fully developed channel flow at low Reynolds number. *J. Fluid Mech.* 177: 133–166, 1987.

Laufer, J. *The Structure of Turbulence in Fully Developed Pipe Flow*. NACA Technical Report No. 1174, 1954.

Lesieur, M., *Turbulence in Fluids*, 4th ed. Springer, Dordrecht, 2008.

McComb, W. D. *The Physics of Fluid Turbulence*. Clarendon Press, Oxford, 1990.

Moin, P. and K. Mahesh. Direct numerical simulation: A tool in turbulence research. *Annu. Rev. Fluid Mech.* 30: 539–578, 1998.

Panchapakesan, N. R. and J. L. Lumley. Turbulence measurements in axisymmetric jets of air and helium. Part 1. Air jet. *J. Fluid Mech.* 246: 197–223, 1993.

Prandtl, L. Bericht über Untersuchungen zur ausgebildeten Turbulenz. *Z. angew. Math. Mech.* 5: 136–139, 1925. [Translated into English as NACA Technical Memorandum No. 1231, 1949.]

Prandtl, L. Neuere Ergebnisse der Turbulenzforschung. *Z. Ver. deutsch. Ing.* 7: 105–114, 1933. [Translated into English as NACA Technical Memorandum No. 720, 1933.]

Reichardt, H. Vollständige Darstellung der turbulenten Geschwindigkeitsverteilung in glatten Leitungen. *Z. Angew. Math. Mech.* 31: 208–219, 1951.

Reynolds, O. An experimental investigation of the circumstances which determine whether the motion of water shall be direct or sinuous, and of the law of resistance in parallel channels. *Philos. Trans. R. Soc. Lond. A* 174: 935–982, 1883.

Reynolds, O. On the dynamical theory of incompressible viscous fluids and the determination of the criterion. *Philos. Trans. R. Soc. Lond. A* 186: 123–164, 1895.

Richardson, L. F. *Weather Prediction by Numerical Process*. Cambridge University Press, Cambridge, 1922.

Sandham, N. D. Turbulence simulation. In *Prediction of Turbulent Flows*, G. F. Hewitt and J. C. Vassillicos (Eds.). Cambridge University Press, Cambridge, 2005, pp. 207–235.

Schlichting, H. *Boundary-Layer Theory*, 6th ed. McGraw-Hill, New York, 1968.

Speziale, C. G. Analytical methods for the development of Reynolds-stress closures in turbulence. *Annu. Rev. Fluid Mech.* 23: 107–157, 1991.

Tennekes, H. and J. L. Lumley. *A First Course in Turbulence*. MIT Press, Cambridge, MA, 1972.

Van Driest, E. R. On turbulent flow near a wall. *J. Aero. Sci.* 23: 1007–1011, 1956.

Van Dyke, M. *An Album of Fluid Motion*. Parabolic Press, Stanford, CA, 1982.

PROBLEMS

10.1. Turbulence scales for water flow in a pipe Calculate the velocity, length, and time scales of the largest and smallest eddies in 20 °C water that is flowing at a mean velocity of 1.0 m/s in a smooth pipe with a diameter of 10 cm.

10.2. Cell damage in turbulent flow It has been suggested that, when turbulence is intense enough to create eddies that are comparable to or smaller than the size of a cell, the fluctuating shear stresses may cause cellular damage. For cells in water at 37 °C in a pipe with a diameter of 2.5 cm, at what mean velocities might this occur? A typical cell diameter is 10 μm.

10.3. Jet velocity from a photograph* Figure P.10.3 provides an instantaneous view of a submerged, circular jet of water. The water coming from the nozzle at the top was visualized using laser-induced fluorescence. In the lower part of the image the Kolmogorov scales are resolved. As will be outlined, length ratios in the photograph can be used to estimate the jet velocity. In particular, the objective is to infer the Reynolds number based on the nozzle diameter and mean velocity.

Figure P10.3 An axisymmetric water jet directed downward into water, visualized using laser-induced fluorescence as described in Dimotakis *et al.* (1983). Image reproduced from Van Dyke (1982, p. 97) with permission from P. E. Dimotakis.

At an axial distance z from the nozzle, let v_m be the maximum smoothed velocity (that at the center of the jet), let u_l be the large-eddy velocity scale (fluctuation in v_z), and let

L_1 and L_2 be the length scales for the largest and smallest eddies, respectively. It is useful to know that the turbulence intensity in a circular jet is $u_1/v_m = 0.24$ (Panchapakesan and Lumley, 1993). The correlation for $v_m(z)$ in Eq. (10.6-24) is applicable far from the nozzle.

(a) Show that Re can be expressed as a function of just two ratios, z/L_1 and L_1/L_2.
(b) Focusing near the bottom of the image, assume that the large-eddy size is that of the largest "vortex" that can be seen and that the small-eddy size (Kolmogorov length) is the thickness of the thinnest "threads." Measurements on an enlargement yield $z = 200$, $L_1 = 20$, and $L_2 = 0.3$, all in arbitrary units. Show that these lengths imply that Re = 1900. This agrees with the reported value of 2300 to one digit, which is perhaps better than might have been expected.
(c) There is some subjectivity in choosing the largest vortex and difficulty in precisely measuring thread widths. A value of L_1 as large as 35 (the approximate jet radius) would be reasonable, as would any L_2 between 0.2 and 0.4. Which of these dimensions is the more important source of uncertainty in estimating Re?

* This problem was suggested by P. S. Virk.

10.4. Reynolds-stress data In addition to the time-smoothed axial velocities for tube flow in Fig. 10.5, Laufer (1954) measured the fluctuations in all three velocity components. Some of those data are shown in Fig. P10.4. The ordinate is the time-smoothed value of the product $u_r u_z$, normalized by the square of the friction velocity. Data are shown for two values of Re. The equation of the line is

$$\frac{\overline{u_r u_z}}{u_\tau^2} = 1 - \frac{y}{R} \qquad \text{(P10.4-1)}$$

where y is the distance from the wall and R is the tube radius. Starting with the definition of the Reynolds stress, derive this equation from first principles. What conclusion can be reached from the fact that the data in Fig. P10.4 closely follow the line?

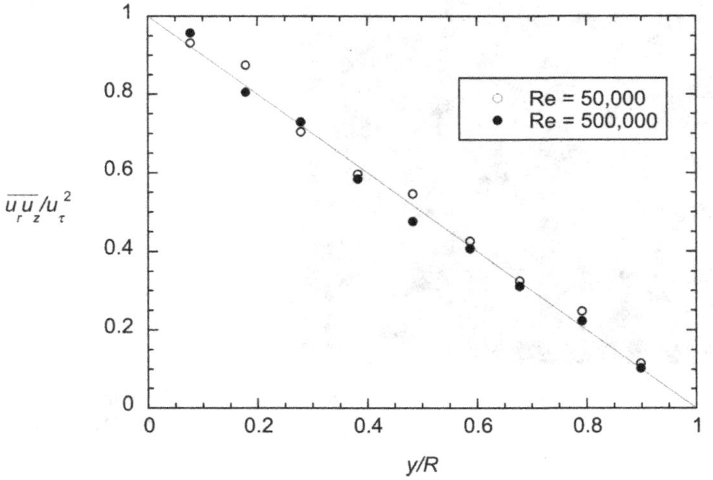

Figure P10.4 Double velocity correlation as a function of radial position in a tube of radius R, based on data in Laufer (1954).

10.5. Eddy diffusivity from near-wall velocity data The velocity profile near a wall is sometimes represented as

$$v_z^+(y^+) = \begin{cases} y^+, & 0 \leq y^+ \leq 5 \\ 5 \ln y^+ - 3.05, & 5 \leq y^+ < 30 \\ 2.5 \ln y^+ + 5.5, & y^+ \geq 30. \end{cases} \quad \text{(P10.5-1)}$$

(a) Plot this piecewise function and, based on the data in Fig. 10.4, comment on its accuracy.
(b) Derive and plot the function $\varepsilon(y^+)/\nu$ that is implied by Eq. (P10.5-1). Does it seem as realistic as Eq. (10.4-10)?

10.6. Mixing length in tube flow* As shown in Fig. 10.5, the velocity profile in a tube at $\text{Re} = 4.1 \times 10^4$ is represented quite well by a 1/7-power relationship. The objective is to see what this implies about the magnitude of the mixing length.

(a) For flow in a tube of radius R at the stated Re, derive the relationship between ℓ/R and y/R that is implied by the 1/7-power velocity profile. Note that this result will be unrealistic near the centerline, as discussed in Problem 10.8.
(b) For the region near the wall, where the 1/7-power profile is inapplicable because it yields an infinite shear rate at the surface, derive the relationship between ℓ/R and y/R that is implied by Eq. (10.4-10).
(c) Use the results from parts (a) and (b) to plot ℓ/R as a function of y/R. It is suggested that the expression from part (a) be restricted to $0.04 < y/R < 0.9$. Your plot should show that $\ell/R \sim 0.1$. That is what supports the assertion in Example 10.2-1 that $0.1D$ is a better estimate of the largest eddy sizes than is D.

* This problem was suggested by P. S. Virk.

10.7. Power-law velocity profile and Blasius friction factor When applied to tube flow, the 1/7-power velocity profile can be written as

$$v_z^+ = C(y^+)^{1/7} \quad \text{(P10.7-1)}$$

where C is a constant. This is analogous to Eq. (10.6-4) for a turbulent boundary layer. Show that this expression implies that the friction factor varies as $\text{Re}^{-1/4}$, as in the Blasius equation for smooth pipes [Eq. (2.2-9)]. How does the value of C implied by the Blasius correlation compare with the constant in Eq. (10.6-4)? (*Hint*: Start by using the 1/7-power expression to evaluate the velocity at the tube centerline.)

10.8. Improved velocity profile for tube flow A weakness of both the power-law and logarithmic velocity profiles is that they give $d\bar{v}_z/dy \neq 0$ at $y = R$. Because symmetry requires that the shear stress vanish at the centerline, this implies that $\varepsilon \to 0$ as $y \to R$. The idea that turbulent eddies somehow disappear at the center of a tube is unrealistic and can create errors in heat and mass transfer calculations. A way around this problem was proposed by Reichardt (1951), who suggested that the eddy diffusivity outside the wall region be approximated as

$$\frac{\varepsilon}{\nu} = \frac{\kappa R^+}{6}(1 - \eta^2)(1 + 2\eta^2) \quad \text{(P10.8-1)}$$

where $\eta = r/R$, R^+ is as defined in Eq. (10.2-6), and κ is an empirical constant.

(a) Show that the Reichardt eddy diffusivity leads to

$$\frac{dv_z^+}{d\eta} = -\frac{6\eta}{\kappa(1-\eta^2)(1+2\eta^2)}. \tag{P10.8-2}$$

Thus, the velocity gradient vanishes at $\eta = 0$, as desired.

(b) Show that the general solution for the velocity is

$$v_z^+(\eta) = \frac{1}{\kappa}\ln\left(\frac{1-\eta^2}{1+2\eta^2}\right) + C \tag{P10.8-3}$$

where C is a constant.

(c) Where the law of the wall applies, $\eta \to 1$. By choosing κ and C for consistency with Eq. (10.5-16), show that

$$v_z^+(\eta) = 2.5\ln\left[\frac{3}{2}y^+\left(\frac{1+\eta}{1+2\eta^2}\right)\right] + 5.5. \tag{P10.8-4}$$

Note that the right-hand side could be rewritten in terms of η alone, if desired, by using $y^+ = R^+(1-\eta)$.

10.9. Friction factor and hydraulic diameter Consider a parallel-plate channel with wall spacing $2H$. If the logarithmic velocity profile given by Eq. (10.5-16) is assumed to apply throughout the channel and if Re is based on the hydraulic diameter, show that the expression for the friction factor is nearly identical to Eq. (10.5-26). This illustrates how the near-universality of the law of the wall underlies the success of the hydraulic-diameter approach for calculating f for turbulent flow in noncircular channels.

10.10. Effects of tube roughness As discussed in Section 2.5, at very high Re the friction factor in a tube becomes independent of Re and is affected only by the relative roughness of the wall. In this "fully rough" regime

$$\frac{1}{\sqrt{f}} = -4\log\left(\frac{k}{3.7D}\right) \tag{P10.10-1}$$

where k is the effective roughness height and D is the tube diameter. Such an expression can be derived using two assumptions. One is to suppose that the mixing length near a rough wall is

$$\ell = \kappa(y + y_0), \quad y_0 = bk. \tag{P10.10-2}$$

In addition to the usual κ, there is now a dimensionless constant b that relates the length y_0 to the roughness height k. Thus, the mixing length is postulated to increase as the wall becomes rougher. The second assumption is that τ_0 is determined now not by viscous shear, but by pressure forces acting on the roughness elements that project in from the wall. This neglect of viscosity is consistent with the finding that f is independent of Re.

(a) Formulate the problem in terms of wall variables and show that

$$v_z^+(y^+) = \frac{1}{\kappa}\ln\left(\frac{y^+}{y_0^+} + 1\right). \tag{P10.10-3}$$

With viscous stresses neglected everywhere, including $y = 0$, this applies even at the wall.

(b) To obtain an expression for f, calculate the mean velocity as in Example 10.5-2. You may simplify the integration by assuming that $y \gg y_0$ over the entire cross-section. That is permissible because $k \ll D$ ordinarily and (as you will find) $y_0 < k$.

(c) What value of b brings your result from part (b) into agreement with Eq. (P10.10-1)? An analogous model for flow past roughened flat plates has a best-fit value of $b = 0.031$ (Kays and Crawford, 1993, p. 231).

10.11. Planar jet Derive OM estimates of the half-thickness (δ), maximum velocity (v_m), and volume flow rate per unit width (q) of a turbulent jet that passes from a slit into a large volume of the same fluid. Assume that the eddy diffusivity is related to δ and v_m as in Eq. (10.6-17). You should find that δ increases linearly with x, which is supported by data for submerged jets (Abramovich, 1963, p. 15). Is ε constant for a planar jet, as for an axisymmetric one, or does it vary with x? How do your results compare with those for a laminar jet in Example 9.3-2?

10.12. Eddy diffusivity in a circular jet Axisymmetric, submerged turbulent jets are conical on average (Example 10.6-2). As shown in Fig. P10.12, let β be the angle at which the velocity at any z is half-maximal. It has been found experimentally that $\tan \beta = 0.096$ (Panchapakesan and Lumley, 1993). The objective is to see what this implies about the magnitude of ε / ν.

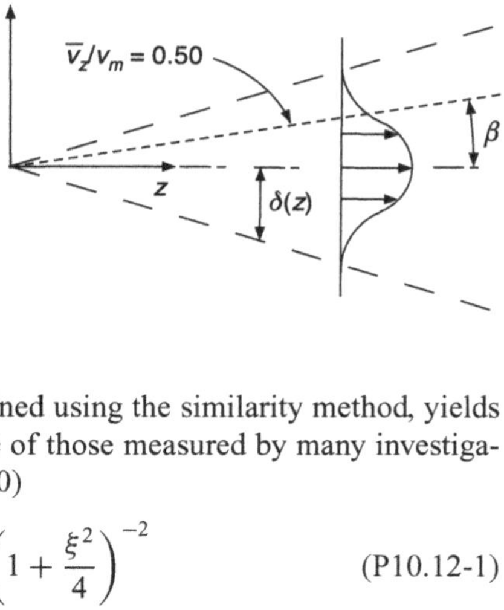

Figure P10.12 Spreading of an axisymmetric turbulent jet. The angle β describes an imaginary conical surface where the velocity is half of $v_m(z)$, its local maximum.

(a) An analytical solution of Eq. (10.6-18), obtained using the similarity method, yields velocity profiles that closely mimic the shape of those measured by many investigators. The solution is (Schlichting, 1968, p. 220)

$$\bar{v}_z(\xi, z) = \frac{3}{8\pi} \frac{K}{\varepsilon z} \left(1 + \frac{\xi^2}{4}\right)^{-2} \qquad \text{(P10.12-1)}$$

with

$$\xi = \left(\frac{3}{16\pi}\right)^{1/2} \frac{\sqrt{K}}{\varepsilon} \frac{r}{z} \qquad \text{(P10.12-2)}$$

where the kinematic momentum K is as defined in Eq. (10.6-20). At what value of ξ is the velocity half-maximal?

(b) Show that the experimental value of β implies that

$$\varepsilon = 0.0182\sqrt{K}. \qquad \text{(P10.12-3)}$$

(c) Use the result from part (b) and the correlation in Eq. (10.6-24) to show that

$$\frac{\varepsilon}{\nu} = 0.017\,\mathrm{Re} \qquad \text{(P10.12-4)}$$

where $\mathrm{Re} = UD/\nu$ is the Reynolds number at the nozzle. This indicates that ε/ν will be large, as assumed in Eq. (10.6-18), whenever Re is at least in the thousands. Because the Reynolds stresses in such a jet greatly exceed the viscous stresses, its spreading angle β is much larger than that of an otherwise-identical laminar jet.

Part IV

Macroscopic analysis

11

Macroscopic balances for mass, momentum, and energy

11.1 INTRODUCTION

Microscopic analysis, as discussed in Part III, is sometimes impractical. Analytical or even numerical solutions of the differential equations can be very hard to obtain for systems with complex shapes or multiple length scales. The difficulty in predicting velocity and pressure fields tends to increase with the Reynolds number and be greatest with turbulent flow. Macroscopic analysis, which employs integral (control volume) rather than differential (point) balances, provides a way to approach such difficult problems. When using only integral balances, much detail and some precision are sacrificed to obtain useful results. Such analysis is valuable even if only the form of a relationship can be predicted, leaving a coefficient to be evaluated experimentally. The benefits of adopting a more macroscopic viewpoint have been illustrated already by the integral analysis of laminar boundary layers (Section 9.4). Conserving momentum only for the boundary layer as a whole, and not necessarily at each point, greatly reduced the computational effort but still yielded acceptable accuracy.

The most direct way to derive a macroscopic balance is to state the volume and surface integrals that represent, respectively, the rate of accumulation of the quantity of interest in a control volume and the net rate at which it enters across the control surface. Any internal formation or loss of the quantity is represented by another volume integral. An alternate method, and sometimes the only option, is to integrate a differential conservation equation over the volume of interest. In this chapter these complementary approaches are used to derive integral balances for mass (Section 11.2), momentum (Section 11.3), and mechanical energy (Section 11.4). The momentum equation, which initially considers only fluid–solid interfaces, is extended then to systems with free surfaces (Section 11.5); no modification of the other integral balances is needed. The use of each kind of macroscopic balance is illustrated by examples.

11.2 CONSERVATION OF MASS

General control volume

The continuity equation was derived in Section 5.2 by imagining a small, cubic volume of fluid and equating the rate of increase of its mass with the net rate of mass entry across its surface. Letting the volume approach zero led to the differential equation that expresses conservation of mass at a point. The objective now is to derive a mass balance for a control volume of any size or shape. A volume $V(t)$ is assumed to be enclosed by a surface $S(t)$, as exemplified by the dashed region in Fig. 11.1. Depending on the system to be analyzed, some or all of S might coincide with interfaces (as with the lower surface

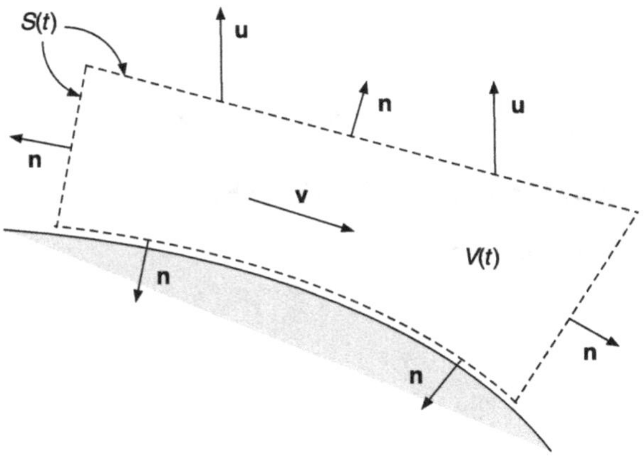

Figure 11.1 Macroscopic control volume.

in the diagram) or S might be entirely imaginary. At each point on S is a unit normal $\mathbf{n}$ that points *outward* from V. On any part of S that is planar, $\mathbf{n}$ will be independent of position, but otherwise the direction of $\mathbf{n}$ will change, as it does along the lower surface in Fig. 11.1. The way in which $\mathbf{n}$ varies over S defines the shape of the control volume. It can be helpful to choose a moving or deforming control volume. At least part of the control surface then has a nonzero velocity $\mathbf{u}$. Any variation in $\mathbf{u}$ over S indicates that the control volume is changing size or shape, making V and S functions of time. In Fig. 11.1 the top surface is moving upward and the others are stationary, indicating that V is increasing.

The total mass in a control volume will be denoted as $m(t)$. The inward component of the fluid velocity at a point on the control surface is $-\mathbf{v} \cdot \mathbf{n}$ (the minus sign because $\mathbf{n}$ points outward) and the inward velocity *relative to the surface* is $-(\mathbf{v} - \mathbf{u}) \cdot \mathbf{n}$. The rate of mass entry across an area element dS is the density times the inward volume flow rate, or $-\rho(\mathbf{v} - \mathbf{u}) \cdot \mathbf{n}\, dS$. Equating the rate of mass increase to the rate of entry gives

$$\frac{dm}{dt} = -\int_{S(t)} \rho(\mathbf{v} - \mathbf{u}) \cdot \mathbf{n}\, dS. \tag{11.2-1}$$

This is the most general statement of conservation of mass for a control volume. Wherever S coincides with an impermeable interface, moving or not, the normal components of $\mathbf{u}$ and $\mathbf{v}$ will be equal [Eq. (6.6-12)], so that $(\mathbf{v} - \mathbf{u}) \cdot \mathbf{n} = 0$. That part of S will not contribute to the integral.

Discrete openings

Equation (11.2-1) is more general than what is often needed. Typically, a device or process unit has discrete inlets and outlets. Suppose that there is a single inlet with open area S_1 and a single outlet with open area S_2, and that both are stationary. The fluid density is assumed to be constant over each opening, although ρ_1 may differ from ρ_2. Angle brackets will be used to denote an average over an opening. The mean velocities entering at S_1 and exiting at S_2 are $\langle v_1 \rangle$ and $\langle v_2 \rangle$, respectively (both positive). The corresponding mass flow rates are $w_1 = \rho_1 \langle v_1 \rangle S_1$ and $w_2 = \rho_2 \langle v_2 \rangle S_2$. Accordingly, the integral in Eq. (11.2-1) has two nonzero parts and

$$\frac{dm}{dt} = w_1 - w_2. \tag{11.2-2}$$

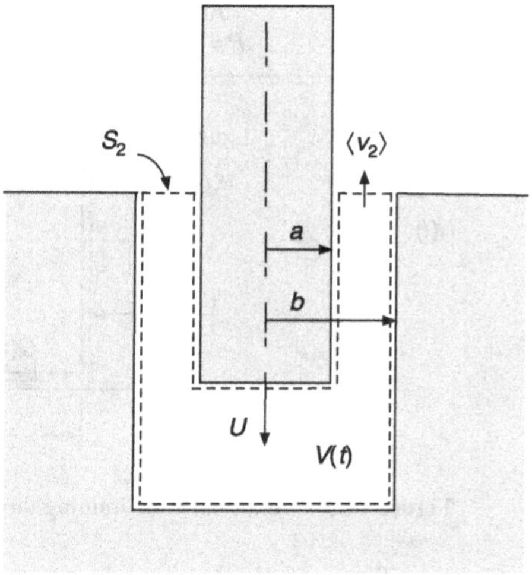

Figure 11.2 Fluid being displaced from a cylindrical cavity by a moving cylinder.

The rate of increase of mass within the control volume is the rate of mass inflow at S_1 minus the rate of mass outflow at S_2. If m is constant, then $w_1 = w_2$, as used in all pipe-flow calculations in Chapter 2. If there are multiple openings, Eq. (11.2-2) may be generalized by adding a term on the right-hand side for each additional inlet and subtracting one for each additional outlet.

Example 11.2-1 Fluid displacement from a cavity. Suppose that a cylinder of radius a is pushed at speed U into a fluid-filled cavity of radius b, as in Fig. 11.2. It is desired to calculate the mean velocity at which the incompressible fluid exits.

The control volume to be used is shown by the dashed lines. There is only an outlet, with a known annular area S_2 and an unknown velocity $\langle v_2 \rangle$ (relative to the cavity). In any short interval Δt the cylinder moves a distance $\Delta z = U \Delta t$ and displaces a volume $\Delta V = \pi a^2 U \Delta t$. Thus, $dV/dt = -\pi a^2 U$. With the fluid density being constant, Eq. (11.2-2) becomes

$$\frac{dm}{dt} = \rho \frac{dV}{dt} = -\rho \pi a^2 U = -\rho \langle v_2 \rangle S_2. \tag{11.2-3}$$

Because ρ can be factored out, conservation of mass is seen to be equivalent here to conservation of volume. That is always the case for incompressible fluids. Setting $S_2 = \pi(b^2 - a^2)$ and solving for $\langle v_2 \rangle$ gives

$$\langle v_2 \rangle = \frac{U}{(b/a)^2 - 1}. \tag{11.2-4}$$

Example 11.2-2 Draining of a tank through a horizontal pipe. Suppose that the bottom of a liquid-filled cylindrical tank of radius R is connected to a horizontal pipe of diameter d and length L, as shown in Fig. 11.3. The top of the tank and the pipe outlet are both open to the atmosphere. At $t = 0$ a valve (not shown) is opened and the tank begins to drain. It is desired to predict the rate of emptying. Because such a tank needs an indefinitely long time to drain fully, what will be calculated is the time $t_{1/2}$ taken to reach half the original volume. The liquid level is $H(t)$ and the volume flow rate through the pipe is $Q(t)$. The initial liquid level is H_0. For simplicity, it will be assumed that the

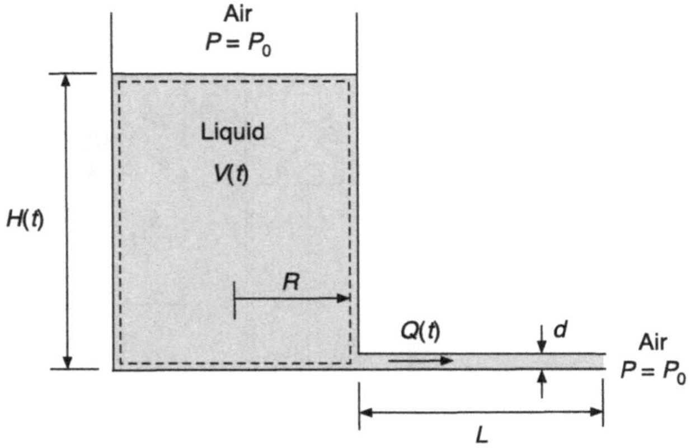

Figure 11.3 Cylindrical tank draining through a horizontal pipe.

pipe flow is fully developed and pseudosteady. Both laminar and turbulent pipe flow will be considered.

Including in the control volume all liquid in the tank, there is only an outlet. The instantaneous liquid volume is $V(t) = \pi R^2 H(t)$ and Eq. (11.2-2) becomes

$$\pi R^2 \frac{dH}{dt} = -Q. \tag{11.2-5}$$

The laminar and turbulent cases differ only in how Q is evaluated.

For laminar pipe flow we use Poiseuille's law [Eq. (2.3-7)],

$$Q = \frac{\pi d^4}{128\mu} \frac{|\Delta \mathcal{P}|}{L}. \tag{11.2-6}$$

In a large tank the fluid will move very slowly, except near the pipe inlet. Accordingly, it is assumed that the upstream pressure in the pipe is approximately equal to the static pressure at the tank bottom, or that $|\Delta \mathcal{P}| = |\Delta P| = \rho g H$. The liquid height is then governed by

$$\frac{dH}{dt} = -\left(\frac{d^4 g}{128 R^2 \nu L}\right) H, \quad H(0) = H_0. \tag{11.2-7}$$

This first-order, linear differential equation has the solution

$$\frac{H(t)}{H_0} = \exp\left(-\frac{d^4 g t}{128 R^2 \nu L}\right). \tag{11.2-8}$$

With the time for half-emptying defined by $H(t_{1/2})/H_0 = 1/2$, the result is

$$t_{1/2} = 128 \ln(2) \frac{R^2 \nu L}{d^4 g} = 88.7 \frac{R^2 \nu L}{d^4 g} \quad \text{(laminar)}. \tag{11.2-9}$$

In turbulent pipe flow Re is often large enough, when combined with wall roughness, to make the friction factor insensitive to changes in velocity. Accordingly, we will assume that f is a known constant. Solving Eq. (2.3-6) for the square of the mean velocity gives

$$U^2 = \frac{d}{2\rho f} \frac{|\Delta \mathcal{P}|}{L}. \tag{11.2-10}$$

Evaluating the pressure drop as before, the flow rate is

$$Q = \frac{\pi d^2}{4} U = \frac{\pi}{4\sqrt{2}} d^2 \left(\frac{gHd}{fL}\right)^{1/2}. \tag{11.2-11}$$

Substitution into Eq. (11.2-5) now gives

$$\frac{dH}{dt} = -\frac{1}{4\sqrt{2}} \left(\frac{d}{R}\right)^2 \left(\frac{gHd}{fL}\right)^{1/2}, \quad H(0) = H_0. \tag{11.2-12}$$

The solution of this separable first-order equation is

$$H_0^{1/2} - H^{1/2}(t) = \frac{1}{8\sqrt{2}} \left(\frac{d}{R}\right)^2 \left(\frac{gd}{fL}\right)^{1/2} t \tag{11.2-13}$$

from which it is found that

$$t_{1/2} = 8(\sqrt{2} - 1)\left(\frac{R}{d}\right)^2 \left(\frac{fLH_0}{gd}\right)^{1/2} = 3.31 \left(\frac{R}{d}\right)^2 \left(\frac{fLH_0}{gd}\right)^{1/2} \quad \text{(turbulent)}. \tag{11.2-14}$$

Whereas for laminar flow $t_{1/2}$ is independent of the initial liquid level in the tank, for turbulent flow it varies as $H_0^{1/2}$.

Integration of the continuity equation

Equation (11.2-1) may be obtained also by integrating the continuity equation,

$$\frac{\partial \rho}{\partial t} = -\nabla \cdot (\rho \mathbf{v}). \tag{11.2-15}$$

Both sides are integrated over $V(t)$, and then the divergence theorem [Eq. (A.4-31)] is used to convert one of the volume integrals into a surface integral, giving

$$\int_{V(t)} \frac{\partial \rho}{\partial t} dV = -\int_{V(t)} \nabla \cdot (\rho \mathbf{v})\, dV = -\int_{S(t)} \rho \mathbf{v} \cdot \mathbf{n}\, dS. \tag{11.2-16}$$

The general formula for differentiating a volume integral with time-dependent limits is

$$\frac{d}{dt} \int_{V(t)} f\, dV = \int_{V(t)} \frac{\partial f}{\partial t} dV + \int_{S(t)} f \mathbf{u} \cdot \mathbf{n}\, dS \tag{11.2-17}$$

where f is a continuous function of position and time and $\mathbf{u}$ is the velocity of the surface that encloses V. (The same equation applies if the scalar f is replaced by a vector or a tensor.) This is a generalization of the Leibniz formula for differentiating integrals that is presented in most calculus books. Equation (11.2-1) is obtained by setting $f = \rho$, using Eq. (11.2-17) to rewrite the left-hand side of Eq. (11.2-16), and recognizing that $m = \int \rho\, dV$. Whenever a conservation principle is expressed as a differential equation that applies throughout a fluid, this procedure can be used to find the corresponding macroscopic balance. Reversing the steps allows the differential equation to be inferred from the integral balance.

11.3 CONSERVATION OF MOMENTUM

General control volume

The force on a control volume equals the rate of change of the momentum in V plus the net rate of momentum outflow across S (Section 6.4). The accumulation and flow terms are evaluated as with conservation of mass (Section 11.2), the mass per unit volume (ρ) being replaced by the linear momentum per unit volume ($\rho\mathbf{v}$). Thus, the total momentum of the fluid in V is $\int \rho\mathbf{v}\,dV$. Letting $\mathbf{F}$ be the force exerted on the control volume by its surroundings,

$$\mathbf{F} = \frac{d}{dt}\int_{V(t)} \rho\mathbf{v}\,dV + \int_{S(t)} \rho\mathbf{v}(\mathbf{v}-\mathbf{u})\cdot\mathbf{n}\,dS. \quad (11.3\text{-}1)$$

With the mass in the control volume denoted as m, and including gravitational, pressure, and viscous contributions, the force is

$$\mathbf{F} = m\mathbf{g} - \int_{S(t)} \mathbf{n}P\,dS + \int_{S(t)} \mathbf{n}\cdot\boldsymbol{\tau}\,dS. \quad (11.3\text{-}2)$$

Equating the two expressions for $\mathbf{F}$ completes the momentum balance,

$$\frac{d}{dt}\int_{V(t)} \rho\mathbf{v}\,dV + \int_{S(t)} \rho\mathbf{v}(\mathbf{v}-\mathbf{u})\cdot\mathbf{n}\,dS = m\mathbf{g} - \int_{S(t)} \mathbf{n}P\,dS + \int_{S(t)} \mathbf{n}\cdot\boldsymbol{\tau}\,dS. \quad (11.3\text{-}3)$$

Discrete openings

Equation (11.3-3) is complete (if gravity is the only body force), but often too general to be convenient. Applications of macroscopic analysis typically involve steady flow through systems of fixed size. To obtain more practical relationships, we assume the following: (i) the total momentum is independent of time; (ii) the control surface is stationary; (iii) S consists of solid boundaries (denoted collectively as S_0) plus a planar inlet of area S_1 with outward normal $\mathbf{n_1}$ and a planar outlet of area S_2 with outward normal $\mathbf{n_2}$; (iv) ρ is uniform over each opening; (v) the viscous forces at the openings are negligible; and (vi) the flow is perpendicular to the openings, such that $\mathbf{v} = \pm v_i\mathbf{n_i}$ on opening i. The velocity v_i is a local value that will usually vary over the opening.[1] Because $\mathbf{n_i}$ points outward, the minus sign is needed for $i = 1$ and the plus sign for $i = 2$.

Assumption (i) requires only that the system be steady in an overall sense. There can be moving internal parts or other time-dependent flow inside a device. What is assumed is that, once turbulent fluctuations in velocity and pressure are smoothed, the volume and surface integrals in Eqs. (11.3-1)–(11.3-3) are constant. Assumption (ii) is self-explanatory and assumptions (iii) and (iv) have been used already to simplify the macroscopic mass balance (Section 11.2). Assumption (v) is motivated by what was found in Part III for unidirectional or nearly unidirectional flows. In such flows the viscous force in the flow direction is usually negligible because the velocity varies slowly in that direction (if at all). Also, the viscous force in the transverse direction is typically zero because of symmetry. Assumption (vi) again presumes nearly unidirectional flow.

The surface forces that contribute to $\mathbf{F}$ are rewritten now as

$$-\int_{S(t)} \mathbf{n}P\,dS + \int_{S(t)} \mathbf{n}\cdot\boldsymbol{\tau}\,dS = -\left(\mathbf{n}_1\langle P_1\rangle S_1 + \mathbf{n}_2\langle P_2\rangle S_2\right) - \mathbf{F}^* \quad (11.3\text{-}4)$$

$$\mathbf{F}^* = \int_{S_0} (\mathbf{n}P - \mathbf{n}\cdot\boldsymbol{\tau})\,dS. \quad (11.3\text{-}5)$$

[1] To simplify the notation, subscripts are used here to identify openings, rather than components of vectors. The directional information for opening i is provided by $\mathbf{n_i}$.

As defined in Eq. (11.3-5), $\mathbf{F}^*$ is the force that the enclosed fluid exerts on S_0, the solid part of the boundary. The signs are consistent with $\mathbf{n}$ pointing outward from the fluid. A minus sign accompanies $\mathbf{F}^*$ in Eq. (11.3-4) because it is a force that the fluid exerts on part of the control surface, whereas $\mathbf{F}$ is the force exerted on the system by its surroundings.

According to assumptions (iii) and (vi), the fluid velocity at a point on opening i is $\pm v_i \mathbf{n_i}$, where $v_i \geq 0$ and $\mathbf{n_i}$ is constant over the area S_i. From assumption (ii), $\mathbf{u} = \mathbf{0}$. The rate of momentum *outflow* through opening i is then

$$\int_{S_i} \rho \mathbf{v}\mathbf{v} \cdot \mathbf{n} \, dS = \rho_i \langle (\pm v_i \mathbf{n}_i)(\pm v_i \mathbf{n}_i) \cdot \mathbf{n}_i \rangle S_i = \rho_i \langle v_i^2 \rangle \mathbf{n}_i S_i. \tag{11.3-6}$$

Notice that the final sign is the same for the inlet and outlet. The steady-state momentum equation is then

$$\mathbf{F}^* = -\left(\langle P_1 \rangle + \rho_1 \langle v_1^2 \rangle\right) \mathbf{n}_1 S_1 - \left(\langle P_2 \rangle + \rho_2 \langle v_2^2 \rangle\right) \mathbf{n}_2 S_2 + m\mathbf{g}. \tag{11.3-7}$$

Because the objective often is to evaluate the force that a fluid exerts on a device, this has been arranged as an expression for $\mathbf{F}^*$.

Equation (11.3-7), which involves mean-square velocities, must be used in conjunction with conservation of mass, which involves just mean velocities. To avoid having mean-square velocities as separate unknowns, we define dimensionless constants a_i such that

$$a_i = \frac{\langle v_i^2 \rangle}{\langle v_i \rangle^2}. \tag{11.3-8}$$

The value of a_i is determined by the velocity profile at opening i, as will be discussed shortly. Introducing these factors, Eq. (11.3-7) becomes

$$\mathbf{F}^* = -(\langle P_1 \rangle + \rho_1 a_1 \langle v_1 \rangle^2)\mathbf{n}_1 S_1 - (\langle P_2 \rangle + \rho_2 a_2 \langle v_2 \rangle^2)\mathbf{n}_2 S_2 + m\mathbf{g} \tag{11.3-9}$$

which is the working form of the momentum equation for a system with one inlet and one outlet.

If there is plug flow (uniform velocity) at an opening, the local and average velocities will be equal, such that $v_i = \langle v_i \rangle$, $\langle v_i^2 \rangle = \langle v_i \rangle^2$, and $a_i = 1$. That this is a good approximation for turbulent flow, but not laminar flow, is suggested by fully developed velocity profiles in circular tubes. The turbulent velocity profile in a tube of radius R is represented well by Eq. (10.5-21), which is rewritten as

$$\frac{v_i}{\langle v_i \rangle} = \frac{60}{49}\left(\frac{y}{R}\right)^{1/7} \tag{11.3-10}$$

where $y = R - r$ is the distance from the wall. Accordingly,

$$a_i = \frac{\langle v_i^2 \rangle}{\langle v_i \rangle^2} = \left(\frac{60}{49}\right)^2 \left(\frac{2}{R^2}\right) \int_0^R \left(\frac{y}{R}\right)^{2/7} (R - y)\, dy = \frac{50}{49}. \tag{11.3-11}$$

Thus, for turbulent pipe flow, the error due to the plug-flow assumption is only 2%. Accordingly, we will set $a_i = 1$ when using Eq. (11.3-9) to analyze any system with turbulent flow, which is the most common kind of application for the macroscopic balances. The analogous result for laminar pipe flow is

$$a_i = \frac{\langle v_i^2 \rangle}{\langle v_i \rangle^2} = 4\left(\frac{2}{R^2}\right) \int_0^R \left[1 - \left(\frac{r}{R}\right)^2\right]^2 r\, dr = \frac{4}{3} \tag{11.3-12}$$

which is too far from unity for us to routinely set $a_i = 1$ in laminar systems.

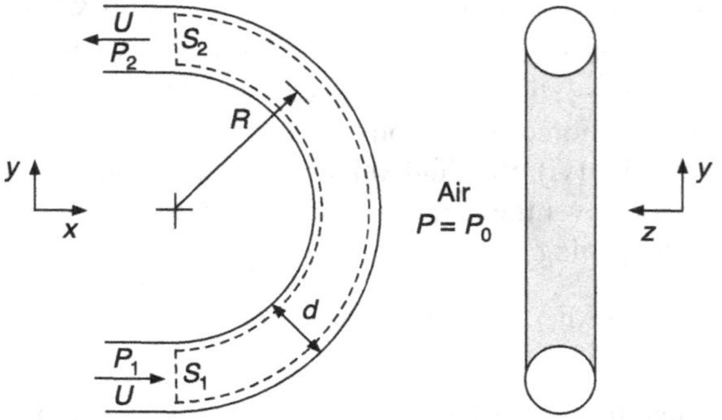

Figure 11.4 Flow through a return bend in a pipe.

It was seen in Part III that $\mathcal{P}$ is uniform over any cross-section in a flow that is either unidirectional or nearly unidirectional. Height differences corresponding to the diameters of pipes or other process equipment are typically small enough that P also will be nearly constant over an opening. Thus, when applying Eq. (11.3-9), the angle brackets around the pressures often will be omitted. When plug flow is assumed, angle brackets around the velocities are also unnecessary.

The momentum balance is easily extended to systems with multiple inlets or outlets. For N openings, Eq. (11.3-9) becomes

$$\mathbf{F}^* = -\sum_{i=1}^{N} (\langle P_i \rangle + \rho_i a_i \langle v_i \rangle^2)\mathbf{n}_i S_i + m\mathbf{g}. \tag{11.3-13}$$

Example 11.3-1 Force on a return bend. The objective is to determine the force that a steady turbulent flow exerts on a 180° pipe bend. As shown on the left in Fig. 11.4, the fluid enters in the $+x$ direction and exits in the $-x$ direction. The inside pipe diameter is d and the bend radius at the centerline is R. The flow path, which is in the x–y plane, is horizontal. Shown on the right is a view of the bend projected onto a plane that is perpendicular to the x axis. Given the pipe dimensions, the mean velocity U, and the inlet and outlet pressures (P_1 and P_2, respectively), it is desired to calculate the force F_x^* that an incompressible fluid of density ρ exerts on the curved section of pipe. Also to be determined is the force F_x^0 that must be supported by the threads or welds that connect the bend to the straight pipe sections.

As shown by the dashed lines, a control volume that encompasses the fluid between the inlet and outlet of the bend is chosen. Given the constant pipe diameter, $S_1 = S_2 = S = \pi d^2/4$. Employing the plug-flow approximation, $v_1 = v_2 = U$. The outward normal vectors at the openings are $\mathbf{n_1} = \mathbf{n_2} = -\mathbf{e_x}$. Gravity need not be considered because $g_x = 0$. Also setting $a_1 = a_2 = 1$ and $\rho_1 = \rho_2 = \rho$ in Eq. (11.3-9), the x component of the internal force is

$$F_x^* = \frac{\pi d^2}{4}(P_1 + P_2 + 2\rho U^2). \tag{11.3-14}$$

It is seen that the pressure forces on the openings and the inertial force due to the change in flow direction all act in the $+x$ direction.

The calculation of F_x^0 must include also the force exerted by the surrounding air, which is at constant pressure P_0. The shaded region on the right-hand side of Fig. 11.4 is

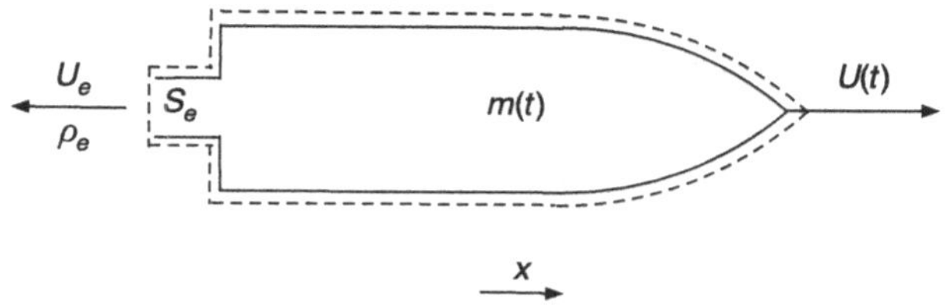

Figure 11.5 Idealized rocket.

the part of the projected area that is exposed to air on both sides, whereas the unshaded, circular areas have air only on one side, the other being inside the pipe. The air exerts no net force on the shaded parts. On the unshaded parts, it exerts a net force of magnitude $2P_0S$ toward the left, and the force on the pipe connections is found to be

$$F_x^0 = F_x^* - 2P_0S = \frac{\pi d^2}{4}[(P_1 - P_0) + (P_2 - P_0) + 2\rho U^2)]. \qquad (11.3\text{-}15)$$

It is seen that F_x^* and F_x^0 differ only in that absolute pressures are needed for the former and gauge pressures for the latter. Neither force depends directly on the bend radius. However, R will influence $P_1 - P_2$, flow resistances being greater for tighter bends. This analysis is extended in Example 12.3-1, where the pressure drop is calculated for a standard-radius bend.

Example 11.3-2 Acceleration of a force-free rocket. Consider the highly idealized rocket in Fig. 11.5, which is assumed to operate where gravity is negligible and there is no atmosphere to create drag. It is desired to determine the rocket velocity $U(t)$ that would result from a constant exhaust-gas velocity U_e and exhaust density ρ_e. The area of the exhaust port is S_e and the total mass of the rocket and its contents is $m(t)$.

As shown, a control volume that encompasses the entire rocket is chosen. Because this is a time-dependent problem with a moving control volume, we must return to Eq. (11.3-1). In the absence of gravity and drag, no force is exerted on the rocket by its surroundings and $\mathbf{F} = \mathbf{0}$. Assuming that the average velocity of the rocket and its contents equals U,

$$\frac{d}{dt}\int_V \rho \mathbf{v}\, dV = \frac{d}{dt}\left(U\mathbf{e_x}\int_V \rho\, dV\right) = \mathbf{e_x}\frac{d}{dt}(mU). \qquad (11.3\text{-}16)$$

At the exhaust, the fluid velocity is $\mathbf{v} = -U_e\mathbf{e_x}$, the surface velocity is $\mathbf{u} = U\mathbf{e_x}$, and the outward normal is $\mathbf{n} = -\mathbf{e_x}$. Thus, the integrand in the momentum outflow term at S_e is

$$\rho\mathbf{v}(\mathbf{v}-\mathbf{u})\cdot\mathbf{n} = \rho_e(-U_e\mathbf{e_x})(-U_e\mathbf{e_x} - U\mathbf{e_x})\cdot(-\mathbf{e_x}) = -\rho_e U_e(U_e + U)\mathbf{e_x} \qquad (11.3\text{-}17)$$

and the x component of Eq. (11.3-1) is

$$\frac{d}{dt}(mU) = \rho_e S_e U_e(U_e + U). \qquad (11.3\text{-}18)$$

This shows that, even though the rocket is force-free, it will accelerate due to the momentum efflux.

If the body of the rocket is massive enough that m remains approximately constant at m_0, and if the rocket is initially at rest, then the solution of Eq. (11.3-18) is

$$U(t) = U_e \left[\exp\left(\frac{\rho_e S_e U_e t}{m_0} \right) - 1 \right]. \qquad (11.3\text{-}19)$$

Thus, as long as U_e and ρ_e could remain constant, the speed would increase exponentially in time.

If there are significant reductions in m, conservation of mass in the form of Eq. (11.2-1) is needed. The integrand in the mass-flow term is

$$\rho(\mathbf{v} - \mathbf{u}) \cdot \mathbf{n} = \rho_e(-U_e \mathbf{e_x} - U \mathbf{e_x}) \cdot (-\mathbf{e_x}) = \rho_e(U_e + U) \qquad (11.3\text{-}20)$$

and the resulting mass balance is

$$\frac{dm}{dt} = -\rho_e S_e (U_e + U). \qquad (11.3\text{-}21)$$

The continual reduction in m would speed the acceleration. Equations (11.3-18) and (11.3-21) are coupled differential equations governing the two unknowns, $m(t)$ and $U(t)$. They could be solved numerically for specified values of ρ_e, S_e, U_e, $m(0)$, and $U(0)$.

11.4 MECHANICAL ENERGY BALANCES

In its most general form, the first law of thermodynamics shows how the total energy of an open system is affected by the flow of internal, kinetic, and potential energy in or out, by heat gain or loss, and by work done on the system by its surroundings [e.g., Deen (2012), p. 423]. The temperature variations that affect the internal energy are crucial for heat transfer, but less important for fluid dynamics. Only kinetic energy and gravitational potential energy are considered in what follows. The kinetic energy per unit mass is $v^2/2$, where $v^2 = \mathbf{v} \cdot \mathbf{v}$, and the potential energy per unit mass is $\Phi = gh$, where h is the height above a reference plane. Thus, the *mechanical energy* per unit volume is $\rho[(v^2/2) + \Phi]$ or $\rho[(v^2/2) + gh]$.

Mechanical energy is not conserved. Unlike mass, it has sources or sinks within a moving fluid, reflecting the interconversion of mechanical and internal energy via compression, expansion, or viscous heating of the fluid. Accordingly, a macroscopic balance for mechanical energy cannot be written as directly as with Eq. (11.2-1) for mass or Eq. (11.3-3) for momentum. Mechanical energy equations are derived instead from the Cauchy momentum equation, as will be explained. They are more a consequence of momentum conservation than they are statements of energy conservation.

Certain mechanical energy balances will be stated first, along with a summary of their origins. Their use will then be illustrated. The detailed derivations, which are more tedious than those for mass or momentum, are deferred until the end of this section.

General control volume

As derived later, the most general macroscopic balance for mechanical energy is

$$\frac{d}{dt} \int_{V(t)} \rho \left(\frac{v^2}{2} + \Phi \right) dV \qquad (11.4\text{-}1)$$

$$= \underbrace{-\int_{S(t)} \rho \left(\frac{v^2}{2} + \Phi \right) (\mathbf{v} - \mathbf{u}) \cdot \mathbf{n}\, dS}_{T1} \underbrace{- \int_{S(t)} P\mathbf{n} \cdot \mathbf{v}\, dS}_{T2} + \underbrace{\int_{S(t)} (\mathbf{n} \cdot \boldsymbol{\tau}) \cdot \mathbf{v}\, dS}_{T3} - E_c - E_v.$$

The left-hand side equals the rate of accumulation of mechanical energy. It is the same as the accumulation term for mass or momentum, except that the concentration variable $\rho[(v^2/2)+\Phi]$ replaces ρ (for mass) or $\rho\mathbf{v}$ (for momentum). On the right-hand side, term T1 (including the minus sign) is the rate of inflow of mechanical energy, $\mathbf{u}$ being the velocity of the control surface and $\mathbf{n}$ the outward unit normal. It is analogous to the flow terms for mass or momentum. Term T2 (including the minus sign) is the rate at which work is done on the system by the surrounding pressure and T3 is the rate at which work is done on the system by viscous forces. Consistent with the idea that a rate of work is a force times a similarly directed velocity, the work terms each involve the dot product of a stress vector with the fluid velocity vector.

The remaining terms in Eq. (11.4-1), E_c and E_v, represent sinks for mechanical energy within a fluid. The rate of loss of mechanical energy due to the compressibility of the fluid is given by

$$E_c = -\int_{V(t)} P(\nabla\cdot\mathbf{v})\,dV. \tag{11.4-2}$$

Continuity in the form of Eq. (5.3-6) indicates that ρ increases when $\nabla\cdot\mathbf{v}<0$, is constant when $\nabla\cdot\mathbf{v}=0$, and decreases when $\nabla\cdot\mathbf{v}>0$. Accordingly, $E_c>0$ when a fluid is compressed, $E_c=0$ when its density is unchanged, and $E_c<0$ when it is expanded. That is, mechanical energy is expended during compression and recovered during expansion. That E_c may have either sign indicates that such interconversions of mechanical and internal energy are reversible.

The loss of mechanical energy due to viscous friction is given by

$$E_v = \int_{V(t)} \boldsymbol{\tau}:\nabla\mathbf{v}\,dV. \tag{11.4-3}$$

For a Newtonian or generalized Newtonian fluid,

$$\boldsymbol{\tau}:\nabla\mathbf{v} = \mu\left[(2\Gamma)^2 - \frac{2}{3}(\nabla\cdot\mathbf{v})^2\right] \tag{11.4-4}$$

where $(2\Gamma)^2$ is evaluated as in Table 6.7. Clearly, $\boldsymbol{\tau}:\nabla\mathbf{v}\geq 0$ for incompressible fluids ($\nabla\cdot\mathbf{v}=0$), and this proves to be true also for compressible fluids (Example A.4-4). Thus, $E_v\geq 0$, reflecting the fact that viscous dissipation of mechanical energy into heat is irreversible.

Discrete openings

Equation (11.4-1) is ordinarily too general to be convenient, but it can be simplified by making assumptions like those in Sections 11.2 and 11.3. We suppose again that there is one inlet (at S_1) and one outlet (at S_2), both of which are stationary. We assume further that the openings are planar, the flow is perpendicular to them, and P, ρ, and Φ are each uniform over a given opening. If there are significant height differences across an opening, then Φ is based on the value of h at the center. The pressure work at each opening can then be evaluated simply and the viscous work there is assumed to be negligible. At any solid surface that is *stationary*, the integrands in terms T2 and T3 vanish and the rates of pressure and viscous work are both zero. The total rate of work (pressure plus viscous) done on the system by *moving* surfaces is denoted as W_m, which is called the *shaft work.* W_m accounts for energy that is transferred into the control volume by such things as pumps or stirrers, and energy that is extracted via turbines or other devices. With these assumptions, Eq. (11.4-1) becomes

$$\begin{aligned}\frac{d}{dt}\int_{V(t)} \rho\left(\frac{v^2}{2}+\Phi\right)dV = &\left[\rho_1\left(\frac{\langle v_1^3\rangle}{2}+\Phi_1\langle v_1\rangle\right)+P_1\langle v_1\rangle\right]S_1 \\ &-\left[\rho_2\left(\frac{\langle v_2^3\rangle}{2}+\Phi_2\langle v_2\rangle\right)+P_2\langle v_2\rangle\right]S_2 \\ &+W_m-E_c-E_v.\end{aligned} \tag{11.4-5}$$

Should a need arise, this mechanical energy balance is readily extended to include additional inlets or outlets.

If the total mass and mechanical energy of the system are both independent of time,[2] Eq. (11.4-5) simplifies to

$$\Delta\left(\frac{\langle v^3\rangle}{2\langle v\rangle}+\frac{P}{\rho}+gh\right)=\Delta\left(\frac{\langle v^3\rangle}{2\langle v\rangle}+\frac{\mathcal{P}}{\rho}\right)=\frac{1}{w}\left(W_m-E_c-E_v\right) \tag{11.4-6}$$

where Δ denotes the difference between outlet (position 2) and inlet (position 1) and $w=\rho_1\langle v_1\rangle S_1=\rho_2\langle v_2\rangle S_2$ is the mass flow rate. It is desirable that the mean-cube velocity not be a separate variable. Accordingly, a factor b (which might differ between inlet and outlet) is defined as

$$b=\frac{\langle v^3\rangle}{\langle v\rangle^3} \tag{11.4-7}$$

and the mechanical energy balance is rewritten as

$$\Delta\left(\frac{b\langle v\rangle^2}{2}+\frac{P}{\rho}+gh\right)=\Delta\left(\frac{b\langle v\rangle^2}{2}+\frac{\mathcal{P}}{\rho}\right)=\frac{1}{w}\left(W_m-E_c-E_v\right). \tag{11.4-8}$$

This widely used expression is called the *engineering Bernoulli equation*.

The engineering Bernoulli equation is applied most often to systems with turbulent flow, where the plug-flow approximation is accurate enough for us to set $b=1$ and replace $\langle v\rangle$ by v. These simplifications are suggested by results for fully developed flow in circular tubes. For turbulent flow, the result analogous to Eq. (11.3-11) is

$$b=\frac{\langle v^3\rangle}{\langle v\rangle^3}=\left(\frac{60}{49}\right)^3\frac{2}{R^2}\int_0^R\left(\frac{y}{R}\right)^{3/7}(R-y)\,dy=1.06. \tag{11.4-9}$$

An error of about 6% is ordinarily negligible, given the other approximations being made. For laminar flow, the result analogous to Eq. (11.3-12) is

$$b=\frac{\langle v^3\rangle}{\langle v\rangle^3}=8\left(\frac{2}{R^2}\right)\int_0^R\left[1-\left(\frac{r}{R}\right)^2\right]^3 r\,dr=2. \tag{11.4-10}$$

[2] The total mechanical energy can be constant without a true steady state, as time-dependent internal motion is not precluded. Also, except for the parts of $S(t)$ that coincide with the inlet and outlet, there can be movement of the control surface; indeed, W_m stems entirely from such motion. As usual, for turbulent flow all quantities are assumed to be time-smoothed.

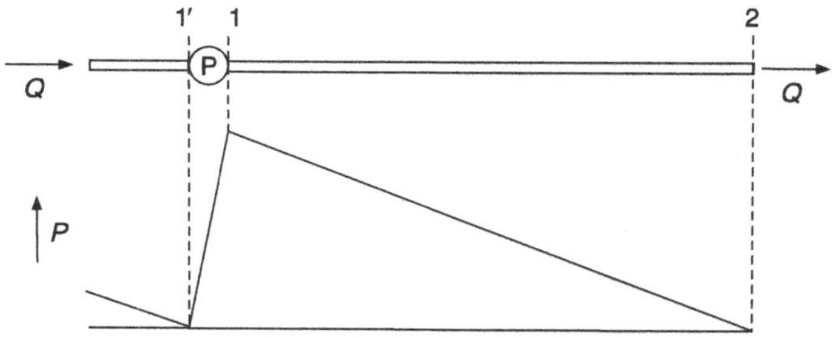

Figure 11.6 A horizontal section of pipe preceded by a pump (P). The pressure variations along the flow path are sketched in the lower part of the diagram. It is assumed here that the pumping power is such that the pressure increase from pump intake to discharge exactly offsets the pressure drop in the pipe.

Thus, as with the momentum equation, the plug-flow assumption is adequate for turbulent flow but generally unsuitable for laminar flow.

The engineering Bernoulli equation is a generalization of the Bernoulli equation derived in Section 9.2,

$$\Delta\left(\frac{v^2}{2}+\frac{P}{\rho}+gh\right)=0. \tag{11.4-11}$$

This applies to any two points on the same streamline in a steady, incompressible, inviscid flow; henceforth, it will be called the *streamline Bernoulli equation*. Shrinking a control volume to a streamline eliminates the distinction between mean and local velocities. It also excludes devices that might yield shaft work, so that $W_m = 0$. Moreover, $E_c = 0$ if ρ is constant and $E_v = 0$ if $\mu = 0$. Thus, the right-hand side of Eq. (11.4-8) vanishes under these conditions and Eq. (11.4-11) is recovered as a special case.

In the hydraulics literature the velocity and pressure terms in the Bernoulli equations are often expressed as equivalent heights, called *heads*. Such heights are obtained by dividing each term by g. Thus, a *velocity head* is $v^2/(2g)$ and a *pressure head* is $P/(\rho g)$, and either may be compared with the height h.

In this chapter the default assumption when applying mechanical energy balances is that the fluid is incompressible, so that $\rho_1 = \rho_2 = \rho$ (a constant) and $E_c = 0$. Compressibility is revisited in Section 12.4.

Example 11.4-1 Viscous loss in pipe flow. The objective is to evaluate E_v for steady, fully developed flow (laminar or turbulent) in a plain section of pipe. The viscous loss will be shown to equal the required pumping power and will be related also to the friction factor. The volume flow rate is Q, the mean velocity is U, and the pipe diameter and length are D and L, respectively.

The control volume that leads most directly to E_v is the fluid between positions 1 and 2 in Fig. 11.6. With D constant, the velocities at the inlet and outlet are the same and there is no change in the kinetic energy. Special devices being absent, $W_m = 0$. With $w = \rho Q$, Eq. (11.4-8) reduces to

$$E_v = Q\,|\Delta\mathcal{P}|\,. \tag{11.4-12}$$

This is the power that a pump must supply, as may be seen by extending the control volume to position 1′ and thereby including the pump that is just upstream. If the pump provides exactly the power needed, the pressure at its intake will equal that at the pipe outlet. In that case there is neither a change in kinetic energy nor a pressure drop from

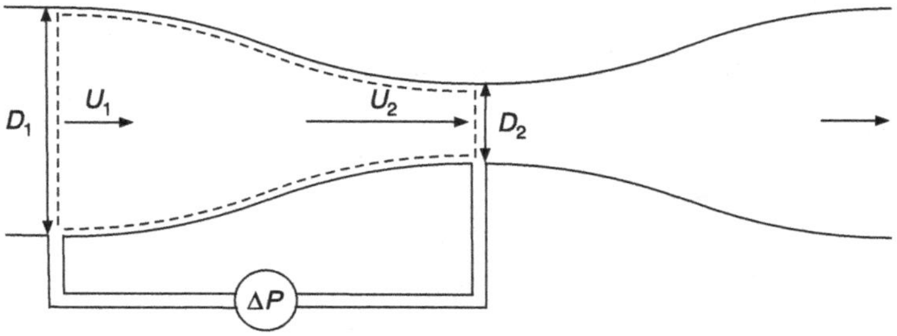

Figure 11.7 Venturi flow meter.

$1'$ to 2, but $W_m \neq 0$ and Eq. (11.4-8) indicates that $W_m = E_v$. Also depicted in Fig. 11.6 is how P for horizontal flow would vary with position. As shown in Chapters 7 and 10, in fully developed flow there is a linear decline in pressure along a pipe. Pumps provide local boosts in pressure to offset the pressure drops in pipe sections. The characteristics of particular types of pumps are discussed in Section 12.3.

The rate of viscous dissipation in any straight pipe can be related also to the friction factor. From Eq. (2.3-6), the pressure drop is

$$|\Delta\mathcal{P}| = \frac{2\rho U^2 L f}{D}. \tag{11.4-13}$$

Substituting this into Eq. (11.4-12) and dividing by the mass flow rate gives

$$\frac{E_v}{w} = \frac{2U^2 L f}{D}. \tag{11.4-14}$$

Notice that E_v is proportional to L. In piping systems there are usually additional contributions to the E_v/w term in Eq. (11.4-8), due to such things as diameter changes, bends, and valves (Section 12.3).

Example 11.4-2 Venturi flow meter. The Venturi tube in Fig. 11.7 provides a way to measure the flow rate in a pipe. In this device the channel diameter is reduced gradually from D_1 to D_2 and then restored to D_1. Taps connected to a gauge allow the pressure difference between the wide and narrow sections to be monitored. The objective is to relate the volume flow rate Q for steady turbulent flow to the pressure difference $|\Delta P| = P_1 - P_2$ for a device that is horizontal. Because Venturi tubes are short and their taper is designed to minimize flow separation and eddying (which can increase viscous losses), it will be assumed that $E_v \cong 0$.

The control volume chosen for the engineering Bernoulli equation encompasses the fluid in the converging part, as shown by the dashed lines. With $b = 1$ (plug flow), $\Delta h = 0$ (horizontal flow), $W_m = 0$ (no moving surfaces), $E_c = 0$ (incompressible fluid), and $E_v = 0$ (no viscous losses), Eq. (11.4-8) reduces to

$$\frac{U_1^2}{2} + \frac{P_1}{\rho} = \frac{U_2^2}{2} + \frac{P_2}{\rho} \tag{11.4-15}$$

where U_1 and U_2 are the mean velocities at the two locations. Because the flow is steady, incompressible, and inviscid, the streamline Bernoulli equation also could have been used. With plug flow, Eq. (11.4-15) is obtained also by applying Eq. (11.4-11) along the center streamline. However it is derived, Eq. (11.4-15) shows that the velocity increase at

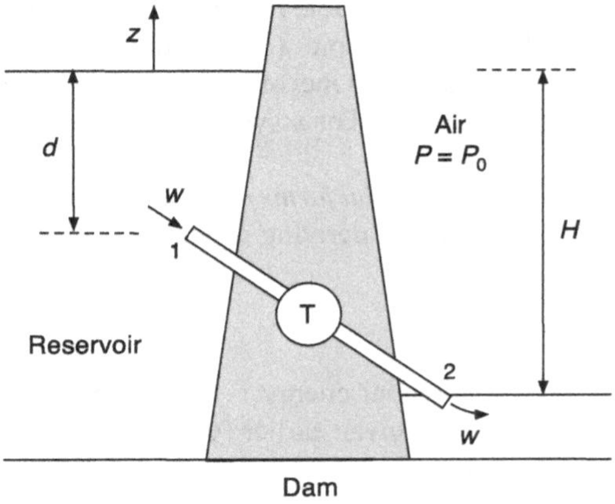

Figure 11.8 Hydroelectric dam and reservoir.

the neck decreases the pressure there. When the channel widens again the pressure recovers, and, if E_v is truly negligible, the pressure recovery will be complete. The pressure reduction caused by a velocity increase is referred to generally as the *Venturi effect*, after the Italian physicist Giovanni Venturi (1746–1822).

Constancy of w requires that $U_2/U_1 = (D_1/D_2)^2$. Using this to eliminate U_2 and relating the volume flow rate to U_1 as $Q = \pi D_1^2 U_1/4$, Eq. (11.4-15) is rearranged to give

$$Q = \frac{\pi D_1^2}{2}\left(\frac{|\Delta P|}{2\rho[(D_1/D_2)^4 - 1]}\right)^{1/2}. \tag{11.4-16}$$

This allows flow rates to be calculated from pressure measurements.

Example 11.4-3 Hydroelectric power. Figure 11.8 is a schematic of a hydroelectric dam and reservoir. It is assumed that water at a constant mass flow rate w enters an intake at a depth d below the top of the reservoir, passes through a turbine (T), and is discharged downstream near the surface. The water in the reservoir is at a height H above the discharge. The objective is to calculate the maximum power that can be generated.

The control volume that is chosen includes the water in the pipe and turbine (between positions 1 and 2 in the diagram). The electric power generated will be greatest if the hydrodynamic losses of mechanical energy are negligible and if all the work the water does on the turbine blades is converted into electricity. Thus, to find the maximum power it is assumed that $E_c = 0 = E_v$. For a constant pipe diameter there is no change in kinetic energy between intake and discharge. Measuring from the top of the reservoir, where $z = 0$, the height difference between outlet and inlet is $\Delta h = -H - (-d) = d - H$. Assuming that the pressure at the intake is nearly static, $\Delta P = P_0 - (P_0 + \rho g d) = -\rho g d$. Equation (11.4-8) then gives

$$W_m = w\left(\frac{\Delta P}{\rho} + g\,\Delta h\right) = -wgH. \tag{11.4-17}$$

The negative value of W_m indicates that the fluid in the control volume is doing work on its surroundings (i.e., the moving blades of the turbine). That is, power is being extracted from the system. Notice that the final result is independent of the intake depth. Large, modern hydroelectric installations generate approximately 90% of the theoretical maximum power, or about 0.9 wgH.

Additional note: mechanical energy derivations[3]

Two microscopic (differential) forms of the mechanical energy equation will be derived first. One of those will be integrated over a finite control volume to obtain Eq. (11.4-1). For a system with one inlet and one outlet, that will be simplified to give Eq. (11.4-5).

Differential forms of the mechanical energy equation. As already mentioned, the starting point for deriving mechanical energy balances is the Cauchy momentum equation,

$$\rho\left(\frac{\partial \mathbf{v}}{\partial t}+\mathbf{v}\cdot\nabla\mathbf{v}\right)=\rho\mathbf{g}-\nabla P+\nabla\cdot\boldsymbol{\tau}. \tag{11.4-18}$$

The potential energy is introduced by setting $\mathbf{g}=-\nabla\Phi$, which is consistent with the definition given earlier ($\Phi = gh$). For example, if the reference plane is at $z = 0$ and the z axis points upward, then $\Phi = gz$ and $\nabla\Phi = g\mathbf{e_z} = -\mathbf{g}$. Putting the potential energy on the left-hand side of the equation and dotting that side with $\mathbf{v}$ gives

$$\begin{aligned}\mathbf{v}\cdot\rho\left(\frac{\partial \mathbf{v}}{\partial t}+\mathbf{v}\cdot\nabla\mathbf{v}+\nabla\Phi\right)&=\rho\left[\frac{\partial}{\partial t}\left(\frac{v^2}{2}+\Phi\right)+\mathbf{v}\cdot\nabla\left(\frac{v^2}{2}+\Phi\right)\right]\\&=\rho\frac{D}{Dt}\left(\frac{v^2}{2}+\Phi\right)\end{aligned} \tag{11.4-19}$$

where Eq. (9.2-3) has been applied to the nonlinear part of the inertial term. The reason why Φ can be included in the time derivative is that $\mathbf{g}$ is constant, which makes $\partial\Phi/\partial t = 0$. Dotting $\mathbf{v}$ also with the other terms in the Cauchy equation gives

$$\rho\frac{D}{Dt}\left(\frac{v^2}{2}+\Phi\right)=-\mathbf{v}\cdot\nabla P+\mathbf{v}\cdot(\nabla\cdot\boldsymbol{\tau}) \tag{11.4-20}$$

which is the most compact form of microscopic mechanical energy balance.

After the microscopic balance has been integrated over a control volume, which will be done shortly, it will be helpful to have ρ inside the time derivative and to express as many terms as possible as divergences of vectors. Toward that end, the left-hand side of Eq. (11.4-20) is expanded as

$$\begin{aligned}\rho\frac{D}{Dt}\left(\frac{v^2}{2}+\Phi\right)=\frac{\partial}{\partial t}\left[\rho\left(\frac{v^2}{2}+\Phi\right)\right]+\nabla\cdot\left[\rho\mathbf{v}\left(\frac{v^2}{2}+\Phi\right)\right]\\-\left(\frac{v^2}{2}+\Phi\right)\left(\frac{\partial\rho}{\partial t}+\nabla\cdot(\rho\mathbf{v})\right)\end{aligned} \tag{11.4-21}$$

where identity (1) of Table A.1 has been used. From the continuity equation, the last term in large parentheses is zero. Again using identity (1) of Table A.1, the pressure term in Eq. (11.4-20) is expanded as

$$\mathbf{v}\cdot\nabla P=\nabla\cdot(P\mathbf{v})-P(\nabla\cdot\mathbf{v}). \tag{11.4-22}$$

Using the identity proven in Example A.4-3, the viscous term is rewritten as

$$\mathbf{v}\cdot(\nabla\cdot\boldsymbol{\tau})=\nabla\cdot(\boldsymbol{\tau}\cdot\mathbf{v})-\boldsymbol{\tau}:\nabla\mathbf{v}. \tag{11.4-23}$$

[3] The approach here generally follows that in Bird *et al.* (2002, pp. 81 and 221–223), although those authors use the opposite sign convention for $\boldsymbol{\tau}$.

The pressure and viscous terms have each now been split in such a way that one part is the divergence of a vector. An alternate form of the mechanical energy equation is then

$$\frac{\partial}{\partial t}\left[\rho\left(\frac{v^2}{2}+\Phi\right)\right] = -\nabla\cdot\left[\rho\mathbf{v}\left(\frac{v^2}{2}+\Phi\right)\right] - \nabla\cdot(P\mathbf{v}) + \nabla\cdot(\boldsymbol{\tau}\cdot\mathbf{v}) + P(\nabla\cdot\mathbf{v}) - \boldsymbol{\tau}:\nabla\mathbf{v}. \tag{11.4-24}$$

Although lengthier than Eq. (11.4-20), this form proves to be more useful.

Integral forms of the mechanical energy equation. Each term in Eq. (11.4-24) will be integrated now over an arbitrary control volume. Using the Leibniz formula [Eq. (11.2-17)], the integral of the accumulation term is rearranged as

$$\int_{V(t)} \frac{\partial}{\partial t}\left[\rho\left(\frac{v^2}{2}+\Phi\right)\right] dV = \frac{d}{dt}\int_{V(t)} \rho\left(\frac{v^2}{2}+\Phi\right) dV - \int_{S(t)} \rho\left(\frac{v^2}{2}+\Phi\right)\mathbf{u}\cdot\mathbf{n}\,dS. \tag{11.4-25}$$

Using Eq. (A.4-31), the volume integrals of the divergences are each converted now to surface integrals:

$$\int_{V(t)} \nabla\cdot\left[\rho\mathbf{v}\left(\frac{v^2}{2}+\Phi\right)\right] dV = \int_{S(t)} \rho\left(\frac{v^2}{2}+\Phi\right)\mathbf{v}\cdot\mathbf{n}\,dS \tag{11.4-26}$$

$$\int_{V(t)} \nabla\cdot(P\mathbf{v})\,dV = \int_{S(t)} P\mathbf{n}\cdot\mathbf{v}\,dS \tag{11.4-27}$$

$$\int_{V(t)} \nabla\cdot(\boldsymbol{\tau}\cdot\mathbf{v})\,dV = \int_{S(t)} (\boldsymbol{\tau}\cdot\mathbf{v})\cdot\mathbf{n}\,dS = \int_{S(t)} (\mathbf{n}\cdot\boldsymbol{\tau})\cdot\mathbf{v}\,dS. \tag{11.4-28}$$

The remaining integrals correspond to E_c and E_v. Assembling all the terms gives Eq. (11.4-1), the most general form of the macroscopic mechanical energy balance.

Obtaining Eq. (11.4-5) from Eq. (11.4-1) involves the evaluation of the energy inflow and work terms under certain restrictions. If the single inlet and single outlet are both fixed (such that $\mathbf{u} = \mathbf{0}$) and planar, then the net rate of inflow of mechanical energy is

$$-\int_{S(t)} \rho\left(\frac{v^2}{2}+\Phi\right)(\mathbf{v}-\mathbf{u})\cdot\mathbf{n}\,dS = \rho_1\left(\frac{\langle v_1\rangle^3}{2}+\Phi_1\langle v_1\rangle\right)S_1 - \rho_2\left(\frac{\langle v_2\rangle^3}{2}+\Phi_2\langle v_2\rangle\right)S_2. \tag{11.4-29}$$

The solid parts of S are distinguished by the fact that $\mathbf{v} = \mathbf{u}$ (from no penetration and no slip), making inflow or outflow absent there. Those surfaces may be subdivided into parts that are fixed and ones that are moving. At fixed solid surfaces, $\mathbf{v} = \mathbf{u} = \mathbf{0}$ and there is neither pressure nor viscous work. The total rate of work done at moving solid surfaces (area S_m) is

$$W_m = \int_{S_m} [-P\mathbf{n} + (\mathbf{n}\cdot\boldsymbol{\tau})]\cdot\mathbf{v}\,dS. \tag{11.4-30}$$

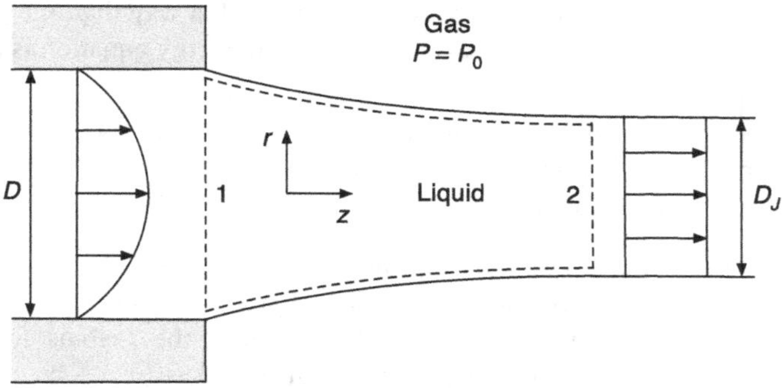

Figure 11.9 Laminar jet with initial diameter D and final diameter D_J.

The integrand is the total stress vector dotted with the fluid velocity vector. At the openings the viscous work is assumed to be negligible, whereas the total pressure work there is

$$-\int_{S_1} P\mathbf{n}\cdot\mathbf{v}\,dS - \int_{S_2} P\mathbf{n}\cdot\mathbf{v}\,dS = P_1\langle v_1\rangle S_1 - P_2\langle v_2\rangle S_2. \qquad (11.4\text{-}31)$$

When the inflow and work terms in Eq. (11.4-1) are evaluated as in Eqs. (11.4-29)–(11.4-31), the result is Eq. (11.4-5).

11.5 SYSTEMS WITH FREE SURFACES

In deriving the simplified momentum balance [Eq. (11.3-9)] the various parts of the control surface S were assumed to be of two kinds, solid surfaces or openings. A solid surface is distinguished both by the absence of inflow or outflow and by the possibility of a significant viscous force. At inlets and outlets flow is present but viscous forces are typically negligible. When part of S coincides with a gas–liquid interface, there is a third kind of surface. As discussed in Example 7.4-1, viscous stresses exerted on a liquid by a gas tend to be negligible. Thus, free surfaces resemble solid ones in that there is no flow across them, but, as with inlets or outlets, there is only a pressure force on the liquid. This presumes that surface tension is negligible, which is a default assumption in macroscopic analyses. To accompany $\mathbf{F}^*$, the force that the enclosed fluid exerts on the solid parts of the control surface, we denote the force the liquid exerts on the gas–liquid interfaces as $\mathbf{F}^{**}$. Free surfaces may have complex shapes, but the pressure there is typically constant. Accordingly, projected areas (Section 4.3) are very useful in evaluating $\mathbf{F}^{**}$.

For a control surface comprised of inlets, outlets, solid boundaries, and gas–liquid interfaces, Eq. (11.3-13) is generalized to

$$\mathbf{F}^* + \mathbf{F}^{**} = -\sum_{i=1}^{N}(\langle P_i\rangle + \rho_i a_i\langle v_i\rangle^2)\mathbf{n}_i S_i + m\mathbf{g}. \qquad (11.5\text{-}1)$$

If the pressure is constant over an opening, then $\langle P_i\rangle = P_i$, and if the plug-flow approximation applies, then $a_i = 1$ and $\langle v_i\rangle = v_i$. The mass and mechanical energy balances remain the same.

Example 11.5-1 Capillary jet. Suppose that a stream of liquid flows steadily from a small, horizontal tube into a gas-filled space, as in Fig. 11.9. The flow is laminar and the liquid emerges at mean velocity U from a tube of diameter D. It is desired to calculate the final jet diameter, D_J. It is assumed that D_J is reached before the jet is bent noticeably by gravity.

As shown, the control volume will extend from the tube opening to where the diameter stops changing. Cylindrical coordinates will be used. If surface tension is negligible and the pressure is constant over any cross-section, then $P = P_0$ throughout the jet. With P constant over the control surface, the net pressure force on the enclosed liquid is zero. Because there is no solid boundary, the viscous force on the liquid is also assumed to be negligible. The absence of a net horizontal force on the control volume suggests the use of a momentum balance. (To apply a mechanical energy balance we would need the viscous loss in the jet, which is unknown.)

The use of projected areas gives $F_z^{**} = P_0(S_1 - S_2)$ and the pressure terms in Eq. (11.5-1) then cancel. With $F_z^* = 0$ also, the momentum balance simplifies to

$$a_1 \langle v_1 \rangle^2 S_1 = a_2 \langle v_2 \rangle^2 S_2. \tag{11.5-2}$$

Thus, the momentum inflow at 1 equals the momentum outflow at 2. The mass inflow and outflow also must be equal, so

$$\langle v_1 \rangle S_1 = \langle v_2 \rangle S_2. \tag{11.5-3}$$

Combining the momentum and mass balances so as to eliminate the velocity ratio gives

$$\frac{S_2}{S_1} = \left(\frac{D_J}{D}\right)^2 = \frac{a_2}{a_1}. \tag{11.5-4}$$

Assuming fully developed flow at the tube end, $a_1 = 4/3$ [Eq. (11.3-12)]. Because the shear stress at the gas–liquid interface is negligible, the radial velocity variations in the tube are not sustained in the jet. Once the radial variations have disappeared, the jet diameter will stop changing and there will be plug flow. Thus, $a_2 = 1$ and

$$\frac{D_J}{D} = \left(\frac{3}{4}\right)^{1/2} = 0.87. \tag{11.5-5}$$

Jet diameters predicted by Eq. (11.5-5) agree well with experimental findings for Re > 150 (Middleman and Gavis, 1961). However, values of $D_J/D > 0.87$, and even exceeding unity, are seen at smaller Re. The simple model fails at smaller Re because (i) τ_{zz} at the tube outlet is no longer negligible, as assumed in Eq. (11.5-1); (ii) the effects of surface tension become evident; and (iii) the velocity profile at the end of the tube is no longer exactly parabolic. For Re $\ll 1$, the effects of a sudden change in boundary condition (such as replacing zero velocity at the wall with zero viscous stress at the gas–liquid interface) propagate approximately one radius upstream. At large Re such exit effects are negligible, as assumed in the analysis, but not at small to moderate Re.

Example 11.5-2 Hydraulic jump. It is desired to determine when an event like that in Fig. 11.10 might occur during steady, turbulent flow in an open, horizontal channel. Such a sudden increase in liquid depth is called a *hydraulic jump*. A hydraulic jump involving radial flow may be observed simply by running water from a faucet into a sink with a flat bottom. The increase in depth that occurs at a certain radius is easy to see. Hydraulic jumps are used to dissipate energy in dam spillways and occur naturally in whitewater rapids and as tidal bores in coastal rivers. Typical of an actual jump are violent surface agitation and an interface that is much more irregular than that depicted in Fig. 11.10. However, the precise shape of the free surface does not affect the analysis.

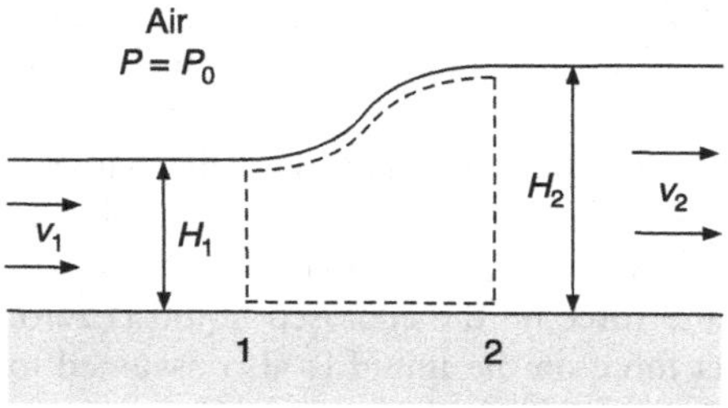

Figure 11.10 Hydraulic jump in an open channel.

The control volume that will be used is indicated by the dashed lines. Aside from plug flow at positions 1 and 2, there are two key assumptions. One is that $F_x^* = 0$, where x is the flow direction. That is, the jump is assumed to occur over a channel length short enough to make the shear force on the bottom negligible. The other key assumption is that the pressure is the same as in a static system. We have often ignored height differences and assumed that P is constant over a given cross-section. However, what is actually constant in unidirectional or nearly unidirectional flow is $\mathcal{P}$. Assuming that the pressure variations are static is the same as assuming that $\mathcal{P}$ is constant.

The known force on the channel bottom (i.e., $F_x^* = 0$) and unknown viscous loss (E_v) make a momentum balance a more attractive starting point than a mechanical energy balance. With P_1 and P_2 each increasing linearly with depth, the height-averaged pressures are $\langle P_1 \rangle = P_0 + \rho g H_1/2$ and $\langle P_2 \rangle = P_0 + \rho g H_2/2$. The pressure force exerted by the liquid on the gas–liquid interface is $F_x^{**} = -P_0(H_2 - H_1)W$, where W is the channel width. The P_0 terms from the right-hand side of Eq. (11.5-1) are canceled by F_x^{**}, leaving

$$v_1^2 H_1 - v_2^2 H_2 = \frac{g}{2}\left(H_2^2 - H_1^2\right). \tag{11.5-6}$$

Also to be satisfied is conservation of mass,

$$v_1 H_1 = v_2 H_2. \tag{11.5-7}$$

If v_1 and H_1 are specified, the momentum and mass balances have the trivial solution $v_2 = v_1$ and $H_2 = H_1$. What will be shown is that they also admit the possibility that $H_2 > H_1$.

Letting $\phi = H_2/H_1$ and $\text{Fr} = v_1^2/(gH_1)$, Eqs. (11.5-6) and (11.5-7) may be combined to give

$$\phi^3 - (1 + 2\,\text{Fr})\phi + 2\,\text{Fr} = 0. \tag{11.5-8}$$

The height ratio ϕ is seen to depend on just one dimensionless parameter, a Froude number (Fr). One root of this cubic equation is the trivial solution, $\phi = 1$. Factoring that out yields a quadratic, $\phi^2 + \phi - 2\,\text{Fr} = 0$, which has one negative and one positive root. The relevant (positive) root is

$$\phi = \frac{\sqrt{1 + 8\,\text{Fr}} - 1}{2} \tag{11.5-9}$$

which indicates that $\phi > 1$ for $\text{Fr} > 1$. That is, the model confirms the possibility of a hydraulic jump and reveals that $v_1^2 > gH_1$ is necessary for such an event. The predicted depth ratios for $\text{Fr} > 1$ are in excellent agreement with observations (Sturm, 2010, p. 76).

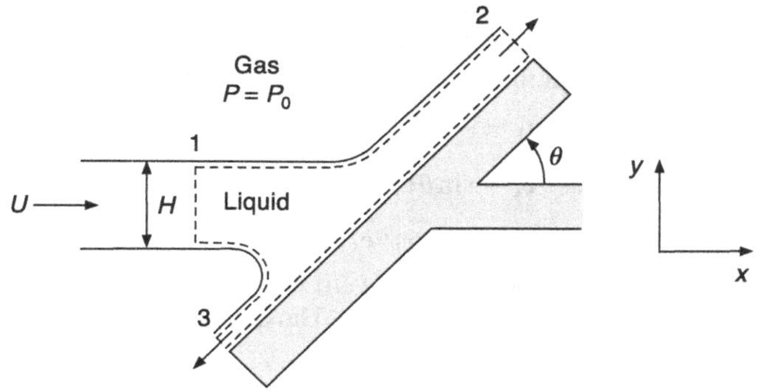

Figure 11.11 A planar liquid jet striking an inclined plate that is held in position.

Equation (11.5-9) also suggests that a "hydraulic drop" might occur ($\phi < 1$ for Fr < 1), but that is excluded by mechanical-energy considerations. Taking care to average the pressures and elevations, the engineering Bernoulli equation becomes

$$\Delta\left(\frac{v^2}{2} + \frac{\langle P\rangle}{\rho} + g\langle h\rangle\right) = -\frac{E_v}{w}. \tag{11.5-10}$$

The difference in the average pressures is $\Delta\langle P\rangle = \rho g(H_2 - H_1)/2$ and the difference in the average heights is $\Delta\langle h\rangle = (H_2 - H_1)/2$. Solving for the viscous loss, we find that

$$\frac{E_v}{w} = \frac{v_1^2 - v_2^2}{2} + g(H_1 - H_2). \tag{11.5-11}$$

Dividing by gH_1, using Eq. (11.5-7) to eliminate v_2, and using Eq. (11.5-6) to relate $v_1^2/(2gH_1)$ to ϕ gives (eventually)

$$\frac{E_v}{wgH_1} = \frac{(\phi - 1)^3}{4\phi}. \tag{11.5-12}$$

The fact that $E_v \geq 0$ in any flow indicates that only $\phi \geq 1$ is possible.

The foregoing does not explain why $\phi > 1$ would ever occur in preference to $\phi = 1$, just as inspection of the Navier–Stokes equation does not reveal why any flow would ever be turbulent instead of laminar. Apparently, Fr > 1 makes undisturbed open-channel flow unstable, just as large values of Re make laminar flow unstable. In that Fr is the ratio of inertial to gravitational forces and Re is the ratio of inertial to viscous forces (Table 1.6), inertia is evidently at the root of both kinds of instabilities, however different they might be otherwise.

Example 11.5-3 Liquid jet striking an inclined plate. Consider a turbulent liquid jet impacting a plate that is supported at a fixed position and angle, as shown in Fig. 11.11. The planar jet has a velocity U and thickness H, and the plate is tilted at an angle θ relative to the approaching flow. The jet and plate are of width W and the linear dimensions are small enough to make gravitational effects negligible. To be determined are the thicknesses and velocities of the exit streams and the force that the jet exerts on the plate, assuming that the flow is nearly inviscid. In particular, what is of interest is how much $\mathbf{F}^*$ exceeds the force from the ambient pressure P_0, as only that excess force is transmitted to the support.

As shown by the dashed lines, the control volume to be used has one inlet (position 1) and two outlets (positions 2 and 3). The solid part of the control surface, coinciding with the plate, is S_0. The outward normals at the openings ($\mathbf{n_1}$ and $\mathbf{n_2}$) and plate ($\mathbf{n_0}$) are

$$\mathbf{n_1} = -\mathbf{e_x} \tag{11.5-13a}$$

$$\mathbf{n_2} = \cos\theta\,\mathbf{e_x} + \sin\theta\,\mathbf{e_y} = -\mathbf{n_3} \tag{11.5-13b}$$

$$\mathbf{n_0} = \sin\theta\,\mathbf{e_x} - \cos\theta\,\mathbf{e_y}. \tag{11.5-13c}$$

Keeping in mind that $\mathbf{n_0}$ points toward the plate (outward from the control volume), the force on the plate due to the ambient pressure is $P_0S_0\mathbf{n_0}$. Thus, the excess force that is to be evaluated is $\mathbf{F}^* - P_0S_0\mathbf{n_0}$.

The pressure force $\mathbf{F}^{**}$ exerted by the liquid on the free surfaces is evaluated indirectly, as follows. The pressure at the free surfaces is constant at P_0, and that may be assumed to be the case also at each opening. If the pressure at S_0 were also P_0, the net pressure force on the control volume would be zero [Eq. (4.3-17)]. Writing such a force balance and solving for $\mathbf{F}^{**}$ gives

$$\mathbf{F}^{**} = -P_0S_0\mathbf{n_0} - \sum_{i=1}^{3} P_0S_i\mathbf{n_i} \tag{11.5-14}$$

for the hypothetical situation. However, this must also be the actual $\mathbf{F}^{**}$, because the free-surface pressure and geometry are the same for both cases (actual and hypothetical).

Neglecting gravity in Eq. (11.5-1) and using Eq. (11.5-14) to evaluate $\mathbf{F}^{**}$, we find that

$$\mathbf{F}^* - P_0S_0\mathbf{n_0} - \sum_{i=1}^{3} P_0S_i\mathbf{n_i} = -\sum_{i=1}^{3}\left(P_0 + \frac{\rho v_i^2}{2}\right)S_i\mathbf{n_i}. \tag{11.5-15}$$

The pressure forces at the openings cancel, leaving

$$\mathbf{F}^* - P_0S_0\mathbf{n_0} = -\sum_{i=1}^{3}\frac{\rho v_i^2}{2}S_i\mathbf{n_i}. \tag{11.5-16}$$

Subsequent relationships are simplified by working with a dimensionless force $\mathbf{f}$. Dividing both sides of Eq. (11.5-16) by $\rho U^2HW/2$ and evaluating the sum gives

$$\mathbf{f} = \frac{\mathbf{F}^* - P_0S_0\mathbf{n_0}}{\rho U^2HW/2} = \mathbf{e_x} + \mathbf{n_2}\left[\left(\frac{v_3}{U}\right)^2\left(\frac{H_3}{H}\right) - \left(\frac{v_2}{U}\right)^2\left(\frac{H_2}{H}\right)\right] \tag{11.5-17}$$

where H_2 and H_3 are the thicknesses of the exiting streams. Those thicknesses are related by conservation of mass as

$$UH = v_2H_2 + v_3H_3. \tag{11.5-18}$$

We have six unknowns ($v_2, v_3, H_2, H_3, f_x, f_y$), but so far only three equations [two from the x and y components of Eq. (11.5-17) and one from Eq. (11.5-18)]. The remaining three relationships all follow from the assumption of inviscid flow. In the absence of viscous losses, the streamline Bernoulli equation can be applied along each gas–liquid interface, from 1 to 2 and 1 to 3. With $P = P_0$ along those interfaces and gravity negligible, that leads to $v_2 = U = v_3$. Conservation of mass indicates then that $H = H_2 + H_3$ and Eq. (11.5-17) simplifies to

$$\mathbf{f} = \mathbf{e_x} + \mathbf{n_2}\left(1 - \frac{2H_2}{H}\right). \tag{11.5-19}$$

The final condition comes from recognizing that there is no shear force in an inviscid flow, or that $\mathbf{n_2} \cdot \mathbf{f} = 0$. Calculating that dot product using Eqs. (11.5-13) and (11.5-19) determines H_2, and the other unknowns follow. The final expressions for the thicknesses and net fluid force are

$$\frac{H_2}{H} = \frac{1+\cos\theta}{2}, \quad \frac{H_3}{H} = \frac{1-\cos\theta}{2} \tag{11.5-20}$$

$$\mathbf{F}^* - P_0 S_0 \mathbf{n_0} = \frac{\rho U^2 H W \sin\theta}{2}(\sin\theta\,\mathbf{e_x} - \cos\theta\,\mathbf{e_y}). \tag{11.5-21}$$

Notice that the force depends on the jet thickness, but not the plate length. If the plate is parallel to the jet ($\theta = 0$), both force components vanish. If it is perpendicular ($\theta = \pi/2$), the y component of the force is zero (as expected from symmetry) and the x component is $\rho U^2 HW/2$, which is the stagnation pressure times the jet cross-section. The y component of the force is greatest when $\theta = \pi/4$. If the plate were a vane of a turbine with an axis of rotation perpendicular to the page, that 45° angle would maximize the torque.

11.6 CONCLUSION

Microscopic analysis is impractical for many systems of engineering interest, especially when the flow is turbulent. An alternative is to adopt a more macroscopic viewpoint and use integral forms of the balance equations. Conservation of mass and momentum for an arbitrary control volume are expressed by Eqs. (11.2-1) and (11.3-3), respectively. Equation (11.4-1) is a similarly general mechanical energy balance. More widely used are macroscopic balances for systems with just one inlet and one outlet. Such relationships for mass and momentum are Eqs. (11.2-2) and (11.3-9), respectively. The one for mechanical energy, Eq. (11.4-8), is the engineering Bernoulli equation. The momentum and mechanical energy equations are simplified by assuming that each opening is planar, that the flow is perpendicular to the opening, and that the total amount of the quantity within the control volume is independent of time. Viscous forces and viscous work at the openings are each assumed to be negligible. In the momentum balance, forces on the system come from the pressures at the openings, pressures and viscous stresses at solid surfaces, and gravity, and it is assumed that the control surface is stationary. Momentum is carried in or out by flow through the openings. In the mechanical energy balance, kinetic and potential energy can flow in or out and there is pressure–volume work at the openings. The rate at which work is done at moving surfaces, called shaft work, is W_m. When $W_m > 0$, work is done on the system by its surroundings, as with a pump; when $W_m < 0$, work is extracted from the system, as with a turbine. Mechanical energy is expended when a fluid is compressed and recovered when it expands. Viscous dissipation is the irreversible conversion of mechanical energy into heat and its rate is E_v.

All macroscopic analyses require that conservation of mass be expressed in some form. For an incompressible fluid in a system with fixed dimensions, the rates of volume inflow and outflow are simply equated. Momentum balances are needed whenever a force is to be calculated or one is given (e.g., no force exerted on the system by its surroundings). The engineering Bernoulli equation arises whenever W_m is to be calculated or is given. Its use is facilitated if viscous dissipation is negligible. Neglecting E_v is favored if the flow path is short and there are no abrupt changes in cross-section. Liquid–gas interfaces resemble solid surfaces in that there is no flow across them (when evaporation or condensation is negligible), but differ from solid surfaces in that the viscous force on the liquid is ordinarily negligible and only the pressure force need be considered. As in microscopic analysis, such free surfaces are distinguished also by the fact that their location may be unknown. Some problems require the use of all three types of balances.

REFERENCES

Bird, R. B., W. E. Stewart, and E. N. Lightfoot. *Transport Phenomena*, 2nd ed. Wiley, New York, 2002.

Deen, W. M. *Analysis of Transport Phenomena*, 2nd ed. Oxford University Press, New York, 2012.

Middleman, S. and J. Gavis. Expansion and contraction of capillary jets of Newtonian liquids. *Phys. Fluids* 4: 355–359, 1961.

Schlichting, H. *Boundary-Layer Theory*, 6th ed. McGraw-Hill, New York, 1968.

Sturm, T. W. *Open Channel Hydraulics*, 2nd ed. McGraw-Hill, New York, 2010.

Townsend, A. A. The fully developed turbulent wake of a circular cylinder. *Austr. J. Sci. Res. A* 2: 451–468, 1949.

PROBLEMS

11.1. Torricelli's law Suppose that the tank in Fig. 11.3 empties through a hole of diameter d_o near its bottom, instead of draining through a long pipe. The opening is smooth and rounded on the inside.

(a) If viscous losses are negligible, the flow is turbulent, and the liquid level changes slowly, show that the outlet velocity is

$$v_o(t) = \sqrt{2gH(t)}. \qquad \text{(P11.1-1)}$$

This relationship between the velocity through a rounded opening and the liquid height above the opening is called *Torricelli's law*, after the Italian scientist Evangelista Torricelli (1608–1647).

(b) Determine $H(t)$ and the time for the tank to become half-empty ($t_{1/2}$).

11.2. Water clock Water clocks were used in Egypt and other parts of the ancient world to monitor the daily passage of time. In the version shown in Fig. P11.2 an open container of varying radius $R(z)$ drains through a small opening of diameter d_o in its bottom. The mean velocity at the opening is $v_o(t)$ and the water level is $H(t)$. If the dimensions are just right, dH/dt will be constant. With each hour yielding the same decrease in H, a linear scale can be used to tell the time. You are asked to design such a clock.

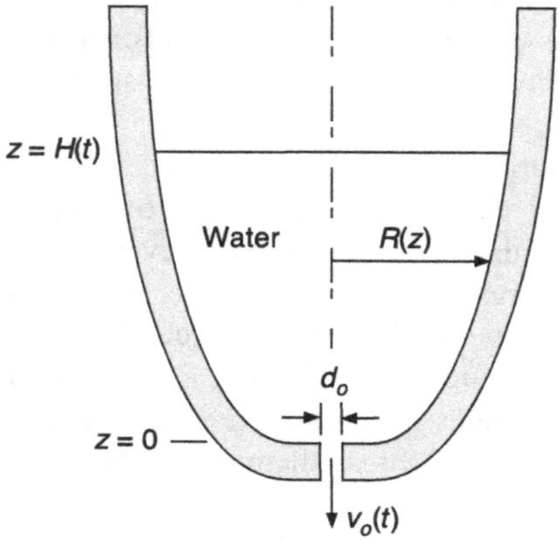

Figure P11.2 Water clock.

(a) Use conservation of mass to relate dH/dt to v_o, d_o, and $R(H)$.
(b) Assume that v_o follows Torricelli's law [Eq. (P11.1-1)] and let $C = |dH/dt|$ be the rate of change in the water level. Identify the function $R(z)$ that will make C constant.
(c) Express the initial water volume V_0 as a function of d_o, C, and the initial height H_0.
(d) As will be apparent in part (e), the clock should not be allowed to drain completely. Let H_f be the water level after operation for a time t_f. For a 12-hour clock that would fit in a room, choose reasonable values for $H_0 - H_f$ and thus C.
(e) For Torricelli's law to apply, at all times the flow must be turbulent and surface tension must be negligible. To complete your design, choose values of H_f and d_o that meet those requirements. Are the resulting values of V_0, $R_{max}[= R(H_0)]$, and $R_{min}[= R(H_f)]$ reasonable? If not, revise your choices for the water heights and outlet diameter.

11.3. Forces on nozzles Figure P11.3 shows two nozzles, one straight and the other curved, in which the diameter is reduced from D_1 to D_2. Suppose that water enters each at a mean velocity U and pressure P_1 and exits to the atmosphere at pressure P_0. The flow is turbulent and the x–y plane is horizontal.

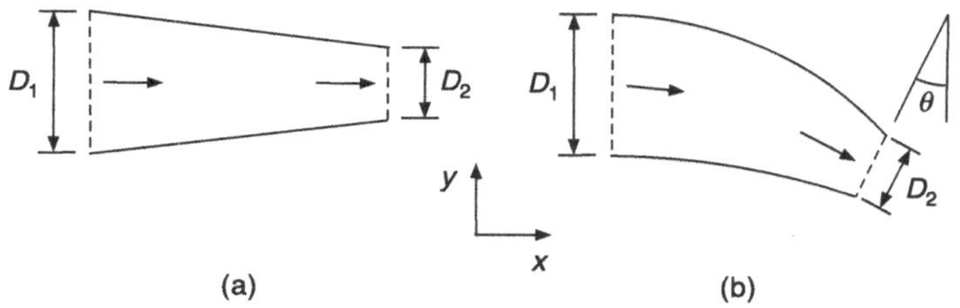

Figure P11.3 Flow through straight and curved nozzles.

(a) Determine the net fluid force exerted on the straight nozzle in Fig. P11.3(a).
(b) Determine the net fluid force exerted on the curved nozzle in Fig. P11.3(b), where the angle between the outlet and inlet planes is θ.

11.4. Drag on a flat plate calculated from the wake velocity The drag on an object can be inferred from the velocity reduction in its wake. Figure P11.4 depicts the boundary layer and wake on one side of a flat plate. The surfaces S_1 through S_4 enclose a control volume that extends from $y = 0$ to $y = H$ and from a position upstream from the leading edge (where $v_x = U$) to one downstream from the trailing edge. To ensure that $v_x = U$ everywhere on S_3, let $H \to \infty$. The plate length is L and its width is W.

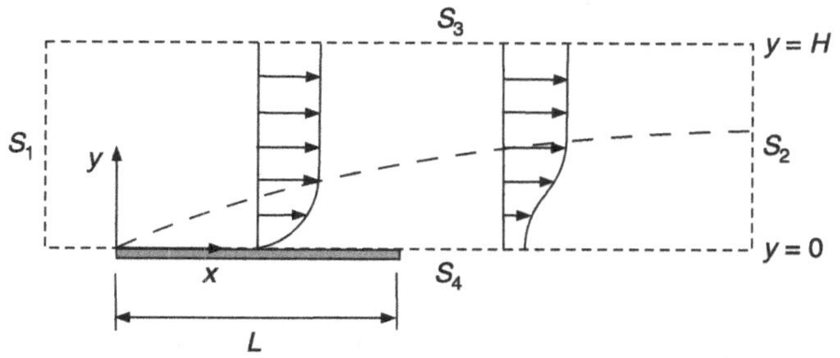

Figure P11.4 Control volume for flow past one side of a flat plate. The boundary-layer and wake thicknesses are greatly exaggerated.

(a) Show that the outward volume flow rate through S_3 is related to U and the velocity v_x evaluated on S_2 as

$$\text{flow rate} = W \int_0^H (U - v_x)\,dy. \tag{P11.4-1}$$

(b) Show that the drag on one side of the plate is

$$F_D = \rho W \int_0^H v_x (U - v_x)\,dy \tag{P11.4-2}$$

where v_x again is evaluated on S_2.

(c) For $x/L > 3$, it has been found that the velocity reduction, $\Lambda(x, y) = U - v_x(x, y)$, is represented accurately by (Schlichting, 1968, p. 169)

$$\frac{\Lambda(x,y)}{U} = 0.375\left(\frac{x}{L}\right)^{-1/2} \exp\left(-\frac{Uy^2}{4\nu x}\right). \tag{P11.4-3}$$

For x large enough that $\Lambda/U \ll 1$, $v_x \cong U$ and the result in part (b) can be simplified to

$$F_D = \rho W U \int_0^\infty \Lambda\,dy. \tag{P11.4-4}$$

Use this to calculate C_f and compare your result with that in Example 9.3-1. A definite integral that is helpful here is

$$\int_0^\infty e^{-a^2x^2}\,dx = \frac{\sqrt{\pi}}{2a}. \tag{P11.4-5}$$

11.5. Drag on a cylinder calculated from the wake velocity* The objective is to estimate the drag on a cylinder from the velocity reduction measured in its wake. The cylinder of diameter d and length L ($\gg d$) is perpendicular to the approaching flow. Figure P11.5 shows a control volume that includes half the system of interest. It extends from a position upstream from the cylinder (S_1, where $v_x = U$) to one many diameters downstream (S_2, where v_x is measured). To ensure that $v_x = U$ everywhere on S_3, let $H \to \infty$. Except for the part that coincides with the top of the cylinder, S_4 is a symmetry plane.

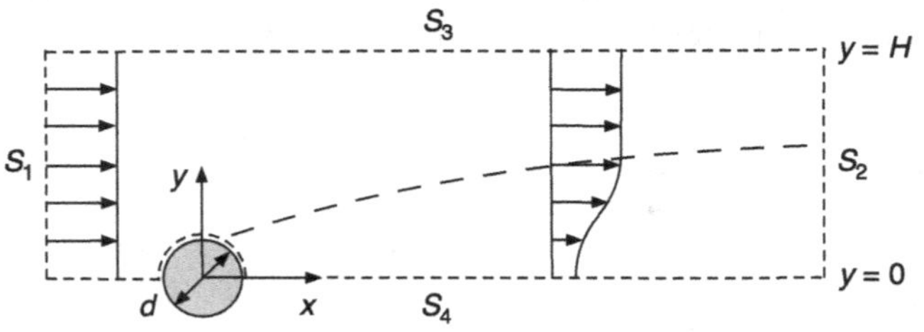

Figure P11.5 Control volume for flow past a cylinder of diameter d, including the wake.

(a) Show that the outward flow rate through S_3 is given by an expression like Eq. (P11.4-1).

(b) Show that the drag can be calculated much as in Eq. (P11.4-2).

(c) Of particular interest is the velocity reduction at S_2, $\Lambda(x,y) = U - v_x(x,y)$. If $\Lambda/U \ll 1$, show that the drag coefficient for the full cylinder is given by

$$C_D = \frac{4}{Ud}\int_0^\infty \Lambda\, dy. \qquad \text{(P11.5-1)}$$

(d) Townsend (1949) mounted a wire with $d = 0.159$ cm in a wind tunnel. For $U = 12.80$ m/s, corresponding to Re $= 1360$, he measured wake velocities at positions within the range $500 \leq x/d \leq 950$. The data were represented fairly well (although not exactly) by

$$\frac{\Lambda}{U} = \frac{1}{2X^{1/2}}\left\{1.835 \exp\left[-\left(\frac{Y}{0.253X^{1/2}}\right)^2\right]\right\} \qquad \text{(P11.5-2)}$$

where $Y = y/d$, $X = (x - x_0)/d$, and $x_0/d = 90$. Evaluate C_D from this information. [Equation (P11.4-5) will be helpful.] How well does it agree with the cylinder results in Chapter 3?

* This problem was suggested by P. S. Virk.

11.6. Jet ejector A jet ejector is a device without moving parts that can be used as a pump. As shown in Fig. P11.6, on the upstream side are concentric tubes of diameters D and λD, where $\lambda < 1$. A liquid at a high velocity v_{jet} in the inner tube entrains another liquid of the same density, giving it a velocity $v_{an} (< v_{jet})$ in the annular region. Where the streams come into contact, the pressure in both is assumed to be P_{jet}. The device discharges to the atmosphere at a pressure P_0 and velocity v_{out}. The outlet is assumed to be just far enough downstream for the mixing to be complete and the plug-flow approximation to be valid again. Supposing that the flow is turbulent and that v_{jet}, P_{jet}, the dimensions, and the fluid properties are known, it is desired to find v_{an}, v_{out}, and the viscous loss E_v.

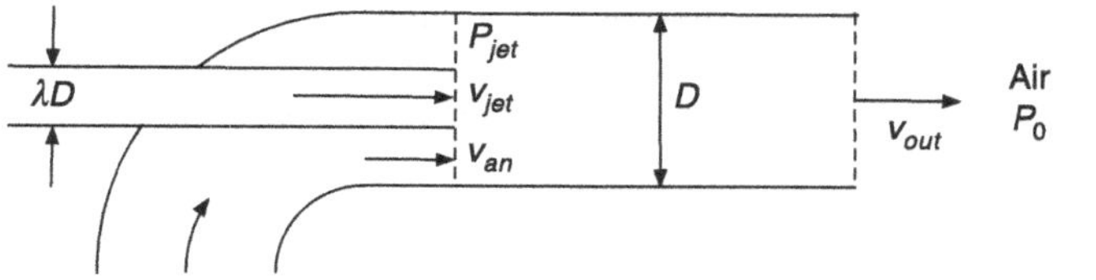

Figure P11.6 Jet ejector.

(a) Assuming that the plug-flow approximation is applicable separately to the jet and annular openings, show how the three velocities must be related.
(b) Supposing that the distance between the two planes shown as dashed lines is small enough to make the shear force on the wall negligible, relate v_{an} to v_{jet} and P_{jet} and show that it is necessary that $\Delta P = P_0 - P_{jet} > 0$.
(c) Derive an expression for E_v.
(d) Calculate v_{an}, v_{out}, and E_v for water with $v_{jet} = 5$ m/s, $\lambda = 0.3$, $D = 0.10$ m, and $P_{jet} = 0.99P_0$.

11.7. Wave tank A laboratory water tank designed to study wave motion is equipped with an end wall that can be moved inward at a desired constant velocity. As shown in Fig. P11.7(a), a wall velocity U creates a wave of height h and velocity c when the resting water level is H. In a reference frame fixed on the tank bottom, the horizontal water velocities behind and ahead of the wave front are $v_x = U$ and 0, respectively. A reference frame that moves with the wave is preferable, because it permits a steady-state analysis. For the control volume in Fig. P11.7(b) the wave velocity is zero, the water velocity on the left-hand side is $U - c$, and the water velocity on the right-hand side is $-c$.

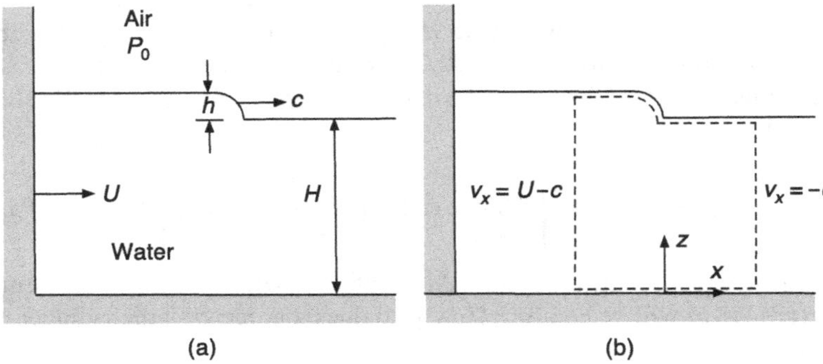

Figure P11.7 Laboratory wave tank: (a) wall and wave velocity relative to the tank bottom; and (b) a control volume moving with the wave.

(a) Relate U to c, h, and H.
(b) Assuming that the shear force on the bottom is negligible, determine c as a function of h and H.

11.8. Force in a syringe pump Syringe pumps such as in Problem 2.4 are limited by the force that the linear motor can apply to the plunger. To guide the design of a new line of pumps, it is desired to predict the forces for syringes of various sizes. Derive a general expression for the force that must be applied to the syringe in Fig. P2.4, assuming that the tubing is horizontal and the plunger itself is frictionless.

11.9. Plate suspended by a water jet Suppose that an upward jet of water strikes an unsupported, horizontal plate of mass m_p, as shown in Fig. P11.9. The diameter of the turbulent jet is $D(z)$ and its velocity is $v(z)$. The diameter and velocity at the nozzle ($z = 0$) are D_0 and v_0, respectively.

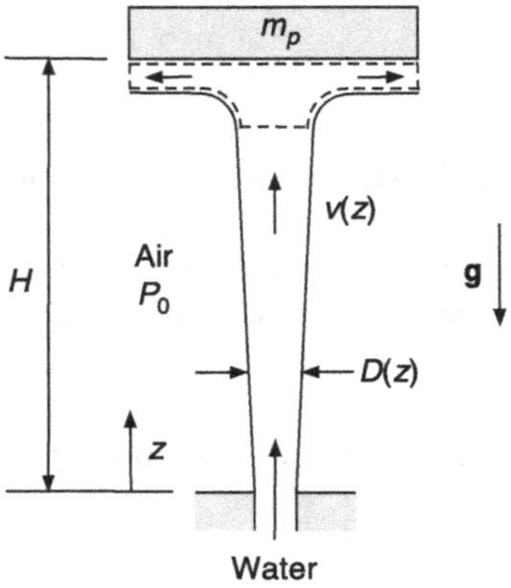

Figure P11.9 A plate of mass m_p suspended at a height H by an upward jet of water.

(a) As it rises the jet will be slowed by gravity, and as it slows it must widen. Determine $D(z)$ and $v(z)$ in the main part of the jet.

(b) Derive an expression for the net fluid force on the plate. A convenient control volume is shown by the dashed lines. You may assume that the actual thickness of the control volume is much less than H and that the mass of water it contains is negligible.

(c) If $m_p = 1$ kg and $D_0 = 0.1$ m, what nozzle velocity v_0 would be needed to suspend the plate at $H = 1$ m?

11.10. Viscous losses in laminar pipe flow When the velocity field in a conduit is completely known, E_v can be calculated directly from Eqs. (11.4-3) and (11.4-4). Confirm that, for fully developed laminar flow of an incompressible fluid in a tube of radius R and length L, the result obtained in this manner is equivalent to Eq. (11.4-12).

11.11. Hydroelectric power Puget Power Plant No. 1 at Snoqualmie Falls, WA is a hydroelectric facility in which water is directed straight down through a pair of steel penstocks located upstream, passes through a set of generator turbines, and exits through a horizontal tunnel that emerges at the base of the falls. The parallel, vertical penstocks are each 8 ft in diameter and 270 ft long and the plant capacity is 11,900 kW.

(a) If viscous losses are negligible and the generator efficiency is 90%, what water flow rate is needed to run at full capacity? You may assume that the river depth at the inflow is 10 ft and that the cross-sectional area of the outflow tunnel equals the total for the two penstocks.

(b) Assuming fully developed flow in the penstocks, show that the viscous loss there is negligible. [Including also entrance effects (Section 12.2) would significantly increase the predicted loss, but it would remain small compared with the shaft work.]

11.12. Pitot tube The *Pitot tube*, invented by the French engineer Henri Pitot (1695–1771) to measure fluid velocities, has various forms. Pitot tubes are used to monitor the airspeed of commercial jets, as anemometers in weather stations, and in industrial or laboratory settings to determine local velocities or velocity profiles. The simplest type of probe, shown in Fig. P11.12(a), is a fluid-filled tube with a tapered end that is pointed at the flow. A connection to a pressure gauge keeps the internal fluid static and creates a stagnation point at the tip (marked by *). The pressure difference between the internal fluid and a nearby reference point where the fluid is static is used to infer the velocity near the tip. The setup in Fig. P11.12(b), where the reference is a tap in the tube wall, could be used to measure the velocity profile by varying the radial position of the probe.

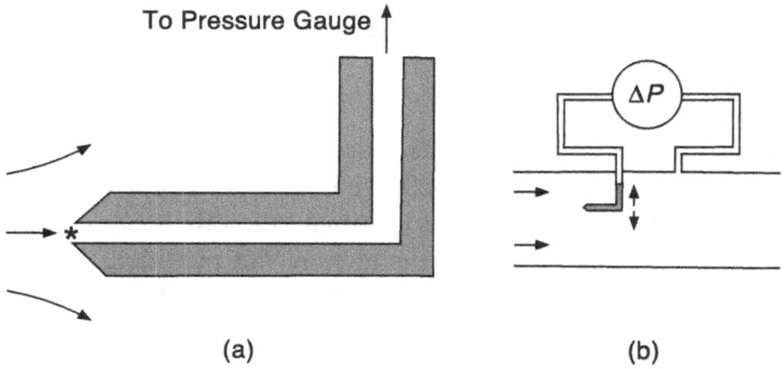

Figure P11.12 Pitot tube: (a) simple probe; and (b) probe with a pressure gauge and a reference line from a tap in a pipe wall. The velocity profile in the pipe could be determined by varying the radial position of the probe.

(In a *Pitot-static tube*, the reference is a second internal channel that is connected to the outside by holes on the *side* of the probe. That eliminates the need for a wall tap or other separate connection to static fluid.)

Derive the relationship between the fluid velocity and the measured pressure difference. You may assume that height differences are negligible.

11.13. Siphon A *siphon* may be used to draw a liquid above its level in an open tank or channel and discharge it below that level. Once initiated, the flow is sustained by the height difference. The simplest configuration is an inverted U-tube, as in Fig. P11.13. Assume here that viscous losses are negligible, although that is not always true in real siphons.

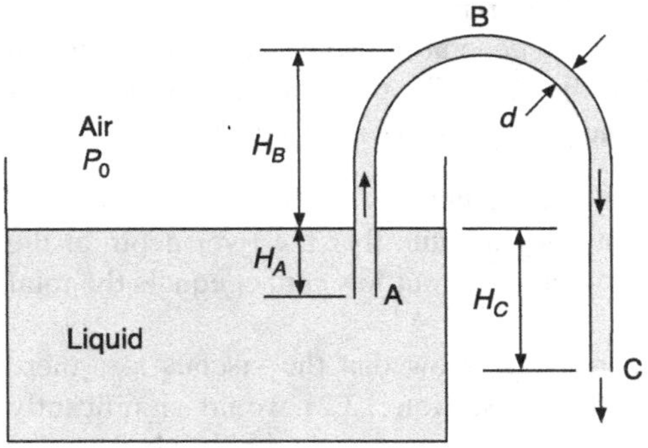

Figure P11.13 Open tank with siphon.

(a) After an empty tube has been inserted into the tank, how much suction must be applied at position C to start the flow?
(b) If the flow, once started, is turbulent, what will the mean velocity U in the tube be?
(c) Show that during operation the pressure P_B at position B is sub-atmospheric.
(d) Lowering the outlet will increase the flow rate. However, cavitation might eventually make the siphon unreliable. That can be avoided if the pressure everywhere exceeds P_V, the vapor pressure of the liquid. What limits would that place on H_C and U?

11.14. Sump pump A nominal 1/2 hp electric sump pump for household use comes with the following data:

Vertical pumping distance (ft)	0	5	10	15	20	25
Flow rate (GPM)	66	61	56	50	43	34

where GPM is gallons per minute. A typical installation is shown in Fig. P11.14. The average water depth at the intake is h and the discharge is through a pipe of diameter D at a height H and mean velocity v. It is desired to estimate the efficiency of the pump from these data. Although the details of the test setup were not provided, the threaded fitting on the pump suggests that the pipe might have been 1 1/2″ Schedule 40 PVC ($D = 4.04$ cm). Assume that the pipe length equaled the height H.

(a) Confirm that the flow was turbulent under all conditions tested and find the range of values of the friction factor. The range of f is small enough that an average value can be used subsequently.

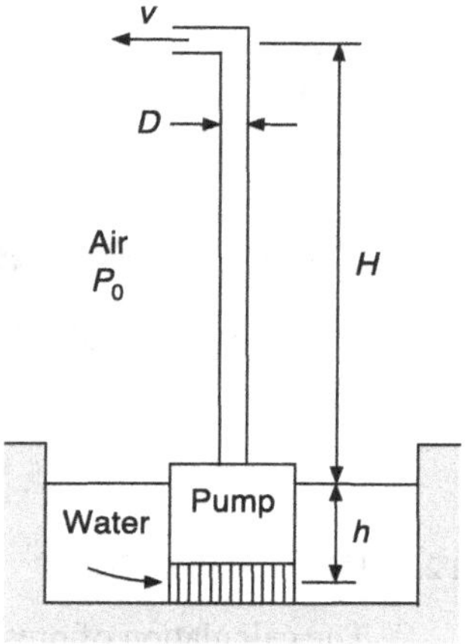

Figure P11.14 Sump pump with discharge pipe.

(b) Relate W_m to H, Q, and the various constants (including f). You may neglect the kinetic energy at the pump intake and consider viscous losses only in the pipe.
(c) Use the data to calculate W_m as a function of H for $H \geq 5$ ft. [Why should the results for $H = 0$ not be used?] Find the nominal efficiency of the pump, which is W_m divided by the rated $1/2$ hp. To obtain a true efficiency (W_m relative to the electric power used), we would need to know how much current was actually drawn during the tests.

11.15. Drainage pipe An underground plastic pipe is to be used to carry water away from a building. As shown in Fig. P11.15, the pipe of diameter D will run downward and have both ends open to the air (pressure P_0). The site requires a drop of $H = 2$ m over a length $L = 40$ m. The volume flow rate Q will depend on the weather, but it is supposed that it will not exceed 1 gal/s, or 3.8×10^{-3} m^3/s ($= Q_{\max}$). It is desired to choose an appropriate value for D.

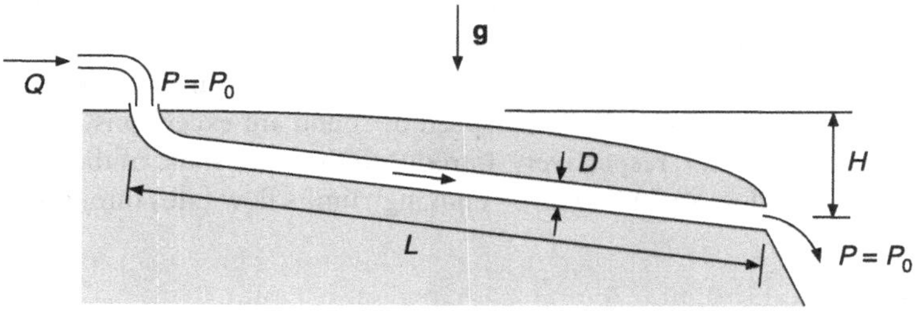

Figure P11.15 Drainage pipe.

(a) Use the engineering Bernoulli equation to calculate the pipe diameter that will exactly accommodate the expected maximum flow. You may assume that the pipe is hydrodynamically smooth.
(b) If D is chosen as in part (a) and $Q < Q_{\max}$, the pipe will not run full. How would you know that mathematically? [*Hint*: Consider the applicable $f(\text{Re})$ relationship from Chapter 2.] What is the viscous loss when $Q = 0.1Q_{\max}$?
(c) For the same D, what would happen if ever $Q > Q_{\max}$? Would the engineering Bernoulli equation still apply to the pipe?

12

Pipe flow: entrance effects, fittings, and compressibility

12.1 INTRODUCTION

The calculation of pressure drops and flow rates for fully developed flow in various kinds of conduits was discussed in Chapter 2. The resistance to such flows is due to a wall shear stress τ_w that is independent of conduit length, and it was seen that information on τ_w could be correlated in terms of the friction factor f. All that is needed for design calculations involving incompressible flow in long pipes is fluid property data and plots or formulas for f. However, any system that distributes a fluid to multiple locations, whether in an oil refinery, a microfluidic device, or the body, consists of conduits of finite length that are connected to ones of different size and to tanks or other dissimilar elements. Additional flow resistances arise from sudden changes in diameter or the presence of valves and fittings. Although the additional resistances are negligible in some systems, in others their total exceeds that from plain sections of pipe. Tube inlets coincide with abrupt reductions in diameter and create entrance regions where the flow adjusts to the new cross-section. The resistance there exceeds that for fully developed flow, as detailed in Section 12.2. Section 12.3 extends the discussion of "other" resistances to pipe fittings and valves and includes basic information on pumps, thereby providing a foundation for practical designs.

Our default assumption for all flow regimes and geometries has been that the fluid density is constant. Assuming a gas to be incompressible is satisfactory if, as is often the case, the change in pressure is a small fraction of the absolute pressure. Gas flows in very long pipes or at velocities comparable to the speed of sound are exceptions, and are the focus of Sections 12.4 and 12.5, respectively. Introduced there are some of the distinctive features of compressible flow, including how "choking" limits flow rates through pipes or nozzles.

12.2 ENTRANCE EFFECTS

Entrance length

Something we have not yet accounted for in calculations is that the velocity profile in a tube or other conduit becomes fully developed only after a certain distance from the inlet. An inlet typically involves a sudden reduction in cross-sectional area, as at position 1 in Fig. 12.1. The radius ratio will be denoted as $\beta = R/R_0$. For a pipe connected to a large tank, β is practically zero. The entrance length L_E is the distance from the inlet at which the flow in the smaller tube becomes fully developed. This is not always a small fraction of its total length L. Because the sudden contraction at position 1 affects the velocity and pressure not just in the first part of the smaller tube, but also for some distance upstream,

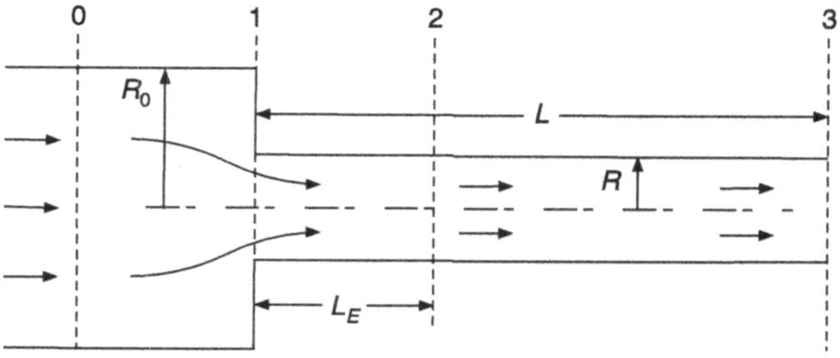

Figure 12.1 Sudden contraction from a tube of radius R_0 to one of radius R. The change in radius at position 1 perturbs the velocity and pressure fields between positions 0 and 2.

the complete entrance region extends from position 0 to position 2. The flow is fully developed from position 2 to where the pipe segment ends at position 3.

The precise point at which the velocity profile in a tube is judged to be fully developed is arbitrary. In laminar flow it is conventional to define L_E as the distance at which the centerline velocity reaches 99% of its final value. As reviewed in Boger (1982), experimental results and numerical simulations for laminar flow of Newtonian fluids through contractions with $\beta \leq 0.5$ indicate that

$$\frac{L_E}{R} = 0.49 + 0.11\,\text{Re} \qquad \text{(laminar)}. \tag{12.2-1}$$

Thus, if laminar flow is maintained up to Re = 2100, the entrance region will be as long as 230 radii.[1] The same formula is approximately correct also for a parallel-plate channel with spacing $2H$, if R is replaced by H and the tube diameter in Re is replaced by $2H$.

Entrance lengths for turbulent flow have been more difficult to establish. Zagarola and Smits (1998) concluded that they are about 160 radii at $\text{Re} = 3 \times 10^5$ and 260 radii at $\text{Re} = 4 \times 10^7$, considerably exceeding the 60–80 radii found in the older literature. However, even the larger values are much less than would be obtained by extrapolating Eq. (12.2-1) to turbulent flow. Velocity profiles develop more quickly in turbulent than in laminar flow because of the more rapid cross-stream transfer of momentum (Chapter 10).

Excess pressure drop in entrance regions

Except when designing an apparatus to study fully developed velocity profiles, the excess pressure drop due to the sudden contraction is more important than L_E. The pumping requirements for the system in Fig. 12.1 are determined by the overall pressure drop, $|\Delta\mathcal{P}| = \mathcal{P}_0 - \mathcal{P}_3$. Knowing how much this exceeds that for fully developed flow permits us to refine the calculations in Chapter 2.

A mechanical energy balance reveals what contributes to $|\Delta\mathcal{P}|$. When applied between positions 0 and 3, the engineering Bernoulli equation [Eq. (11.4-8)] becomes

$$\Delta\left(\frac{b\langle v\rangle^2}{2} + \frac{\mathcal{P}}{\rho}\right) = -\frac{E_v}{w} \tag{12.2-2}$$

[1] A slightly different expression (with 1.10 instead of 0.49) is obtained for the less realistic but computationally more convenient situation in which plug flow is assumed at position 1 and the upstream region is ignored (Boger, 1982).

where $b = 2$ for laminar flow and $b = 1$ for turbulent flow. If $\beta \ll 1$, the upstream kinetic energy is negligible because $\langle v_0 \rangle = \beta^2 \langle v_3 \rangle \ll U$, where U is the mean velocity in the smaller tube. Accordingly, Eq. (12.2-2) simplifies to

$$\left| \frac{\Delta \mathcal{P}}{\rho} \right| = \frac{E_v}{w} + \frac{bU^2}{2}. \tag{12.2-3}$$

It is seen that the overall pressure drop arises from a combination of the viscous loss and the gain in kinetic energy.

The overall pressure drop may be viewed as having two parts: $|\Delta \mathcal{P}|_{FD}$, that for fully developed flow in a tube of radius R and length L; and $|\Delta \mathcal{P}|_{EN}$ an increment due to the entrance region. Using Eq. (2.3-4) to relate $|\Delta \mathcal{P}|_{FD}$ to the wall shear stress for fully developed flow (τ_w),

$$|\Delta \mathcal{P}| = \frac{2\tau_w L}{R} + |\Delta \mathcal{P}|_{EN}. \tag{12.2-4}$$

It is convenient to think of $|\Delta \mathcal{P}|_{EN}$ as increasing the effective length of the smaller tube. Dividing by $2\tau_w$ gives

$$\frac{|\Delta \mathcal{P}|}{2\tau_w} = \frac{L}{R} + \frac{|\Delta \mathcal{P}|_{EN}}{2\tau_w} = \frac{L + L^*}{R} \tag{12.2-5}$$

where L^* is the apparent increase in length due to the entrance region. For laminar flow with $\beta \leq 0.25$, experimental and computational results (Boger, 1982) indicate that

$$\frac{|\Delta \mathcal{P}|_{EN}}{2\tau_w} = \frac{L^*}{R} = 0.0709\,\mathrm{Re} + 0.589 \quad \text{(laminar)}. \tag{12.2-6}$$

The increase in effective tube length for turbulent flow is inferred from results for fully developed flow and sudden contractions. The respective contributions to the viscous loss are $E_v^{(FD)}$ and $E_v^{(EN)}$. From Eq. (11.4-14), $E_v^{(FD)}/w = U^2 L f/R$. From results for sudden contractions with $\beta \to 0$, $E_v{}^{(EN)}/w = 0.22U^2$ (Martin, 1974). Thus,

$$\frac{E_v}{w} = \frac{U^2 L f}{R} + 0.22U^2. \tag{12.2-7}$$

Substituting this and $b = 1$ into Eq. (12.2-3) gives

$$\frac{|\Delta \mathcal{P}|}{\rho} = \frac{E_v}{w} + \frac{U^2}{2} = \frac{U^2 L f}{R} + 0.72U^2. \tag{12.2-8}$$

The effective increase in tube length is defined now using

$$\frac{|\Delta \mathcal{P}|}{\rho} = \frac{U^2 f (L + L^*)}{R} \tag{12.2-9}$$

and a comparison with Eq. (12.2-8) gives

$$\frac{L^*}{R} = \frac{0.72}{f} \quad \text{(turbulent)}. \tag{12.2-10}$$

In a smooth pipe with $\mathrm{Re} = 3 \times 10^5$, where $f = 3.6 \times 10^{-3}$, this gives $L^*/R = 200$, which happens to coincide fairly well with L_E/R. The entrance length and apparent length increase are comparable also for laminar flow. From Eqs. (12.2-1) and (12.2-6), L^*/L_E ranges from 1.20 at $\mathrm{Re} = 0$ to 0.65 at $\mathrm{Re} = 2000$.

Example 12.2-1 Entrance correction for a process pipe. It is desired to determine the effect of the entrance region on the pressure drop for water in the horizontal pipe of Example 2.3-1, assuming that at the inlet there is a sudden contraction with $\beta \ll 1$.

Because Q, the pipe dimensions, and the fluid properties are fixed, the additional resistance of the entrance region will increase $|\Delta\mathcal{P}|$. As determined previously, the flow is turbulent, with $\text{Re} = 1.02 \times 10^5$ and $f = 4.43 \times 10^{-3}$. Equation (12.2-10) gives

$$\frac{L^*}{R} = \frac{0.72}{4.43 \times 10^{-3}} = 163.$$

From the pipe dimensions, $L = 30$ m, $D = 0.1$ m, and $L/R = 30/0.05 = 600$. The fact that L^*/R is a significant fraction of L/R indicates that the additional pressure drop is not negligible. The required pumping power is $W_m = Q|\Delta\mathcal{P}|$ (Example 11.4-1). Thus, the pressure drop and pumping power are each $763/600 = 1.27$ times that for fully developed flow.

Example 12.2-2 Entrance correction for a capillary viscometer. The objective is to see whether a correction factor is needed to interpret measurements with the viscometer in Example 2.3-4, assuming that there is a sudden contraction at the start of the test section.

With $|\Delta\mathcal{P}|$ for the viscometer fixed, the entrance resistance must have decreased the measured Q and caused μ to be overestimated. In the test section, $L = 5$ cm, $D =$ 0.5 mm, and $L/R = 5/0.025 = 200$. The flow is laminar and it was found that $\text{Re} = 34.0$. From Eq. (12.2-6),

$$\frac{L^*}{R} = 0.0709\,(34.0) + 0.589 = 3.00.$$

This is 1.5% of L/R, which is enough to be of concern for precise work. Letting μ_{app} be the apparent (uncorrected) viscosity found before and μ the true value,

$$\mu_{app} L = \mu(L + L^*)$$

$$\mu = 1.35 \times 10^{-3} \left(\frac{200}{203}\right) = 1.33 \times 10^{-3} \text{ Pa} \cdot \text{s}.$$

The interpretation of data from capillary viscometers is discussed in Kestin *et al.* (1973).

12.3 FITTINGS, VALVES, AND PUMPS

It was seen in Section 12.2 that a sudden reduction in pipe diameter creates an additional viscous loss, which helps increase the pressure drop. That is true also for sudden expansions, pipe bends, and obstructions due to valves or fittings, irrespective of whether the flow is laminar or turbulent. These "other" losses are not necessarily minor. We will focus on turbulent flow, where the effects tend to be of greater practical importance and much more data is available.

Loss coefficients

The loss $E_{v,i}$ due to event or device i is expressed as

$$\frac{E_{v,i}}{w} = K_i \frac{U^2}{2} \tag{12.3-1}$$

where K_i is a dimensionless *loss coefficient* and U is the mean velocity. Where there is a change in diameter, the convention we will use is that U is the value in the narrower

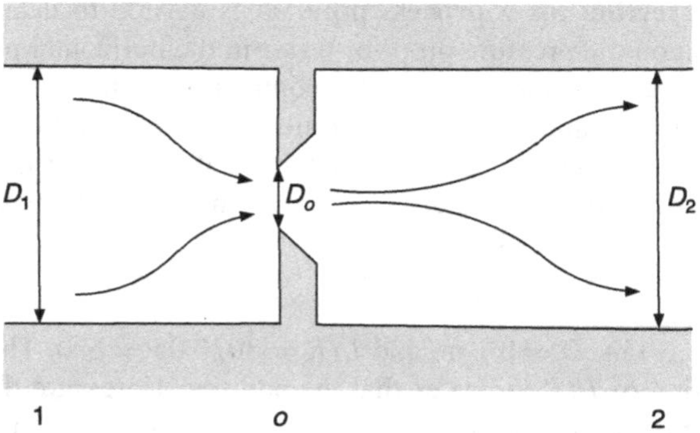

Figure 12.2 Flow through an orifice plate in a tube.

channel and β is the smaller diameter divided by the larger one ($\beta \leq 1$). When Re is large enough to justify the plug-flow approximation (greater than about 10^5), K_i is typically independent of Re and depends only on the flow geometry. Methods for measuring K_i are described, for example, in Crane (2012). In a few instances, expressions for K_i can be derived by combining macroscopic balances with one or two key assumptions, as will be shown.

The loss coefficient for turbulent flow through a *sudden contraction* is (Martin, 1974)

$$K_c = \left(\frac{2}{m} - \beta^2 - 1\right)^2 \tag{12.3-2a}$$

$$m = \frac{1.2\sqrt{1 - 0.64\beta^4} - 0.72\beta^2}{1 - \beta^4}. \tag{12.3-2b}$$

For $\beta \to 0$, $m \to 1.2$ and $K_c \to 0.44$, as used already in Eq. (12.2-7).

The loss coefficient for a *sudden expansion* is given by the *Borda–Carnot equation*,

$$K_e = (1 - \beta^2)^2 \tag{12.3-3}$$

which is derived in Example 12.3-2. This applies only to a stream passing into a larger-diameter vessel that is filled with the same fluid. For a liquid jet entering a gas, $K \cong 0$.

An *orifice* is a hole in a thin plate or sheet of material; mathematically, it is a tube of diameter D_0 and length L_0 in which $L_0/D_0 \to 0$. Flow through an orifice involves a sudden contraction followed immediately by a sudden expansion, as shown in Fig. 12.2 for an orifice plate in a tube. The flow remains undisturbed up to position 1 (about one diameter upstream) and becomes fully developed again at position 2 (several diameters downstream). For equal upstream and downstream diameters ($D_1 = D_2$), the volume flow rate Q is related to the overall pressure drop as (Tilton, 2007)

$$Q = v_o S_o = C_o S_o \left[\frac{2\,(P_1 - P_2)}{\rho\,(1 - \beta^2)\,(1 - \beta^4)}\right]^{1/2} \tag{12.3-4}$$

where $S_0 = \pi D_0^2/4$, $\beta = D_0/D_1$, and C_0 is the *orifice coefficient*. Equation (12.3-4) is used to interpret data from *orifice flow meters*, which are equipped with pressure taps at positions 1 and 2 (Crane, 2012). With $C_0 = 0.62$ (Tilton, 2007), the loss coefficient based

Table 12.1 Selected loss coefficients for turbulent flow[a]

Object	K
Sudden contraction ($\beta \to 0$)	0.44
Sudden expansion ($\beta \to 0$)	1.0
Orifice ($\beta \to 0$)	2.6
Rounded tube entrance[b]	0.1
90° bend (standard elbow)	0.75
180° bend (close return)	1.5
Diaphragm valve	
Open	2.3
Half open	4.3
Gate valve	
Open	0.17
Half open	4.5
Globe valve (bevel seat)	
Open	6.0
Half open	9.5
Long pipe	$\frac{4Lf}{D}$

[a] The long-pipe result is from Eq. (11.4-14). The other values are from Eqs. (12.3-2), (12.3-3), and (12.3-5), or Tilton (2007).

[b] Trumpet-shaped entrance rounded with a radius that is >0.15D.

on v_0 is

$$K_o = \frac{(1-\beta^2)(1-\beta^4)}{C_o^2} = 2.6(1-\beta^2)(1-\beta^4). \tag{12.3-5}$$

If $D_2 \gg D_1$ (e.g., if a tube ends at an orifice), the $(1-\beta^2)$ term in each equation is omitted.

Selected loss coefficients are given in Table 12.1. The values for sudden contractions, sudden expansions, and orifices are shown only for $\beta \to 0$. Included also are a rounded tube entrance, two kinds of standard bends, three types of valves, and fully developed flow in a pipe of diameter D and length L. Loss coefficients differ greatly among valve types, as shown. More complete results are given in Blevins (2003), Crane (2012), and Tilton (2007). Variations in design, especially of valves, make it prudent to obtain the manufacturer's data for specific devices.

Pump characteristics

A pump increases the pressure at a particular point along a pipe or other conduit, as was illustrated in Fig. 11.6. Pumps are of two general types, *positive displacement* (PD) and *centrifugal* (CF). Positive-displacement pumps maintain a constant volume flow rate Q and can be used with gases or liquids; PD pumps for gases are called *compressors*. The power output of a PD pump varies according to the pressure needed to achieve a set Q. There are both reciprocating and rotary types for industrial applications (Wilkes, 2006, pp. 188–194). A syringe pump (Problem 2.4) is a simple type of PD pump that is used to give small, constant values of Q in laboratory settings. Centrifugal pumps also can be

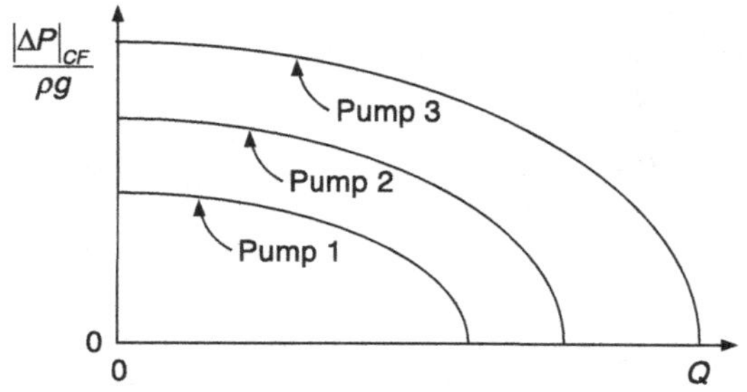

Figure 12.3 Hypothetical operating curves for centrifugal pumps of differing size.

used with gases or liquids; CF pumps for gases are called *blowers*, a household example of which is a hair dryer. Centrifugal pumps are better suited for large flow rates and are more common than PD pumps. In general, the fluid enters a CF pump through an inlet at the center, is rotated by an impeller, and is discharged at the periphery. If the motor speed is fixed, Q is not, which merits some discussion.

Manufacturers of CF pumps provide performance data in plots like Fig. 12.3. The ordinate is the pressure increase across the pump, $|\Delta P|_{CF}$, which is expressed as a head by dividing it by ρg, and the abscissa is Q. Shown are hypothetical performance curves for three sizes of CF pumps. The head developed by each is greatest for small Q, gradually declines as Q increases, and reaches zero at a certain maximum flow rate (which might be beyond the range of the data). The decline in the head is due largely to the increasing viscous losses within the pump. A typical performance curve is of the form

$$\frac{|\Delta P|_{CF}}{\rho g} = A - BQ^n \tag{12.3-6}$$

where A and B (dimensional) and n (dimensionless) are empirical constants. Typically, $n \cong 2$.

Guidance on how to construct a more universal performance curve is provided by dimensional analysis. It may be supposed that CF pumps of a given design are characterized mainly by a linear dimension d (e.g., their overall diameter) and the angular velocity ω of their impeller. If the pumps are geometrically similar, the diameters of the inlet and outlet, the radius of the impeller, the sizes of impeller vanes, and other internal dimensions will all be proportional to d. The various geometric ratios that would be the same within such a family of pumps need not be enumerated. Assuming that the Reynolds number will always be large enough for us to ignore the viscosity, the relevant quantities aside from d and ω are ρ, Q, and $|\Delta P|_{CF}$. With these five quantities involving three independent dimensions (M, L, T) there will be two groups, a dimensionless flow rate N_1 and a dimensionless pump head N_2. Proceeding as in Section 1.4, we find that

$$N_1 = \frac{Q}{\omega d^3}, \quad N_2 = \frac{|\Delta P|_{CF}}{\rho \omega^2 d^2}. \tag{12.3-7}$$

A plot of N_2 vs. N_1 would have the same shape as any of the curves in Fig. 12.3. However, a single curve would apply now to all CF pumps of a given design.

Example 12.3-1 Force on a return bend (revisited). In Example 11.3-1 a momentum balance was used to show that the net fluid force on a 180° pipe bend with turbulent flow is

$$F_x^0 = S(P_1 + P_2 - 2P_0 + 2\rho U^2) \tag{12.3-8}$$

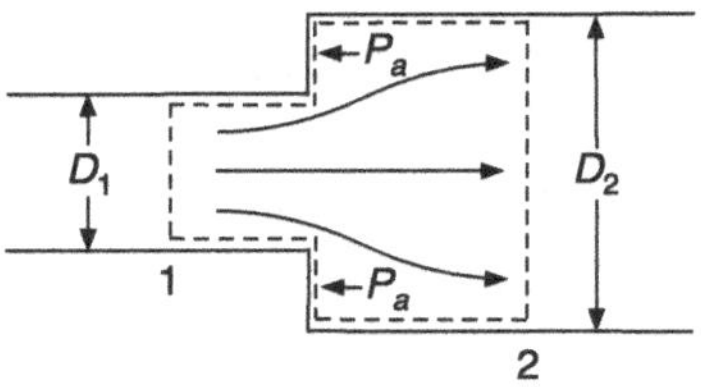

Figure 12.4 Sudden expansion in tube diameter.

where the flow enters in the $+x$ direction, S is the pipe cross-sectional area, P_1 and P_2 are the inlet and outlet pressures, respectively, and P_0 is the ambient pressure. It is desired to complete the force calculation by relating P_2 to P_1.

The engineering Bernoulli equation [Eq. (11.4-8)] is applied by setting $b = 1$, $\Delta(v^2) = 0$, $\Delta h = 0$, $W_m = 0$, and $E_c = 0$. Solving for the outlet pressure gives

$$P_2 = P_1 - \rho \frac{E_{v,b}}{w} \tag{12.3-9}$$

where $E_{v,b}$ is the viscous loss in the bend. Using $K_b = 1.5$ (Table 12.1), the loss is

$$\frac{E_{v,b}}{w} = K_b \frac{U^2}{2} = 0.75U^2. \tag{12.3-10}$$

Substitution of these results into Eq. (12.3-8) gives

$$F_x^0 = S[2(P_1 - P_0) + 1.25\rho U^2]. \tag{12.3-11}$$

The only pressure that must be specified now to calculate the force is the gauge value at the inlet.

Example 12.3-2 Borda–Carnot equation. The objective is to derive Eq. (12.3-3), which gives the viscous loss for turbulent flow through an expansion like that in Fig. 12.4.

The dashed lines indicate the control volume, which encompasses the fluid between planes 1 and 2. Assuming the tubes to be horizontal, the engineering Bernoulli equation indicates that the pressure drop in the expansion is

$$\frac{P_1 - P_2}{\rho} = \frac{v_2^2 - v_1^2}{2} + \frac{E_{v,e}}{w}. \tag{12.3-12}$$

A momentum balance based on Eq. (11.3-9) will be used to find another expression for the pressure drop. If the overall flow is in the $+x$ direction, the outward normals at the openings are $\mathbf{n_1} = -\mathbf{e_x}$ and $\mathbf{n_2} = \mathbf{e_x}$ and

$$F_x^* = (P_1 + \rho v_1^2)S_1 - (P_2 + \rho v_2^2)S_2. \tag{12.3-13}$$

We suppose now that the distance between positions 1 and 2 is short enough to make the shear forces on the walls of both tube segments negligible. The fluid force on the solid part of the control surface is then due only to the pressure P_a acting on the annular area and

$$F_x^* = -P_a(S_2 - S_1). \tag{12.3-14}$$

The key assumption needed to complete the analysis is that $P_a \cong P_1$. Why the annular pressure should nearly equal the upstream value is not obvious, but this approximation is

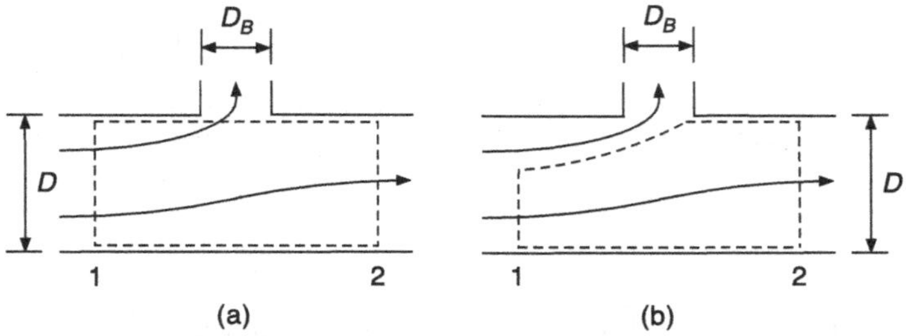

Figure 12.5 Alternative control volumes for calculating the pressure change near a pipe branch.

validated by the success of the final result, as is the neglect of the shear forces. Evaluating F_x^* in this manner, Eq. (12.3-13) is rearranged to give

$$\frac{P_1 - P_2}{\rho} = v_2^2 - \left(\frac{S_1}{S_2}\right) v_1^2. \tag{12.3-15}$$

Equating this pressure drop with that in Eq. (12.3-12) and solving for the viscous loss, we obtain

$$\frac{E_{v,e}}{w} = K_e \frac{U^2}{2} = \frac{v_2^2}{2} + \frac{v_1^2}{2}\left(1 - \frac{2S_1}{S_2}\right). \tag{12.3-16}$$

Here $v_1 = U$ and conservation of mass gives $v_2 = v_1 S_1 / S_2 = \beta^2 U$. Substitution of those velocities into Eq. (12.3-16) leads to Eq. (12.3-3).

Example 12.3-3 Pressure increase at a diverging branch. Figure 12.5 shows a location in a pipe of diameter D where some of the fluid exits via a side port or branch of diameter D_B. The opening in the wall might either be an orifice or the inlet of another pipe. A *manifold* that distributes fluid to multiple outlets often consists of a pipe with a series of such openings along its length. A distribution section in which the flow is outward might be accompanied by a return section in which the flow is inward. A burner in a kitchen gas range is an example of a distribution manifold with orifices. The objective is to predict the pressure change along the main channel in the vicinity of such a branch, for turbulent flow.

The exiting fluid reduces the average velocity in the main pipe. If the flow is nearly inviscid the pressure will then increase, as in the expanding part of a Venturi tube. The key geometric parameters are the area ratio κ and diameter ratio β, which are related as $\kappa = S_B/S = (D_B/D)^2 = \beta^2$. It will be assumed that $\beta \ll 1$ and the plug-flow approximation will be used.

One approach for calculating the pressure increase would be to apply a momentum balance to the control volume in Fig. 12.5(a) and neglect the shear force on the walls. However, the fact that the flow into the branch is not perpendicular to that part of the control surface creates a problem: an unknown amount of axial momentum is convected into the branch. A more promising approach is to apply the engineering Bernoulli equation to the control volume in Fig. 12.5(b). The top surface is now a hypothetical streamline that separates the flow continuing in the main pipe from that entering the branch. Fortunately, neither the precise location of that imaginary surface (which is far from planar) nor the cross-sectional area at position 1 (which is far from circular) is needed for the analysis. It will be assumed that the viscous loss between the "inlet" (position 1) and "outlet" (position 2) is negligible.

For plug flow in a horizontal channel without a viscous loss, the engineering Bernoulli equation indicates that

$$\frac{P_2 - P_1}{\rho} = \frac{1}{2}\left(v_1^2 - v_2^2\right). \tag{12.3-17}$$

Because $v_2 < v_1$ here, there is a pressure increase or "recovery," as has already been stated. From conservation of mass, the velocities in the main pipe are related to that in the branch (v_B) as

$$v_2 = v_1 - \kappa v_B. \tag{12.3-18}$$

Using this to eliminate v_2 from Eq. (12.3-17), the pressure recovery is

$$\frac{P_2 - P_1}{\rho} = \frac{\kappa v_B}{2}(2v_1 - \kappa v_B). \tag{12.3-19}$$

To evaluate v_B, we assume now that the side port is an orifice that opens to a pressure P_0. Given the indefinitely large area beyond the orifice, the $(1 - \beta^2)$ term in Eq. (12.3-4) is omitted. Moreover, with $\beta < 0.3$, $(1 - \beta^4) > 0.99 \cong 1$. Using $(P_1 + P_2)/2$ to approximate the pressure upstream from the orifice,

$$v_B = C_o\left(\frac{P_1 + P_2 - 2P_0}{\rho}\right)^{1/2} = C_o\left(\frac{P_2 - P_1}{\rho} + \frac{2(P_1 - P_0)}{\rho}\right)^{1/2}. \tag{12.3-20}$$

Because only the product κv_B is needed in Eq. (12.3-19), the area ratio and orifice coefficient will be combined as $\phi = \kappa C_0$. With $C_0 = 0.62$ and $\beta < 0.3$, $\phi = \beta^2 C_0 < 0.06$. The likelihood that $\phi \ll 1$ will be used to simplify the final results. Given $P_1, v_1, \phi, \rho,$ and P_0, Eqs. (12.3-19) and (12.3-20) are sufficient to determine P_2.

The algebra needed to evaluate the pressure increase and branch flow rate is simplified by using the dimensionless quantities

$$X = \frac{P_2 - P_1}{\rho v_1^2}, \quad F = \frac{\kappa v_B}{v_1}, \quad G = \frac{P_1 - P_0}{\rho v_1^2} \tag{12.3-21}$$

where X is the pressure recovery relative to the inertial pressure scale, F is the fraction of the flow that enters the branch, and G is the applied pressure difference relative to the inertial pressure scale. The two unknowns are X and F, whereas G ordinarily would be specified. Converting Eqs. (12.3-19) and (12.3-20) to the new variables gives, respectively,

$$X = \frac{F}{2}(2 - F), \quad F = \phi(X + 2G)^{1/2}. \tag{12.3-22}$$

The second equation suggests another variable change, such that $Y = (X + 2G)^{1/2}$ or $X = Y^2 - 2G$. Transforming X to Y, and using the second equation to evaluate F in the first, yields a single, quadratic equation in Y. That equation has only one physically realistic (positive) root. Choosing that root and changing back to X gives

$$X = \left(\frac{\phi + \sqrt{\phi^2 + 4(2 + \phi^2)G}}{2 + \phi^2}\right) - 2G \tag{12.3-23}$$

$$F = \phi\left(\frac{\phi + \sqrt{\phi^2 + 4(2 + \phi^2)G}}{2 + \phi^2}\right) \tag{12.3-24}$$

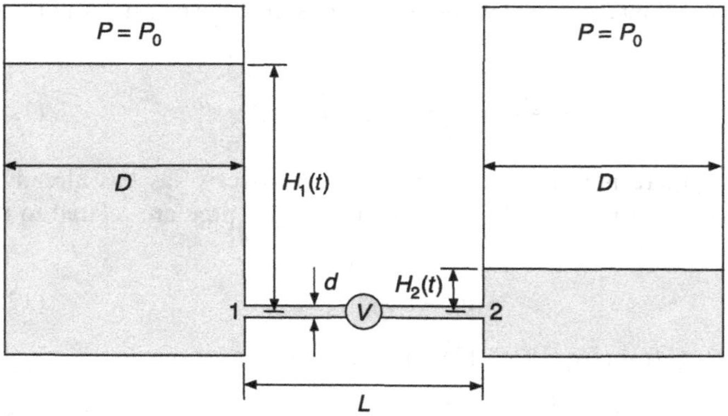

Figure 12.6 Two water tanks connected by a pipe, with a valve that is opened at $t = 0$.

for the pressure recovery and the branch flow rate. These results can be simplified when $\phi \ll 1$. In that case, using the expansions $(1+x)^{1/2} = 1 + (x/2) + \cdots$ and $(1+x)^{-1} = 1 - x + \cdots$ (each valid for $x \to 0$) and retaining only the leading terms gives

$$X = \phi\sqrt{2G} = F. \tag{12.3-25}$$

For $G = 1$ and $\phi = 0.1$, this gives $X = 0.141 = F$. That is only 4% higher than the result for X from Eq. (12.3-23) and 3% lower than that for F from Eq. (12.3-24). Returning to dimensional quantities, the small-orifice results are

$$\frac{P_2 - P_1}{\rho v_1^2} = \kappa C_o \left[\frac{2(P_1 - P_0)}{\rho v_1^2}\right]^{1/2} = \frac{\kappa v_B}{v_1} \quad (\kappa C_o \ll 1). \tag{12.3-26}$$

Additional discussion of manifolds may be found in Denn (1980, pp. 125–129).

Example 12.3-4 Draining of one tank into another. Suppose that two large water tanks of diameter D are connected by a horizontal pipe of length L and diameter d, as shown in Fig. 12.6. The water heights $H_1(t)$ and $H_2(t)$ differ initially by $H_0 = H_1(0) - H_2(0)$, and at $t = 0$ the valve is opened. Vents maintain a constant gas pressure P_0 in both tanks. It is desired to predict how long it will take for the levels to equalize. The dimensions are $D = 10$ m, $L = 10$ m, $d = 0.05$ m, and $H_0 = 5$ m. The pipe friction factor is estimated as $f = 5 \times 10^{-3}$.

The mean pipe velocity $U(t)$ will decrease as the tank levels approach one another. However, we will assume that the changes are slow enough for us to use the engineering Bernoulli equation, which assumes steady flow. As shown in the additional note that follows, this pseudosteady approximation requires only that the total volume of water transferred greatly exceed the volume of the pipe.

The water surfaces in the tanks are good reference positions. The pressures there are known and, given the large tank diameters, the kinetic energies are likely to be negligible. In any event, the velocities are equal. With $b = 1$, $\Delta\langle v\rangle^2 = 0$, $\Delta P = 0$, $\Delta h = H_2 - H_1$, $W_m = 0$, and $E_c = 0$, Eq. (11.4-8) reduces to

$$g(H_1 - H_2) = \frac{E_v}{w}. \tag{12.3-27}$$

The total viscous loss, which includes the pipe, the contraction at the inlet, the expansion at the outlet, and the valve, is

$$\frac{E_v}{w} = K\frac{U^2}{2} = \left(\frac{4Lf}{d} + K_c + K_e + K_v\right)\frac{U^2}{2}. \quad (12.3\text{-}28)$$

Here K is the sum of the loss coefficients, as shown. Evaluating the viscous loss in Eq. (12.3-27) and solving for U gives

$$U = \left[\frac{2g(H_1 - H_2)}{K}\right]^{1/2}. \quad (12.3\text{-}29)$$

The overall loss coefficient K will vary with time because of the dependence of f on Re. That variation is small, as will be discussed, and for simplicity we will regard K as a constant.

Mass balances for each tank indicate that $dV_1/dt = -Q = -dV_2/dt$, where V_1 and V_2 are the respective water volumes and Q is the volume flow rate in the pipe. Subtracting one balance from the other gives $d(V_1 - V_2)/dt = -2Q$. Introducing the tank and pipe diameters and using Eq. (12.3-29) to evaluate U, the combined mass balance is

$$\frac{\pi D^2}{4}\frac{d(H_1 - H_2)}{dt} = -2\left(\frac{\pi d^2}{4}\right)\left[\frac{2g(H_1 - H_2)}{K}\right]^{1/2}. \quad (12.3\text{-}30)$$

A convenient independent variable is $\theta(t) = [H_1(t) - H_2(t)]/H_0$, the height difference as a fraction of its initial value. Multiplying both sides of Eq. (12.3-30) by $4/(\pi D^2 H_0)$ gives

$$\frac{d\theta}{dt} = -2\left(\frac{d}{D}\right)^2\left(\frac{2g}{KH_0}\right)^{1/2}\theta^{1/2}, \quad \theta(0) = 1. \quad (12.3\text{-}31)$$

The time needed for the height difference to reach 1% of its initial value will be defined as the process time and denoted as t_p. Integrating from $t = 0$ (when $\theta = 1$) to $t = t_p$ (when $\theta = 0.01$),

$$\int_1^{0.01} \theta^{-1/2}\, d\theta = 2\theta^{1/2}\Big|_1^{0.01} = -1.8 = -2\left(\frac{d}{D}\right)^2\left(\frac{2g}{KH_0}\right)^{1/2} t_p. \quad (12.3\text{-}32)$$

The process time is then

$$t_p = 0.64\left(\frac{D}{d}\right)^2\left(\frac{KH_0}{g}\right)^{1/2}. \quad (12.3\text{-}33)$$

Using the given value of f and the coefficients in Table 12.1 for a contraction, an expansion, and an open globe valve,

$$K = \frac{4(10)(5 \times 10^{-3})}{0.05} + 0.44 + 1 + 6.0 = 11.44.$$

It is seen that the frictional loss in the pipe ($4Lf/d = 4.00$) is only about one-third of the total viscous loss. Thus, although f will increase gradually as Re falls, K will be nearly constant for most of the process, as assumed. The process time is evaluated as

$$t_p = 0.64\left(\frac{10}{0.05}\right)^2\left(\frac{11.44(5)}{9.81}\right)^{1/2} = 6.18 \times 10^4 \text{ s} = 17.2 \text{ h}.$$

The accuracy of f, and thus K, may be checked as follows. With $K = 11.44$, Eq. (12.3-29) gives an initial velocity of $U(0) = 2.93$ m/s and an initial Reynolds number of $\mathrm{Re}(0) = 1.46 \times 10^5$. Using also the effective roughness for commercial steel ($k =$ 0.046 mm, Table 2.2), it is found from Eq. (2.5-3) that $f = 5.25 \times 10^{-3}$. This improved value of f increases the loss coefficient in the plain pipe from 4.0 to 4.2 and the overall loss coefficient from 11.44 to 11.64. Those refinements barely affect Re(0), which decreases to 1.45×10^5. It is worth noting that the friction factor is close to the "fully rough" value of $f = 4.81 \times 10^{-3}$ obtained from Eq. (2.5-1). In that regime, where f is independent of Re, K would indeed be constant.

Additional note: pseudosteady approximation for tank filling or emptying
Calculating liquid flow rates to or from tanks is greatly simplified if the engineering Bernoulli equation is applicable to the pipes, even though the velocities are time-dependent. What is needed to justify such pseudosteady approximations is examined now for turbulent flow in a stationary, liquid-filled pipe of constant diameter.

To include contraction and expansion losses, the transient mechanical energy balance in Eq. (11.4-5) will be applied to the liquid from just outside the pipe inlet to just outside the outlet. The external volumes and that within any valves or fittings will be assumed to be negligible relative to the volume V_p in plain sections of pipe. The energy-accumulation term is then

$$\frac{d}{dt}\int_{V_p} \rho\left(\frac{v^2}{2} + gh\right) dV = \rho V_p \frac{d}{dt}\left(\frac{U^2}{2}\right). \tag{12.3-34}$$

The kinetic-energy integral may be evaluated simply, as shown, because the mean velocity U is the same throughout the pipe. The time derivative of the potential energy is zero because the pipe is filled and stationary. Neglecting the difference in kinetic energy between the reference positions and evaluating the remaining terms as in Eq. (11.4-8) gives

$$\rho V_p \frac{d}{dt}\left(\frac{U^2}{2}\right) = -w\left[\Delta\left(gh + \frac{P}{\rho}\right) - \frac{W_m}{w} + \frac{E_v}{w}\right]. \tag{12.3-35}$$

How to decide when the more exact relationship in Eq. (12.3-35) can be replaced by Eq. (11.4-8) will be illustrated by returning to Example 12.3-4. For the pipe in Fig. 12.6, $V_p = SL$, where $S = \pi d^2/4$. Assuming that the pressure difference equals the static difference between the tanks, $\Delta P/\rho = g(H_2 - H_1)$. Also, $w = \rho U S$, $\Delta h = 0$, and $E_v/w = KU^2/2$. Equation (12.3-35) then simplifies to

$$\underset{\mathrm{T1}}{\frac{dU}{dt}} = \underset{\mathrm{T2}}{\frac{g(H_1 - H_2)}{L}} - \underset{\mathrm{T3}}{\frac{KU^2}{2L}}, \qquad U(0) = 0. \tag{12.3-36}$$

The pseudosteady approximation assumes that term T1 is negligible during most of the process and that T2 ≅ T3. Indeed, setting T2 = T3 is what gives Eq. (12.3-29). Although it may possibly be accurate after a certain time, this cannot hold initially because T3 = 0 at $t = 0$. The pseudosteady approximation might give a result like the dashed curve in Fig. 12.7, whereas the actual velocity might more nearly resemble the solid curve. The two differ significantly for $t \le t_0$. Using the pseudosteady approximation to predict t_p requires that $t_0 \ll t_p$.

The duration of the initial transient can be estimated by noting that, during a time t_0, U increases from 0 to roughly $U_0 = (2gH_0/K)^{1/2}$, the initial velocity obtained from

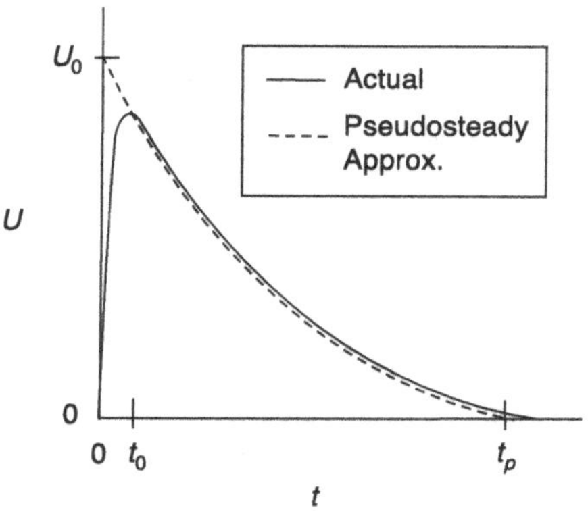

Figure 12.7 Qualitative comparison of the actual velocity in a pipe connecting two large tanks (as in Fig. 12.6) with that obtained using a pseudosteady approximation in the mechanical energy balance.

Eq. (12.3-29). During this period the water levels will change little, so that $H_1 - H_2 \sim H_0$. Also, T1 $\sim$ T2 during this phase. Thus, $dU/dt \sim U_0/t_0 \sim gH_0/L$, or $t_0 \sim U_0L/(gH_0)$. A comparison with the process time given by Eq. (12.3-33) indicates that a necessary condition for the pseudosteady analysis is

$$\frac{t_0}{t_p} = \frac{\sqrt{2}}{0.64}\left(\frac{L}{KH_0}\right)\left(\frac{d}{D}\right)^2 = \frac{1.1}{K}\left(\frac{V_p}{V_t}\right) \ll 1 \qquad (12.3\text{-}37)$$

where $V_t = \pi D^2 H_0/8$ is the total volume that is ultimately transferred. The last expression is particularly informative. In that K always includes at least entrance and exit losses, $1.1/K$ will never be large. Thus, it is sufficient that the pipe volume be much smaller than the volume transferred. For the data in Example 12.3-4,

$$\frac{t_0}{t_p} = \frac{2.2}{K}\left(\frac{L}{H_0}\right)\left(\frac{d}{D}\right)^2 = \frac{2.2}{11.44}\left(\frac{10}{5}\right)\left(\frac{0.05}{10}\right)^2 = 9.7 \times 10^{-6}$$

which easily satisfies the pseudosteady requirement. Indeed, $t_0 = (9.7 \times 10^{-6}) \times (6.2 \times 10^4\ \text{s}) = 0.6\ \text{s}$ is less than the time it takes to open a typical valve.

12.4 COMPRESSIBLE FLOW IN LONG PIPES

Compressible-fluid dynamics is one of the foundations of acoustics and aeronautics, and among the many books on this subject are Liepmann and Roshko (2001), Shapiro (1953), and Thompson (1972). The present introduction to compressible flow is necessarily limited in scope and focuses mainly on pipe flow. Middleman (1998, pp. 477–495) has a similar emphasis.

The key velocity scale in compressible flow is the speed of sound, c. As shown by Laplace in 1825, in an ideal gas it is

$$c = \left(\frac{\gamma RT}{M}\right)^{1/2}, \quad \gamma = \frac{\hat{C}_P}{\hat{C}_V} \qquad (12.4\text{-}1)$$

where M is the molar mass, $\hat{C}_P$ is the heat capacity at constant pressure (per unit mass), and $\hat{C}_V$ is the heat capacity at constant volume (also per unit mass). For air at 27 °C and atmospheric pressure, $\gamma = 1.40$ and $c = 347$ m/s. Hereafter, R [$= 8314\ \text{Pa m}^3(\text{kg} - \text{mol})^{-1}\ \text{K}^{-1}$] is always the universal gas constant (not a radius) and γ is always the heat-capacity ratio (not a surface tension). A derivation of Eq. (12.4-1) is outlined in Problem 12.19.

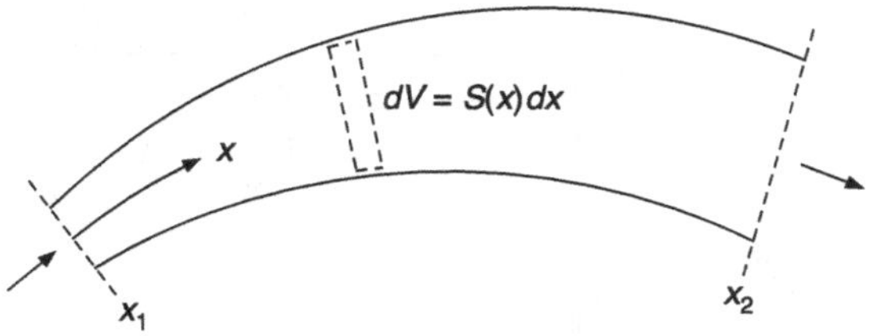

Figure 12.8 Schematic for evaluating the compression term in the engineering Bernoulli equation.

As mentioned already, compressibility is important for gas flows either in very long pipes or at velocities comparable to c. Turbulent flow in long pipes is the present focus and flow at near-sonic velocities is the subject of Section 12.5.

Engineering Bernoulli equation for variable density

To apply the engineering Bernoulli equation to gas flows in very long pipes, we need to know the rate at which mechanical energy is recovered as the pressure decreases and the gas expands. As derived in Section 11.4, the rate of mechanical energy loss from compressing the fluid within an arbitrary control volume is

$$E_c = -\int_V P(\nabla \cdot \mathbf{v})\, dV. \tag{12.4-2}$$

This loss can be estimated fairly simply for steady flow in a system like that in Fig. 12.8. The local cross-sectional area is $S(x)$, where x is an arc length. As usual in the macroscopic analysis of turbulent flows, it will be assumed that the flow is perpendicular to the openings, the plug-flow approximation can be used, and P and ρ are each uniform over any cross-section. Such a one-dimensional model is best for straight or gently curved conduits of constant or slowly varying cross-section.

For steady flow, the continuity equation reduces to $\nabla \cdot (\rho \mathbf{v}) = 0$, which may be rearranged as $\nabla \cdot \mathbf{v} = -(\mathbf{v} \cdot \nabla \rho)/\rho$. In one dimension, $\mathbf{v} \cdot \nabla \rho = v\, d\rho/dx$ and Eq. (12.4-2) becomes

$$E_c = \int_{x_1}^{x_2} \frac{Pv}{\rho} \frac{d\rho}{dx} S\, dx = w \int_{x_1}^{x_2} \frac{P}{\rho^2} \frac{d\rho}{dx}\, dx \tag{12.4-3}$$

where use has been made of the fact that $w = \rho v S$ is constant. Using integration by parts,

$$\int_{x_1}^{x_2} \frac{P}{\rho^2} \frac{d\rho}{dx}\, dx = -\int_{x_1}^{x_2} P \frac{d}{dx}\left(\frac{1}{\rho}\right) dx = -\frac{P}{\rho}\bigg|_{x_1}^{x_2} + \int_{x_1}^{x_2} \frac{1}{\rho} \frac{dP}{dx}\, dx. \tag{12.4-4}$$

Thus, the compression term is

$$\frac{E_c}{w} = -\Delta\left(\frac{P}{\rho}\right) + \int_{P_1}^{P_2} \frac{dP}{\rho}. \tag{12.4-5}$$

When this is substituted into Eq. (11.4-8), the $\Delta(P/\rho)$ term from E_c/w cancels that on the left-hand side. The engineering Bernoulli equation for turbulent flow with variable ρ

is then

$$\Delta\left(\frac{v^2}{2}+gh\right)+\int_{P_1}^{P_2}\frac{dP}{\rho}=\frac{1}{w}\left(W_m-E_v\right). \quad (12.4\text{-}6)$$

If ρ is constant, the integral equals $\Delta P/\rho$ and the simpler form used previously is recovered. Otherwise, an equation of state that relates ρ to P and T is needed to evaluate the integral, along with information about the temperature.

We will consider only ideal gases, for which

$$\rho=\frac{MP}{RT}. \quad (12.4\text{-}7)$$

If the gas is isothermal, then

$$\int_{P_1}^{P_2}\frac{dP}{\rho}=\frac{RT}{M}\int_{P_1}^{P_2}\frac{dP}{P}=\frac{RT}{M}\ln\left(\frac{P_2}{P_1}\right)=\frac{P_1}{\rho_1}\ln\left(\frac{P_2}{P_1}\right). \quad (12.4\text{-}8)$$

Noticing that $P_2/P_1=1+(\Delta P/P_1)$ and using $\ln(1+x)=x-x^2/2+\cdots$, which is valid for small x, this becomes

$$\int_{P_1}^{P_2}\frac{dP}{\rho}=\frac{\Delta P}{\rho_1}\left(1-\frac{1}{2}\frac{\Delta P}{P_1}+\ldots\right). \quad (12.4\text{-}9)$$

The first term is the result for incompressible flow. Thus, the corrections associated with density changes can be ignored if $|\Delta P|\ll 2P_1$. That requirement is often met. For instance, if the air duct of Example 2.4-1 is horizontal and the inlet pressure is atmospheric, then $|\Delta P|=4.9$ Pa, $P_1=1.0\times10^5$ Pa, and $|\Delta P|/(2P_1)=2.5\times10^{-5}$.

For flow past an object at high Re, the pressure variations are on the order of ρv^2, the inertial pressure scale. For those variations to be small relative to the absolute pressure, the velocity must be such that $\rho v^2/P\ll 1$. For ambient air ($\rho=1.2$ kg m^{-3} and $P=1\times10^5$ Pa), $\rho v^2/P=0.1$ when $v=91$ m/s, which is roughly one-fourth the speed of sound. Thus, for compressibility to be important in fluid dynamics, velocities comparable to the speed of sound are often needed. Very long conduits, which will be discussed next, are an exception to that rule.

Isothermal pipe flow

Assuming the gas to be isothermal yields the simplest model for compressible flow in a pipe. The longer the transit time, the greater the opportunity for the fluid temperature to reach that of the pipe wall. Thus, the isothermal assumption is best suited for moderate velocities in long pipes. Although the following derivation is for a cylindrical tube of diameter D and length L, the results are readily extended to other conduit shapes by using hydraulic diameters (Section 2.4).

In an isothermal, one-dimensional, compressible flow there are three unknowns: $v(x)$, $P(x)$, and $\rho(x)$. They are governed by conservation of mass, conservation of momentum (as a mechanical energy balance), and the equation of state. For steady flow in a horizontal pipe ($W_m=0$ and $\Delta h=0$), Eq. (12.4-6) may be rewritten in differential form as

$$\frac{d}{dx}\left(\frac{v^2}{2}\right)+\frac{1}{\rho}\frac{dP}{dx}=-\frac{2v^2 f}{D} \quad (12.4\text{-}10)$$

where Eq. (11.4-14) has been used to express the viscous loss as

$$dE_v = \frac{2wv^2 f}{D}\,dx. \tag{12.4-11}$$

The variable velocity notwithstanding, the friction factor f may be assumed to be constant because $\mathrm{Re} = Dv\rho/\mu = 4w/(\pi D\mu)$ is independent of position; recall from Section 1.2 that μ depends mainly on temperature, which is assumed here to be constant. For a uniform diameter, conservation of mass requires that ρv be constant. Differentiating that product gives

$$\frac{1}{\rho}\frac{d\rho}{dx} + \frac{1}{v}\frac{dv}{dx} = 0. \tag{12.4-12}$$

If the gas is ideal, the pressure and density gradients are related as

$$\frac{dP}{dx} = \left(\frac{\partial P}{\partial \rho}\right)_T \frac{d\rho}{dx} = \frac{RT}{M}\frac{d\rho}{dx} \equiv b^2\frac{d\rho}{dx}. \tag{12.4-13}$$

The constant b defined by the last equality has the dimension of velocity. As will be seen, it is the scale for $v(x)$ in isothermal pipe flow. For air at 27 °C, $b = 293$ m/s, which is about 85% of the speed of sound.

The governing equations can be combined in various ways to give a single differential equation that involves just one of the three unknowns. We will work with $v(x)$. From Eqs. (12.4-12) and (12.4-13),

$$\frac{1}{\rho}\frac{dP}{dx} = \frac{b^2}{\rho}\frac{d\rho}{dx} = -\frac{b^2}{v}\frac{dv}{dx} = -\left(\frac{b}{v}\right)^2 \frac{d}{dx}\left(\frac{v^2}{2}\right). \tag{12.4-14}$$

Using this to eliminate ρ and P from Eq. (12.4-10) gives

$$\frac{d}{dx}\left(\frac{v^2}{2}\right)\left[1 - \left(\frac{b}{v}\right)^2\right] = -\frac{2v^2 f}{D} \tag{12.4-15}$$

or, multiplying by $-2/b^2$,

$$\frac{d}{dx}\left(\frac{v}{b}\right)^2\left[\left(\frac{b}{v}\right)^2 - 1\right] = \frac{4f}{D}\left(\frac{v}{b}\right)^2. \tag{12.4-16}$$

It is seen that a convenient dependent variable is $\theta(x) = [v(x)/b]^2$, the square of the velocity ratio. Another simplification is to normalize the coordinate as $X = x/L$. The problem to be solved is then

$$\frac{1}{\theta}\left(\frac{1}{\theta} - 1\right)\frac{d\theta}{dX} = F, \quad \theta(0) = \theta_1 \tag{12.4-17}$$

for $0 \le X \le 1$. The two dimensionless parameters are

$$F = \frac{4Lf}{D} \tag{12.4-18}$$

$$\theta_1 = \left(\frac{v_1}{b}\right)^2 = \frac{1}{\rho_1 P_1}\left(\frac{w}{S}\right)^2 \tag{12.4-19}$$

where the subscript 1 refers to the inlet and $S = \pi D^2/4$. The friction factor and pipe dimensions determine F and the inlet conditions and mass flow rate are embedded in θ_1.

Equation (12.4-17) reveals some remarkable features of this type of flow. In particular, it indicates that $d\theta/dX > 0$ when $\theta < 1$ and $d\theta/dX < 0$ when $\theta > 1$. Consequently, $\theta \to 1$ at large x, for any value of θ_1. In other words, whatever the initial velocity, eventually $v(x) \to b$ if the pipe is long enough and the gas temperature is constant.

Equation (12.4-17) is nonlinear but separable. It has the implicit solution

$$\frac{1}{\theta_1} - \frac{1}{\theta} - \ln\left(\frac{\theta}{\theta_1}\right) = FX. \tag{12.4-20}$$

Once $\theta(X)$ is calculated, the local velocity, density, and pressure are evaluated as

$$\frac{v(x)}{v_1} = \frac{b}{v_1}\theta^{1/2}(x) = \frac{\rho_1}{\rho(x)} = \frac{P_1}{P(x)} \tag{12.4-21}$$

provided that the inlet values (v_1, ρ_1, and P_1) are known.

Example 12.4-1 Natural-gas pipeline. A natural-gas pipeline is an example of a system in which compressibility is made important by the pipe length. It is desired to find the power that must be delivered by a compressor serving a segment of such a pipeline. Representative of US practice is a steel pipe of diameter $D = 1.0$ m and length $L = 8.0 \times 10^4$ m (50 miles), with an inlet pressure of $P_1 = 7.0 \times 10^6$ Pa and a mass flow rate of $w = 400$ kg/s. It is assumed that the temperature is constant at 27 °C and the pipe is horizontal. After initial processing, natural gas is almost pure methane (CH_4), with $M = 16.0$ kg (kg-mol)$^{-1}$ and $\mu = 1.12 \times 10^{-5}$ Pa · s (Table 1.1). Using the ideal-gas law, $\rho_1 = 44.9$ kg m^{-3}.

The control volume chosen for calculating W_m extends from the compressor inlet to the pipe outlet, as from 1′ to 2 in Fig. 11.6. To exactly compensate for the viscous loss in the pipe (i.e., to have equal pressures at the compressor inlet and pipe outlet),

$$W_m = E_v = w\int_0^L \frac{2v^2 f}{D}\,dx = \frac{wFb^2}{2}\langle\theta\rangle \tag{12.4-22}$$

where $\langle\theta\rangle$ is the average of θ over the pipe length. The differential form of the viscous loss [Eq. (12.4-11)] was used because v depends on x.

The inlet value θ_1 is needed to determine $\theta(X)$. From Eq. (12.4-19),

$$\theta_1 = \frac{1}{\rho_1 P_1}\left(\frac{4w}{\pi D^2}\right)^2 = \frac{1}{(44.9)(7.0\times 10^6)}\left(\frac{4(400)}{\pi(1)^2}\right)^2 = 8.253\times 10^{-4}.$$

The other input parameter is F, which contains the friction factor. The Reynolds number is

$$\mathrm{Re} = \frac{4w}{\pi D\mu} = \frac{4(450)}{\pi(1)(1.12\times 10^{-5})} = 5.12\times 10^7.$$

Commercial steel has an effective roughness of $k = 0.046$ mm (Table 2.2), so that

$$\frac{k}{D} = \frac{0.046}{1000} = 4.6\times 10^{-5}.$$

Although k/D is small, Re is just large enough for the flow to be in the "fully rough" regime, where f is independent of Re. Using Eq. (2.5-1),

$$f = \left[4\log\left(\frac{k}{3.7D}\right)\right]^{-2} = \left[4\log\left(\frac{4.6\times 10^{-5}}{3.7}\right)\right]^{-2} = 2.60\times 10^{-3}.$$

Accordingly, from Eq. (12.4-18),

$$F = \frac{4Lf}{D} = \frac{4(8.0 \times 10^4)(2.60 \times 10^{-3})}{(1)} = 832.$$

With θ_1 and F now specified, $\theta(X)$ could be determined by solving Eq. (12.4-20) iteratively. However, because θ_1 is very small, we suppose (subject to confirmation) that $\theta \ll 1$ throughout the pipe. This suggests that Eq. (12.4-17) be simplified to

$$\frac{1}{\theta^2}\frac{d\theta}{dX} = F, \quad \theta(0) = \theta_1. \tag{12.4-23}$$

This has the explicit solution

$$\theta(X) = \left(\frac{1}{\theta_1} - FX\right)^{-1}. \tag{12.4-24}$$

The outlet velocity ratio so obtained is $\theta_2 = \theta(1) = 2.633 \times 10^{-3}$, which confirms that $\theta \ll 1$ everywhere. It follows that the average of θ over the pipe length is

$$\langle\theta\rangle = \int_0^1 \theta\, dX = -\frac{\ln(1 - F\theta_1)}{F}. \tag{12.4-25}$$

Using the pipeline parameter values,

$$\langle\theta\rangle = -\frac{\ln[1 - 832(8.253 \times 10^{-4})]}{832} = 1.395 \times 10^{-3}.$$

This differs by $<0.1\%$ from what is obtained by solving Eq. (12.4-20) for $\theta(X)$ and integrating numerically to find $\langle\theta\rangle$.

The remaining input needed to calculate the compressor power is

$$b^2 = \frac{RT}{M} = \frac{(8314)(300)}{16} = 1.559 \times 10^5 \text{ m}^2\text{ s}^{-2}$$

which corresponds to $b = 395$ m/s. The power requirement is

$$W_m = \frac{wFb^2}{2}\langle\theta\rangle = \frac{(400)(832)(1.559 \times 10^5)}{2}(1.395 \times 10^{-3}) = 3.62 \times 10^7 \text{ W}$$

or about 36 megawatts. This and the other results could be refined somewhat by using an equation of state that is more accurate for high-pressure methane than is the ideal-gas law.

From b, θ_1, and θ_2, the inlet and outlet velocities are $v_1 = 11.3$ m/s and $v_2 = 20.3$ m/s, respectively. These are much smaller than the speed of sound in methane, which is about 450 m/s. This confirms that it is the pipe length that makes compressibility important here, not the gas velocity. Corresponding to the nearly two-fold increase in velocity along the pipe are roughly two-fold reductions each in density and pressure. The velocity and pressure variations are shown in Fig. 12.9, where the velocity has been normalized by its outlet value and the pressure by its inlet value. These profiles are quite different than for incompressible flow, where the velocity would be constant and the pressure decline linear.

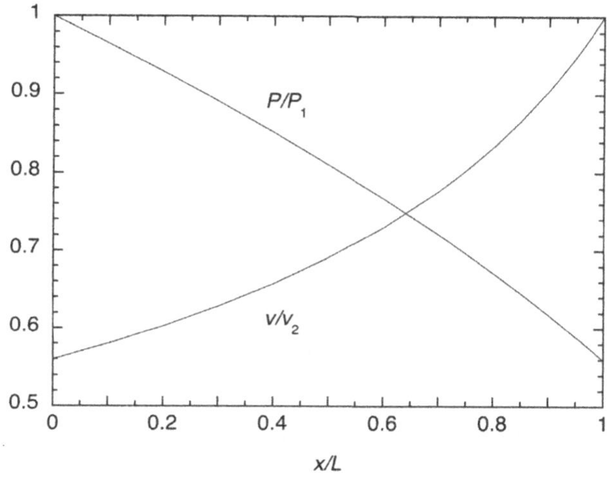

Figure 12.9 Velocity and pressure profiles for isothermal flow with $\theta_1 = 8.253 \times 10^{-4}$ and $F = 832$, representative of a natural-gas pipeline.

12.5 COMPRESSIBLE FLOW NEAR THE SPEED OF SOUND

Adiabatic pipe flow

As mentioned in Section 1.3, the fluid velocity divided by the speed of sound is the *Mach number*,

$$\mathrm{Ma} = \frac{v}{c}. \tag{12.5-1}$$

This is ordinarily not a constant, because v varies with position and, if there are temperature variations, so does c. In a truly "high-speed" gas flow, Ma $\sim$ 1. At such velocities there is little time for heat transfer between the gas and a tube of moderate length, making adiabatic flow a more realistic approximation than isothermal flow.

In adiabatic flow, T is an additional variable and one more governing equation is needed. That is provided by conservation of energy. On a per-unit-mass basis, the total energy is the sum of the internal ($\hat{U}$), kinetic ($v^2/2$), and potential (Φ) energies. For steady flow with one inlet and one *outlet*,

$$\Delta\left(\hat{U} + \frac{v^2}{2} + \Phi\right) = \hat{Q} + \hat{W} \tag{12.5-2}$$

where $\hat{Q}$ is the heat added to the fluid and $\hat{W}$ is the work done on it, each per unit mass. For adiabatic pipe flow $\hat{Q} = 0$, and the only contribution to $\hat{W}$ is the pressure–volume work at the inlet and outlet. Evaluating that work as in Section 11.4 and neglecting the potential energy change, conservation of energy becomes

$$\Delta\left(\hat{U} + \frac{v^2}{2} + \frac{P}{\rho}\right) = \Delta\left(\hat{H} + \frac{v^2}{2}\right) = 0 \tag{12.5-3}$$

where $\hat{H}$ is the enthalpy per unit mass. Thus, what is constant in adiabatic pipe flow is the sum of the enthalpy and kinetic energy. In an ideal gas $d\hat{H} = \hat{C}_P\, dT$ and the energy equation in one-dimension (for constant $\hat{C}_P$) is

$$\hat{C}_P \frac{dT}{dx} + \frac{d}{dx}\left(\frac{v^2}{2}\right) = 0. \tag{12.5-4}$$

In compressible-flow analysis it is customary to express $\hat{C}_P$ in terms of R, M, and γ. In an ideal gas,

$$\hat{C}_P - \hat{C}_V = \frac{R}{M} \tag{12.5-5}$$

from which it follows that

$$\hat{C}_P = \left(\frac{\hat{C}_P}{\hat{C}_P - \hat{C}_V}\right)\frac{R}{M} = \left(\frac{\gamma}{\gamma - 1}\right)\frac{R}{M}. \tag{12.5-6}$$

The energy equation for steady, one-dimensional, adiabatic flow is then

$$\left(\frac{\gamma}{\gamma - 1}\right)\frac{R}{M}\frac{dT}{dx} + \frac{d}{dx}\left(\frac{v^2}{2}\right) = 0. \tag{12.5-7}$$

Also governing the four primary unknowns (v, P, ρ, T) in adiabatic pipe flow are conservation of mass, conservation of momentum, and the equation of state. The mass and momentum (mechanical energy) equations are as in Section 12.4. With ρ, P, and T all variable, writing the ideal-gas equation as $P/(\rho T) =$ constant and differentiating gives

$$\frac{1}{P}\frac{dP}{dx} - \frac{1}{\rho}\frac{d\rho}{dx} - \frac{1}{T}\frac{dT}{dx} = 0. \tag{12.5-8}$$

In summary, the four differential equations that must be solved simultaneously are Eqs. (12.4-12) (mass), (12.4-10) (momentum), (12.5-7) (energy), and (12.5-8) (equation of state). The speed of sound (c) is a fifth unknown, but it can be calculated readily from T using Eq. (12.4-1).

As in the isothermal analysis of Section 12.4, we will combine all the information into a single differential equation. The manipulations are made lengthier by the additional unknowns. Briefly, Ma is introduced into Eq. (12.4-10) by multiplying everything by γ/c^2. The result is a relationship between dP/dx and dv/dx that involves Ma. A second such relationship is obtained by using Eqs. (12.4-12) and (12.5-7) to eliminate $d\rho/dx$ and dT/dx from Eq. (12.5-8). Equating the two expressions for dP/dx yields a single equation involving dv/dx and Ma. Differentiating $\text{Ma}(x) = v(x)/c(x)$ and evaluating $c(x)$ using Eq. (12.4-1) allows dv/dx to be rewritten in terms of $d\text{Ma}/dx$. Retaining Ma(x) as the primary dependent variable gives

$$\frac{1 - \text{Ma}^2}{\text{Ma}^4\{1 + [(\gamma - 1)/2]\text{Ma}^2\}}\frac{d}{dx}(\text{Ma}^2) = \frac{4\gamma f}{D}. \tag{12.5-9}$$

Recalling that $\gamma > 1$, it is seen that $d\text{Ma}/dx > 0$ when $\text{Ma} < 1$ and $d\text{Ma}/dx < 0$ when $\text{Ma} > 1$. The important conclusion is that the gas velocity in a long, adiabatic tube can reach the speed of sound, but not pass that limit. That is true irrespective of whether the flow is initially subsonic ($\text{Ma} < 1$ at the inlet) or supersonic ($\text{Ma} > 1$ at the inlet).

Setting $\phi = \text{Ma}^2$ and normalizing the coordinate ($X = x/L$), Eq. (12.5-9) becomes

$$\frac{1 - \phi}{\phi^2(1 + \Gamma\phi)}\frac{d\phi}{dX} = \gamma F, \quad \phi(0) = \phi_1 \tag{12.5-10}$$

where $\Gamma = (\gamma - 1)/2$ and F is given again by Eq. (12.4-18).[2] For air, $\Gamma = (1.40 - 1)/2 = 0.20$. The inlet value of ϕ (the square of the inlet Mach number) is

$$\phi_1 = \frac{1}{\rho_1 P_1 \gamma}\left(\frac{w}{S}\right)^2 = \text{Ma}_1^2. \tag{12.5-11}$$

Equation (12.5-10) is separable and has the implicit solution

$$\frac{1}{\phi_1} - \frac{1}{\phi} + (1+\Gamma)\ln\left[\frac{\phi_1(1+\Gamma\phi)}{\phi(1+\Gamma\phi_1)}\right] = \gamma F X. \tag{12.5-12}$$

The gas velocity, $v(x) = c(x)\phi(x)^{1/2}$, is not fully determined until $c(x)$ is known, which requires knowledge of $T(x)$. From Eq. (12.5-7),

$$\left(\frac{\gamma}{\gamma-1}\right)\frac{R}{M}\frac{dT}{dx} = -\frac{d}{dx}\left(\frac{v^2}{2}\right) = -\frac{d}{dx}\left(\frac{\gamma R T \phi}{2M}\right). \tag{12.5-13}$$

Canceling the common factor $\gamma R/M$ in the first and last terms and expanding $d(T\phi)/dx$, this is rearranged as

$$\frac{1}{T}\frac{dT}{dx} = -\frac{\Gamma}{1+\Gamma\phi}\frac{d\phi}{dx}. \tag{12.5-14}$$

Thus, in a flow that is initially subsonic, where $\phi_1 < 1$ and ϕ increases toward unity, T decreases along the pipe. In one that is initially supersonic, T increases. Noticing that Eq. (12.5-14) is equivalent to $d\ln T = -d\ln(1+\Gamma\phi)$, it is integrated to give

$$\frac{T(x)}{T_1} = \frac{1+\Gamma\phi_1}{1+\Gamma\phi(x)}. \tag{12.5-15}$$

It follows that the speed of sound, gas velocity, density, and pressure are given by

$$\frac{c(x)}{c_1} = \left[\frac{T(x)}{T_1}\right]^{1/2} = \left[\frac{1+\Gamma\phi_1}{1+\Gamma\phi(x)}\right]^{1/2} \tag{12.5-16}$$

$$\frac{v(x)}{v_1} = \frac{c(x)}{c_1}\left[\frac{\phi(x)}{\phi_1}\right]^{1/2} = \left[\frac{[1+\Gamma\phi_1]\phi(x)}{[1+\Gamma\phi(x)]\phi_1}\right]^{1/2} = \frac{\rho_1}{\rho(x)} \tag{12.5-17}$$

$$\frac{P(x)}{P_1} = \frac{\rho(x)T(x)}{\rho_1 T_1} = \left[\frac{[1+\Gamma\phi_1]\phi_1}{[1+\Gamma\phi(x)]\phi(x)}\right]^{1/2}. \tag{12.5-18}$$

In an initially subsonic flow, where T decreases along the pipe, c decreases and v increases, both contributing to the increase in Ma. Accompanying those changes are decreases in both ρ and P. In an initially supersonic flow, all variables change in the opposite direction.

[2] In a smooth tube, $f = f(\text{Re}, \text{Ma})$ from dimensional analysis. In a high-speed flow, temperature variations affect μ and thus Re, and of course Ma also changes along the tube. Thus, f is not independent of x and Eq. (12.4-18) is made more precise by replacing f by $\langle f \rangle$, the length-averaged value. However, in turbulent flows the dependence of f on Re is weak enough that the effects of temperature variations on μ tend to be negligible. Also, in fully developed subsonic flow f is insensitive to Ma (Shapiro, 1953, pp. 185 and 1131). Thus, the use of $\langle f \rangle$ is important mainly when entrance effects are prominent or the flow is supersonic.

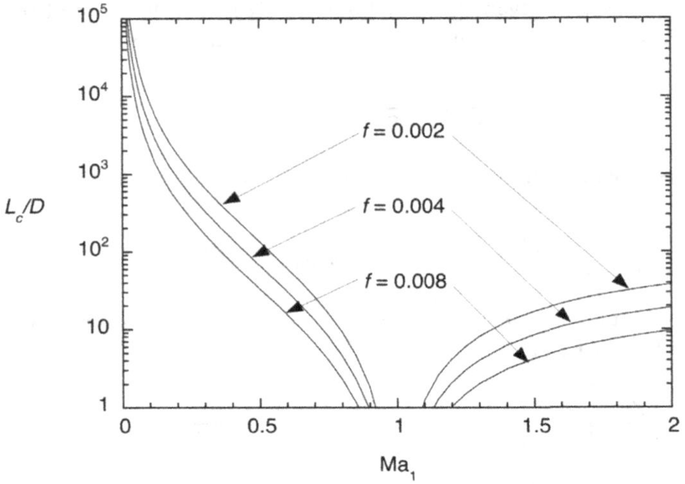

Figure 12.10 Pipe lengths for the onset of choking in adiabatic flow with $\gamma = 1.40$. Such gas flows are choked if $L \geq L_c$.

Choked flow

In a sufficiently long pipe, the velocity will reach the speed of sound [$c(x)$, Eq. (12.4-1)] if the system is adiabatic, or an analogous speed [b, Eq. (12.4-13)] if it is isothermal.[3] When this occurs the flow is said to be *choked*. The minimum pipe length for choked flow (L_c) is found by setting $\theta(1) = 1$ in Eq. (12.4-20) or $\phi(1) = 1$ in Eq. (12.5-12). The adiabatic result is

$$\frac{L_c}{D} = \frac{1}{4\gamma f}\left\{\frac{1}{\phi_1} - 1 + (1+\Gamma)\ln\left[\frac{\phi_1(1+\Gamma)}{1+\Gamma\phi_1}\right]\right\} \quad \text{(adiabatic)}. \quad (12.5\text{-}19)$$

These critical lengths are shown in Fig. 12.10, where Ma_1 is the inlet Mach number. The more Ma_1 deviates from unity and the smaller the friction factor, the greater the pipe length needed for choking to occur. Initially supersonic flows ($\text{Ma}_1 > 1$) are much more prone to choking than initially subsonic ones ($\text{Ma}_1 < 1$). Extremely long lengths are needed when Ma_1 is small.

The peculiarity of choked flow may be seen by plotting Eq. (12.5-19) in another way and imagining a set of experiments in which $\text{Ma}_1 < 1$. What was just described as a critical length for the onset of choking may be viewed more generally as a critical value of the parameter F, namely $F_c = 4fL_c/D$. The curve in Fig. 12.11 shows Ma_1 as a function of γF when $F = F_c$. Suppose now that we were to measure the mass flow rate w as a function of pipe length L and outlet pressure P_2, for the same pipe diameter, gas, and inlet conditions (ρ_1 and P_1). The experimental variations in w and L would cause proportional changes in the calculated values of Ma_1 and γF, respectively [Eqs. (12.5-11) and (12.4-18)]. In other words, in this plot Ma_1 is a proxy for w and γF is a proxy for L.[4]

The labeled points in Fig. 12.11 represent the results of a set of hypothetical experiments. Point A corresponds to a value of L too small for choking. If P_2 is reduced at constant L, w can be increased up to the value at point B. Or, reducing P_2 while increasing

[3] For a flow to remain isothermal as $v \to b$, the rate of heat transfer from the pipe wall to the gas must become infinite (Middleman, 1998, p. 492). Thus, the limit $v = b$ is never actually reached.

[4] It is assumed here that the Reynolds number and pipe roughness are large enough to make f nearly constant.

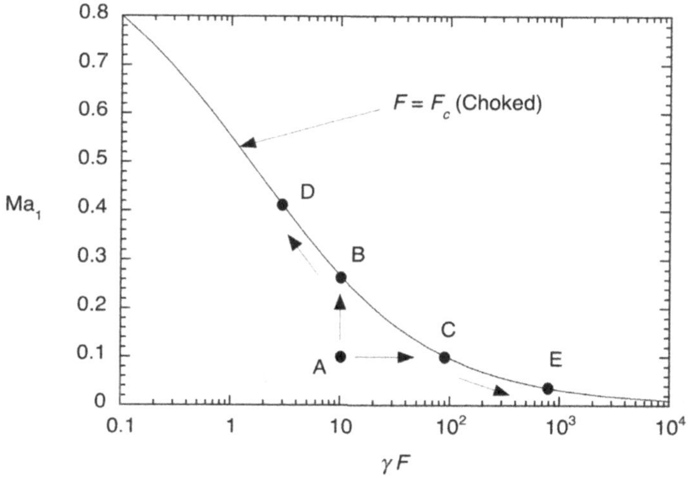

Figure 12.11 Relationship between the inlet Mach number (Ma_1, proportional to the mass flow rate w) and the product γF (proportional to the tube length L). The curve is the constraint present when the flow is choked. The points labeled A–E are the results of hypothetical experiments (see the text).

L could keep w constant, until point C. However, once either B or C is reached, the flow is choked and any further changes follow the curve. From point B, w can be increased more only by reducing L, as in going to D. From point C, if L is increased further the flow rate must fall, as at E. In other words, choking makes the region above and to the right of the curve inaccessible. As will be shown in Example 12.5-2, the underlying cause of this is that the pressure just inside the downstream end of the pipe cannot be controlled independently when the flow is choked.

Choking is a key consideration in the design of pressure-relief devices, where failing to recognize the limit placed on the discharge velocity could seriously impact boiler or reactor safety. Although it makes flow rates more difficult to adjust, choked flow can be used to advantage if it is desired that a mass flow rate be insensitive to the downstream pressure, as in a mass-flow controller (Middleman, 1998, pp. 486–488).

Example 12.5-1 Absence of choking in a natural-gas pipeline. How long would the pipe in Example 12.4-1 need to be for the flow to become choked?

The result for isothermal flow that is analogous to Eq. (12.5-19) is

$$\frac{L_c}{D} = \frac{1}{4f}\left(\frac{1}{\theta_1} - 1 + \ln\theta_1\right) \quad \text{(isothermal)}. \tag{12.5-20}$$

Using the values of D, f, and θ_1 from Example 12.4-1,

$$L_c = \frac{1.0}{4(2.60\times10^{-3})}\left[\frac{1}{8.253\times10^{-4}} - 1 + \ln(8.253\times10^{-4})\right] = 1.16\times10^5\ \mathrm{m}$$

or 72 miles. Segments of natural-gas pipelines in the US are typically not more than 50–60 miles in length.

Example 12.5-2 Choked air flow. Suppose that ambient air is drawn through a pipe by partially evacuating a chamber connected to its outlet. Let position 1 be just inside the inlet, position 2 be just inside the outlet, and position 3 be in the downstream chamber.

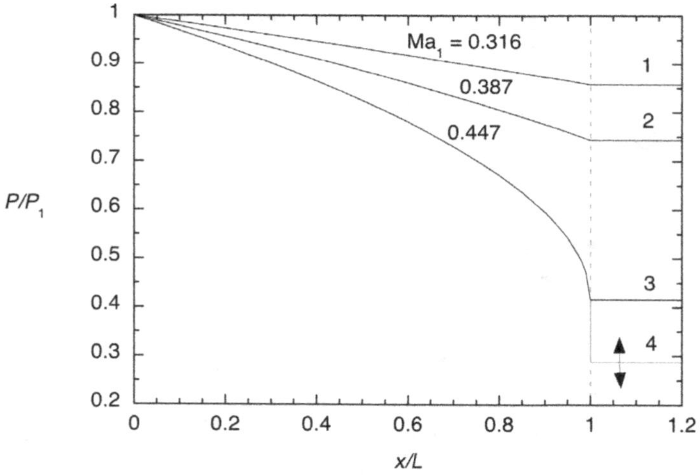

Figure 12.12 Pressure profiles for adiabatic air flow in a pipe. The three inlet Mach numbers (Ma_1) correspond to $\phi_1 = 0.10$, 0.15, and 0.20, and in all cases $F = 1.60$. The values for $x/L > 1$ correspond to controlled pressures in the downstream chamber.

Varying the rate at which air is pumped out allows the chamber pressure (P_3) to be controlled. The inlet conditions (ρ_1, P_1, T_1) are fixed. It is desired to determine the effect of P_3 on w for initially subsonic, adiabatic flow when $F = 1.60$ (e.g., for $f = 2.5 \times 10^{-3}$ and $L/D = 160$).

How P_3 affects w may be seen by examining the pressure variations along the pipe. The pressure profiles are determined by solving Eq. (12.5-12) iteratively for $\phi(X)$ and then using Eq. (12.5-18) to find $P(X)/P_1$. Representative plots are shown in Fig. 12.12. Each curve corresponds to a specified value of P_3/P_1 and the mass flow rates are expressed as values of Ma_1. For curves 1–3, increasing the vacuum increases the pressure drop within the pipe because $P_3 = P_2$. Curve 3 is special, in that it corresponds to a value of P_3/P_1 just small enough for the flow to be choked. Because the velocity at the outlet is then maximal (and the pressure gradient there is infinite), further decreases in P_3 (as in curve 4) do not influence either P_2 or w. In choked flow the air pressure falls from P_2 to P_3 somewhere beyond the pipe outlet. That last drop in pressure might occur over a negligible distance, as shown, making the function $P(X)$ practically discontinuous. Thin regions with such abrupt changes in P and the other state variables are called *shocks*.

The inability of decreases in P_3 to influence a choked flow may be viewed in terms of information transfer. The speed of sound is the speed at which pressure disturbances move *relative to the fluid*. Pressure information does not move upstream when $\mathrm{Ma}_2 = 1$ because the speed of sound is exactly offset by the air velocity. Choked flow does not occur with liquids because Ma is always small. The speed of sound in ambient water is about 1500 m/s, which is much larger than practical water velocities. Thus, if a piston increases the pressure at one end of a column of liquid, the increase in pressure at the other end is almost immediate. Only in gases can the fluid travel as fast as, or faster than, the pressure wave.

Varying cross-section: nozzles and diffusers

The analysis of high-speed gas flow through nozzles or diffusers requires that their varying cross-sectional area $S(x)$ be taken into account. Such devices are short enough that viscous losses tend to be small, in addition to the flow being nearly adiabatic. In a frictionless, adiabatic flow the specific entropy remains constant, so that this idealization is called *isentropic flow*. Diameter changes notwithstanding, one-dimensional analyses of isentropic flow in nozzles or diffusers are accurate enough to be informative, yet simpler than what has already been discussed for tube flow with losses. As will be seen, predicting the mass flow rate of a gas through a nozzle requires only algebraic equations. The development given below generally follows that in Thompson (1972).

In nozzle or diffuser problems $S(x)$ is generally specified and there are five unknown functions of position, namely, v, P, ρ, T, and c. Needed again are forms of the continuity equation, momentum (mechanical energy) equation, energy equation, equation of state, and the relationship between c and T. It is expeditious to focus first on the thermodynamic variables and leave continuity and the variation in S for last.

The governing equations used previously can be rewritten in terms of differential changes in the variables, instead of their x derivatives. The adiabatic energy equation and inviscid momentum equation may then be combined as

$$\left(\frac{\gamma}{\gamma-1}\right)\frac{R}{M}dT=-d\left(\frac{v^2}{2}\right)=\frac{1}{\rho}dP. \tag{12.5-21}$$

The first relationship is integrated by setting $T=T_0$ in a static gas, which gives

$$\left(\frac{\gamma}{\gamma-1}\right)\frac{R}{M}(T-T_0)=-\frac{v^2}{2}. \tag{12.5-22}$$

The parameter T_0, the *stagnation temperature*, is a measure of the total specific energy of the gas and is meaningful even if the gas is not static anywhere in the system of interest. The subscript "0" will be used hereafter to denote quantities evaluated at a possibly hypothetical position where $v=0$. Rearranging Eq. (12.5-22) and using Eq. (12.4-1) to introduce c yields

$$\frac{T_0}{T}=1+\left(\frac{\gamma-1}{2}\right)\mathrm{Ma}^2 \tag{12.5-23}$$

which relates the local temperature to the local Mach number.

Considering now the temperature–pressure relationship in Eq. (12.5-21), ρ is eliminated using the ideal-gas equation and the result rearranged as

$$\frac{dT}{T}=\left(\frac{\gamma-1}{\gamma}\right)\frac{dP}{P} \tag{12.5-24}$$

or $d\ln T=d\ln P^{(\gamma-1)/\gamma}$. Integration of this gives another expression for T_0/T,

$$\frac{T_0}{T}=\left(\frac{P_0}{P}\right)^{(\gamma-1)/\gamma} \tag{12.5-25}$$

in which P_0 is the *stagnation pressure*. Using this relationship between P and T, it follows from the ideal-gas equation that ρ and T are related as

$$\frac{T_0}{T}=\left(\frac{\rho_0}{\rho}\right)^{\gamma-1} \tag{12.5-26}$$

where ρ_0 is the *stagnation density*. From Eq. (12.4-1) comes yet another expression for T_0/T,

$$\frac{T_0}{T}=\left(\frac{c_0}{c}\right)^2 \tag{12.5-27}$$

where c_0 is the speed of sound at the stagnation temperature. All of the thermodynamic variables have been related now to the temperature, and thus the Mach number. In summary,

$$1+\left(\frac{\gamma-1}{2}\right)\mathrm{Ma}^2=\frac{T_0}{T}=\left(\frac{P_0}{P}\right)^{(\gamma-1)/\gamma}=\left(\frac{\rho_0}{\rho}\right)^{\gamma-1}=\left(\frac{c_0}{c}\right)^2. \quad (12.5\text{-}28)$$

What remains is to connect Ma(x) with $S(x)$. For steady flow, conservation of mass requires that $\rho v S$ be constant. We now define a second reference condition (identified by the subscript "$*$"), in which ρ and S are such that $v = c$ or Ma = 1. As with the stagnation values, these sonic values may or may not be realized in the system of interest. Setting $\rho v S = \rho_* v_* S_*$, recalling that Ma = v/c in general, and noting that $v_* = c_*$ by definition, we find that

$$\frac{S}{S_*}=\frac{\rho_* c_*}{\rho v}=\left(\frac{\rho_*}{\rho_0}\right)\left(\frac{\rho_0}{\rho}\right)\frac{1}{\mathrm{Ma}}\left(\frac{c_*}{c_0}\right)\left(\frac{c_0}{c}\right). \quad (12.5\text{-}29)$$

The ratios of sonic to stagnation values that appear here are evaluated by setting Ma = 1 in Eq. (12.5-28). Including also the pressure ratio (which is needed later),

$$\frac{\rho_*}{\rho_0}=\left(1+\frac{\gamma-1}{2}\right)^{-1/(\gamma-1)},\quad \frac{c_*}{c_0}=\left(1+\frac{\gamma-1}{2}\right)^{-1/2},\quad \frac{P_*}{P_0}=\left(1+\frac{\gamma-1}{2}\right)^{-\gamma/(\gamma-1)}. \quad (12.5\text{-}30)$$

For air, where $\gamma = 1.40$, the values are $\rho_*/\rho_0 = 0.634$, $c_*/c_0 = 0.913$, and $P_*/P_0 = 0.528$. Using Eqs. (12.5-28) and (12.5-30) to evaluate the four ratios on the right-hand side of Eq. (12.5-29) gives the relationship between S and Ma,

$$\frac{S}{S_*}=\frac{1}{\mathrm{Ma}}\left(\frac{2}{\gamma+1}+\frac{\gamma-1}{\gamma+1}\mathrm{Ma}^2\right)^{(\gamma+1)/[2(\gamma-1)]}. \quad (12.5\text{-}31)$$

For air this simplifies to

$$\frac{S}{S_*}=\frac{1}{\mathrm{Ma}}\left(0.833+0.167\,\mathrm{Ma}^2\right)^3. \quad (12.5\text{-}32)$$

The use of these relationships is illustrated in the example that follows.

Example 12.5-3 Converging nozzle. Suppose that a large air tank discharges through a nozzle that narrows to an area S_t at its throat, where it contacts external air. Because the tank air is nearly static, its pressure and temperature will nearly equal P_0 and T_0 respectively. The external air pressure is P_e. For $P_0 = 3.0$ atm, it is desired to compare the mass flow rates for $P_e = 2.0$ atm and $P_e = 1.0$ atm. To obtain only relative rates, values of S_t and T_0 are not needed.

As with a pipe, an initially subsonic flow in a converging nozzle can become sonic at most (Problem 12.20). Thus, the key consideration is how P_e compares with the pressure P_* for Ma = 1 at the throat. If $P_e > P_*$, the throat pressure will be $P_t = P_e$. If $P_e < P_*$, then $P_t = P_*$ and the flow will be choked. For air at $P_0 = 3.0$ atm, Eq. (12.5-30) gives $P_* = (3.0)(0.528) = 1.58$ atm. Thus, the flow will be choked for $P_e = 1.0$ atm, but not 2.0 atm.

For $P_e = 2.0$ atm, $P_t/P_0 = P_e/P_0 = 0.667$. Rearranging Eq. (12.5-28) to relate the throat Mach number to this pressure ratio (and using $\gamma = 1.40$) gives

$$\text{Ma}_t = \left\{ \frac{2}{\gamma - 1} \left[\left(\frac{P_t}{P_0} \right)^{-(\gamma-1)/\gamma} - 1 \right] \right\}^{1/2} = 0.784. \qquad (12.5\text{-}33)$$

Returning to Eq. (12.5-28),

$$\frac{c_t}{c_0} = \left(1 + \frac{\gamma - 1}{2} \text{Ma}_t^2 \right)^{-1/2} = 0.944 \qquad (12.5\text{-}34)$$

$$\frac{\rho_t}{\rho_0} = \left(1 + \frac{\gamma - 1}{2} \text{Ma}_t^2 \right)^{-1/(\gamma-1)} = 0.749. \qquad (12.5\text{-}35)$$

It follows that the outlet velocity and mass flow rate are

$$v_t = \text{Ma}_t c_t = \text{Ma}_t \left(\frac{c_t}{c_0} \right) c_0 = (0.784)\,(0.944)\, c_0 = 0.740 c_0 \qquad (12.5\text{-}36)$$

$$w_t = \rho_t v_t S_t = \left(\frac{\rho_t}{\rho_0} \right) \left(\frac{v_t}{c_0} \right) \rho_0 c_0 S_t = (0.749)(0.740)\rho_0 c_0 S_t = 0.554 \rho_0 c_0 S_t. \qquad (12.5\text{-}37)$$

For $P_e = 1.0$ atm (or any other value ≤ 1.58 atm), $P_t/P_0 = P_*/P_0 = 0.528$ from Eq. (12.5-30). With $\text{Ma}_t = 1$, $c_t/c_0 = c_*/c_0 = 0.913$ and $\rho_t/\rho_0 = \rho_*/\rho_0 = 0.634$. Thus, $v_t = 0.913 c_0$ and $w_t = (0.634)(0.913)\rho_0 c_0 S_t = 0.579 \rho_0 c_0 S_t$. This mass flow rate is only 4.5% higher than for $P_e = 2.0$ atm, and no additional increase in w_t could be achieved by reducing P_e further.

12.6 CONCLUSION

Although friction factors for fully developed flow are central to the analysis and design of piping systems, flow resistances other than those in straight pipes are not necessarily negligible. Resistances due to diameter changes, bends, and valves can even exceed those in the plain sections of pipe. The additional pressure drop in an entrance region can be expressed as an effective increase in tube length, as given by Eq. (12.2-6) for laminar flow and Eq. (12.2-10) for turbulent flow. In more complex systems, the viscous losses due to individual components can be calculated from their loss coefficients, as defined in Eq. (12.3-1) and listed in Table 12.1. Loss coefficients in pipes and fittings distinguish real flows from inviscid ones, and are important when using the engineering Bernoulli equation in design. In pipes connecting large tanks or reservoirs, time-dependent flows are often pseudosteady, allowing the engineering Bernoulli equation (derived in Section 11.4 for steady flow) to be applied.

The incompressible idealization breaks down for gas flows with sizable pressure variations, as in very long pipes or with Mach numbers approaching unity. The Mach number (Ma) is the local velocity relative to the speed of sound, which is given by Eq. (12.4-1). A form of the engineering Bernoulli equation for compressible fluids is Eq. (12.4-6). In very long pipes, the pressure drop can be a significant fraction of the inlet pressure, even when velocities are moderate. Assuming isothermal flow can be useful in such systems and leads to Eq. (12.4-20) for the velocity variation along a pipe. At high speeds, adiabatic flow is a more realistic assumption and velocity variations are described then by Eq. (12.5-12). A special aspect of adiabatic pipe flow is that Ma $\rightarrow$ 1 if there is sufficient pipe length, independent of the inlet conditions. Once the speed of sound is reached at the outlet, decreasing the external pressure has no effect on the flow rate and the flow is said to be choked. Choking occurs also in nozzles. A useful first approximation for high-speed, compressible flow in nozzles and diffusers is that it is inviscid as well as adiabatic, making it isentropic. The assumption of isentropic flow leads to relatively simple relationships among the state variables, the gas velocity, and the varying cross-sectional area, as given by Eqs. (12.5-28), (12.5-30), and (12.5-31).

REFERENCES

Atkinson, B., M. P. Brocklebank, C. C. H. Card, and J. M. Smith. Low Reynolds number developing flows. *AIChE J.* 15: 548–553, 1969.

Blevins, R. D. *Applied Fluid Dynamics Handbook*. Krieger Publishing, Malabar, FL, 2003.

Boger, D. V. Circular entry flows of inelastic and viscoelastic fluids. *Adv. Transp. Processes* 2: 43–104, 1982.

Crane Company, *Flow of Fluids through Valves, Fittings and Pipe*. Technical Paper No. 410M, Crane Co., Stamford, CT, 2012.

Deen, W. M. *Analysis of Transport Phenomena*, 2nd ed. Oxford University Press, New York, 2012.

Denn, M. M. *Process Fluid Mechanics*. Prentice-Hall, Englewood Cliffs, NJ, 1980.

Epstein, N. B. Hindered transport through porous membranes. S.M. thesis, Department of Chemical Engineering, Massachusetts Institute of Technology, Cambridge, MA, 1979.

Kestin, J., M. Sokolov, and W. Wakeham. Theory of capillary viscometers. *Appl. Sci. Res.* 27: 241–264, 1973.

Liepmann, H. W. and A. Roshko. *Elements of Gasdynamics*. Dover, Mineola, NY, 2001.

Martin, J. J. Expansion and contraction losses in fluid flow. *Chem. Eng. Educ.*, Summer, 138–140 and 148, 1974.

Middleman, S. *An Introduction to Fluid Dynamics*. Wiley, New York, 1998.

Pritchard, P. J. *Fox and McDonald's Introduction to Fluid Mechanics*, 8th ed. Wiley, New York, 2011.

Shapiro, A. H. *The Dynamics and Thermodynamics of Compressible Fluid Flow*. Ronald Press, New York, 1953 (2 volumes).

Thompson, P. A. *Compressible-Fluid Dynamics*. McGraw-Hill, New York, 1972.

Tilton, J. N. Fluid and Particle Dynamics. In *Perry's Chemical Engineers' Handbook*, D. E. Green and R. H. Perry (Eds.), 8th ed. McGraw-Hill, New York, 2007.

Wilkes, J. O. *Fluid Mechanics for Chemical Engineers*, 2nd ed. Prentice Hall, Upper Saddle River, NJ, 2006.

Zagarola, M. V. and A. J. Smits. Mean-flow scaling of turbulent pipe flow. *J. Fluid Mech.* 373: 33–79, 1998.

PROBLEMS

12.1. Entrance effects with air flow Consider air flow at a mean velocity of 5.0 m/s through a smooth pipe of diameter 0.10 m and length 10 m, at 27 °C. The air comes from a tank at atmospheric pressure.

(a) If there is a sudden contraction at the inlet, what is the length of the entrance region?
(b) How does the actual pressure drop (from just outside the inlet to the end of the pipe segment) compare with that for fully developed flow?
(c) Confirm that ρ is nearly constant.

12.2. Entrance-region model It is desired to estimate the entrance length L_E for laminar flow at large Re in a parallel-plate channel with wall spacing $2H$ and mean velocity U. In such a channel the velocity profile develops much as in Fig. P12.2. During the transition from plug flow at the inlet to the final parabolic profile, "shear layers" next to the walls accompany a "core" where the profile is flat. The shear layers are like laminar boundary layers and the core resembles an outer region. Let $\delta(x)$ and $u(x)$ be the shear-layer thickness and core velocity, respectively.

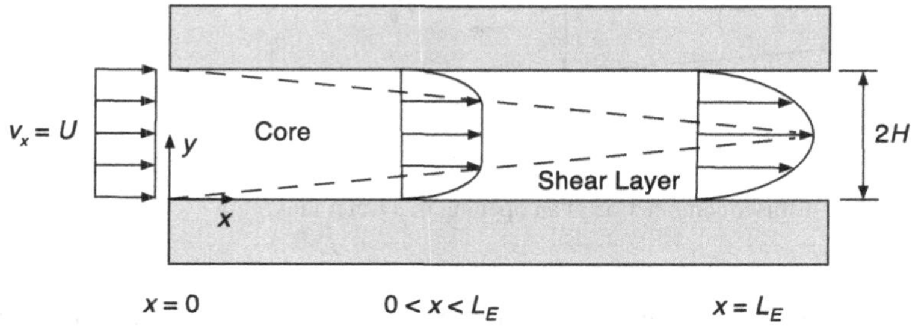

Figure P12.2 Developing velocity profile in a parallel-plate channel.

(a) As a first approximation, assume that $\delta(x)$ equals the boundary-layer thickness for flow past a flat plate. Show that, consistent with Eq. (12.2-1) for Re $\gg$ 1, this gives a relationship of the form

$$\frac{L_E}{H} = a\,\mathrm{Re} \qquad \text{(P12.2-1)}$$

where a is a constant and $\mathrm{Re} = 2UH/v$.

(b) Using the integral analysis in Example 9.4-1, show that a ranges from 0.017 to 0.042, depending on whether velocity profile (1), (2), or (3) of Table 9.2 is used.

(c) Whereas for a flat plate the outer velocity is constant, in the channel $u(x)$ must increase as $\delta(x)$ increases. Show that, for profile (2) in the shear layers and $v_x = u(x)$ in the core,

$$\frac{u(x)}{U} = \left(1 - \frac{\delta(x)}{3H}\right)^{-1}. \qquad \text{(P12.2-2)}$$

A somewhat lengthy calculation that takes this into account (which you are not asked to reproduce) yields $a = 0.052$ (Deen, 2012, p. 287). For comparison, if $x = L_E$ is considered to be where the velocity change at the channel center is 99% complete, a numerical solution of the Navier–Stokes equation gives $a = 0.088$ (Atkinson *et al.*, 1969). The discrepancy between the integral-momentum and numerical results is less than it might seem, as $a = 0.052$ actually corresponds to about 93% of the velocity change and the precise percentage used to define L_E is arbitrary.

12.3. Nozzle with diffuser A nozzle increases the fluid velocity by tapering inward, whereas a *diffuser* decreases it by tapering outward. Figure P12.3 shows a rounded nozzle followed by a diffuser whose diameter varies from D_1 to D_2 over a length L. Suppose this combination is installed in the wall of a large, open tank at a depth H below the water level. Although it might seem that adding any tube (tapered or not) to a nozzle would reduce the flow by increasing the resistance, the ability of such an arrangement to increase the flow was known even in ancient Rome.[5]

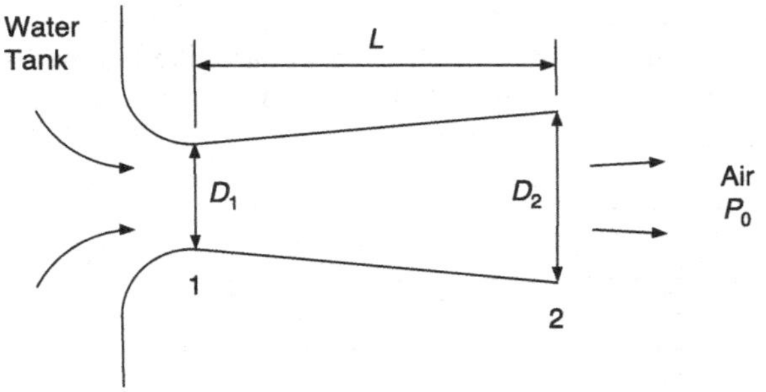

Figure P12.3 Nozzle–diffuser combination at an opening in a water tank.

(a) Explain qualitatively why, under near-inviscid conditions, adding the diffuser will increase the volume flow rate.
(b) Derive an expression to predict the flow rate with the diffuser relative to that with the nozzle alone. Let K_e and K_d be the loss coefficients for the rounded entrance and diffuser, respectively.
(c) For $D_2 = 2D_1$ and $L = 5D_1$, data plotted in Blevins (2003, p. 153) indicate that $K_d = 0.32$. What then would be the percentage increase in flow rate due to the diffuser?

12.4. Water siphon For the siphon in Fig. P11.13, suppose that $d = 0.10\,\mathrm{m}$, $H_A = 0.50\,\mathrm{m}$, $H_B = 1.25\,\mathrm{m}$, $H_C = 1.00\,\mathrm{m}$, and the liquid is water at 20 °C. Assume that the tube is smooth and has a total length $L = 5.0$ m.

(a) If the viscous losses were negligible, what would be the flow rate Q (in $\mathrm{m^3/s}$)?
(b) Taking all the losses into account, predict the actual Q. You may assume that the tube curvature is gradual enough to neglect the loss due to the bend.

12.5. Pumping from a lower to a higher reservoir Suppose that water is to be pumped upward at a volume flow rate Q from one reservoir to another, as in Fig. P12.5. The difference in the water levels is nearly constant at H. The connection is a steel pipe of diameter D and total length L that has four 90° bends. The depths at its inlet and outlet are d_1 and d_2, respectively. Find the pump power (in kW) that would be required if $Q = 0.13\,\mathrm{m^3/s}$, $H = 10\,\mathrm{m}$, $D = 0.40\,\mathrm{m}$, $L = 1000\,\mathrm{m}$, $d_1 = 5\,\mathrm{m}$, and $d_2 = 4\,\mathrm{m}$.

12.6. Water transfer from a higher to a lower reservoir When it is necessary to reverse the flow in the system in Fig. P12.5, the pump is bypassed by a short section of pipe. The valves involved have negligible resistance when open. If the downward flow were not assisted by a pump, what would be the flow rate (in $\mathrm{m^3/s}$)?

[5] As discussed in Pritchard (2011, p. 380), around 97 A.D. some citizens used this strategy to circumvent restrictions on water use. Flow rates had been standardized by the issuance of calibrated bronze nozzles. To eliminate abuses, the water regulations were amended to require that the nozzle discharge into a long, constant-diameter pipe.

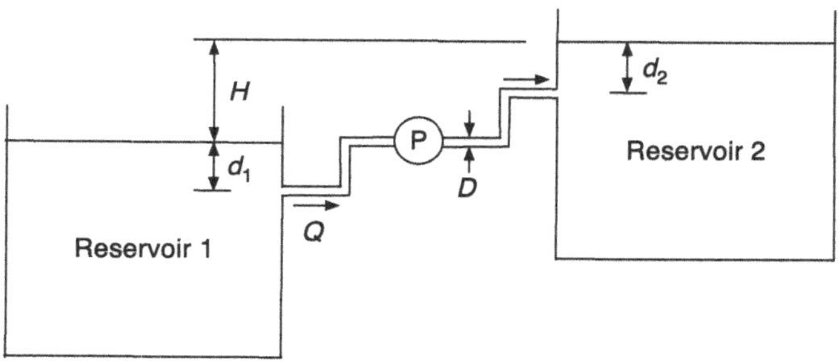

Figure P12.5 Pumping from a lower reservoir to a higher one.

12.7. Home plumbing It is desired to infer the value of the loss coefficient of the valve–spigot combination in a household bathtub. As shown in Fig. P12.7, leading from the underground water main to the house is pipe of diameter D_1 and total length L_1, with one 90° bend. Inside the house are pipe segments of diameter D_2 and total length L_2. The interior pipe leading to the tub has seven 90° bends and an open gate valve. There is a total rise H from water main to tub spigot. The main and atmospheric pressures are P_m and P_0, respectively.

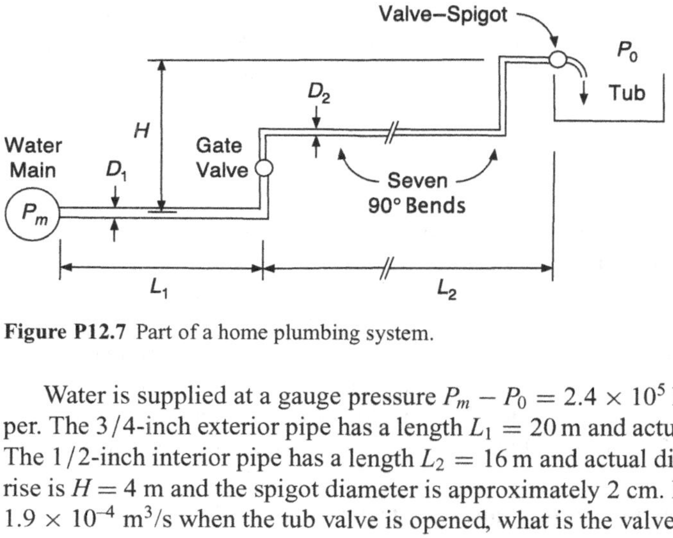

Figure P12.7 Part of a home plumbing system.

Water is supplied at a gauge pressure $P_m - P_0 = 2.4 \times 10^5$ Pa. All of the pipe is copper. The 3/4-inch exterior pipe has a length $L_1 = 20$ m and actual diameter $D_1 = 1.9$ cm. The 1/2-inch interior pipe has a length $L_2 = 16$ m and actual diameter $D_2 = 1.4$ cm. The rise is $H = 4$ m and the spigot diameter is approximately 2 cm. If the flow rate at 20 °C is 1.9×10^{-4} m^3/s when the tub valve is opened, what is the valve–spigot loss coefficient?

12.8. Membrane hydraulic permeability The setup in Fig. P12.8 was used to help characterize porous membranes employed in diffusion studies (Epstein, 1979). The membrane hydraulic permeability k_m obtained using this apparatus was used to find the pore diameter d. If a transmembrane pressure difference $|\Delta P|_m$ gives a volume flow rate Q through a membrane of area S_m, then $k_m = Q/(S_m|\Delta P|_m)$. After a reservoir was placed at a measured height H, Q was found by timing the movement of the meniscus in a pipette placed at the outlet. The cross-sectional area of the outlet was S_o.

(a) The membranes had straight cylindrical pores and the pore number density n (number of pores per unit area) and pore length L for each membrane were measured separately. If Poiseuille's law applies to each pore, show how to calculate d from

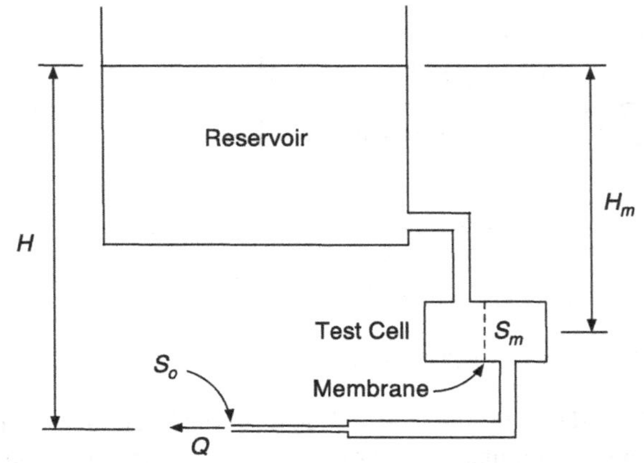

Figure P12.8 Apparatus for measuring membrane hydraulic permeabilities (Epstein, 1979).

k_m, n, and L. With $H = 20$ cm, $d = 80$ nm, and $L = 6.0$ μm being typical, why was it reasonable to assume fully developed laminar flow in the pores?

(b) If the membrane were the only resistance to flow, how could k_m be calculated from the dependence of Q on H?

(c) It was suspected that resistances in other parts of the apparatus were not always negligible, especially for relatively large pore sizes and high flow rates. Accordingly, Q was measured as a function of H with no membrane present. The extramembrane loss $E_{v,o}$ was expressed as

$$\frac{E_{v,o}}{w} = Fv_o^2 \qquad \text{(P12.8-1)}$$

where F is dimensionless and $v_o = Q/S_o$ is the mean velocity at the outlet. Show how to evaluate F from such data. In contrast to turbulent flow, where loss coefficients are nearly constant, in this laminar-flow apparatus F was sensitive to Q. (It was found to vary as $\text{Re}^{-1.24}$, where Re is the outlet Reynolds number.)

(d) With F a known function of Re, show how to find k_m from the original experiments. (*Hint*: Add the loss in going from the reservoir to the upstream side of the membrane to the loss in going from the downstream side of the membrane to the outlet.) How might a plot of Q vs. H reveal that there were extramembrane resistances?

12.9. Design of distribution manifolds In distribution manifolds it is often desired to have the same flow rate Q_B in each branch. If the branches are identical, the inlet pressure at each would need to be the same. There are at least two strategies for making those pressures uniform. If viscous losses along the main pipe are negligible, a series of diameter reductions could keep the velocity from falling and the pressure from rising, as shown in Fig. P12.9(a). If the losses are not negligible, a constant pipe diameter D_0 and a certain length L between branches could be chosen so that the pressure recovery at a branch is exactly balanced by the pressure drop in the next pipe segment, as shown in Fig. P12.9(b). Suppose that the flow is turbulent everywhere and let the initial diameter, flow rate, and velocity for both cases be D_0, Q_0, and v_0, respectively.

(a) For the case with negligible losses and variable pipe diameter, let D_n be the diameter after branch n. How should D_1 differ from D_0? What should D_n be in general ($n \geq 1$)?

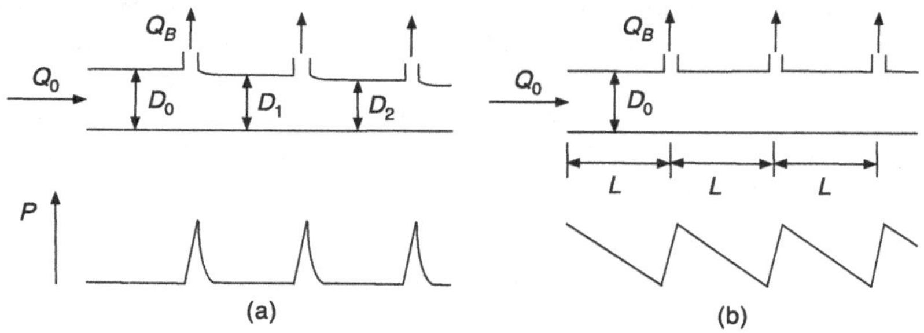

Figure P12.9 Distribution manifolds in which the pressure recoveries at the side openings are offset by: (a) diameter reductions in the main pipe (if viscous losses are negligible); or (b) viscous losses along a constant-diameter main pipe. The approximate pressure variations along the main pipe are shown below each design. An experimental pressure profile like that in (b) is shown in Denn (1980, p. 129).

(b) For the case where the viscous pressure drops balance the pressure recoveries, derive a relationship to guide the choices of D_0 and L. Explain how it could be used.
(c) For air flow with $v_0 = 12.3$ m/s, $D_0 = 100$ mm, $D_B = 25$ mm, $L = 1.8$ m, $P_1 = 1.025$ atm, and $P_0 = 1.000$ atm, how would the pressure drop in the first segment compare with the pressure recovery at the first branch? Which of the two approaches would be most applicable in optimizing such a manifold?

12.10. Tubular reactors in parallel To boost production in a chemical plant without enlarging any of the tubular reactors, several might be arranged in parallel, as shown in Fig. P12.10. Each might be an open tube of diameter D_R and length L_R. Or, each might be a tube of those dimensions that is packed with spherical catalyst pellets of diameter d with void fraction ε. The mean velocity in an open tube and the superficial velocity in a packed tube are each denoted as v_R. The reactants are supplied by a distribution manifold and the products collected via a return manifold. It is desired to derive pressure–velocity relationships that would help in designing such systems.

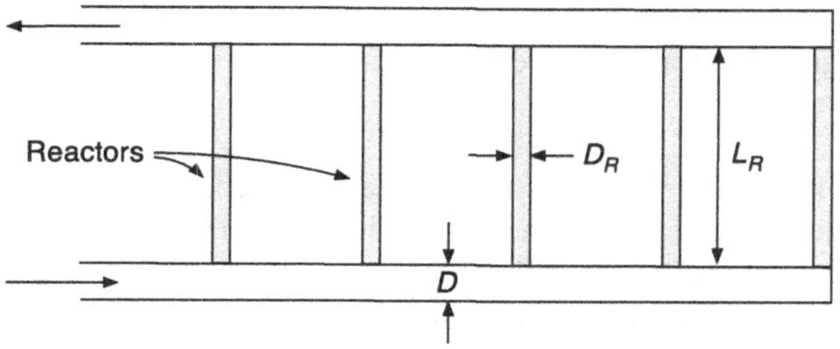

Figure P12.10 Tubular reactors in parallel.

(a) Modify the derivation in Example 12.3-3 as needed to find $P_2 - P_1$ and v_R for a tubular reactor without packing. You may assume that the flow is turbulent and that P_0 is given. (Evaluating the inlet and outlet pressures at a given branch would require analysis of the entire flow network, manifolds and reactors.)
(b) Repeat part (a) for a packed-bed reactor in which $\mathrm{Re}_p > 1000$.
(c) If v_R for either type of reactor is given, find the pressure change at a branch point in the *return* manifold. Does the pressure there in the main pipe increase or decrease in the direction of flow?

12.11. Pumping between tanks All of the liquid in a tank is to be transferred to another tank of the same size. As shown in Fig. P12.11, both are of radius R and at first they are slightly less than half full, with a liquid height H_0 in each. They are open at the top and are connected at the bottom by a pipe with a constant-speed centrifugal pump. The pump performs as in Eq. (12.3-6) with $n = 2$. For simplicity, assume that viscous losses outside the pump itself are negligible.

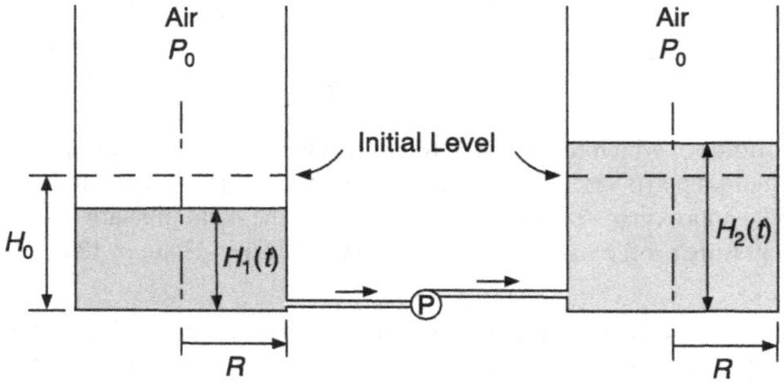

Figure P12.11 Pumped transfer of a liquid between tanks.

(a) What is the minimum value of the parameter A that is needed? In other words, how large must the pump be?
(b) Use the engineering Bernoulli equation and pump characteristics to show that

$$Q = \left(\frac{A - \Delta H}{B}\right)^{1/2} \tag{P12.11-1}$$

where $\Delta H = H_2 - H_1$ is the difference in liquid levels. Combine this with a mass balance to obtain the differential equation that governs ΔH.
(c) A more convenient dependent variable than ΔH is $\phi = (A - \Delta H)/H_0$. Show that

$$\phi_0^{1/2} - \phi^{1/2}(t) = \left(\frac{H_0}{B}\right)^{1/2} \frac{t}{V_0} \tag{P12.11-2}$$

where $\phi_0 = \phi(0) = A/H_0$ and $V_0 = \pi R^2 H_0$ is the initial liquid volume in either tank.
(d) What is the time t_p required to transfer the entire volume?
(e) How will Q vary with time?

12.12. Pumps in series or parallel A tank is to be added to a refinery to increase the storage capacity for a product such as home heating oil. The oil level in the new tank may be as much as a height H above that in the tank leading to it, which will have the same pressure in the head space (i.e., the same gas pressure). Thus, pumping will be required. Two identical centrifugal pumps, whose performance is described by Eq. (12.3-6) with $n = 2$, are available for this purpose. You are asked to decide how best to configure them.

(a) If the pumps are arranged in *series* and the loss in the pipe that connects them is negligible, how will the total pressure increase $|\Delta P|_T$ generated by them be related to the flow rate Q? How large would H need to be to make both pumps necessary? When would the two pumps not be enough?
(b) Repeat part (a) for pumps connected in *parallel*, assuming again that losses in the pipes and fittings that connect them are negligible. If H is small enough that either a single pump or two in parallel could be used, what would be the functional advantage of the parallel arrangement?

12.13. Conical diffuser Figure P12.13 shows a *conical diffuser*, a tube segment that tapers outward linearly in the direction of flow. Its local diameter is $D(z) = D_1 + (D_2 - D_1)(z/L)$. The rate of diameter increase corresponds to an angle θ between the tube wall and centerline. It is desired to predict the loss coefficient K for small values of θ.

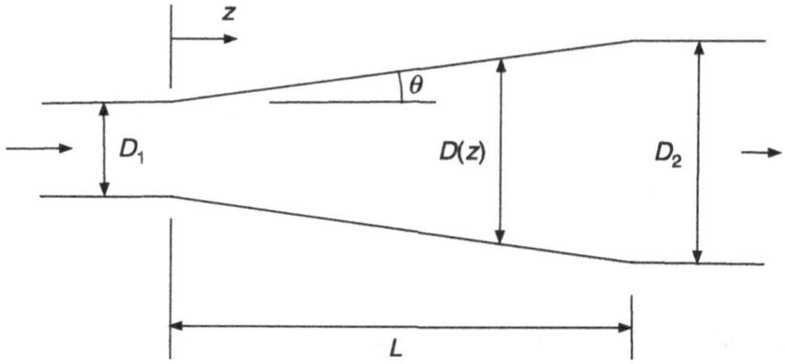

Figure P12.13 Conical diffuser.

(a) For turbulent flow, show that the pressure increase in the diffuser is

$$P_2 - P_1 = \frac{\rho v_1^2}{2}(1 - \beta^4 - K) \tag{P12.13-1}$$

where $\beta = D_1/D_2$.

(b) For small θ, the loss for incompressible flow can be estimated using Eq. (12.4-11). By integrating the differential loss over the length of the diffuser, show that, if the friction factor f is approximately constant, then

$$K = \frac{Lf}{D_1}\beta(1+\beta)(1+\beta^2). \tag{P12.13-2}$$

Notice that this reduces to the loss coefficient in a plain tube, $K = 4Lf/D$, for $\beta \to 1$.

(c) Let $\mathrm{Re}_1 = 1 \times 10^5$ be the inlet Reynolds number. Assuming the wall is smooth and basing f on Re_1, calculate K for $D_2 = 2D_1$ and $L = 10D_1$. What percentage reduction in $P_2 - P_1$ is predicted to be caused by the viscous loss? How much would that be affected if f were based on Re_2? The wall angle for these dimensions, $\theta = \tan^{-1}(0.05) = 2.9°$, is said to be within the range where the approach in part (b) is accurate ($\theta < 3.5°$) (Tilton, 2007, pp. 6–17).

12.14. Balloon inflation A large balloon is to be inflated from radius a_1 to a_2 using a piston pump that delivers air at a constant mass flow rate w. If the wall tension is proportional to how much the radius a exceeds a_0, then the balloon pressure P and ambient pressure P_0 will be related as

$$P - P_0 = \frac{2\chi\,(a - a_0)}{a} \tag{P12.14-1}$$

where χ is a measure of the elasticity of the balloon material. (This is analogous to the Young–Laplace equation, the constant surface tension in that relationship being replaced here by a variable wall tension.) For simplicity, assume that χ is constant and that the air temperature in the balloon remains at the ambient value T_0.

(a) Explain qualitatively why a larger χ will increase the time t_p that is needed for the inflation.
(b) Derive an expression that could be used to predict t_p if χ and a_0 were known.

12.15. Discharge of a compressed-air tank A rigid steel tank of finite volume V is filled with compressed air and at $t = 0$ a valve is to be opened to allow the tank to discharge into the atmosphere. It is desired to predict the resulting tank pressure $P(t)$, if the temperature remains constant at T_0.

(a) Assuming that $w(t) = \alpha P(t)$, where w is the mass flow rate out of the tank and α is a constant, determine $P(t)$.
(b) For what range of P is the assumed proportionality between w and P likely to be a good approximation, and how then could you calculate α? (*Hint*: Would choked flow favor or disfavor this model?)

12.16. Automobile tire inflation Suppose that the pressure in a car tire is discovered to be 20 psig, which is well below the 30 psig recommended for that type of tire. A nearby service station provides compressed air at $P_1 = 50$ psig. The air is delivered through a smooth rubber hose of diameter $D = 6$ mm and length $L = 5$ m. It is desired to estimate the time needed to correct the tire inflation, if the flow is isothermal at 20 °C. The air properties at that temperature and the average tire pressure ($P_2 = 25$ psig) are $\rho_2 = 3.3$ kg/m^3 and $\nu_2 = 5.5 \times 10^{-6}$ m^2/s.

(a) Show how to calculate the mass flow rate w for specified values of P_1 and P_2. [*Hint*: Assume that the air velocity is small enough to use the approximate solution given by Eq. (12.4-24), and that the Blasius equation can be used to find the friction factor.]
(b) As an estimate of the average mass flow rate, calculate w for steady flow at $P_2 = 25$ psig. Confirm that Eq. (12.4-24) and the Blasius equation are both applicable.
(c) Car tires are rigid enough to have a nearly constant air volume. If $V = 600$ in^3, what is the approximate inflation time?

12.17. Comparison of isothermal and adiabatic pipe flow Assume that an air tank at $T_1 = 27$ °C and $P_1 = 2$ atm discharges into the atmosphere through a steel pipe of diameter $D = 2.5$ cm. The tank is large enough that T_1 and P_1 remain nearly constant. The ambient temperature is also 27 °C.

(a) If the flow were isothermal, what pipe length L_c would just suffice to create choking? What then would be the mass flow rate w?
(b) If the flow were isothermal, what would w be if $L = 0.5L_c$?
(c) Repeat part (a) for adiabatic flow. [*Hint*: Note that the gas exiting the pipe will now be cooler than 27 °C. Thus, the Reynolds numbers based on the inlet and outlet conditions will differ. However, show that the corresponding friction factors are nearly equal to one another and to the isothermal f.]

12.18. Gas-cylinder hazard Fatalities have resulted from the careless handling of compressed-gas cylinders, both during transportation and during use. The objective of this problem is to illustrate the consequences of a large, improperly secured, inert-gas cylinder falling over in a lab and shearing off the pressure regulator at its top.

A standard 44-liter steel gas cylinder is 9 inches in diameter, 51 inches tall, and has a tare weight of 133 pounds, corresponding to a mass when empty of 60.4 kg. Suppose that such a cylinder is filled with argon [$M = 40$ kg/(kg-mol), $\gamma = 1.67$] at 2000 psig. In SI units, $V = 0.044$ m^3, $P = 1.38 \times 10^7$ Pa (absolute), and $T = 293$ K. The inside diameter of the regulator connection is 5 mm. Thus, breaking off the regulator at $t = 0$ creates a nozzle with that throat diameter.

(a) Assuming that the gas expands isentropically, show that the initial throat velocity v_t will be sonic and find its value. Explain why v_t will be constant during almost the entire discharge, if the gas in the cylinder remains at 293 K.
(b) Find the initial mass flow rate $w(0)$.
(c) Let $U(t)$ be the velocity of the *cylinder* relative to the lab. If after tipping over it makes little contact with the floor, derive an expression for dU/dt. (*Hint*: Part of the rocket analysis in Example 11.3-2 is applicable.)
(d) If the initial dU/dt is sustained and no obstacles are encountered, show that $U = 9.4$ m/s at $t = 3$ s. That is more than enough to severely damage a masonry wall, not to mention a person! (Accounting for both the decreasing w and the decreasing dU/dt as the cylinder empties gives $U = 8.4$ m/s at 3 s.)

12.19. Speed of sound Sounds are the result of small changes in pressure. To predict the speed at which such disturbances travel, assume that P, ρ, v, and T in a fluid are each perturbed slightly from their constant values at rest. Denoting the static values by a subscript zero and the time- and position-dependent perturbations by primes, and setting $v_0 = 0$,

$$P = P_0 + P', \quad \rho = \rho_0 + \rho', \quad v = v', \quad T = T_0 + T'. \qquad \text{(P12.19-1)}$$

In sound transmission the velocity and temperature gradients are small enough that viscous stresses and heat conduction are each negligible, making the process isentropic. As outlined below, this leads both to Eq. (12.4-1) for c and to the identification of c as a wave speed.

(a) Starting with the time-dependent, one-dimensional continuity and inviscid momentum equations, and assuming that products of primed quantities are negligible, show that

$$\frac{\partial \rho}{\partial t} = -\rho_0 \frac{\partial v}{\partial x} \qquad \text{(P12.19-2)}$$

$$\frac{\partial v}{\partial t} = -\frac{1}{\rho_0}\frac{\partial P}{\partial x} + g_x. \qquad \text{(P12.19-3)}$$

(b) Under isentropic conditions the pressure and density gradients are related as

$$\frac{\partial P}{\partial x} = \left(\frac{\partial P}{\partial \rho}\right)_S \frac{\partial \rho}{\partial x} \qquad \text{(P12.19-4)}$$

where the subscript S denotes constant entropy. Use Eq. (12.5-28) and the ideal-gas law to show that $(\partial P/\partial \rho)_S = \gamma RT/M$, which has the dimension of velocity squared ($L^2 T^{-2}$). Because this quantity is positive, it may be denoted as c^2.
(c) The physical significance of c is revealed by returning to the continuity and momentum equations. Show that they can be combined now to give

$$\frac{\partial^2 \rho}{\partial t^2} = c^2 \frac{\partial^2 \rho}{\partial x^2}. \qquad \text{(P12.19-5)}$$

A partial differential equation of this form is a *wave equation*. (The same equation governs v.)

(d) Wave equations are among the most thoroughly studied partial differential equations, but detailed solutions are not needed to interpret c. Show by substitution that Eq. (P12.19-5) has solutions of the form

$$\rho(x, t) = A(x - ct) + B(x + ct) \tag{P12.19-6}$$

where A and B are *any functions that are differentiable*. That x and t can be combined into a single variable as $x \pm ct$ indicates that c is a wave speed. For example, suppose that $\rho(x, t) = A(x - ct)$ and $B = 0$. At time t_1 the density at position x_1 is $A(x_1 - ct_1)$. At a later time t_2 that density will be replicated at the position x_2 that keeps the functional argument the same. That is, $x_2 - ct_2 = x_1 - ct_1$ or $\Delta x = x_2 - x_1 = c(t_2 - t_1) = c\,\Delta t$. Because each value of ρ migrates a distance $\Delta x = c\,\Delta t$ in a time Δt, there must be a density wave traveling at speed c in the $+x$ direction, as depicted in Fig. P12.19. Pressure variations accompany density disturbances and therefore travel at the same speed. Similar reasoning shows that $B(x + ct)$ represents waves traveling at speed c in the $-x$ direction.

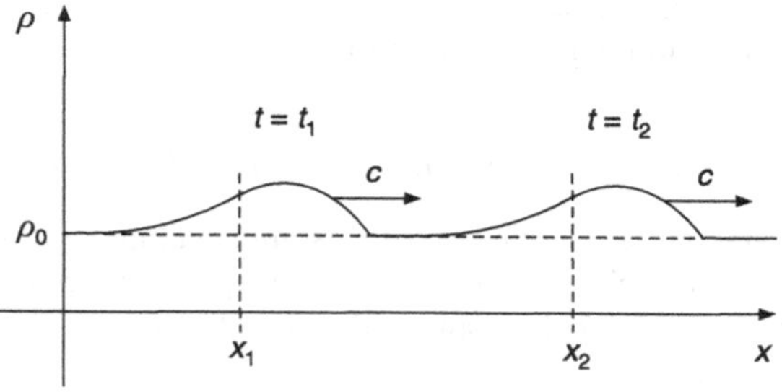

Figure P12.19 A density wave moving in the $+x$ direction at speed c, at two instants in time.

Speeds of sound predicted from Eq. (12.4-1) are typically within 1% of values measured for various gases at atmospheric pressure (Thompson, 1972, p. 165). This confirms that sound transmission is isentropic, rather than isothermal. In a perfectly incompressible fluid, $c = \infty$. In the above derivation the bulk fluid was assumed to be stationary. More generally, c is the speed of sound relative to the fluid velocity.

12.20. Transonic flow A flow in which the velocity changes somewhere from subsonic (Ma < 1) to supersonic (Ma > 1), or the reverse, is called *transonic*. Equation (12.5-9) shows that this cannot occur in a constant-diameter tube. The objective is to see what variation in cross-sectional area is needed to make transonic flows possible.

(a) Show that the continuity and inviscid momentum equations can be written as

$$\frac{1}{\rho}\frac{d\rho}{dx} + \frac{1}{v}\frac{dv}{dx} + \frac{1}{S}\frac{dS}{dx} = 0 \tag{P12.20-1}$$

$$\rho v \frac{dv}{dx} = -\frac{dP}{dx} \tag{P12.20-2}$$

where $S(x)$ is the cross-sectional area.

(b) In isentropic (adiabatic and inviscid) flow the pressure and density gradients are related as in Eq. (P12.19-4). Show then that

$$\frac{1}{v}\frac{dv}{dx} = \left(\frac{1}{\mathrm{Ma}^2 - 1}\right)\frac{1}{S}\frac{dS}{dx}. \qquad \text{(P12.20-3)}$$

This implies that v and S vary inversely for subsonic conditions, but in parallel for supersonic conditions. It indicates also that a finite acceleration can be sustained when $\mathrm{Ma} = 1$ only if $dS/dx = 0$, thereby making both the numerator and denominator on the right-hand side zero. Thus, either a throat or bulge is necessary for adiabatic, inviscid flow to pass through $\mathrm{Ma} = 1$.

(c) To prove that what is actually required for transonic flow is a throat, it is necessary to relate dv/dx to $d\mathrm{Ma}/dx$. Using conservation of energy for adiabatic flow [Eq. (12.5-7)] and evaluating the speed of sound as in Eq. (12.4-1), show that

$$\frac{1}{v}\frac{dv}{dx}(1 + \Gamma \mathrm{Ma}^2) = \frac{1}{\mathrm{Ma}}\frac{d\mathrm{Ma}}{dx} \qquad \text{(P12.20-4)}$$

where $\Gamma = (\gamma - 1)/2$. Substitution of this into Eq. (P12.20-3) gives

$$\frac{1}{\mathrm{Ma}}\frac{d\mathrm{Ma}}{dx} = \left(\frac{1 + \Gamma \mathrm{Ma}^2}{\mathrm{Ma}^2 - 1}\right)\frac{1}{S}\frac{dS}{dx}. \qquad \text{(P12.20-5)}$$

(d) Using l'Hôpital's rule, show that

$$\frac{2}{1+\Gamma}\left(\frac{d\mathrm{Ma}}{dx}\right)^2\Bigg|_{\mathrm{Ma}=1} = \left(\frac{1}{S}\frac{d^2 S}{dx^2}\right)_{\mathrm{Ma}=1}. \qquad \text{(P12.20-6)}$$

Thus, $d^2S/dx^2 > 0$ at $\mathrm{Ma} = 1$, indicating that the gas velocity will pass through the speed of sound only where S is a *minimum*. Such a nozzle–diffuser combination, called a *de Laval nozzle*, is sketched in Fig. P12.20. Developed about 1890 for steam turbines by the Swedish engineer Gustaf de Laval (1845–1913), it is a standard part of modern rocket engines.

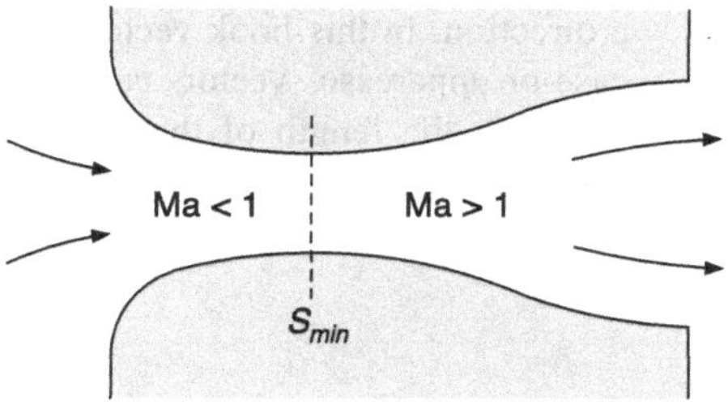

Figure P12.20 De Laval nozzle. The minimum cross-sectional area, where $\mathrm{Ma} = 1$, is denoted as S_{min}.

Appendix

Vectors, tensors, and coordinate systems

A.1 INTRODUCTION

What follows is background information on vectors, tensors, and analytic geometry that is needed for various derivations in this book. This is largely a condensation of an appendix in Deen (2012), which in turn was influenced by Bird *et al.* (2002), Brand (1947), and Hildebrand (1976). Section A.2 describes the notation and reviews some basic concepts. Gibbs notation[1] (Wilson, 1901) is favored, in which vectors and tensors are shown as boldface characters. This allows equations to be written in an uncluttered and coordinate-independent form. Vector and tensor multiplications are examined in Section A.3, including dot, cross, and dyadic products. Differential operators such as the gradient and Laplacian are discussed in Section A.4, along with operations such as the divergence and curl. Certain integral transformations are also presented. Section A.5 begins with a general discussion of orthogonal coordinate systems, and then specializes the results to the Cartesian, cylindrical, and spherical systems that find greatest use.

A.2 NOTATION AND FUNDAMENTALS

Representation of vectors and tensors

A *scalar* in physics has only a magnitude, whereas a *vector* has both a magnitude and a direction. In this book vectors are identified by boldface Roman letters, either lowercase or uppercase. Vectors can be depicted as arrows and diagrams can be constructed in which the length of the arrow corresponds to the magnitude of the vector. However, such graphics have limited utility for analysis, and instead we use expressions such as

$$\mathbf{v} = v_x \mathbf{e_x} + v_y \mathbf{e_y} + v_z \mathbf{e_z}. \tag{A.2-1}$$

Here $\mathbf{e_x}$, $\mathbf{e_y}$, and $\mathbf{e_z}$ are vectors of unit magnitude that parallel the Cartesian axes[2], and v_x, v_y, and v_z are the corresponding scalar components of $\mathbf{v}$. The magnitude of a vector is denoted usually by an italic letter, although absolute value signs are sometimes used for clarity (e.g., $\mathbf{v}$ for the vector and either v or $|\mathbf{v}|$ for its magnitude).

[1] In his career at Yale University, Josiah W. Gibbs (1839–1903) established much of the basic theory of chemical thermodynamics.

[2] The reader may be accustomed to **i**, **j**, and **k** as the Cartesian unit vectors. What makes the e-with-subscript notation preferable is that it allows many relationships to be generalized immediately to other coordinate systems.

It is desirable to have more compact representations of vectors in component form. Two alternatives to Eq. (A.2-1) are

$$\mathbf{v} = (v_1, v_2, v_3) = \sum_{i=1}^{3} v_i \mathbf{e_i} \tag{A.2-2}$$

in which x, y, and z (or the three coordinates in another system) are coordinates 1, 2, and 3. The components of $\mathbf{v}$ are displayed here either as an ordered set of three numbers or in a sum; the latter will be preferred. Hereafter, the limits of such sums are omitted, with the understanding that in three-dimensional space the index (which could also be any letter other than i) varies always from 1 to 3. In Cartesian tensor notation (Jeffreys, 1963) the summation symbol and unit vectors are omitted and the entire vector is represented as v_i. The present system is less compact but more explicit.

A *tensor* in physics generalizes the concept of a vector in a way that conveys additional directional information. In this book tensors are represented by boldface Greek letters, either lowercase or uppercase. Each component of a *second-order* tensor has two subscripts, corresponding to a pair of directions. Such a tensor $\boldsymbol{\tau}$, which has 9 components τ_{ij}, can be represented either as a 3×3 matrix or as a double sum,

$$\boldsymbol{\tau} = \begin{pmatrix} \tau_{11} & \tau_{12} & \tau_{13} \\ \tau_{21} & \tau_{22} & \tau_{23} \\ \tau_{31} & \tau_{32} & \tau_{33} \end{pmatrix} = \sum_i \sum_j \tau_{ij} \mathbf{e_i} \mathbf{e_j}. \tag{A.2-3}$$

The summation notation again is preferred. The quantity $\mathbf{e_i e_j}$, which is a *unit dyad*, should not be confused with the dot product of the unit vectors $\mathbf{e_i}$ and $\mathbf{e_j}$ (Section A.3). It is a kind of dual-directional marker or placeholder. Tensors lack the simple graphical representation that arrows provide for vectors. In Cartesian tensor notation the sums and unit dyads are omitted and the entire tensor is represented as τ_{ij}.

A third-order tensor has three subscripts and 27 components, a fourth-order tensor four subscripts and 81 components, and so on. A scalar can be classified as a zeroth-order tensor and a vector as a first-order tensor. However, because the 9 components that result from two directions usually suffice, "tensor" is used in this book as shorthand for "second-order tensor."

Basic operations

Adding vectors or tensors and multiplying either by a scalar is the same as with matrices in linear algebra. That is, when adding or subtracting two vectors or two tensors, the corresponding components are added or subtracted. When multiplying or dividing a vector or tensor by a scalar, each component is multiplied or divided by that factor. Scalar multiplication changes the magnitude of a vector but not its direction. Likewise, scalar multiplication of a tensor changes its magnitude but not the directional information therein. Derivatives with respect to a single variable, such as time, are calculated by differentiating each component. Derivatives involving multiple spatial variables, as with gradient or Laplacian operators, require more attention and are discussed in Section A.4. The zero vector and zero tensor, all of whose components are zero, are each denoted as $\mathbf{0}$. For two vectors or two tensors to be equal, each of their components must match. Put another way, the difference between such vectors or tensors must be $\mathbf{0}$.

With one exception, vectors do not have specific positions in space. Thus, equality of two vectors requires only that they have the same direction and magnitude; they need not be collinear. Position vectors, which extend from an origin to a point in space, are the exception.

The *transpose* of a tensor is like that of a square matrix. Thus, the transpose of $\boldsymbol{\tau}$ is

$$\boldsymbol{\tau}^t = \begin{pmatrix} \tau_{11} & \tau_{21} & \tau_{31} \\ \tau_{12} & \tau_{22} & \tau_{32} \\ \tau_{13} & \tau_{23} & \tau_{33} \end{pmatrix} = \sum_i \sum_j \tau_{ji} \mathbf{e}_i \mathbf{e}_j \tag{A.2-4}$$

in which the off-diagonal components have been interchanged. A tensor is *symmetric* if $\tau_{ij} = \tau_{ji}$, in which case $\boldsymbol{\tau} = \boldsymbol{\tau}^t$. It is *antisymmetric* if $\tau_{ij} = -\tau_{ji}$. A tensor can be antisymmetric only if each of its diagonal components is zero (i.e., $\tau_{ii} = 0$). By adding and subtracting its transpose, any tensor can be decomposed into a symmetric one plus an antisymmetric one. That is,

$$\boldsymbol{\tau} = \frac{1}{2}(\boldsymbol{\tau} + \boldsymbol{\tau}^t) + \frac{1}{2}(\boldsymbol{\tau} - \boldsymbol{\tau}^t). \tag{A.2-5}$$

Starting with either type of representation in Eqs. (A.2-3) and (A.2-4), it is readily confirmed that $\boldsymbol{\tau} + \boldsymbol{\tau}^t$ is symmetric and $\boldsymbol{\tau} - \boldsymbol{\tau}^t$ is antisymmetric.

Coordinate independence

While of limited help in analysis, the concept of an arrow in space is a reminder that a vector has properties that are independent of any coordinate system that might be chosen for calculations. That is, no matter how it is decomposed into components, its magnitude and direction must be preserved. Accordingly, vector components obey certain rules in coordinate transformations, and the components of a vector in one coordinate system uniquely determine them in other systems. A tensor also has a magnitude that is coordinate-independent, as well as other invariant properties. Thus, the most general way to express any physical law that involves directional quantities is with vector–tensor notation.

The multiplication rules explained in Section A.3 apply to any orthogonal coordinate system. The coordinate-independence of vector and tensor relationships is exploited further in Section A.4, where identities involving differential operators are derived. Among the many systems in use, Cartesian coordinates are unique in that all three base vectors ($\mathbf{e}_x$, $\mathbf{e}_y$, $\mathbf{e}_z$) are independent of position. That constancy greatly facilitates derivations. Even if it was derived using Cartesian operators, if an expression can be converted to Gibbs form it is valid for any coordinate system. However, that requires that the differential operators be tailored to the coordinate system, as described in Section A.5.

A.3 VECTOR AND TENSOR PRODUCTS

Vector dot product

The *dot product* of $\mathbf{a}$ and $\mathbf{b}$ is the scalar given by

$$\mathbf{a} \cdot \mathbf{b} = ab \cos \phi \tag{A.3-1}$$

where ϕ is the angle between the vectors ($\leq 180°$ or π radians), as shown in Fig. A.1(a). Because the result is a scalar, $\mathbf{a} \cdot \mathbf{b}$ is also called the *scalar product*. In that $b \cos \phi$ is the magnitude of $\mathbf{b}$ as projected onto $\mathbf{a}$, the dot product may be interpreted as the projection of $\mathbf{b}$ times the magnitude of $\mathbf{a}$. It may be viewed also as the magnitude of $\mathbf{b}$ times the projection of $\mathbf{a}$ onto $\mathbf{b}$.

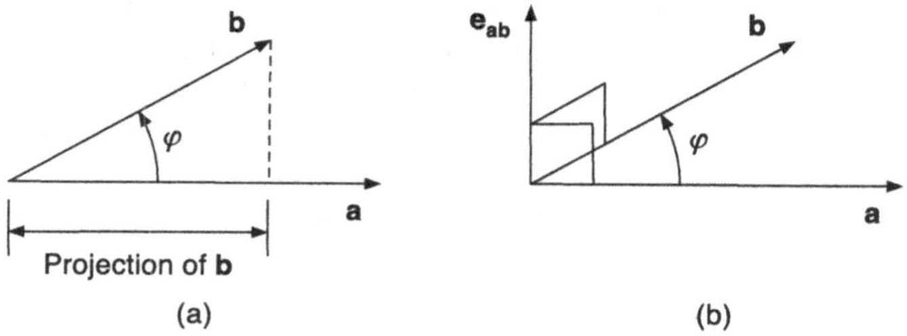

Figure A.1 Geometry of (a) the dot product $\mathbf{a} \cdot \mathbf{b}$ and (b) the cross product $\mathbf{a} \times \mathbf{b}$.

If the vectors are perpendicular to one another, then $\phi = \pi/2$, $\cos\phi = 0$, and $\mathbf{a} \cdot \mathbf{b} = 0$. If a vector is dotted with itself, then $\phi = 0$, $\cos\phi = 1$, and $\mathbf{a} \cdot \mathbf{a} = a^2$. Thus, the *magnitude* of a vector is given by

$$a = |\mathbf{a}| = (\mathbf{a} \cdot \mathbf{a})^{1/2}. \tag{A.3-2}$$

It follows also from Eq. (A.3-1) that dot products of vectors are commutative,

$$\mathbf{a} \cdot \mathbf{b} = \mathbf{b} \cdot \mathbf{a}. \tag{A.3-3}$$

They are also distributive

$$\mathbf{a} \cdot (\mathbf{b} + \mathbf{c}) = (\mathbf{a} \cdot \mathbf{b}) + (\mathbf{a} \cdot \mathbf{c}). \tag{A.3-4}$$

Although the geometric interpretation of dot products is good to keep in mind, it is usually more efficient to dispense with angles and work directly with vector components. That is facilitated by the properties of the base vectors. For any *orthogonal* (mutually perpendicular) set of unit vectors, it follows from Eq. (A.3-1) that the independent dot products are

$$\begin{array}{lll} \mathbf{e}_1 \cdot \mathbf{e}_1 = 1, & \mathbf{e}_2 \cdot \mathbf{e}_2 = 1, & \mathbf{e}_3 \cdot \mathbf{e}_3 = 1, \\ \mathbf{e}_1 \cdot \mathbf{e}_2 = 0, & \mathbf{e}_2 \cdot \mathbf{e}_3 = 0, & \mathbf{e}_3 \cdot \mathbf{e}_1 = 0, \\ \mathbf{e}_2 \cdot \mathbf{e}_1 = 0, & \mathbf{e}_3 \cdot \mathbf{e}_2 = 0, & \mathbf{e}_1 \cdot \mathbf{e}_3 = 0. \end{array} \tag{A.3-5}$$

The *Kronecker delta* (δ_{ij}) allows these nine results to be summarized as

$$\mathbf{e_i} \cdot \mathbf{e_j} = \delta_{ij} = \begin{cases} 1, & i = j \\ 0, & i \neq j. \end{cases} \tag{A.3-6}$$

Thus, the dot product of orthogonal base vectors equals 1 if the subscripts are the same and 0 if they are different.

Using components, the dot product of **a** and **b** is calculated as

$$\mathbf{a} \cdot \mathbf{b} = \left(\sum_i a_i \mathbf{e_i} \right) \cdot \left(\sum_j b_j \mathbf{e_j} \right) = \sum_i \sum_j a_i b_j \mathbf{e_i} \cdot \mathbf{e_j} = \sum_i \sum_j a_i b_j \delta_{ij} = \sum_i a_i b_i. \tag{A.3-7}$$

The intermediate expressions illustrate the basic steps in evaluating any vector product. First, each vector is represented as a sum. Second, the sums are combined, keeping in mind that anything can be moved inside or outside a sum that involves a different index.[3] Third, the product of the base vectors is evaluated. Finally, any redundancy is eliminated.

[3] This is analogous to $\int a(x)\,dx \int b(y)\,dy = \int \int a(x)\,b(y)\,dx\,dy$.

Because nonzero terms are obtained here only for $i = j$, a single sum suffices in the end. At any step, other letters could have been substituted for either i or j. That is because the indices disappear once the summations are performed, just as the integration variable disappears once a definite integral is evaluated. In Cartesian tensor notation the final result is written simply as $a_i b_i$, the repeated index implying a sum over i.

From Eqs. (A.3-2) and (A.3-7), the magnitude of **a** in terms of its components is

$$a = |\mathbf{a}| = \left(\sum_i a_i^2\right)^{1/2}. \tag{A.3-8}$$

A *unit* vector that parallels **a** is $\mathbf{a}/a$ or $\mathbf{a}/|\mathbf{a}|$. A component of a vector is its projection onto the corresponding coordinate axis. The ith component of **a** can be evaluated as

$$a_i = \mathbf{e_i} \cdot \mathbf{a} = \mathbf{a} \cdot \mathbf{e_i}. \tag{A.3-9}$$

Vector cross product

The *cross product* (or *vector product*) of **a** and **b** is the vector given by

$$\mathbf{a} \times \mathbf{b} = ab \sin\phi \, \mathbf{e_{ab}} \tag{A.3-10}$$

where $\mathbf{e_{ab}}$ is a unit vector that is perpendicular to both **a** and **b**, as shown in Fig. A.1(b). If the fingers of the right hand are curled from **a** toward **b**, then $\mathbf{e_{ab}}$ points in the direction of the extended thumb. The magnitude of $\mathbf{a} \times \mathbf{b}$ equals the area of a parallelogram that has **a** and **b** as adjacent sides. Crossing any vector with itself or with a parallel vector gives **0**.

The right-hand rule implies that

$$\mathbf{a} \times \mathbf{b} = -\mathbf{b} \times \mathbf{a}. \tag{A.3-11}$$

Thus, the cross product is *not* commutative. However, it is distributive:

$$\mathbf{a} \times (\mathbf{b} + \mathbf{c}) = (\mathbf{a} \times \mathbf{b}) + (\mathbf{a} \times \mathbf{c}). \tag{A.3-12}$$

As with dot products, the evaluation of cross products in terms of components relies on the multiplication properties of the orthogonal base vectors. If follows from Eq. (A.3-10) that the independent cross products are

$$\begin{aligned} &\mathbf{e_1} \times \mathbf{e_1} = \mathbf{0}, \quad &&\mathbf{e_2} \times \mathbf{e_2} = \mathbf{0}, \quad &&\mathbf{e_3} \times \mathbf{e_3} = \mathbf{0}, \\ &\mathbf{e_1} \times \mathbf{e_2} = \mathbf{e_3}, \quad &&\mathbf{e_2} \times \mathbf{e_3} = \mathbf{e_1}, \quad &&\mathbf{e_3} \times \mathbf{e_1} = \mathbf{e_2}, \\ &\mathbf{e_2} \times \mathbf{e_1} = -\mathbf{e_3}, \quad &&\mathbf{e_3} \times \mathbf{e_2} = -\mathbf{e_1}, \quad &&\mathbf{e_1} \times \mathbf{e_3} = -\mathbf{e_2}. \end{aligned} \tag{A.3-13}$$

These results are summarized as

$$\mathbf{e_i} \times \mathbf{e_j} = \sum_k \varepsilon_{ijk} \mathbf{e_k} \tag{A.3-14}$$

where ε_{ijk} is the *permutation symbol*, defined as

$$\varepsilon_{ijk} = \begin{cases} 0, & i = j, \ j = k, \text{ or } k = i \\ 1, & ijk = 123, 231, \text{ or } 312 \\ -1, & ijk = 132, 213, \text{ or } 321. \end{cases} \tag{A.3-15}$$

Thus, $\varepsilon_{ijk} = 0$ if any indices are repeated, $\varepsilon_{ijk} = 1$ if they increase in cyclic order, and $\varepsilon_{ijk} = -1$ if they decrease in cyclic order. Whatever the values of the indices, $\varepsilon_{ijk} = \varepsilon_{jki} = \varepsilon_{kij}$ because the same cyclic order is maintained.

In terms of components, the cross product of **a** and **b** is

$$\mathbf{a}\times\mathbf{b}=\sum_i\sum_j\sum_k a_i b_j \varepsilon_{ijk}\mathbf{e_k}. \tag{A.3-16}$$

Written out, this is

$$\mathbf{a}\times\mathbf{b}=(a_2b_3-a_3b_2)\mathbf{e_1}-(a_1b_3-a_3b_1)\mathbf{e_2}+(a_1b_2-a_2b_1)\mathbf{e_3}. \tag{A.3-17}$$

Only 6 of 27 terms in the triple sum survive because the others involve repeated indices. A way to remember this result is to notice that it is equivalent to a determinant:

$$\mathbf{a}\times\mathbf{b}=\begin{vmatrix}\mathbf{e_1} & \mathbf{e_2} & \mathbf{e_3}\\ a_1 & a_2 & a_3\\ b_1 & b_2 & b_3\end{vmatrix}. \tag{A.3-18}$$

Expressions in which one cross product follows another require that a pair of permutation symbols be multiplied and then summed over a common index. A helpful identity is

$$\sum_k \varepsilon_{ijk}\varepsilon_{mnk}=\delta_{im}\delta_{jn}-\delta_{in}\delta_{jm}. \tag{A.3-19}$$

Dyadic product

A tensor constructed from a pair of vectors is called a *dyad*. Multiplying the vectors **a** and **b** without a dot or cross gives the dyad

$$\mathbf{ab}=\begin{pmatrix}a_1b_1 & a_1b_2 & a_1b_3\\ a_2b_1 & a_2b_2 & a_2b_3\\ a_3b_1 & a_3b_2 & a_3b_3\end{pmatrix}=\sum_i\sum_j a_i b_j \mathbf{e_i}\mathbf{e_j}. \tag{A.3-20}$$

In general, dyadic products are not commutative, but they are distributive. Dot products involving dyads are evaluated like those involving tensors, as discussed next.

Tensor products

The evaluation of vector–tensor and tensor–tensor dot products is based on the following four operations involving unit vectors and unit dyads:

$$\mathbf{e_k}\cdot\mathbf{e_i}\mathbf{e_j}=\delta_{ik}\mathbf{e_j} \tag{A.3-21}$$

$$\mathbf{e_i}\mathbf{e_j}\cdot\mathbf{e_k}=\delta_{jk}\mathbf{e_i} \tag{A.3-22}$$

$$\mathbf{e_i}\mathbf{e_j}\cdot\mathbf{e_m}\mathbf{e_n}=\delta_{jm}\mathbf{e_i}\mathbf{e_n} \tag{A.3-23}$$

$$\mathbf{e_i}\mathbf{e_j}:\mathbf{e_m}\mathbf{e_n}=\delta_{jm}\delta_{in}. \tag{A.3-24}$$

It is seen that dotting a vector with a dyad (in either order) yields a vector, and that dotting two dyads yields another dyad. The double-dot product of two dyads (written with a colon) is a scalar. A mnemonic is that each relationship may be obtained by evaluating the product of the base vectors next to the dot and retaining any "unused" base vectors. The order in which the base vectors appear is preserved, as shown in Eq. (A.3-23). Equation (A.3-24) may be viewed as two dot products in succession, the first involving $\mathbf{e_j}$ and $\mathbf{e_m}$ and the second involving $\mathbf{e_i}$ and $\mathbf{e_n}$.[4]

[4] Using the original definition (Wilson, 1901), the result in Eq. (A.3-24) would be $\delta_{im}\delta_{jn}$. The present definition, preferred now by a number of authors, makes the double-dot product a more natural extension of Eq. (A.3-23).

In general, vector–tensor dot products are *not* commutative. Evaluating the two possible dot products of **a** and $\boldsymbol{\tau}$ using Eqs. (A.3-21) and (A.3-22) gives

$$\mathbf{a} \cdot \boldsymbol{\tau} = \sum_i \sum_j a_i \tau_{ij} \mathbf{e_j} \neq \sum_i \sum_j a_i \tau_{ji} \mathbf{e_j} = \boldsymbol{\tau} \cdot \mathbf{a}. \tag{A.3-25}$$

The two expressions will be equal only if $\boldsymbol{\tau}$ is symmetric ($\tau_{ij} = \tau_{ji}$). Likewise, the single-dot product of two tensors is commutative only if both tensors are symmetric. Single-dot products are formally the same as matrix multiplication, which also is commutative only for symmetric matrices. Although the single-dot product of two tensors is not commutative in general, the double-dot product is. For tensors $\boldsymbol{\tau}$ and $\boldsymbol{\kappa}$ that product is

$$\boldsymbol{\tau} : \boldsymbol{\kappa} = \sum_i \sum_j \tau_{ij} \kappa_{ji} = \sum_i \sum_j \kappa_{ij} \tau_{ji} = \boldsymbol{\kappa} : \boldsymbol{\tau}. \tag{A.3-26}$$

The middle equality is confirmed by noticing that either double sum may be obtained from the other by interchanging the arbitrary indices i and j.

The double-dot product of a tensor with its transpose leads to the *magnitude* of the tensor. The magnitude τ of the tensor $\boldsymbol{\tau}$ is

$$\tau = \left(\frac{1}{2} \boldsymbol{\tau} : \boldsymbol{\tau}^t \right)^{1/2} = \left(\frac{1}{2} \sum_i \sum_j \tau_{ij}^2 \right)^{1/2}. \tag{A.3-27}$$

As with the magnitude of a vector, τ is an *invariant*, meaning that it is unaffected by the coordinate system. If $\boldsymbol{\tau}$ is symmetric and its only nonzero components are τ_{12} and τ_{21}, then $\tau = |\tau_{12}| = |\tau_{21}|$. This is analogous to a vector **v** with just one nonzero component, v_1, in which case $v = |v_1|$. The factor 1/2 in Eq. (A.3-27) is included to preserve that analogy. Another tensor invariant is the *trace*, which is the sum of the elements on the main diagonal. The trace of $\boldsymbol{\tau}$ is

$$\operatorname{tr} \boldsymbol{\tau} = \sum_i \tau_{ii}. \tag{A.3-28}$$

Dotting a tensor with vectors that precede and follow it is associative. That is,

$$(\mathbf{a} \cdot \boldsymbol{\tau}) \cdot \mathbf{b} = \sum_i \sum_j a_i \tau_{ij} b_j = \mathbf{a} \cdot (\boldsymbol{\tau} \cdot \mathbf{b}). \tag{A.3-29}$$

Accordingly, the parentheses can be omitted and the overall product written simply as $\mathbf{a} \cdot \boldsymbol{\tau} \cdot \mathbf{b}$. Because both products within parentheses are vectors and vector–vector dot products are commutative, two other equivalent expressions are $\mathbf{b} \cdot (\mathbf{a} \cdot \boldsymbol{\tau})$ and $(\boldsymbol{\tau} \cdot \mathbf{b}) \cdot \mathbf{a}$. In these cases the parentheses are needed to avoid ambiguities.

The mnemonic described after Eq. (A.3-24) leads also to dot-product rules for *unit polyads* such as $\mathbf{e_i e_j e_k}$ and $\mathbf{e_i e_j e_m e_n}$. Cross products involving tensors can be calculated as well. Cross-product rules for unit dyads are discussed in Brand (1947, pp. 187–189). These other kinds of products are not used in this book.

Identity tensor

The identity tensor, which is analogous to an identity matrix, is

$$\boldsymbol{\delta} = \begin{pmatrix} 1 & 0 & 0 \\ 0 & 1 & 0 \\ 0 & 0 & 1 \end{pmatrix} = \sum_i \sum_j \delta_{ij} \mathbf{e_i e_j} = \sum_i \mathbf{e_i e_i}. \tag{A.3-30}$$

Each component is the Kronecker delta; hence, the symbol $\boldsymbol{\delta}$ for the tensor. The essential property of the identity tensor is that dotting it with any vector or tensor, in either order, returns that vector or tensor. That is,

$$\boldsymbol{\delta} \cdot \mathbf{a} = \mathbf{a} = \mathbf{a} \cdot \boldsymbol{\delta} \tag{A.3-31}$$

$$\boldsymbol{\delta} \cdot \boldsymbol{\tau} = \boldsymbol{\tau} = \boldsymbol{\tau} \cdot \boldsymbol{\delta}. \tag{A.3-32}$$

The trace of a tensor is obtained by double-dotting it with the identity tensor, in either order,

$$\operatorname{tr} \boldsymbol{\tau} = \boldsymbol{\delta} : \boldsymbol{\tau} = \boldsymbol{\tau} : \boldsymbol{\delta} \tag{A.3-33}$$

and the trace of the identity tensor itself is

$$\operatorname{tr} \boldsymbol{\delta} = \boldsymbol{\delta} : \boldsymbol{\delta} = \sum_i \delta_{ii} = 3. \tag{A.3-34}$$

Example A.3-1 Repeated dot products of a vector with an antisymmetric tensor. The objective is to show that, for any vector $\mathbf{p}$ and any *antisymmetric* tensor $\boldsymbol{\Omega}$,

$$\mathbf{p} \cdot \boldsymbol{\Omega} \cdot \mathbf{p} = \sum_i \sum_j p_i \Omega_{ij} p_j = 0. \tag{A.3-35}$$

The component representation that is shown follows from Eq. (A.3-29).

Because $\Omega_{ii} = 0$ if $\boldsymbol{\Omega}$ is antisymmetric, the double sum will give at most six nonzero terms. Grouping each Ω_{ij} with the corresponding Ω_{ji} $(i \neq j)$, they are

$$\mathbf{p} \cdot \boldsymbol{\Omega} \cdot \mathbf{p} = p_1 p_2 (\Omega_{12} + \Omega_{21}) + p_1 p_3 (\Omega_{13} + \Omega_{31}) + p_2 p_3 (\Omega_{23} + \Omega_{32}). \tag{A.3-36}$$

Each sum in parentheses vanishes because $\Omega_{ij} = -\Omega_{ji}$. This completes the proof.

Example A.3-2 Scalar triple products. It is desired to show that

$$\mathbf{a} \cdot (\mathbf{b} \times \mathbf{c}) = \mathbf{b} \cdot (\mathbf{c} \times \mathbf{a}) = \mathbf{c} \cdot (\mathbf{a} \times \mathbf{b}). \tag{A.3-37}$$

These expressions are called *scalar triple products*. They give the volume of a parallelepiped that has $\mathbf{a}$, $\mathbf{b}$, and $\mathbf{c}$ as adjacent sides.

The first expression in Eq. (A.3-37) is expanded as

$$\mathbf{a} \cdot (\mathbf{b} \times \mathbf{c}) = \sum_n a_n \mathbf{e_n} \cdot \sum_i \sum_j \sum_k b_i c_j \varepsilon_{ijk} \mathbf{e_k} = \sum_i \sum_j \sum_k a_k b_i c_j \varepsilon_{ijk}. \tag{A.3-38}$$

By analogy, the next expression is

$$\mathbf{b} \cdot (\mathbf{c} \times \mathbf{a}) = \sum_i \sum_j \sum_k b_k c_i a_j \varepsilon_{ijk} = \sum_i \sum_j \sum_k a_k b_i c_j \varepsilon_{jk_i}. \tag{A.3-39}$$

The second equality was obtained by re-labeling the indices as follows: $i \to j$, $j \to k$, and $k \to i$. The three vector components were also rearranged to match Eq. (A.3-38). Because $\varepsilon_{ijk} = \varepsilon_{jki}$ (same cyclic order), the result is the same as before. The third expression in Eq. (A.3-37) is written as

$$\mathbf{c} \cdot (\mathbf{a} \times \mathbf{b}) = \sum_i \sum_j \sum_k c_k a_i b_j \varepsilon_{ijk} = \sum_i \sum_j \sum_k a_k b_i c_j \varepsilon_{kij}. \tag{A.3-40}$$

The last step again is simply a cyclic shift in the labels $(i \to k,\ j \to i,\ k \to j)$, and $\varepsilon_{kij} = \varepsilon_{ijk}$. This completes the verification that the three expressions in Eq. (A.3-37) are equal.

A.4 DIFFERENTIAL AND INTEGRAL IDENTITIES

Gradient

The rate at which a quantity varies with position is calculated using the *gradient operator*, the Cartesian form of which is

$$\nabla = \mathbf{e_x}\frac{\partial}{\partial x} + \mathbf{e_y}\frac{\partial}{\partial y} + \mathbf{e_z}\frac{\partial}{\partial z} = \sum_i \mathbf{e_i}\frac{\partial}{\partial x_i}. \tag{A.4-1}$$

In the sum the coordinates (x, y, z) are replaced by (x_1, x_2, x_3). This and other component representations in this section are restricted to Cartesian coordinates. In dot, cross, or dyadic products the gradient operator may be treated as an ordinary vector, except that the derivatives require that its order of appearance not be changed. For example, unlike an ordinary dot product, $\nabla \cdot \mathbf{a} \neq \mathbf{a} \cdot \nabla$. In Eq. (A.4-1) the base vectors are placed ahead of the derivatives to emphasize that they are not being differentiated. However, because the *Cartesian* base vectors are each constant, $\mathbf{e_i}$ could be written here after $\partial/\partial x_i$.

The gradient of any function is the operation of ∇ on it, without a dot or cross. Thus, the gradient of the scalar f is the vector

$$\nabla f = \frac{\partial f}{\partial x}\mathbf{e_x} + \frac{\partial f}{\partial y}\mathbf{e_y} + \frac{\partial f}{\partial z}\mathbf{e_z} = \sum_i \frac{\partial f}{\partial x_i}\mathbf{e_i}. \tag{A.4-2}$$

Hereafter, only summation symbols will be used. The gradient of the vector $\mathbf{a}$ is the dyad

$$\nabla \mathbf{a} = \sum_i \sum_j \frac{\partial a_j}{\partial x_i}\mathbf{e_i e_j}. \tag{A.4-3}$$

Divergence

The *divergence* of a vector or tensor is the dot product of ∇ with that vector or tensor. Thus, the divergence of $\mathbf{a}$ is the scalar

$$\nabla \cdot \mathbf{a} = \sum_i \frac{\partial a_i}{\partial x_i} \tag{A.4-4}$$

and the divergence of $\boldsymbol{\tau}$ is the vector

$$\nabla \cdot \boldsymbol{\tau} = \sum_i \sum_j \frac{\partial \tau_{ij}}{\partial x_i}\mathbf{e_j}. \tag{A.4-5}$$

Curl

The cross product of ∇ with a vector is called the *curl*. The curl of $\mathbf{a}$ is the vector

$$\nabla \times \mathbf{a} = \sum_i \sum_j \sum_k \frac{\partial a_j}{\partial x_i}\varepsilon_{ijk}\mathbf{e_k}. \tag{A.4-6}$$

Laplacian

The *Laplacian operator* (∇^2) is formed by dotting ∇ with itself. That is,

$$\nabla^2 = \nabla \cdot \nabla = \sum_i \frac{\partial^2}{\partial x_i^2}. \tag{A.4-7}$$

A.4 Differential and integral identities

Table A.1 Identities involving differential operators[a]

(1)	$\nabla \cdot (f\mathbf{a}) = (\nabla f) \cdot \mathbf{a} + f\nabla \cdot \mathbf{a}$
(2)	$\nabla \times (f\mathbf{a}) = (\nabla f) \times \mathbf{a} + f\nabla \times \mathbf{a}$
(3)	$\nabla \cdot (\mathbf{a} \times \mathbf{b}) = (\nabla \times \mathbf{a}) \cdot \mathbf{b} - (\nabla \times \mathbf{b}) \cdot \mathbf{a}$
(4)	$\nabla \times (\mathbf{a} \times \mathbf{b}) = \mathbf{b} \cdot \nabla\mathbf{a} - \mathbf{a} \cdot \nabla\mathbf{b} + \mathbf{a}(\nabla \cdot \mathbf{b}) - \mathbf{b}(\nabla \cdot \mathbf{a})$
(5)	$\nabla \times \nabla f = \mathbf{0}$
(6)	$\nabla \cdot (\nabla f \times \nabla g) = 0$
(7)	$\nabla \cdot (\nabla \times \mathbf{a}) = 0$
(8)	$\nabla \times (\nabla \times \mathbf{a}) = \nabla(\nabla \cdot \mathbf{a}) - \nabla^2\mathbf{a}$
(9)	$\nabla(\mathbf{a} \cdot \mathbf{b}) = (\nabla\mathbf{a}) \cdot \mathbf{b} + (\nabla\mathbf{b}) \cdot \mathbf{a} = \mathbf{a} \cdot \nabla\mathbf{b} + \mathbf{b} \cdot \nabla\mathbf{a} + \mathbf{a} \times (\nabla \times \mathbf{b}) + \mathbf{b} \times (\nabla \times \mathbf{a})$
(10)	$\nabla \cdot (\mathbf{ab}) = \mathbf{a} \cdot \nabla\mathbf{b} + \mathbf{b}(\nabla \cdot \mathbf{a})$
(11)	$\nabla \cdot (f\boldsymbol{\delta}) = \nabla f$

[a] These relationships are for any differentiable scalar functions f and g and any differentiable vector functions **a** and **b**.

This sum of second derivatives can operate on any function of position. The Laplacian of the scalar f is the scalar

$$\nabla^2 f = \sum_i \frac{\partial^2 f}{\partial x_i^2} \qquad \text{(A.4-8)}$$

and the Laplacian of the vector **a** is the vector

$$\nabla^2 \mathbf{a} = \sum_i \sum_j \frac{\partial^2 a_j}{\partial x_i^2} \mathbf{e_j}. \qquad \text{(A.4-9)}$$

Differential identities

As emphasized already, each of the foregoing expressions involving ∇ is valid only for Cartesian coordinates. The forms of the gradient, divergence, curl, and Laplacian in other coordinate systems are discussed in Section A.5. Although the component representations in this section are Cartesian, they can be used to prove vector–tensor identities that are valid in general. A number of identities involving differential operators are shown in Table A.1. Many others may be found in Brand (1947) and Hay (1953).

Example A.4-1 Proof of a differential identity. It is desired to confirm that

$$\nabla \times (\nabla \times \mathbf{a}) = \nabla(\nabla \cdot \mathbf{a}) - \nabla^2 \mathbf{a} \qquad \text{(A.4-10)}$$

for any vector **a**, which is identity (8) of Table A.1.

Representing ∇ and $\nabla \times \mathbf{a}$ each as sums [as in Eqs. (A.4-1) and (A.4-6), respectively], the left-hand side of Eq. (A.4-10) is expanded as

$$\begin{aligned} \nabla \times (\nabla \times \mathbf{a}) &= \left(\sum_m \mathbf{e_m} \frac{\partial}{\partial x_m} \right) \times \left(\sum_i \sum_j \sum_k \frac{\partial a_j}{\partial x_i} \varepsilon_{ijk} \mathbf{e_k} \right) \\ &= \sum_i \sum_j \sum_k \sum_m \sum_n \frac{\partial^2 a_j}{\partial x_i \partial x_m} \varepsilon_{ijk} \varepsilon_{mkn} \mathbf{e_n}. \end{aligned} \qquad \text{(A.4-11)}$$

The second equality was obtained by combining the original sums, forming the mixed second derivatives, and evaluating the unit-vector cross product as

$$\mathbf{e_m} \times \mathbf{e_k} = \sum_n \varepsilon_{mkn} \mathbf{e_n}. \tag{A.4-12}$$

Recognizing that $\varepsilon_{mkn} = \varepsilon_{nmk}$ and [from Eq. (A.3-19)] that

$$\sum_k \varepsilon_{ijk} \varepsilon_{nmk} = \delta_{in} \delta_{jm} - \delta_{im} \delta_{jn} \tag{A.4-13}$$

the last expression in Eq. (A.4-11) is simplified to

$$\sum_i \sum_j \frac{\partial^2 a_j}{\partial x_i \, \partial x_j} \mathbf{e_i} - \sum_i \sum_j \frac{\partial^2 a_j}{\partial x_i^2} \mathbf{e_j}. \tag{A.4-14}$$

The first term on the right-hand side of Eq. (A.4-10) is expanded now as

$$\nabla(\nabla \cdot \mathbf{a}) = \sum_i \mathbf{e_i} \frac{\partial}{\partial x_i} \sum_j \frac{\partial a_j}{\partial x_j} = \sum_i \sum_j \frac{\partial^2 a_j}{\partial x_i \, \partial x_j} \mathbf{e_i} \tag{A.4-15}$$

which matches the first term in Eq. (A.4-14). The second term on the right-hand side of Eq. (A.4-10) is expanded as in Eq. (A.4-9), which matches the second term in Eq. (A.4-14). This completes the proof.

Example A.4-2 Proof of a differential identity. The objective is to show that

$$\nabla \cdot [\nabla \mathbf{v} + (\nabla \mathbf{v})^t] = \nabla^2 \mathbf{v} + \nabla(\nabla \cdot \mathbf{v}) \tag{A.4-16}$$

which is used in Section 6.6 in deriving the Navier–Stokes equation.

The first term on the left-hand side of Eq. (A.4-16) is expanded as

$$\nabla \cdot (\nabla \mathbf{v}) = \sum_i \mathbf{e_i} \frac{\partial}{\partial x_i} \cdot \sum_j \sum_k \frac{\partial v_k}{\partial x_j} \mathbf{e_j} \mathbf{e_k} = \sum_i \sum_k \frac{\partial^2 v_k}{\partial x_i^2} \mathbf{e_k} = \nabla^2 \mathbf{v} \tag{A.4-17}$$

where the last equality is from Eq. (A.4-9). In that $\nabla \cdot \nabla = \nabla^2$, this is actually just a special case of $\mathbf{a} \cdot \mathbf{bc} = (\mathbf{a} \cdot \mathbf{b})\mathbf{c}$, which holds for any vector $\mathbf{a}$ and dyad $\mathbf{bc}$. The second term on the left-hand side of Eq. (A.4-16) is

$$\begin{aligned} \nabla \cdot (\nabla \mathbf{v})^t &= \sum_i \mathbf{e_i} \frac{\partial}{\partial x_i} \cdot \sum_j \sum_k \frac{\partial v_j}{\partial x_k} \mathbf{e_j} \mathbf{e_k} = \sum_i \sum_k \frac{\partial^2 v_i}{\partial x_i \partial x_k} \mathbf{e_k} \\ &= \sum_k \mathbf{e_k} \frac{\partial}{\partial x_k} \sum_i \frac{\partial v_i}{\partial x_i} = \nabla(\nabla \cdot \mathbf{v}) \end{aligned} \tag{A.4-18}$$

which completes the proof.

Example A.4-3 Proof of a differential identity. It is desired to show that

$$\mathbf{v} \cdot (\nabla \cdot \boldsymbol{\tau}) = \nabla \cdot (\boldsymbol{\tau} \cdot \mathbf{v}) - \boldsymbol{\tau} : \nabla \mathbf{v} \tag{A.4-19}$$

if $\boldsymbol{\tau}$ is symmetric. This is used in deriving the mechanical energy balances in Section 11.4.

The left-hand side is evaluated first as

$$\mathbf{v} \cdot (\nabla \cdot \boldsymbol{\tau}) = \sum_i \sum_j \sum_m \sum_n v_i \frac{\partial \tau_{mn}}{\partial x_j} \mathbf{e_i} \cdot (\mathbf{e_j} \cdot \mathbf{e_m} \mathbf{e_n}) = \sum_i \sum_j v_i \frac{\partial \tau_{ji}}{\partial x_j}. \tag{A.4-20}$$

The first term on the right-hand side of Eq. (A.4-19) is expanded as

$$\begin{aligned}\nabla\cdot(\boldsymbol{\tau}\cdot\mathbf{v}) &= \sum_i\sum_j\sum_m\sum_n \frac{\partial}{\partial x_i}(\tau_{jm}v_n)\mathbf{e_i}\cdot(\mathbf{e_j e_m}\cdot\mathbf{e_n}) \\ &= \sum_i\sum_m \frac{\partial}{\partial x_i}(\tau_{im}v_m) = \sum_i\sum_j\left(v_i\frac{\partial\tau_{ji}}{\partial x_j}+\tau_{ji}\frac{\partial v_i}{\partial x_j}\right).\end{aligned} \tag{A.4-21}$$

In the last expression the indices have been re-labeled ($i\to j$ and $m\to i$) to clarify that the first term equals the result in Eq. (A.4-20). The last term in Eq. (A.4-19) is

$$\boldsymbol{\tau}:\nabla\mathbf{v} = \sum_i\sum_j\sum_m\sum_n \tau_{ij}\frac{\partial v_n}{\partial x_m}\mathbf{e_i e_j}:\mathbf{e_m e_n} = \sum_i\sum_j \tau_{ij}\frac{\partial v_i}{\partial x_j}. \tag{A.4-22}$$

When $\tau_{ij}=\tau_{ji}$, this cancels the remaining part of Eq. (A.4-21), thereby completing the proof.

Example A.4-4 Proof of a differential identity.[5] The objective is to prove that

$$(2\Gamma)^2-\frac{2}{3}(\nabla\cdot\mathbf{v})^2\geq 0 \tag{A.4-23}$$

where $\boldsymbol{\Gamma}=[\nabla\mathbf{v}+(\nabla\mathbf{v})^t]/2$, a symmetric tensor. This is used in Section 11.4 to show that viscous dissipation of mechanical energy is irreversible.

It is helpful to first rewrite the equation in terms of a double-dot product and a trace. From the definition of tensor magnitude in Eq. (A.3-27), $(2\Gamma)^2=2\boldsymbol{\Gamma}:\boldsymbol{\Gamma}$. Also,

$$\nabla\cdot\mathbf{v}=\sum_i\frac{\partial v_i}{\partial x_i}=\sum_i\Gamma_{ii}=\operatorname{tr}\boldsymbol{\Gamma}. \tag{A.4-24}$$

Thus, what is to be shown is that

$$\boldsymbol{\Gamma}:\boldsymbol{\Gamma}-\frac{1}{3}(\operatorname{tr}\boldsymbol{\Gamma})^2\geq 0. \tag{A.4-25}$$

A tensor $\boldsymbol{\Gamma}'$ is defined now by

$$\boldsymbol{\Gamma}\equiv\boldsymbol{\Gamma}'+\frac{1}{3}(\operatorname{tr}\boldsymbol{\Gamma})\boldsymbol{\delta}. \tag{A.4-26}$$

Because $\boldsymbol{\Gamma}$ and $\boldsymbol{\delta}$ are symmetric, so is $\boldsymbol{\Gamma}'$. Moreover, its construction makes $\boldsymbol{\Gamma}'$ "traceless." That is, using Eqs. (A.3-33) and (A.3-34),

$$\operatorname{tr}\boldsymbol{\Gamma}=\boldsymbol{\delta}:\boldsymbol{\Gamma}=\boldsymbol{\delta}:\boldsymbol{\Gamma}'+\left(\frac{1}{3}\right)(\operatorname{tr}\boldsymbol{\Gamma})\boldsymbol{\delta}:\boldsymbol{\delta}=\operatorname{tr}\boldsymbol{\Gamma}'+\operatorname{tr}\boldsymbol{\Gamma} \tag{A.4-27}$$

which shows that $\operatorname{tr}\boldsymbol{\Gamma}'=\boldsymbol{\delta}:\boldsymbol{\Gamma}'=0$. The first term in Eq. (A.4-25) is evaluated now as

$$\boldsymbol{\Gamma}:\boldsymbol{\Gamma}=\boldsymbol{\Gamma}':\boldsymbol{\Gamma}'+\frac{2}{3}(\operatorname{tr}\boldsymbol{\Gamma})\boldsymbol{\delta}:\boldsymbol{\Gamma}'+\frac{1}{9}(\operatorname{tr}\boldsymbol{\Gamma})^2\boldsymbol{\delta}:\boldsymbol{\delta}=\boldsymbol{\Gamma}':\boldsymbol{\Gamma}'+\frac{1}{3}(\operatorname{tr}\boldsymbol{\Gamma})^2. \tag{A.4-28}$$

Using the symmetry of $\boldsymbol{\Gamma}'$, it follows that

$$\boldsymbol{\Gamma}:\boldsymbol{\Gamma}-\frac{1}{3}(\operatorname{tr}\boldsymbol{\Gamma})^2=\boldsymbol{\Gamma}':\boldsymbol{\Gamma}'=\sum_i\sum_j\Gamma'_{ij}\Gamma'_{ji}=\sum_i\sum_j(\Gamma'_{ij})^2\geq 0 \tag{A.4-29}$$

which completes the proof.

[5] This proof was provided by J. W. Swan.

Integral transformations

It can be helpful to transform an integral over a closed surface S to an integral over the volume V that S encloses. The relationships of interest involve the *unit normal* $\mathbf{n}$, which is a vector of unit magnitude that is normal (perpendicular) to the surface and points *outward* from the enclosed volume. The shape of the region under consideration is defined by how $\mathbf{n}$ varies with position on S. Incidentally, S may have more than one piece and the pieces need not be adjacent. For example, if V is the volume between concentric spheres, the surfaces of the inner and outer spheres are distinct parts of S.

Three closely related integral transformations are

$$\int_S \mathbf{n} f \, dS = \int_V \nabla f \, dV \tag{A.4-30}$$

$$\int_S \mathbf{n} \cdot \mathbf{a} \, dS = \int_V \nabla \cdot \mathbf{a} \, dV \tag{A.4-31}$$

$$\int_S \mathbf{n} \cdot \boldsymbol{\tau} \, dS = \int_V \nabla \cdot \boldsymbol{\tau} \, dV \tag{A.4-32}$$

where surface and volume integrals are denoted as $\int_S dS$ and $\int_V dV$, respectively. The corresponding double or triple integrals would depend on the coordinate system adopted, which would be influenced in turn by the shape of the domain. Equation (A.4-31), the *divergence theorem*, is derived in books on vector calculus (e.g., Thomas and Finney, 1984, pp. 976–979). Equations (A.4-30) and (A.4-32) are the corresponding theorems for scalars or tensors, respectively.

Another family of theorems, not used in this book, relates integrals over closed contours to integrals over surfaces enclosed by such curves. Several that are helpful when delving further into the analytic geometry of surfaces are discussed in Deen (2012). Many other integral theorems (with proofs), may be found in Brand (1947).

Unit normal and unit tangent vectors

Equations (A.4-30)–(A.4-32) all involve unit normals. If the equation that defines a surface is written as $G(x, y, z) = 0$, the unit normal is

$$\mathbf{n} = \frac{\nabla G}{|\nabla G|}. \tag{A.4-33}$$

The function G is chosen according to the desired direction of $\mathbf{n}$. For example, if $G(x, y, z) = z - F(x, y)$, then $\mathbf{n}$ will have a positive z component; if $G(x, y, z) = F(x, y) - z$, then $\mathbf{n}$ will have a negative z component.

Also needed on occasion are unit vectors that are tangent to a surface. Any such vector is necessarily orthogonal to $\mathbf{n}$. Thus, two requirements for any *unit tangent* $\mathbf{t}$ are that $\mathbf{t} \cdot \mathbf{t} = 1$ and $\mathbf{t} \cdot \mathbf{n} = 0$. These conditions do not fully define $\mathbf{t}$ for a given $\mathbf{n}$, because one of its three components can still be chosen freely, corresponding to rotations of $\mathbf{t}$ about $\mathbf{n}$. The choice for that component is dictated by the problem at hand (e.g., $t_z = 0$ if $\mathbf{t}$ is to lie in an x–y plane). Once $\mathbf{n}$ and a first unit tangent $\mathbf{t}_1$ have been found, a second unit tangent can be calculated as $\mathbf{t}_2 = \mathbf{n} \times \mathbf{t}_1$. The vectors $(\mathbf{n}, \mathbf{t}_1, \mathbf{t}_2)$ form a right-handed orthonormal set.

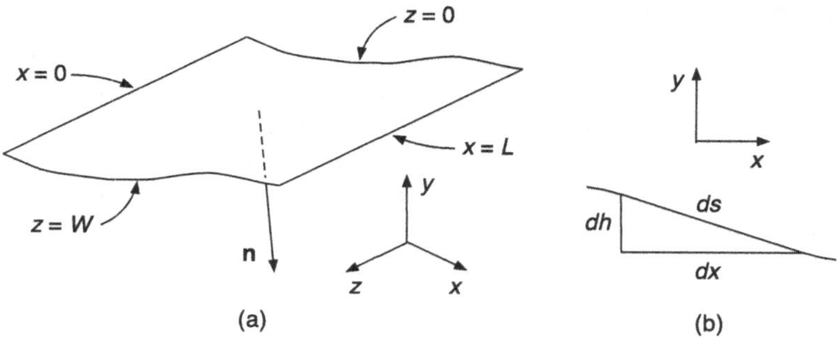

Figure A.2 Surface at $y = h(x)$ with downward-pointing unit normal $\mathbf{n}$: (a) the entire surface S, and (b) the differential arc length ds along a curve where S intersects an x–y plane.

Example A.4-5 Integration of a unit normal over a surface. Consider a surface S that is located at $y = h(x)$. As shown in Fig. A.2(a), in this case S is not a closed surface. It extends from $x = 0$ to L and from $z = 0$ to W. It is desired to find the unit normal $\mathbf{n}$ that points toward smaller y and to evaluate the integral $\int_S \mathbf{n}\, dS$.

Letting $G = h(x) - y$, the unit normal is calculated as

$$\nabla G = \mathbf{e_x}\frac{\partial G}{\partial x} + \mathbf{e_y}\frac{\partial G}{\partial y} + \mathbf{e_z}\frac{\partial G}{\partial z} = \mathbf{e_x}\frac{dh}{dx} - \mathbf{e_y} \tag{A.4-34}$$

$$|\nabla G| = \left[\left(\frac{dh}{dx}\right)^2 + 1\right]^{1/2} \equiv g(x) \tag{A.4-35}$$

$$\mathbf{n} = \frac{1}{g}\left(\frac{dh}{dx}\mathbf{e_x} - \mathbf{e_y}\right). \tag{A.4-36}$$

Notice that $\mathbf{n}$ points down, as desired (i.e., $n_y = -1/g < 0$). For a planar surface, dh/dx will be constant and $\mathbf{n}$ will be independent of position; otherwise, the direction of $\mathbf{n}$ will depend on x. Referring to Fig. A.2(b), the differential arc length ds along the surface in an x–y plane is related to variations in x and h as $(ds)^2 = (dx)^2 + (dh)^2$. It follows that

$$ds = [(dx)^2 + (dh)^2]^{1/2} = \left[1 + \left(\frac{dh}{dx}\right)^2\right]^{1/2} dx = g\,dx. \tag{A.4-37}$$

Because the surface location is independent of z, the arc length in a y–z plane is simply dz. Thus, the differential area is $dS = g\,dx\,dz$. In the surface integral the g factors in $\mathbf{n}$ and dS cancel and

$$\int_S \mathbf{n}\,dS = \int_0^W\int_0^L \left(\frac{dh}{dx}\mathbf{e_x} - \mathbf{e_y}\right) dx\,dz = W\int_0^L \left(\frac{dh}{dx}\mathbf{e_x} - \mathbf{e_y}\right) dx = W(\Delta h\,\mathbf{e_x} - L\,\mathbf{e_y}) \tag{A.4-38}$$

where $\Delta h = h(L) - h(0)$.

A.5 ORTHOGONAL CURVILINEAR COORDINATES

The most useful coordinate systems are *orthogonal*. That is, at any point in space, vectors aligned with the three coordinate directions are mutually perpendicular. Varying a single coordinate typically generates a curve in space, rather than a straight line; hence the term *curvilinear*. The general properties of such coordinate systems will be discussed first, and then the details for Cartesian, cylindrical, and spherical systems will be presented. A reader seeking only the specifics for those coordinate systems may skip ahead to Tables A.2–A.4 and to the text relationships that begin with Eq. (A.5-33).

Base vectors

To distinguish them from the Cartesian coordinates (x_1, x_2, x_3) of Section A.4, the coordinates in a general orthogonal system will be denoted as (u_1, u_2, u_3). The unit vector $\mathbf{e_i}$ points in the direction of increasing u_i $(i = 1, 2, \text{ or } 3)$. A curve generated by varying u_i while holding the other coordinates constant is called a " u_i curve." Although $\mathbf{e_i}$ has a constant (unit) magnitude, it may change direction along a u_i curve and must be viewed as a variable. When a distinction is necessary, the Cartesian unit vectors will be denoted as $\mathbf{e_i}^{(x)}$. As mentioned already, that set of base vectors is unique in that each is independent of position in space.

The only coordinate systems to be considered are right-handed and normalized. Even when their base vectors vary with position, they have the properties used in Section A.3, namely

$$\mathbf{e_i} \cdot \mathbf{e_j} = \delta_{ij} \tag{A.5-1}$$

$$\mathbf{e_i} \times \mathbf{e_j} = \sum_k \varepsilon_{ijk} \mathbf{e_k}. \tag{A.5-2}$$

Thus, the product formulas in Section A.3 are valid for any such coordinate system. The main task now is to show how the differential operators in Section A.4 must be modified. In the process, expressions for differential volumes and surface areas will be obtained.

Position vectors and scale factors

A position vector $\mathbf{r}$ extends from the origin to a point in space, and therefore locates that point. The Cartesian representation of such a vector is

$$\mathbf{r} = \sum_i x_i \mathbf{e_i}^{(x)}. \tag{A.5-3}$$

That is, its Cartesian components equal the respective coordinates. A differential change in three-dimensional position is then

$$d\mathbf{r} = \sum_i \mathbf{e_i}^{(x)} \, dx_i. \tag{A.5-4}$$

Using general orthogonal coordinates, such a displacement is expressed as

$$d\mathbf{r} = \sum_i \mathbf{e_i} h_i \, du_i \tag{A.5-5}$$

where h_i is the *scale factor* for the ith coordinate. Scale factors are needed because du_i does not necessarily equal the corresponding change in length. For example, the cylindrical coordinate θ, being an angle, does not even have the dimension of length. Scale factors may vary with position, so that in general $h_i = h_i(u_1, u_2, u_3)$.

The differential change in **r** given in Eq. (A.5-5) implies that

$$\frac{\partial \mathbf{r}}{\partial u_i} = h_i \mathbf{e_i}. \tag{A.5-6}$$

This is a vector of magnitude h_i that is tangent to a u_i curve, the tangency being indicated by the unit vector $\mathbf{e_i}$. With **r** written in terms of Cartesian components as in Eq. (A.5-3),

$$\frac{\partial \mathbf{r}}{\partial u_i} = \sum_j \frac{\partial x_j}{\partial u_i} \mathbf{e_j}^{(x)}. \tag{A.5-7}$$

Accordingly, if the relationship between the Cartesian coordinates (x_1, x_2, x_3) and other coordinates (u_1, u_2, u_3) is known, the scale factors can be calculated as

$$h_i = \left| \frac{\partial \mathbf{r}}{\partial u_i} \right| = \left[\sum_j \left(\frac{\partial x_j}{\partial u_i} \right)^2 \right]^{1/2}. \tag{A.5-8}$$

From Eqs. (A.5-6) and (A.5-7), the general base vectors are related to the Cartesian ones as

$$\mathbf{e_i} = \frac{1}{h_i} \sum_j \frac{\partial x_j}{\partial u_i} \mathbf{e_j}^{(x)}. \tag{A.5-9}$$

Replacing (x_1, x_2, x_3) by (x, y, z) and writing out the sums, the recipes for finding scale factors and base vectors are

$$h_i = \left[\left(\frac{\partial x}{\partial u_i} \right)^2 + \left(\frac{\partial y}{\partial u_i} \right)^2 + \left(\frac{\partial z}{\partial u_i} \right)^2 \right]^{1/2} \tag{A.5-10}$$

$$\mathbf{e_i} = \frac{1}{h_i} \left(\frac{\partial x}{\partial u_i} \mathbf{e_x} + \frac{\partial y}{\partial u_i} \mathbf{e_y} + \frac{\partial z}{\partial u_i} \mathbf{e_z} \right). \tag{A.5-11}$$

Volumes and surface areas

The scale factors indicate how to calculate volumes and surface areas. Differential changes in all three coordinates define a rectangular prism of volume

$$dV = h_1 h_2 h_3 \, du_1 \, du_2 \, du_3. \tag{A.5-12}$$

A *coordinate surface* is generated by holding one coordinate constant and varying the other two. If the constant coordinate is u_i, this is a "u_i surface." Denoting the differential area of a u_i surface as dS_i, the expressions for the three kinds of coordinate surfaces are

$$dS_1 = h_2 h_3 \, du_2 \, du_3 \tag{A.5-13a}$$

$$dS_2 = h_1 h_3 \, du_1 \, du_3 \tag{A.5-13b}$$

$$dS_3 = h_1 h_2 \, du_1 \, du_2. \tag{A.5-13c}$$

Gradient

A general expression for the gradient operator is obtained by examining the differential change in a function f that corresponds to a differential change in position. In Cartesian coordinates,

$$df = \sum_i \frac{\partial f}{\partial x_i} dx_i = \sum_i \mathbf{e_i^{(x)}} \, dx_i \cdot \sum_j \mathbf{e_j^{(x)}} \frac{\partial f}{\partial x_j} = d\mathbf{r} \cdot \nabla f. \tag{A.5-14}$$

The first equality holds for any function $f(x_1, x_2, x_3)$ and the rest follows from Eqs. (A.5-4) and (A.4-2). What is sought now is a definition of ∇ that will ensure that $df = d\mathbf{r} \cdot \nabla f$ for *any* orthogonal coordinate system. The unknown vector ∇f is represented as

$$\nabla f = \sum_j \lambda_j \mathbf{e_j} \tag{A.5-15}$$

where the components λ_j are to be determined. Using Eqs. (A.5-5) and (A.5-15), the differential change in f is

$$df = \sum_i \frac{\partial f}{\partial u_i} du_i = d\mathbf{r} \cdot \nabla f = \sum_i \mathbf{e_i} h_i \, du_i \cdot \sum_j \lambda_j \mathbf{e_j} = \sum_i h_i \lambda_i \, du_i. \tag{A.5-16}$$

Thus, $\lambda_i = (1/h_i)(\partial f/\partial u_i)$ and

$$\nabla = \sum_i \frac{\mathbf{e_i}}{h_i} \frac{\partial}{\partial u_i}. \tag{A.5-17}$$

This is the desired generalization of the gradient operator.

Using Eq. (A.5-17), the gradients of the scalar f and vector $\mathbf{v}$ are

$$\nabla f = \sum_i \frac{\mathbf{e_i}}{h_i} \frac{\partial f}{\partial u_i} \tag{A.5-18}$$

$$\nabla \mathbf{v} = \sum_i \sum_j \frac{\mathbf{e_i}}{h_i} \frac{\partial}{\partial u_i} (v_j \mathbf{e_j}) = \sum_i \sum_j \frac{1}{h_i} \left(\frac{\partial v_j}{\partial u_i} \mathbf{e_i e_j} + v_j \mathbf{e_i} \frac{\partial \mathbf{e_j}}{\partial u_i} \right). \tag{A.5-19}$$

The results are a vector and a tensor, respectively. Equation (A.5-19) and many subsequent expressions are complicated by the need to calculate partial derivatives of the base vectors themselves. How to evaluate such partial derivatives is described later, using cylindrical and spherical coordinates as examples.

Dotting $\mathbf{v}$ with its gradient gives the vector

$$\mathbf{v} \cdot \nabla \mathbf{v} = \sum_i \sum_j \frac{v_i}{h_i} \left(\frac{\partial v_j}{\partial u_i} \mathbf{e_j} + v_j \frac{\partial \mathbf{e_j}}{\partial u_i} \right). \tag{A.5-20}$$

If $\mathbf{v}$ is the velocity of a fluid and t is time, then $D\mathbf{v}/Dt = \partial \mathbf{v}/\partial t + \mathbf{v} \cdot \nabla \mathbf{v}$ is the *material derivative* of the velocity (Section 5.3), which is needed to describe the inertia of the fluid (Section 6.4).

Scale-factor identities

Two identities involving scale factors and base vectors are very helpful when deriving general expressions for the divergence and curl. One is

$$\nabla \times \frac{\mathbf{e_i}}{h_i} = 0. \tag{A.5-21}$$

Equation (A.5-17) is used to obtain

$$\nabla u_i = \sum_j \frac{\mathbf{e_j}}{h_j}\frac{\partial u_i}{\partial u_j} = \sum_j \frac{\mathbf{e_j}}{h_j}\delta_{ij} = \frac{\mathbf{e_i}}{h_i} \quad \text{(A.5-22)}$$

and identity (5) of Table A.1 then gives Eq. (A.5-21).

The other helpful identity is

$$\nabla \cdot \frac{\mathbf{e_1}}{h_2 h_3} = \nabla \cdot \frac{\mathbf{e_2}}{h_3 h_1} = \nabla \cdot \frac{\mathbf{e_3}}{h_1 h_2} = 0. \quad \text{(A.5-23)}$$

Starting with the first vector and using Eq. (A.5-22),

$$\frac{\mathbf{e_1}}{h_2 h_3} = \frac{\mathbf{e_2}}{h_2} \times \frac{\mathbf{e_3}}{h_3} = \nabla u_2 \times \nabla u_3. \quad \text{(A.5-24)}$$

Identity (6) of Table A.1 indicates that the divergence of this vector is zero. Verification of the other parts of Eq. (A.5-23) requires only that the indices be rotated.

Divergence

The divergence of the first part of the vector $\mathbf{v}$ is

$$\begin{aligned}\nabla \cdot (v_1 \mathbf{e_1}) &= \nabla \cdot \left[h_2 h_3 v_1 \left(\frac{\mathbf{e_1}}{h_2 h_3} \right) \right] = \nabla (h_2 h_3 v_1) \cdot \left(\frac{\mathbf{e_1}}{h_2 h_3} \right) \\ &= \sum_i \frac{\mathbf{e_i}}{h_i}\frac{\partial}{\partial u_i}(h_2 h_3 v_1) \cdot \left(\frac{\mathbf{e_1}}{h_2 h_3} \right) = \frac{1}{h_1 h_2 h_3}\frac{\partial}{\partial u_1}(h_2 h_3 v_1).\end{aligned} \quad \text{(A.5-25)}$$

To obtain the second equality, the divergence of the scalar–vector product was expanded using identity (1) of Table A.1 and a term not shown was eliminated using Eq. (A.5-23). Combining Eq. (A.5-25) with the analogous results for the second and third parts of $\mathbf{v}$ (obtained by rotating subscripts) gives

$$\nabla \cdot \mathbf{v} = \frac{1}{h_1 h_2 h_3}\left[\frac{\partial}{\partial u_1}(h_2 h_3 v_1) + \frac{\partial}{\partial u_2}(h_3 h_1 v_2) + \frac{\partial}{\partial u_3}(h_1 h_2 v_3) \right]. \quad \text{(A.5-26)}$$

The general expression for the divergence of the tensor $\boldsymbol{\tau}$ is

$$\begin{aligned}\nabla \cdot \boldsymbol{\tau} &= \sum_i \sum_j \sum_k \frac{\mathbf{e_k}}{h_k} \cdot \frac{\partial}{\partial u_k}(\tau_{ij}\mathbf{e_i}\mathbf{e_j}) \\ &= \sum_i \sum_j \left(\frac{1}{h_i}\frac{\partial \tau_{ij}}{\partial u_i}\mathbf{e_j} + \sum_k \frac{\tau_{ij}}{h_k}\mathbf{e_k} \cdot \frac{\partial \mathbf{e_i}}{\partial u_k}\mathbf{e_j} + \frac{\tau_{ij}}{h_i}\frac{\partial \mathbf{e_j}}{\partial u_i} \right).\end{aligned} \quad \text{(A.5-27)}$$

Curl

The procedure for the curl is similar to that for the divergence. The curl of the first part of the vector $\mathbf{v}$ is

$$\begin{aligned}\nabla \times (v_1 \mathbf{e_1}) &= \nabla \times \left[h_1 v_1 \left(\frac{\mathbf{e_1}}{h_1} \right) \right] = \nabla (h_1 v_1) \times \frac{\mathbf{e_1}}{h_1} = \sum_i \frac{\mathbf{e_i}}{h_i}\frac{\partial}{\partial u_i}(h_1 v_1) \times \frac{\mathbf{e_1}}{h_1} \\ &= \frac{1}{h_1 h_2 h_3}\left(h_2 \mathbf{e_2}\frac{\partial}{\partial u_3} - h_3 \mathbf{e_3}\frac{\partial}{\partial u_2} \right)(h_1 v_1).\end{aligned} \quad \text{(A.5-28)}$$

To obtain the second equality, the curl of the vector–tensor product was expanded using identity (2) of Table A.1 and a term not shown was eliminated using Eq. (A.5-21). Combining Eq. (A.5-28) with the analogous results for the other parts of $\mathbf{v}$ gives

$$\begin{aligned}\nabla \times \mathbf{v} = {} & \frac{\mathbf{e_1}}{h_2 h_3}\left[\frac{\partial}{\partial u_2}(h_3 v_3) - \frac{\partial}{\partial u_3}(h_2 v_2)\right] \\ & + \frac{\mathbf{e_2}}{h_3 h_1}\left[\frac{\partial}{\partial u_3}(h_1 v_1) - \frac{\partial}{\partial u_1}(h_3 v_3)\right] \\ & + \frac{\mathbf{e_3}}{h_1 h_2}\left[\frac{\partial}{\partial u_1}(h_2 v_2) - \frac{\partial}{\partial u_2}(h_1 v_1)\right].\end{aligned} \tag{A.5-29}$$

This may be written also as a determinant:

$$\nabla \times \mathbf{v} = \frac{1}{h_1 h_2 h_3}\begin{vmatrix} h_1 \mathbf{e_1} & h_2 \mathbf{e_2} & h_3 \mathbf{e_3} \\ \partial/\partial u_1 & \partial/\partial u_2 & \partial/\partial u_3 \\ h_1 v_1 & h_2 v_2 & h_3 v_3 \end{vmatrix}. \tag{A.5-30}$$

Laplacian

To obtain the Laplacian ($\nabla^2 = \nabla \cdot \nabla$) it is necessary only to set $\mathbf{v} = \nabla$ or $v_i = (1/h_i)\partial/\partial u_i$ in Eq. (A.5-26). The resulting operator is applied to a scalar f or vector $\mathbf{v}$ to give

$$\nabla^2 f = \frac{1}{h_1 h_2 h_3}\left[\frac{\partial}{\partial u_1}\left(\frac{h_2 h_3}{h_1}\frac{\partial f}{\partial u_1}\right) + \frac{\partial}{\partial u_2}\left(\frac{h_3 h_1}{h_2}\frac{\partial f}{\partial u_2}\right) + \frac{\partial}{\partial u_3}\left(\frac{h_1 h_2}{h_3}\frac{\partial f}{\partial u_3}\right)\right] \tag{A.5-31}$$

$$\nabla^2 \mathbf{v} = \frac{1}{h_1 h_2 h_3}\left[\frac{\partial}{\partial u_1}\left(\frac{h_2 h_3}{h_1}\frac{\partial}{\partial u_1}\right) + \frac{\partial}{\partial u_2}\left(\frac{h_3 h_1}{h_2}\frac{\partial}{\partial u_2}\right) + \frac{\partial}{\partial u_3}\left(\frac{h_1 h_2}{h_3}\frac{\partial}{\partial u_3}\right)\right]\sum_i v_i \mathbf{e_i}. \tag{A.5-32}$$

Cartesian coordinates

In Cartesian (rectangular) coordinates the scale factors are $h_x = h_y = h_z = 1$ and the general expressions for the differential operators simplify to those in Section A.4. A number of results for a scalar f and vector $\mathbf{v}$ are summarized in Table A.2. Included are ∇f, $\nabla \cdot \mathbf{v}$, $\nabla \times \mathbf{v}$, $\nabla^2 f$, and the components of $\nabla \mathbf{v}$. Not shown here but available in Chapter 6 are the components of $\mathbf{v} \cdot \nabla \mathbf{v}$ (Tables 6.1 and 6.8), $\nabla \cdot \boldsymbol{\tau}$ (Table 6.1), and $\nabla^2 \mathbf{v}$ (Table 6.8).

Cylindrical coordinates

The cylindrical coordinate system is shown in Fig. A.3. The cylindrical (r, θ, z) and Cartesian (x, y, z) coordinates are related as

$$x = r\cos\theta, \quad y = r\sin\theta, \quad z = z. \tag{A.5-33}$$

Using Eqs. (A.5-10) and (A.5-11), the scale factors and base vectors are found to be

$$h_r = 1, \quad h_\theta = r, \quad h_z = 1 \tag{A.5-34}$$

$$\mathbf{e_r} = \cos\theta\,\mathbf{e_x} + \sin\theta\,\mathbf{e_y} \tag{A.5-35a}$$

$$\mathbf{e_\theta} = -\sin\theta\,\mathbf{e_x} + \cos\theta\,\mathbf{e_y} \tag{A.5-35b}$$

$$\mathbf{e_z} = \mathbf{e_z}. \tag{A.5-35c}$$

Table A.2 Differential operations in Cartesian coordinates[a]

(1) $\nabla f = \frac{\partial f}{\partial x}\mathbf{e_x} + \frac{\partial f}{\partial y}\mathbf{e_y} + \frac{\partial f}{\partial z}\mathbf{e_z}$
(2) $\nabla \cdot \mathbf{v} = \frac{\partial v_x}{\partial x} + \frac{\partial v_y}{\partial y} + \frac{\partial v_z}{\partial z}$
(3) $\nabla \times \mathbf{v} = \left[\frac{\partial v_z}{\partial y} - \frac{\partial v_y}{\partial z}\right]\mathbf{e_x} + \left[\frac{\partial v_x}{\partial z} - \frac{\partial v_z}{\partial x}\right]\mathbf{e_y} + \left[\frac{\partial v_y}{\partial x} - \frac{\partial v_x}{\partial y}\right]\mathbf{e_z}$
(4) $\nabla^2 f = \frac{\partial^2 f}{\partial x^2} + \frac{\partial^2 f}{\partial y^2} + \frac{\partial^2 f}{\partial z^2}$
(5) $(\nabla \mathbf{v})_{xx} = \frac{\partial v_x}{\partial x}$
(6) $(\nabla \mathbf{v})_{xy} = \frac{\partial v_y}{\partial x}$
(7) $(\nabla \mathbf{v})_{xz} = \frac{\partial v_z}{\partial x}$
(8) $(\nabla \mathbf{v})_{yx} = \frac{\partial v_x}{\partial y}$
(9) $(\nabla \mathbf{v})_{yy} = \frac{\partial v_y}{\partial y}$
(10) $(\nabla \mathbf{v})_{yz} = \frac{\partial v_z}{\partial y}$
(11) $(\nabla \mathbf{v})_{zx} = \frac{\partial v_x}{\partial z}$
(12) $(\nabla \mathbf{v})_{zy} = \frac{\partial v_y}{\partial z}$
(13) $(\nabla \mathbf{v})_{zz} = \frac{\partial v_z}{\partial z}$

[a] In these relationships f is any differentiable scalar and $\mathbf{v}$ is any differentiable vector.

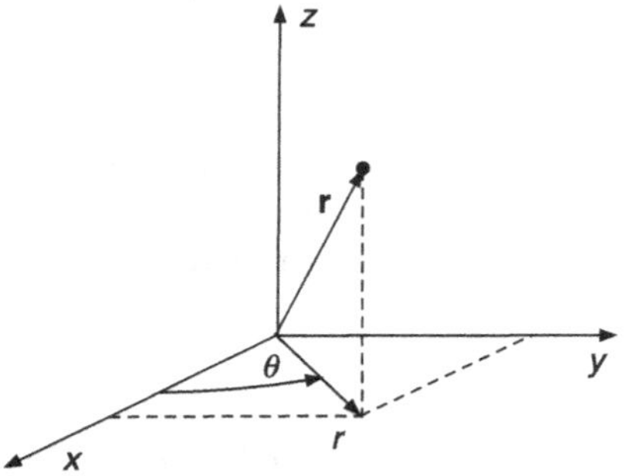

Figure A.3 Relationship between cylindrical (r, θ, z) and Cartesian (x, y, z) coordinates. The range of the cylindrical angle is $0 \le \theta \le 2\pi$. The position vector for an arbitrary point above the x–y plane is shown as $\mathbf{r}$. There is an irregularity in notation, such that $r \ne |\mathbf{r}|$. Rather, r is the distance from the origin projected onto the x–y plane, as indicated by the dashed lines.

Solving Eqs. (A.5-35) for the Cartesian base vectors gives

$$\mathbf{e_x} = \cos\theta\,\mathbf{e_r} - \sin\theta\,\mathbf{e_\theta} \qquad \text{(A.5-36a)}$$

$$\mathbf{e_y} = \sin\theta\,\mathbf{e_r} + \cos\theta\,\mathbf{e_\theta} \qquad \text{(A.5-36b)}$$

$$\mathbf{e_z} = \mathbf{e_z}. \qquad \text{(A.5-36c)}$$

Table A.3 Differential operations in cylindrical coordinates[a]

$$(1)\ \nabla f = \frac{\partial f}{\partial r}\mathbf{e_r} + \frac{1}{r}\frac{\partial f}{\partial \theta}\mathbf{e_\theta} + \frac{\partial f}{\partial z}\mathbf{e_z}$$

$$(2)\ \nabla\cdot\mathbf{v} = \frac{1}{r}\frac{\partial}{\partial r}(rv_r) + \frac{1}{r}\frac{\partial v_\theta}{\partial \theta} + \frac{\partial v_z}{\partial z}$$

$$(3)\ \nabla\times\mathbf{v} = \left[\frac{1}{r}\frac{\partial v_z}{\partial \theta} - \frac{\partial v_\theta}{\partial z}\right]\mathbf{e_r} + \left[\frac{\partial v_r}{\partial z} - \frac{\partial v_z}{\partial r}\right]\mathbf{e_\theta} + \left[\frac{1}{r}\frac{\partial}{\partial r}(rv_\theta) - \frac{1}{r}\frac{\partial v_r}{\partial \theta}\right]\mathbf{e_z}$$

$$(4)\ \nabla^2 f = \frac{1}{r}\frac{\partial}{\partial r}\left(r\frac{\partial f}{\partial r}\right) + \frac{1}{r^2}\frac{\partial^2 f}{\partial \theta^2} + \frac{\partial^2 f}{\partial z^2}$$

$$(5)\ (\nabla\mathbf{v})_{rr} = \frac{\partial v_r}{\partial r}$$

$$(6)\ (\nabla\mathbf{v})_{r\theta} = \frac{\partial v_\theta}{\partial r}$$

$$(7)\ (\nabla\mathbf{v})_{rz} = \frac{\partial v_z}{\partial r}$$

$$(8)\ (\nabla\mathbf{v})_{\theta r} = \frac{1}{r}\frac{\partial v_r}{\partial \theta} - \frac{v_\theta}{r}$$

$$(9)\ (\nabla\mathbf{v})_{\theta\theta} = \frac{1}{r}\frac{\partial v_\theta}{\partial \theta} + \frac{v_r}{r}$$

$$(10)\ (\nabla\mathbf{v})_{\theta z} = \frac{1}{r}\frac{\partial v_z}{\partial \theta}$$

$$(11)\ (\nabla\mathbf{v})_{zr} = \frac{\partial v_r}{\partial z}$$

$$(12)\ (\nabla\mathbf{v})_{z\theta} = \frac{\partial v_\theta}{\partial z}$$

$$(13)\ (\nabla\mathbf{v})_{zz} = \frac{\partial v_z}{\partial z}$$

[a] In these relationships f is any differentiable scalar and $\mathbf{v}$ is any differentiable vector.

The vectors $\mathbf{e_r}$ and $\mathbf{e_\theta}$ each depend only on θ, whereas $\mathbf{e_z}$ is independent of position. Differentiating Eqs. (A.5-35) and using Eqs. (A.5-36) to return to the cylindrical base vectors, it is found that

$$\frac{\partial \mathbf{e_r}}{\partial \theta} = \mathbf{e_\theta}, \qquad \frac{\partial \mathbf{e_\theta}}{\partial \theta} = -\mathbf{e_r}. \tag{A.5-37}$$

The other seven partial derivatives of the cylindrical base vectors with respect to cylindrical coordinates are all zero. The differential volume and the differential areas of the coordinate surfaces are

$$dV = r\,dr\,d\theta\,dz \tag{A.5-38}$$

$$dS_r = r\,d\theta\,dz, \quad dS_\theta = dr\,dz, \quad dS_z = r\,dr\,d\theta. \tag{A.5-39}$$

Various differential operations in cylindrical coordinates are shown in Table A.3. Not included here but available in Chapter 6 are the components of $\mathbf{v}\cdot\nabla\mathbf{v}$ (Table 6.2 or 6.9), $\nabla\cdot\boldsymbol{\tau}$ (Table 6.2), and $\nabla^2\mathbf{v}$ (Table 6.9).

Spherical coordinates

The spherical coordinate system is shown in Fig. A.4. The spherical (r, θ, ϕ) and Cartesian (x, y, z) coordinates are related as

$$x = r\sin\theta\cos\phi, \quad y = r\sin\theta\sin\phi, \quad z = r\cos\theta. \tag{A.5-40}$$

Table A.4 Differential operations in spherical coordinates[a]

(1) $$\nabla f = \frac{\partial f}{\partial r}\mathbf{e_r} + \frac{1}{r}\frac{\partial f}{\partial \theta}\mathbf{e_\theta} + \frac{1}{r\sin\theta}\frac{\partial f}{\partial \phi}\mathbf{e_\phi}$$

(2) $$\nabla \cdot \mathbf{v} = \frac{1}{r^2}\frac{\partial}{\partial r}(r^2 v_r) + \frac{1}{r\sin\theta}\frac{\partial}{\partial \theta}(v_\theta \sin\theta) + \frac{1}{r\sin\theta}\frac{\partial v_\phi}{\partial \phi}$$

(3) $$\nabla \times \mathbf{v} = \frac{1}{r\sin\theta}\left[\frac{\partial}{\partial \theta}(v_\phi \sin\theta) - \frac{\partial v_\theta}{\partial \phi}\right]\mathbf{e_r} + \left[\frac{1}{r\sin\theta}\frac{\partial v_r}{\partial \phi} - \frac{1}{r}\frac{\partial}{\partial r}(r v_\phi)\right]\mathbf{e_\theta} + \left[\frac{1}{r}\frac{\partial}{\partial r}(r v_\theta) - \frac{1}{r}\frac{\partial v_r}{\partial \theta}\right]\mathbf{e_\phi}$$

(4) $$\nabla^2 f = \frac{1}{r^2}\frac{\partial}{\partial r}\left(r^2\frac{\partial f}{\partial r}\right) + \frac{1}{r^2\sin\theta}\frac{\partial}{\partial \theta}\left(\sin\theta\frac{\partial f}{\partial \theta}\right) + \frac{1}{r^2\sin^2\theta}\frac{\partial^2 f}{\partial \phi^2}$$

(5) $$(\nabla \mathbf{v})_{rr} = \frac{\partial v_r}{\partial r}$$

(6) $$(\nabla \mathbf{v})_{r\theta} = \frac{\partial v_\theta}{\partial r}$$

(7) $$(\nabla \mathbf{v})_{r\phi} = \frac{\partial v_\phi}{\partial r}$$

(8) $$(\nabla \mathbf{v})_{\theta r} = \frac{1}{r}\frac{\partial v_r}{\partial \theta} - \frac{v_\theta}{r}$$

(9) $$(\nabla \mathbf{v})_{\theta\theta} = \frac{1}{r}\frac{\partial v_\theta}{\partial \theta} + \frac{v_r}{r}$$

(10) $$(\nabla \mathbf{v})_{\theta\phi} = \frac{1}{r}\frac{\partial v_\phi}{\partial \theta}$$

(11) $$(\nabla \mathbf{v})_{\phi r} = \frac{1}{r\sin\theta}\frac{\partial v_r}{\partial \phi} - \frac{v_\phi}{r}$$

(12) $$(\nabla \mathbf{v})_{\phi\theta} = \frac{1}{r\sin\theta}\frac{\partial v_\theta}{\partial \phi} - \frac{v_\phi \cot\theta}{r}$$

(13) $$(\nabla \mathbf{v})_{\phi\phi} = \frac{1}{r\sin\theta}\frac{\partial v_\phi}{\partial \phi} + \frac{v_r}{r} + \frac{v_\theta \cot\theta}{r}$$

[a] In these relationships f is any differentiable scalar and $\mathbf{v}$ is any differentiable vector.

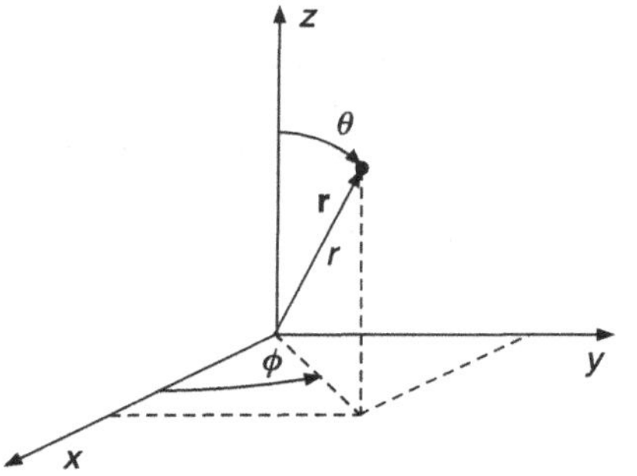

Figure A.4 Relationship between spherical (r, θ, ϕ) and Cartesian (x, y, z) coordinates. The ranges of the spherical angles are $0 \le \theta \le \pi$ and $0 \le \phi \le 2\pi$. The position vector $\mathbf{r}$ for an arbitrary point above the x–y plane is shown. Its magnitude is r, the distance from the origin.

The scale factors and base vectors obtained from Eqs. (A.5-10) and (A.5-11) are

$$h_r = 1, \quad h_\theta = r, \quad h_\phi = r\sin\theta \tag{A.5-41}$$

$$\mathbf{e_r} = \sin\theta\cos\phi\,\mathbf{e_x} + \sin\theta\sin\phi\,\mathbf{e_y} + \cos\theta\,\mathbf{e_z} \tag{A.5-42a}$$

$$\mathbf{e_\theta} = \cos\theta\cos\phi\,\mathbf{e_x} + \cos\theta\sin\phi\,\mathbf{e_y} - \sin\theta\,\mathbf{e_z} \tag{A.5-42b}$$

$$\mathbf{e_\phi} = -\sin\phi\,\mathbf{e_x} + \cos\phi\,\mathbf{e_y}. \tag{A.5-42c}$$

Solving for the Cartesian base vectors gives

$$\mathbf{e_x} = \sin\theta\cos\phi\,\mathbf{e_r} + \cos\theta\cos\phi\,\mathbf{e_\theta} - \sin\phi\,\mathbf{e_\phi} \tag{A.5-43a}$$

$$\mathbf{e_y} = \sin\theta\sin\phi\,\mathbf{e_r} + \cos\theta\sin\phi\,\mathbf{e_\theta} + \cos\phi\,\mathbf{e_\phi} \tag{A.5-43b}$$

$$\mathbf{e_z} = \cos\theta\,\mathbf{e_r} - \sin\theta\,\mathbf{e_\theta}. \tag{A.5-43c}$$

The partial derivatives of the spherical base vectors are

$$\frac{\partial \mathbf{e_r}}{\partial r} = \mathbf{0}, \quad \frac{\partial \mathbf{e_r}}{\partial \theta} = \mathbf{e_\theta}, \quad \frac{\partial \mathbf{e_r}}{\partial \phi} = \sin\theta\,\mathbf{e_\phi} \tag{A.5-44a}$$

$$\frac{\partial \mathbf{e_\theta}}{\partial r} = \mathbf{0}, \quad \frac{\partial \mathbf{e_\theta}}{\partial \theta} = -\mathbf{e_r}, \quad \frac{\partial \mathbf{e_\theta}}{\partial \phi} = \cos\theta\,\mathbf{e_\phi} \tag{A.5-44b}$$

$$\frac{\partial \mathbf{e_\phi}}{\partial r} = \mathbf{0}, \quad \frac{\partial \mathbf{e_\phi}}{\partial \theta} = \mathbf{0}, \quad \frac{\partial \mathbf{e_\phi}}{\partial \phi} = -(\sin\theta\,\mathbf{e_r} + \cos\theta\,\mathbf{e_\theta}). \tag{A.5-44c}$$

The differential volume and the differential areas of the coordinate surfaces are

$$dV = r^2 \sin\theta\, dr\, d\theta\, d\phi \tag{A.5-45}$$

$$dS_r = r^2 \sin\theta\, d\theta\, d\phi, \quad dS_\theta = r\sin\theta\, dr\, d\phi, \quad dS_\phi = r\, dr\, d\theta. \tag{A.5-46}$$

Table A.4 shows a number of differential operations in spherical coordinates. Not included, but available in Chapter 6, are the components of $\mathbf{v} \cdot \nabla \mathbf{v}$ (Table 6.3 or 6.10), $\nabla \cdot \boldsymbol{\tau}$ (Table 6.3), and $\nabla^2 \mathbf{v}$ (Table 6.10).

Many other orthogonal coordinate systems have been developed for special applications. Scale factors, differential operators, and solutions of Laplace's equation for 40 such systems are compiled in Moon and Spencer (1961).

REFERENCES

Bird, R. B., W. E. Stewart, and E. N. Lightfoot. *Transport Phenomena*, 2nd ed. Wiley, New York, 2002.

Brand, L. *Vector and Tensor Analysis*. Wiley, New York, 1947.

Deen, W. M. *Analysis of Transport Phenomena*, 2nd ed. Oxford University Press, New York, 2012.

Hay, G. B. *Vector and Tensor Analysis*. Dover, New York, 1953.

Hildebrand, F. B. *Advanced Calculus for Applications*, 2nd ed. Prentice-Hall, Englewood Cliffs, NJ, 1976.

Jeffreys, H. *Cartesian Tensors*. Cambridge University Press, Cambridge, 1963.

Moon, P. and D. E. Spencer. *Field Theory Handbook*. Springer-Verlag, Berlin, 1961.

Thomas, G.B., Jr. and R. L. Finney. *Calculus and Analytic Geometry*, 6th ed. Addison-Wesley, Reading, MA, 1984.

Wilson, E. B. *Vector Analysis*. Yale University Press, New Haven, 1901.

Author index

Page numbers are given only for first authors. Bold numbers indicate where full citations or biographical data appear.

Subject index

Printed in the United States
by Baker & Taylor Publisher Services